Automation and Control

Edited by Constantin Voloşencu,
Serdar Küçük, José Guerrero and Oscar Valero

Published in London, United Kingdom

IntechOpen

Supporting open minds since 2005

Automation and Control
http://dx.doi.org/10.5772/intechopen.87702
Edited by Constantin Voloşencu, Serdar Küçük, José Guerrero and Oscar Valero

Contributors
Aamir Shahzad, Yasir Al-Yasir, Ameer Saleh, Adel Obed, Hamza Qasim, Waleed Breesam, Naser Ojaroudi Parchin, Raed. Alhameed, Igor N. Sinitsyn, Vladimir I. Sinitsyn, Edward R. Korepanov, Kuo-Chi Chang, Andrey Petrovich Nikiforov, Kangsoo Kim, Marcio A.F. Martins, Odilon S.L. de Abreu, Leizer Schnitman, Mangayarkarasi Ramaiah, Dilip Kumar Prasad, Wesam Jasim, Jozef Ritonja, Noor Hafizah Amer, Khisbullah Hudha, Hairi Zamzuri, Vimal Rau Aparow, Zulkiffli Abd. Kadir, Muhamad Murrad, Amar Faiz Zainal Abidin, Flabio Dario Mirelez-Delgado, Miguel Abraham Gallardo-Carreón, José Ronaldo Díaz-Paredes, Ivan Virgala, Michal Kelemen, Erik Prada, İhsan Omur Ömür Bucak, Olalekan Samuel Ogunleye, Billy Mathias Kalema, Dongbing Gu, Kai-Chun Chu, Yuh-Chung Lin, Jeng-Shyang Pan, Muhammad Kashif, Tariq Munir, Maogang He

First published in London, United Kingdom, 2021 by IntechOpen
IntechOpen is the global imprint of INTECHOPEN LIMITED, registered in England and Wales, registration number: 11086078, 5 Princes Gate Court, London, SW7 2QJ, United Kingdom
Printed in Croatia

British Library Cataloguing-in-Publication Data
A catalogue record for this book is available from the British Library

Additional hard and PDF copies can be obtained from orders@intechopen.com

Automation and Control
Edited by Constantin Voloşencu, Serdar Küçük, José Guerrero and Oscar Valero
p. cm.
Print ISBN 978-1-83962-713-2
Online ISBN 978-1-83962-714-9
eBook (PDF) ISBN 978-1-83962-715-6

For EU product safety concerns:
IN TECH d.o.o., Prolaz Marije Krucifikse Kozulić 3, 51000 Rijeka, Croatia, info@intechopen.com or visit our website at intechopen.com.

Meet the editors

Constantin Volosencu is a professor in the Department of Automation, "Politehnica" University Timisoara, Romania. He is the editor of nine books and author of ten books, five book chapters, more than 170 scientific papers published in journals and conference proceedings, and twenty-seven patents. He is also a manager of research grants, a member of international journal editorial boards, a former plenary speaker, a member of scientific committees, and chair at international conferences. His research is in the fields of control systems, control of electric drives, fuzzy control systems, neural network applications, fault detection and diagnosis, sensor network applications, monitoring of distributed parameter systems, and power ultrasound applications. He has developed electrical equipment for machine tools, spooling machines, high-power ultrasound processes, and more.

Dr. Serdar Kucuk received a BA and MSc from Marmara University, Istanbul, Turkey, in 1995 and 1998, respectively. He received a Ph.D. from Kocaeli University, Turkey, in 2004. He is currently a full professor in the Department of Biomedical Engineering, Kocaeli University. He has published several international conference papers, journal papers, books, and book chapters. He serves as a reviewer for several well-known robotics journals. He has also edited a number of scientific books. His research interests include optimization, control, kinematics, and dynamics modeling of serial and parallel robotic manipulators. Lately, Dr. Kucuk has also been interested in designing electrical controlled above-the-knee prosthetics and hand–wrist rehabilitation robots, surgical robots, and biomedical robotic devices.

Dr. José Guerrero Sastre received a degree in Computer Science from the University of the Balearic Islands (UIB). In 2012, he completed his Ph.D. in Computer Science at UIB with a thesis entitled "New Methodologies for Allocating Tasks and Coalition Formation in Multi-Robot Systems." He has been a member of the Department of Mathematics and Computer Science, UIB, since 2002, where he is currently a lecturer and postdoctoral researcher. He is also a member of the Systems, Robotics and Vision Group (SRV) and a collaborator with the Models for Information Processing (MOTIBO) research group. His research interests include multi-robot and multi-agent task allocation mechanisms with auction, swarm-like coordination mechanisms, and possibility theory.

Dr. Oscar Valero graduated in Mathematical Sciences from the University of Valencia and obtained a Ph.D. in Mathematical Sciences from the Polytechnic University of Valencia in 20003. Since then, he has worked as an associate professor in the Department of Mathematics and Computer Science, University of the Balearic Islands (UBI). His research focuses on the theoretical study of generalized metric structures, measures of similarity, fixed point theory, functional analysis, asymmetric topology, fuzzy logic, and the application of theoretical results to engineering, economics, and medicine. Dr. Valero has published more than 140 international research articles in these areas. In addition, he has presented fifty-two communications at several international conferences, seven of which he served as the main speaker. Moreover, he has been president and member of the organizing and scientific committees of several international congresses. He is currently a member of the editorial board of nine international research journals.

Contents

Preface

This book presents recent information on automation and control, a field that deals with a wide range of technologies that reduce human intervention in processes using predetermined criteria and actuator relationships and embodying this predetermination in machines based on digital equipment and algorithms. Automatic control includes the use of various control systems for operating equipment such as machinery, processes in factories, telephone networks, and ships, aircraft, and vehicles with reduced intervention, intelligent process monitoring, and optimal decision making based on practical measurements. Automation and control applications range from home control applications to large industrial control systems with tens of thousands of input measurements and output control signals, process monitoring, and centralized management policies.

Researchers in automation and control are continually developing new concepts and tools that enhance human understanding and improve the specialist's ability to design and implement high-performance solutions. As such, this book presents practical automation and control applications across various domains that emphasize methodologies used with implementation and commissioning issues. It also discusses new methodologies and techniques in the field.

The readers get new solutions and answers to questions related to the emerging automation and control in various applications and their implementation.

The book is structured in seven sections, each of which examines one of the following thematic areas: control theory, which deals with the control of dynamical systems in engineered processes and machines; autonomous or self-driving vehicles, which are vehicles capable of sensing the environment and moving safely with little or no human input via a variety of sensors and advanced control systems; mechatronics, which is an interdisciplinary branch of engineering that focuses on the engineering of electrical and mechanical systems and includes a combination of robotics, electronics, computers, telecommunications, and control; digital image processing, which is the use of digital computers to process digital images using algorithms, electrical grids, which are networks that provide power to homes and industries within an extended area; artificial intelligence, which is a field of research that seeks to develop machines, programs, systems, algorithms, and methods to mimic natural intelligence; and electric motor drives, which are equipment, systems, control algorithms, and programs used to control the speed of machinery.

This book contents fifteen chapters. Chapter 1 presents a theoretic study on ellipsoidal analysis and filtering methods for nonlinear control stochastic systems with some basic applications. Stochastic control is a subfield of control theory that deals with the existence of uncertainty either in observation or in the noise that drives the evolution of the system. Methods of stochastic control based on the parametrization of distributions allow the design of practical software tools. Chapter 2 presents an interesting approach of reconfigurable minimum-time guidance of autonomous vehicles moving in variable sea currents, characterized by robustness and fault tolerance to environmental uncertainties. Chapter 3 discusses the development of an adaptive path-tracking controller equipped with a knowledge-based supervisory algorithm for an autonomous heavy vehicle, with optimal parameters depending on the maneuvering and vehicle conditions. Chapter 4 investigates an integral backstepping control strategy for unmanned aerial vehicle team formation to improve the performance indicators of the control. Chapter 5 presents a stabilizing model-predictive control strategy for inverted pendulum-like behaved systems that has an offset-free optimization control law. The Lyapunov stability of the closed-loop system is achieved by adopting an infinite prediction horizon. Chapter 6 presents the design of a digital PID control for a Stewart-Gough platform with six degrees of freedom. The authors develop a prototype for the study, design, and control of parallel mechanisms. Chapter 7 is a theoretic study of kinematic modeling of serial manipulators, as multibody systems. Chapter 8 analyzes a sliding mode solution applied to robotic manipulators. Chapter 9 presents an iterative technique for polygonal approximation of digital image boundary, with application in parallel manipulators in detecting target boundary of an image with varying scale. Chapter 10 presents a study on selecting the appropriate robust and adaptive control theories for power system stabilizer implementation. Chapter 11 is a study in the field of control and protection algorithms for the electric power industry. Chapter 12 evaluates the algorithmic management of a digital work platform in developing countries. Chapter 13 is a short overview of some intelligent control structures and dedicated algorithms. Chapter 14 presents a technical solution for speed control of a DC motor based on a microcontroller. Finally, Chapter 15 presents a study on using wavelet neural networks for speed control of BLDC motors.

The chapters were edited and published following a rigorous selection process, with only a small number of the proposed chapters selected for publication.

The editor thanks the authors for their excellent contributions in the field as well as their understanding during the editing process. The editor also wishes to thank Josip Knapic, assistants to the editor, and the staff at IntechOpen.

Constantin Voloşencu
Department of Automation and Applied Informatics,
"Politehnica" University Timisoara,
Timisoara, Romania

Dr. Oscar Valero and Dr. José Guerrero
Department of Mathematics and Computer Science,
University of the Balearic Islands,
Spain

Dr. Serdar Küçük
University of Kocaeli,
Turkey

Section 1

Control Theory

Chapter 1

Development of Ellipsoidal Analysis and Filtering Methods for Nonlinear Control Stochastic Systems

Igor N. Sinitsyn, Vladimir I. Sinitsyn and Edward R. Korepanov

Abstract

The methods of the control stochastic systems (CStS) research based on the parametrization of the distributions permit to design practically simple software tools. These methods give the rapid increase of the number of equations for the moments, the semiinvariants, coefficients of the truncated orthogonal expansions of the state vector Y, and the maximal order of the moments involved. For structural parametrization of the probability (normalized and nonnormalized) densities, we shall apply the ellipsoidal densities. A normal distribution has an ellipsoidal structure. The distinctive characteristics of such distributions consist in the fact that their densities are the functions of positively determined quadratic form of the centered state vector. Ellipsoidal approximation method (EAM) cardinally reduces the number of parameters. For ellipsoidal linearization method (ELM), the number of equations coincides with normal approximation method (NAM). The development of EAM (ELM) for CStS analysis and CStS filtering are considered. Based on nonnormalized densities, new types of filters are designed. The theory of ellipsoidal Pugachev conditionally optimal control is presented. Basic applications are considered.

Keywords: conditionally optimal filtering and control, control stochastic system, ellipsoidal approximation method (EAM), ellipsoidal linearization method (ELM)

1. Introduction

The methods for the control stochastic systems (CStS) research based on the parametrization of the distributions permit to design practically simple software tools [1–6]. These methods give the rapid increase of the number of equations for the moments, the semiinvariants, and coefficients of the truncated orthogonal expansions of the state vector Y for the maximal order of the moments involved. For structural parametrization of the probability (normalized and nonnormalized) densities, we shall apply the ellipsoidal densities. A normal distribution has an ellipsoidal structure. The distinctive characteristics of such distributions consist in the fact that their densities are the functions of positively determined quadratic form $u = u(y) = (y^T - m^T)C(y - m)$ where m is an expectation of Y, C is some positively determined matrix. Ellipsoidal approximation method (EAM) cardinally reduces the number of parameters till $Q^{EAM} = Q^{NAM} + n_m - 1$ and $Q^{NAM} = r(r+3)/2$ where $2n_m$ being the number of probabilistic moments. For ellipsoidal linearization method (ELM), we get $Q^{ELM} = Q^{NAM}$.

The theory of conditionally optimal filters (COF) is described in [7, 8] on the basis of methods of normal approximation (NAM), methods of statistical linearization (SLM), and methods of orthogonal expansions (OEM) for the differential stochastic systems on smooth manifolds with Wiener noise in the equations of observation and Wiener and Poisson noises in the state equations. The COF theory relies on the exact nonlinear equations for the normalized one-dimensional a posteriori distribution. The paper [9] considers extension of [7, 8] to the case where the a posteriori one-dimensional distribution of the filtration error admits the ellipsoidal approximation [4]. The exact filtration equations are obtained, as well as the OEM-based equation of accuracy and sensitivity, the elements of ellipsoidal analysis of distributions are given, and the equations of ellipsoidal COF (ECOF) using EAM and ELM are derived. The theory of analytical design of the modified ellipsoidal suboptimal filters was developed in [10, 11] on the basis of the approximate solution by EAM (ELM) of the filtration equation for the nonnormalized a posteriori characteristic function. The modified ellipsoidal conditionally optimal filters (MECOF) were constructed in [12] on the basis of the equations for nonnormalized distributions. It is assumed that there exist the Wiener and Poisson noises in the state equations and only Wiener noise being in the observation equations. At that, the observation noise can be non-Gaussian.

Special attention is paid to the conditional generalization of Pugachev optimal control [13] based on EAM (ELM).

Let us consider the development of EAM (ELM) for solving problems of ellipsoidal analysis and optimal, suboptimal, and conditionally optimal filtering and control in continuous CStS with non-Gaussian noises and stochastic factors.

2. Ellipsoidal approximation method

This method was worked out in [1–4] for analytical modeling of stochastic process (StP) in multidimensional nonlinear continuous, discrete and continuous-discrete (CStS). Let us consider elements of EAM.

Following [1–4] let us find ellipsoidal approximation (EA) for the density of r-dimensional random vector by means of the truncated expansion based on biorthogonal polynomials $\{p_{r,\nu}(u(y)), q_{r,\nu}(u(y))\}$, depending only on the quadratic form $u = u(y)$ $u = u(y)$ for which some probability density of the ellipsoidal structure $w(u(y))$ serves as the weight:

$$\int_{-\infty}^{\infty} w(u(y)) p_{r,\nu}(u(y)) q_{r,\mu}(u(y)) dy = \delta_{\nu\mu}. \tag{1}$$

The indexes ν and μ at the polynomials mean their degrees relative to the variable u. The concrete form and the properties of the polynomials are determined further. But without the loss of generality, we may assume that $q_{r,0}(u) = p_{r,0}(u) = 1$. Then the probability density of the vector Y may be approximately presented by the expression of the form:

$$f(y) \approx f^*(u) = w(u)\left[1 + \sum_{\nu=2}^{N} c_{r,\nu} p_{r,\nu}(u)\right]. \tag{2}$$

Here the coefficients $c_{r,\nu}$ are determined by the formula:

$$c_{r,\nu} = \int_{-\infty}^{\infty} f(y) q_{r,\nu}(u) dy = E q_{r,\nu}(U), \quad (\nu = 1, \dots, N). \tag{3}$$

As $p_{r,0}(u)$ and $q_{r,0}(u)$ are reciprocal constants (the polynomials of zero degree), then always $c_{r,0}p_{r,0} = 1$ and we come to the following results.

Statement 1. Formulae (2) and (3) express the essence of the EA of the probability density of the random vector$Y$.

For the control problems, the case when the normal distribution is chosen as the distribution $w(u)$ is of great importance

$$w(u) = w\left(x^T C x\right) = \frac{1}{\sqrt{(2\pi)^r |K|}} \exp\left(-x^T K^{-1} x/2\right); \tag{4}$$

accounting that $C = K^{-1}$, we reduce the condition of the biorthonormality (1) to the form

$$\frac{1}{2^{r/2}\Gamma(r/2)} \int_0^{\infty} p_{r,\nu}(u) q_{r,\mu}(u) u^{r/2-1} e^{-u/2} du = \delta_{\nu\mu}, \tag{5}$$

where $\Gamma(\cdot)$ is gamma function [5].

Statement 2. The problem of the choosing of the polynomial system $\left\{p_{r,\nu}(u) q_{r,\mu}(u)\right\}$ which is used at the EA of the densities (4) and (5) is reduced to finding a biorthonormal system of the polynomials for which the χ^2-distribution with r degrees of the freedom serves as the weigh.

A system of the polynomials which are relatively orthogonal to χ^2-distribution with r degrees of the freedom is described by series:

$$S_{r,\nu}(u) = \sum_{\mu=0}^{\nu} (-1)^{\nu+\mu} C_{\nu}^{\mu} \frac{(r + 2\nu - 2)!!}{(r + 2\mu - 2)!!} u^{\mu}. \tag{6}$$

The main properties of polynomials $S_{r,\nu}$ are given in [2–4]. Between the polynomials $S_{r,\nu}(u)$ and the system of the polynomials $\left\{p_{r,\nu}(u),\ q_{r,\mu}(u)\right\}$, the following relations exist:

$$p_{r,\nu}(u) = S_{r,\nu}(u), q_{r,\nu}(u) = \frac{(r-2)!!}{(r + 2\nu - 2)!!(2\nu)!!} S_{r,\nu}(u), \quad r \ge 2. \tag{7}$$

Example 1. Formulae for polynomials $p_{r,\nu}(u)$ and $q_{r,\nu}(u)$ and its derivatives for some r and ν are as follows [4]:

- At $r = 2$, $\nu \ge 2$,

$$p_{2,\nu}(u) = u^{\nu}, \quad q_{2,\nu}(u) \equiv 0, \quad q'_{2,\nu}(u) \equiv 0, \quad q''_{2,\nu}(u) \equiv 0;$$

- At $r \geq 2$, $\nu = 2$

$$p_{r,2}(u) = u^2, \quad q_{r,2}(u) = \frac{1}{8}u^2, \quad q'_{r,2}(u) = \frac{1}{4}u, \quad q''_{r,2}(u) = \frac{1}{4}.$$

For $r = 2$ at $\nu = 3$ we have

$$p_{2,3}(u) = u^3, \quad q_{2,3}(u) \equiv 0, \quad q'_{2,3}(u) \equiv 0, \quad q''_{2,3}(u) \equiv 0;$$

at $r = 3$

$$\left.\begin{aligned} &p_{3,3}(u) = S_{3,3}(u), \quad q_{3,3}(u) = \frac{1}{5040}S_{3,3}(u), \\ &q'_{3,3}(u) = \frac{1}{5040}S'_{3,3}(u), \quad q''_{3,3}(u) = \frac{1}{5040}S''_{3,3}(u), \\ &S_{3,3}(u) = -105 + 105u - 21u^2 + u^3, \\ &S'_{3,3}(u) = 105 - 42u + 3u^2, \quad S''_{3,3}(u) = -42 + 6u; \end{aligned}\right\}$$

at $r = 4$:

$$\left.\begin{aligned} &p_{4,3}(u) = S_{4,3}(u), \quad q_{4,3}(u) = \frac{1}{9216}S_{4,3}(u), \\ &q'_{4,3}(u) = \frac{1}{9216}S'_{4,3}(u), \quad q''_{4,3}(u) = \frac{1}{9216}S'_{4,3}(u), \\ &S_{4,3}(u) = -197 + 144u - 24u^2 + u^3, \\ &S'_{4,3}(u) = 144 - 48u + 3u^2, \quad S''_{4,3}(u) = -48 + 6u. \end{aligned}\right\}$$

Following [5] we consider the H-space $L_2(R^r)$ and the orthogonal system of the functions in them where the polynomials $S_{r,\nu}(u)$ are given by Formula (6), and $w(u)$ is a normal distribution of the r-dimensional random vector (4). This system is not complete in $L_2(R^r)$. But the expansion of the probability density $f(u) = f((y^T - m^T)C(y - m))$ of the random vector Y which has an ellipsoidal structure over the polynomials $p_{r,\nu}(u) = S_{r,\nu}(u)$, m.s. converges to the function $f(u)$ itself. The coefficients of the expansion in this case are determined by relation:

$$c_{r,\nu} = \int_{-\infty}^{\infty} f(u)p_{r,\nu}(u)dy \Big/ \frac{(2\nu)!!(r + 2\nu - 2)!!}{(r - 2)!!}. \tag{8}$$

Statement 3. The system of the functions $\{\sqrt{w(u)}S_{r,\nu}(u)\}$ forms the basis in the subspace of the space $L_2(R^r)$ generated by the functions $f(u)$ of the quadratic form $u = (y - m)^T C(y - m)$.

At the probability density expansion over the polynomial $S_{r,\nu}(u)$, the probability densities of the random vector Y and all its possible projections are consistent. In other words, at integrating the expansions over the polynomials $S_{h+l,\nu}(u)$ and $h + l = r$, of the probability densities of the r-dimensional vector Y,

$$f(y) = \frac{1}{\sqrt{(2\pi)^{h+l}|K|}}e^{-u/2}\left[1 + \sum_{\nu=2}^{N} c_{h+l,\nu}S_{h+l,\nu}(u)\right], \quad u = (y - m)^T K^{-1}(y - m),$$

$$y = \left[y'^T y''^T\right]^T, \tag{9}$$

on all the components of the l-dimensional vector y'', we obtain the expansion over the polynomials $S_{h,\nu}(u_1)$ of the probability density of the h-dimensional vector Y' with the same coefficients

$$f(y') = \frac{1}{\sqrt{(2\pi)^h |K_{11}|}} e^{-u_1/2} \left[1 + \sum_{\nu=2}^{N} c_{h,\nu} S_{h,\nu}(u_1)\right], \quad u_1 = (y' - m')^T K_{11}^{-1} (y' - m'),$$
$$c_{h,\nu} = c_{h+l,\nu}, \tag{10}$$

where K_{11} is a covariance matrix of the vector Y'.

But in approximation (10) the probability density of h-dimensional random vector Y' obtained by the integration of expansion (9) the density of $(h + l)$-dimensional vector is not optimal EA of the density.

For the random r-dimensional vector with an arbitrary distribution, the EA (2) of its distribution determines exactly the moments till the Nth order inclusively of the quadratic form $U = (Y - m)^T K^{-1} (Y - m)$, i.e.,

$$EU^\mu = E^{EA} U^\mu, \quad \mu \le N. \tag{11}$$

(E^{EA} stands for expectation relative to EA distribution).

In this case the initial moments of the order s and $s = s_1 + \dots + s_r$ of the random vector Y at the approximation (4) are determined by the formula:

$$\alpha_{s_1,\dots,s_r} = \alpha_s = EY_1^{s_1} \dots Y_r^{s_r} \approx \int_{-\infty}^{\infty} y_1^{s_1} \dots y_r^{s_r} w(u) dy + \sum_{v=2}^{N} c_{r,v} \int_{-\infty}^{\infty} y_1^{s_1} \dots y_r^{s_r} p_{r,v}(u) w(u) dy \tag{12}$$

Statement 4. At the EA of the distribution of the random vector, its moments are combined as the sums of the correspondent moments of the normal distribution and the expectations of the products of the polynomials $p_{r,\nu}(u)$ by the degrees of the components of the vector Y at the normal density $w(u)$.

3. EAM accuracy

For control problems the weak convergence of the probability measures generated by the segments of the density expansion to the probability measure generated by the density itself is more important than m.s. convergence of the segments of the density expansion over the polynomials $S_{r,\nu}(u)$ to the density, namely,

$$\int_A w(u) \left[1 + \sum_{\nu=2}^{N} c_{r,\nu} p_{r,\nu}(u)\right] \to \int_A f(u) dy$$

uniformly relative to A at $N \to \infty$ on the σ-algebra of Borel sets of the space R^r. Thus the partial sums of series (2) give the approximation of the distribution, i.e., the probability of any event A determined by the density $f(u)$ with any degree of the accuracy. The finite segment of this expansion may be practically used for an approximate presentation of $f(u)$ with any degree of the accuracy even in those cases when $f(u)/\sqrt{w(u)}$ does not belong to $L_2(R^r)$. In this case it is sufficient to substitute $f(u)$ by the truncated density. Expansion (2) is valid only for the densities which have the ellipsoidal structure. It is impossible in principal to approximate with any degree of the accuracy by means of the EA (2) the densities which arbitrarily depend on the vector y.

One is the way of the estimate of the accuracy of the distribution approximation in the comparison of the probability characteristics calculated by means of the known density and its approximate expression. The most complete estimate of the accuracy of the approximation may be obtained by the comparison of the probability occurrence on the sets of some given class. Besides that taking into consideration that the probability density is usually approximated by a finite segment of its orthogonal expansion for instance, over Hermite polynomials or by a finite segment of the Edgeworth series [1–5] which contain the moments till the fourth order, the accuracy may be characterized by the accuracy of the definition of the moments of the random vector or its separate components, in particular, of the fourth order moments.

Corresponding estimates for these two ways of approximation are given in [2, 3].

4. Ellipsoidal linearization method

Now we consider ellipsoidal linearization of nonlinear transforms of random vectors Y using mean square error (m.s.e.) criterion optimal m.s.e. regression of vector $Z = \varphi(Y)$ on vector Y is determined by the formula [4, 6]:

$$m_z(Y) = h_2 Y, \quad h_2 = \Gamma_{zy}\Gamma_y^{-1} \tag{13}$$

or

$$m_z(Y) = h_1 Y + a, \quad h_1 = K_{zy}K_y^{-1}, \quad a = m_z - h_1 m_y. \tag{14}$$

where h_1 and h_2 are equivalent linearization matrices and m_y and K_y are mathematical expectation and covariance matrix $(\det|\ K_y| \neq 0)$. In case (14) coefficient h_1 is equal to

$$\begin{aligned} h_1 = K_{zy}K_y^{-1} &= \int_{-\infty}^{\infty}\int_{-\infty}^{\infty} (z - m_z)(y - m_y)^T K_y^{-1} f(z,y)\,dz\,dy \\ &= \int_{-\infty}^{\infty} [m_z(y) - m_z](y - m_y)^T K_y^{-1} f_1(y)\,dy \end{aligned} \tag{15}$$

where $f_1(y)$ is the density of random vector Y.
For ellipsoidal density $f_1(y)$ in (15) is defined by

$$f_1(y) = f_1^{\mathrm{EL}}(y) = \tilde{f}_1^{\mathrm{EL}}(u(y), m_y, K_y, c). \tag{16}$$

In case (14) we get Statement 5 for ELM:

$$m_z(Y) \approx m_{1z}^{\mathrm{EL}} + h_1^{\mathrm{EL}}(m_y, K_y, c) Y^0, \tag{17}$$

where

$$h_1^{\mathrm{EL}} = h_1^{\mathrm{EL}}(m_y, K_y, c) = \int_{-\infty}^{\infty} [m_z(y) - m_z](y - m_y)^T K_y^{-1} f_1^{\mathrm{EL}}(y) dy$$
$$= \int_{-\infty}^{\infty} [m_z(y) - m_z](y - m_y)^T K_y^{-1} \tilde{f}_1^{\mathrm{EL}}(u(y), m_y, K_y, c) dy. \quad (18)$$

In case (13) we have Statement 6 for ELM:

$$m_z(Y) \approx h_2^{\mathrm{EL}}(\Gamma_y, c) Y, \quad (19)$$

$$h_2^{\mathrm{EL}}(\Gamma_y, c) = \int_{-\infty}^{\infty} m_z(y) y^T \Gamma_y^{-1} f_1^{\mathrm{EL}}(y) dy = \int_{-\infty}^{\infty} m_z(y) y^T \Gamma_y^{-1} \tilde{f}_1(u(y), m_y, \Gamma_y, c) dy. \quad (20)$$

For control problems the following ELM new generalizations are useful:

1. Let us consider for fixed dimension $p = \dim y$ and N in (2) with normal $w(u)$ distinguish modifications of various orders $\mathrm{ELM}_w^{p,2}$, $\mathrm{ELM}_w^{p,3}$, …, $\mathrm{ELM}_w^{p,N}$. In this case $c = \{c_{p,v}\}$ characterizes partial deviations from normal distributions of various orders v jointly for all p components of vector Y (be part of quadratic form $U(Y)$.

2. At decomposition of vector Y on $l_1, l_2, \dots, l_r$ random subvectors, $Y = \left[Y_{l_1}^T Y_{l_2}^T \dots Y_{l_r}^T\right]^T$, we distinguish $\mathrm{ELM}_w^{l_1,\dots,l_r,N}$. Coefficients $c_{l_1,v}, \dots, c_{l_2,v}$ characterize partial deviations of subvectors from normal distribution.

3. For matrix transforms $Z = \varphi(Y) = \left[\varphi_1(Y) \dots \varphi_q(Y)\right]^T$, $\varphi_i(Y) = \left[\varphi_{i1}(Y) \dots \varphi_{ip}(Y)\right]^T$ $\left(i = \overline{1,q}\right)$, $\dim\varphi = p \times q$, we have the following formulae for ELM:

$$m_z(Y) \approx m_{1z}^{\mathrm{EL}} + H_1^{\mathrm{EL}}(m_y, K_y, c)(Y - m); \quad (21)$$
$$m_z(Y) \approx H_2^{\mathrm{EL}} Y. \quad (22)$$

where

$$H_1^{\mathrm{EL}}(m_y, K_y, c) = \left[h_{11}^{\mathrm{EL}}(m_y, K_y, c) \dots h_{1q}^{\mathrm{EL}}(m_y, K_y, c)\right] \quad (23)$$
$$H_2^{\mathrm{EL}}(\Gamma_y, c) = \left[h_{21}^{\mathrm{EL}}(\Gamma_y, c) \dots h_{2q}^{\mathrm{EL}}(\Gamma_y, c)\right], \quad (24)$$

(h_{1i}^{EL} and h_{2i}^{EL} $\left(i = \overline{1,q}\right)$ are determined by formulae (18) and (19)).

4. For transforms depending on time process t, it is useful to work with overage ELM coefficients $\langle m_{iz} \rangle$ and $\langle h_i^{\mathrm{EL}} \rangle$ for time intervals.

5. EAM and ELM for nonlinear CStS analysis

Let us consider nonlinear CStS defined by the following Ito vector stochastic differential equation:

$$dY_t = a(Y_t,t)dt + b(Y_t,t)dW_0 + \int_{R_0^q} c(Y_t,t,v)P^0(dt,dv), \quad Y(t_0) = Y_0. \tag{25}$$

Here $Y_t \in \Delta^y$ is (Δ^y is a smooth state manifold) $W_0 = W_0(t)$ is an r – dimensional Wiener StP of intensity $v_0 = v_0(t)$, $P(\Delta, \mathscr{A})$ is simple Poisson StP for any set $\mathscr{A}$, $\Delta = (t_1, t_2]$, $P^0(\Delta, \mathscr{A}) = P^0(\Delta, \mathscr{A}) - \mu_P(\Delta, \mathscr{A})$, $\mu_P(\Delta, \mathscr{A}) = \mathrm{E}P^0(\Delta, \mathscr{A}) =$

$\int_\Delta v_P(\tau, \mathscr{A})d\tau$. Integration by v extends to the entire space R^q with deleted origin, a and b are certain functions mapping $R^p \times R$, respectively, into R^p, R^{pr}, and c is for $R^p \times R^q$ into R^p.

Following [4] we use for finding the one-dimensional probability density $f_1(y;t)$ of the r-dimensional $Y(t)$ which is determined by Eq. (25). Suppose that we know a distribution of the initial value $Y_0 = Y(t_0)$ of the StP $Y(t)$. Following the idea of EAM, we present the one-dimensional density in the form of a segment of the orthogonal expansion in terms of the polynomials dependent on the quadratic form $u = (y^T - m^T)C(y - m)$ where m and $K = C^{-1}$ are the expectation and the covariance matrix of the StP$Y(t)$:

$$f_1(y;t) \cong f_1^{EAM}(u) = w_1(u)\left[1 + \sum_{\nu=2}^{N} c_{p,\upsilon} p_{p,\nu}(u)\right]. \tag{26}$$

Here $w_1(u)$ is the normal density of the p-dimensional random vector which is chosen in correspondence with the requirement $c_{p,1} = 0$. The optimal coefficients of the expansion $c_{p,\upsilon}$ are determined by the relation

$$c_{p,\upsilon} = \int_{-\infty}^{\infty} f_1(y;t) q_{p,\nu}(u)dy = Eq_{p,\nu}(U), \qquad (\nu = 1, \dots, N). \tag{27}$$

The set of the polynomials $\left\{p_{p,\nu}(u), q_{p,\nu}(u)\right\}$ is constructed on the base of the orthogonal set of the polynomials $\left\{S_{p,\upsilon}(u)\right\}$ according to the following rule which provides the biorthonormality of system at $p \geq 2$ given by (5). Thus the solution of the problem of finding the one-dimensional probability density by EAM is reduced to finding the expectation m, the covariance matrix K of the state vector, and the coefficients of the correspondent expansion $c_{p,\upsilon}$ also.

So we get the equations

$$\dot{m} = \varphi_{10}(m,K,t) + \sum_{\nu=2}^{N} c_{p,\upsilon}\varphi_{1\nu}(m,K,t), \tag{28}$$

$$\dot{K} = \varphi_{20}(m,K,t) + \sum_{\nu=2}^{N} c_{p,\upsilon}\varphi_{2\nu}(m,K,t), \tag{29}$$

$$\dot{c}_{p,\kappa} = -\left(\frac{c_{p,\kappa-1}}{2p} - \frac{\kappa c_{p,\kappa}}{p}\right) \times \mathrm{tr}\left\{K^{-1}\varphi_{20}(m,K,t) + K^{-1}\sum_{\nu=2}^{N} c_{p,\upsilon}\varphi_{2\nu}(m,K,t)\right\}$$
$$+\,\psi_{\kappa 0}(m,K,t) + \sum_{\nu=2}^{N} c_{p,\upsilon}\psi_{\kappa\nu}(m,K,t), \qquad \kappa = 2, \dots, N, \tag{30}$$

where the following indications are introduced:

$$\varphi_{10}(m,K,t) = \int_{-\infty}^{\infty} a(y,t)w_1(u)dy, \varphi_{1\nu}(m,K,t) = \int_{-\infty}^{\infty} a(y,t)p_{p,\nu}(u)w_1(u)dy,$$

$$\varphi_{20}(m,K,t) = \int_{-\infty}^{\infty} \left[a(y,t)\left(y^T - m^T\right) + (y-m)a(y,t)^T + \overline{\sigma}(y,t)\right]w_1(u)dy,$$

$$\varphi_{2\nu}(m,K,t) = \int_{-\infty}^{\infty} \left[a(y,t)\left(y^T - m^T\right) + (y-m)a(y,t)^T + \overline{\sigma}(y,t)\right]p_{p,\nu}(u)w_1(u)dy,$$

$$\overline{\sigma}(y,t) = \overline{\sigma}(y,t) + \int_{R_0^q} c(y,t,v)c(y,t,v)^T \nu_P(t,dv), \quad \sigma(y,t) = b(y,t)\nu_0(t)b(y,t)^T, \tag{31}$$

$$\begin{aligned} \psi_{\kappa 0}(m,K,t) &= \int_{-\infty}^{\infty} \Big[q'_{p,\kappa}(u)\left(2(y-m)^T K^{-1}a(y,t) + \operatorname{tr} K^{-1}\sigma(y,t)\right) \\ &\quad + 2q''_{\kappa}(u)(y-m)^T K^{-1}\sigma(y,t)(y-m)\Big]w_1(u)dy \\ \psi_{\kappa\nu}(m,K,t) &= \int_{-\infty}^{\infty} q'_{p,\kappa}(u)\left[2(y-m)^T K^{-1}a(y,t) + \operatorname{tr} K^{-1}\sigma(y,t)\right] \\ &\quad + 2q''_{\kappa}(u)(y-m)^T K^{-1}\sigma(y,t)(y-m)\Big\}p_{p,\nu}(u)w_1(u)dy. \end{aligned} \tag{32}$$

Eqs. (28)–(30) at the initial conditions

$$m(t_0) = m_0, \quad K(t_0) = K_0, \quad c_{p,\kappa}(t_0) = c^0_{p,\kappa} \quad (\kappa = 2, \dots, N) \tag{33}$$

determine $m, K, c_{p,2}, \dots, c_{p,N}$ as time functions. For finding the variables $c^0_{p,\kappa}$, the density of the initial value Y_0 of the state vector should be approximated by Formula (26).

So we get the following result.

Statement 7. At sufficient conditions of existence and uniqueness of StP in Eq. (25), Eqs. (28)–(33) define EAM.

For stationary CStS we get the corresponding EAM equations putting in Eqs. (28)–(30) right-hand equal to zero.

Example 2. Following [4, 14, 15] in case of vibroprotection Duffing StS:

$$\ddot{X} + \delta\dot{X} + \omega^2 X + \mu X^3 = U + V, \quad X(t_0) = X_0, \quad \dot{X}(t_0) = \dot{X}_0,$$

(δ, ω^2, μ, U are constants, V is the white noise with intensity v) with accuracy till 4th probabilistic moments, ellipsoidal approximation of one-dimensional density is described by the set of parameters:

$$m_1 = EX, \quad m_2 = \dot{X}, \quad K_{11} = EX^{0^2}, \quad K_{12} = EX^0\dot{X}^0, \quad K_{22} = E\dot{X}^{0^2} \quad \text{and} \quad c_{2,2}.$$

These parameters satisfy the following ordinary differential equations:

$$\dot{m}_1 = m_2, \quad m_1(t_0) = m_1^0, \quad \dot{m}_2 = U - \omega^2 m_1 + \mu\left(m_1^3 + 3m_1K_{11}\right) - \delta m_2,$$
$$m_2(t_0) = m_2^0;$$

$$\dot{K}_{11} = 2K_{12},$$
$$\dot{K}_{12} = K_{22} - \omega^2 K_{11} + 3\mu K_{11}\left(K_{11} + m_1^2\right) - \delta K_{12} + 24\mu c_{2,2} K_{11}^2,$$
$$\dot{K}_{22} = \nu - 2\left(\omega^2 K_{12} - 3\mu K_{12}\left(K_{11} + m_1^2\right) + \delta K_{22}\right) + 48\mu c_{2,2} K_{11} K_{12},$$
$$K_{11}(t_0) = K_{11}^0, \quad K_{12}(t_0) = K_{12}^0, \quad K_{22}(t_0) = K_{22}^0;$$
$$\dot{c}_{2,2} = c_{2,2}\left(\frac{K_{11}\nu}{|K|} - 5\delta\right) \quad \left(|K| = K_{11}K_{22} - K_{12}^2\right), \quad c_{2,2}(t_0) = c_{2,2}^0.$$

At $U = 0$ stationary values are as follows:

$$\overline{m}_1 = 0, \quad \overline{m}_2 = 0, \quad \overline{K}_{11} = \frac{\omega^2\sqrt{\omega^4 - 6\mu\nu/\delta}}{6\mu}, \quad \overline{K}_{12} = 0, \quad \overline{K}_{22} = \frac{\nu}{2\delta}, \quad \overline{c}_{2,2} = 0.$$

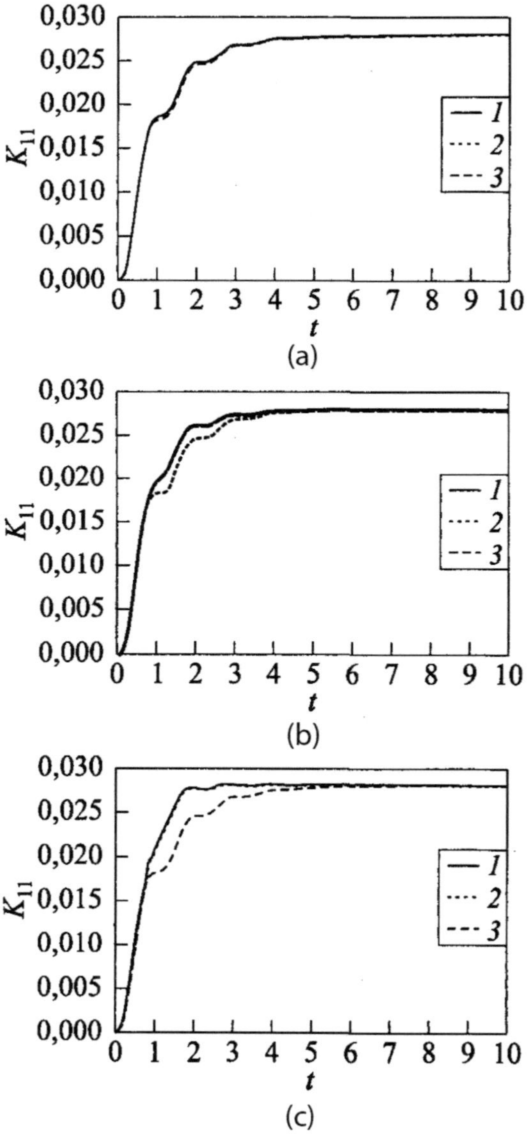

Figure 1.
K_{11} graphs for at 0,1 (a); 0,5 (b); 1,0 (c).

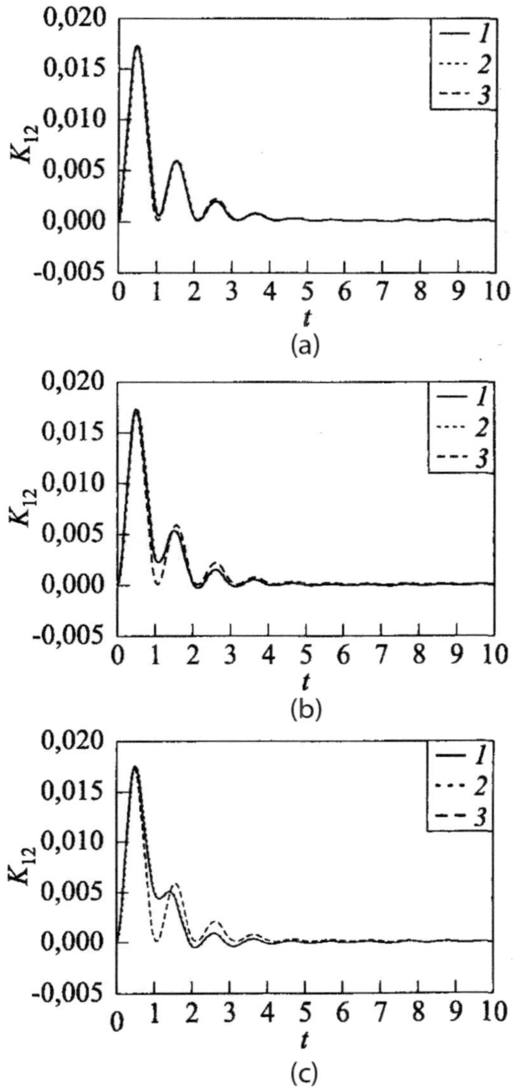

Figure 2.
K_{12} graphs for at 0,1 (a); 0,5 (b); 1,0 (c).

At conditions

$$1.\ U = 0;\quad \mu = 0.1;\quad \omega = 3;\quad \delta = 1;\quad \nu = 0.5;$$

$$2.\ U = 0;\quad \mu = 0.5;\quad \omega = 3;\quad \delta = 1;\quad \nu = 0.5;$$

$$3.\ U = 0;\quad \mu = 1;\quad \omega = 3;\quad \delta = 1;\quad \nu = 0.5$$

And at zero initial conditions, the results of analytical modeling for K_{11}, K_{12}, K_{22} are given in **Figures 1–3**. Mathematical expectations m_1 and m_n are equal to zero.

Graphs (1) are the results of integration of NAM equations at initial stage. Then for nongenerated covariance matrix K integration of EAM equations (2). Graphs are the results of EAM equation integration at the whole stage.

The results of investigations for $c_{2,2}$ are given in **Figure 4** for the following sets of conditions:

1. $U = 0$; $\mu = 1$; $\omega = 3$; $\delta = 0,5$; $\nu = 0,5$; $T = [0, 20]$ zero initial conditions;
2. $U = 0$; $\mu = 1$; $\omega = 3$; $\delta = 0,5$; $\nu = 1$; $T = [0, 20]$ zero initial conditions;
3. $U = 0$; $\mu = 1$; $\omega = 3$; $\delta = 0,5$; $\nu = 1$; $T = [0, 20]$ zero initial conditions except $m_1(0) = 0,2$;
4. $U = 0$; $\mu = 1$; $\omega = 3$; $\delta = 1$; $\nu = 1$; $T = [0, 20]$ zero initial conditions.

For the stationary CStS regimes, EAM gives the same results as NAM (MSL). EAM describes non-Gaussian transient vibro StP at initial stage.

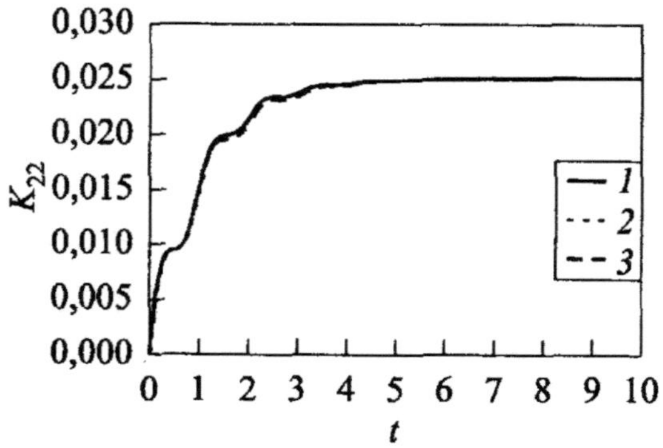

(a)

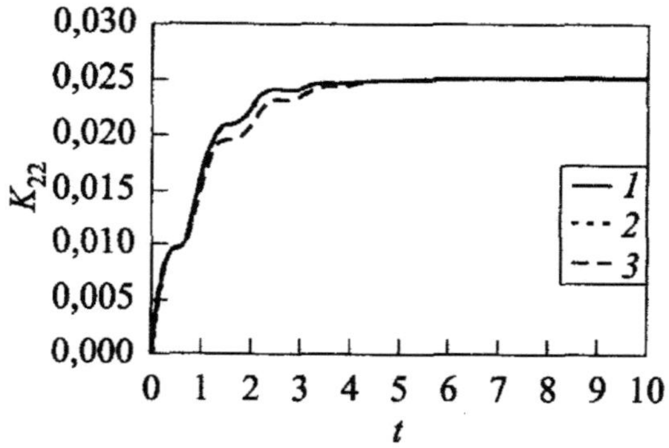

(b)

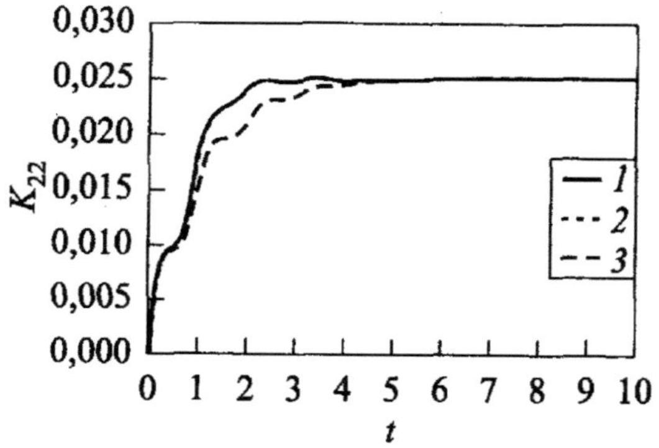

(c)

Figure 3.
K_{22} graphs for at 0,1 (a); 0,5 (b); 1,0 (c).

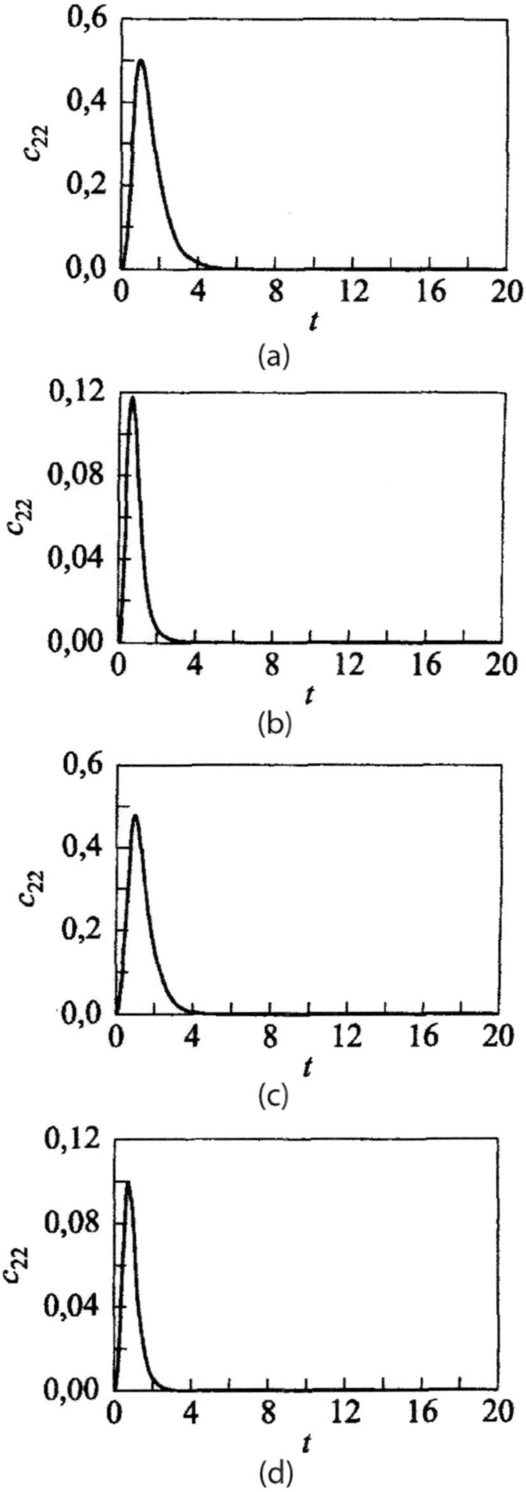

Figure 4.
C_{22} graphs for at sets № 1 (a); 2 (b); 3 (c); 4 (d).

Methodological and software support for analysis and filtering problem CStS for shock and vibroprotection is given in [4, 14].

6. Exact filtering equations for continuous a posteriori distribution

Following [7–9, 15], let the vector StP $\left[X_t^{\mathrm{T}} Y_t^{\mathrm{T}}\right]^{\mathrm{T}}$ be defined by a system on vector stochastic differential Ito equations:

$$dX_t = \varphi(X_t, Y_t, \Theta, t)dt + \psi'(X_t, Y_t, \Theta, t)dW_0 \\ + \int_{R_0^q} \psi''(X_t, Y_t, \Theta, t, v)P^0(dt, dv), \quad X(t_0) = X_0, \tag{34}$$

$$dY_t = \varphi_1(X_t, Y_t, \Theta, t)dt + \psi_1'(X_t, Y_t, \Theta, t)dW_0 \\ + \int_{R_0^q} \psi_1''(X_t, Y_t, \Theta, t, v)P^0(dt, dv), \quad Y(t_0) = Y_0. \tag{35}$$

where $Y_t = Y(t)$ is an n_y-dimensional observed StP $Y_t \in \Delta^y$ (Δ^y is a smooth manifold of observations); $X_t \in \Delta^x$ (Δ^x is a smooth state manifold), $W_0 = W_0(t)$ is an n_w-dimensional Wiener StP $(n_w \geq n_y)$ of intensity $\nu_0 = \nu_0(\Theta, t)$; $P^0(\Delta, A) = P(\Delta, A) - \mu_P(\Delta, A)$, $P(\Delta, A)$ for any set A represents a simple Poisson StP, and $\mu_P(\Delta, A)$ is its expectation,

$$\mu_P(\Delta, A) = \mathrm{E}P(\Delta, A) = \int_{\Delta} \nu_P(\tau, A)d\tau;$$

$v_P(\Delta, A)$ is the intensity of the corresponding Poisson flow of events, $\Delta = (t_1, t_2]$; integration by v extends to the entire space R^q with deleted origin; Θ is the vector of random parameters of size n_Θ; $\varphi = \varphi(X_t, Y_t, \Theta, t)$, $\varphi_1 = \varphi_1(X_t, Y_t, \Theta, t)$, $\psi' = \psi'(X_t, Y_t, \Theta, t)$, and $\psi_1' = \psi_1'(X_t, Y_t, \Theta, t)$ are certain functions mapping $R^{n_x} \times R^{n_y} \times R$, respectively, into $R^{n_x}, R^{n_y}, R^{n_x n_w}$, $R^{n_y n_w}$; $\psi'' = \psi''(X_t, Y_t, \Theta, t, v)$, and $\psi_1'' = \psi_1''(X_t, Y_t, \Theta, t, v)$ are certain functions mapping $R^{n_x} \times R^{n_y} \times R^q$ into R^{n_x}, R^{n_y}. Determine the estimate $\hat{X}_t$ StP X_t at each time instant t from the results of observation of StP $Y(\tau)$ until the instant t, $Y_{t_0}^t = \{Y(\tau) : t_0 \leq \tau < t\}$.

Let us assume that the state equation has the form (34); the observation Eq. (35), first, contains no Poisson noise ($\psi''_1 \equiv 0$); and, second, the coefficient at the Wiener noise ψ_1' in the observation equations is independent of the state $\left(\psi_1'(X_t, Y_t, \Theta, t) = \psi_1'(Y_t, \Theta, t)\right)$, and then the equations of the problem of nonlinear filtration are given by

$$dX_t = \varphi(X_t, Y_t, \Theta, t)dt + \psi'(X_t, Y_t, \Theta, t)dW_0 \\ + \int_{R_0^q} \psi''(X_t, Y_t, \Theta, t, v)P^0(dt, dv), \quad X(t_0) = X_0, \tag{36}$$

$$dY_t = \varphi_1(X_t, Y_t, \Theta, t)dt + \psi_1(Y_t, \Theta, t)dW_0, \quad Y(t_0) = Y_0. \tag{37}$$

The known sufficient conditions for the existence and uniqueness of StP defined by (36) and (37) under the corresponding initial conditions [1, 3, 16] are satisfied.

The optimal estimate $\hat{X}_t$ minimizing the mean square of the error at each time instant t is known [10–14] to represent for any StP X_t and Y_t.

An a posteriori expectation StP X_t: $\hat{X}_t = \mathrm{E}\left[X_t | Y_{t_0}^t\right]$. To determine this conditional expectation, one needs to know $p_t = p_t(x)$ and $g_t = g_t(\lambda)$, the a posteriori one-dimensional density, and the characteristic function of the distribution StP X_t.

Introduce the nonnormalized one-dimensional a posteriori density $\tilde{p}_t(x, \Theta)$ and a characteristic function $\tilde{g}_t(\lambda, \Theta)$ according to

$$\tilde{p}_t(x, \Theta) = \mu_t p_t(x, \Theta), \ \tilde{g}_t(\lambda, \Theta) = \mathrm{E}_{\Delta^x}^{p_t}\left[e^{i\lambda^{\mathrm{T}} X_t} \mu_t\right] = \mu_t g_t(\lambda, \Theta), \tag{38}$$

where μ_t is a normalizing function and $\mathrm{E}_{\Delta^x}^{p_t}$ is the symbol of expectation on the manifold Δ^x on the basis of density $p_t(x)$. Then, by generalizing [11] to the case of Eqs. (36) and (37), we get the following exact equation of the rms optimal nonlinear filtration:

$$d\tilde{g}_t(\lambda, \Theta) = \mathrm{E}_{\Delta^x}^{\tilde{p}_t}\left\{\left[i\lambda^{\mathrm{T}} \varphi(X, Y_t, \Theta, t) - \frac{1}{2}\left(\psi' \nu_0 \psi'^{\mathrm{T}}\right)(X, Y_t, \Theta, t)\right.\right.$$

$$+ \int_{R_0^q}\left[e^{i\lambda^{\mathrm{T}} \psi''(X, Y_t, \Theta, t, v)} - 1 - i\lambda^{\mathrm{T}} \psi''(X, Y_t, \Theta, t, v)\right] \nu_P(\Theta, t, dv)\Big] e^{i\lambda^{\mathrm{T}} X}\Big\} dt$$

$$+ \mathrm{E}_{\Delta^x}^{\tilde{p}_t}\left\{\left[\varphi_1(X, Y_t, \Theta, t)^{\mathrm{T}} + i\lambda^{\mathrm{T}}\left(\psi' \nu_0 \psi'^{\mathrm{T}}\right)(X, Y_t, \Theta, t)\right] e^{i\lambda^{\mathrm{T}} X}\right\}\left(\psi' \nu_0 \psi'^{\mathrm{T}}\right)^{-1}(Y_t, \Theta, t) dY_t. \tag{39}$$

If by following [15, 17] the function ψ'' in (36) admits the representation

$$\psi'' = \psi' \omega(\Theta, v), \tag{40}$$

where $P^0(\Delta, A) = P^0((0, t], dv)$, then Eqs. (36) and (37) take the form

$$\dot{X}_t = \varphi(X_t, Y_t, \Theta, t) + \psi'(X_t, Y_t, \Theta, t) V(\Theta, t), \quad X(t_0) = X_0, \tag{41}$$

$$\dot{Y}_t = \varphi(X_t, Y_t, \Theta, t) + \psi_1(Y_t, \Theta, t) V_0(\Theta, t), \quad Y(t_0) = Y_0. \tag{42}$$

with $V_0(\Theta, t) = \dot{W}_0(\Theta, t); \quad V(\Theta, t) = \dot{\overline{W}}(\Theta, t)$,

$$\overline{W}(\Theta, t) = W_0(\Theta, t) + \int_{R_0^q} \omega(\Theta, v) P^0((0, t], dv), \tag{43}$$

where $\nu_P(\Theta, t, v) dv = [\partial \mu(\Theta, t, v)/\partial t] dv$ is the intensity of the Poisson flow of discontinuities equal to $\omega(\Theta, t)$; the logarithmic derivatives of the one-dimensional characteristic functions obey certain formulas

$$\chi^{\overline{W}}(\rho; \Theta, t) = \chi^{W_0}(\rho; \Theta, t) + \int_{R_0^q}\left[e^{i\rho^{\mathrm{T}} \omega(\Theta, v)} - 1 - i\rho^{\mathrm{T}} \omega(\Theta, v)\right] \nu_P(\Theta, t, v) dv, \tag{44}$$

where

$$\chi^{W_0}(\rho; \Theta, t) = -\frac{1}{2} \rho^{\mathrm{T}} \nu_0(\Theta, t) \rho.$$

In this case, the integral term in (39) admits the following notation:

$$\gamma = \int_{R_0^q} \left[e^{i\lambda^{\mathrm{T}}\psi''(X_t,Y_t,\Theta,t)\omega(\Theta,v)} - 1 - i\lambda^{\mathrm{T}}\psi''(X_t,Y_t,\Theta,t)\omega(\Theta,v)\right]\nu_P(\Theta,t,v)dv. \qquad (45)$$

For the Gaussian CStS, the condition $\gamma \equiv 0$ is, obviously, true, and we come to the well-known statements [11, 15, 17].

Statement 8. Let the conditions for existence and uniqueness be satisfied for the non-Gaussian CStS (36) and (37). Then, the equation with a continuous rms of the optimal nonlinear filtration for the nonnormalized characteristic function (38) is given by (39).

Statement 9. Let the non-Gaussian CStS (41) and (42) the conditions for existence and uniqueness be satisfied. Then, the equation with continuous rms of optimal nonlinear filtration for the nonnormalized characteristic function is given by (39) provided that (45).

7. EAM (ELM) for nonlinear CStS filtering

EAM (ELM) for approximate conditionally optimal and suboptimal filtering (COF and SOF) in continuous CStS for normalized one-dimensional density is given in [11]. Let us consider the case of nonnormalized densities:

$$\tilde{p}_t(x,\Theta) \approx p_t^*(u,\Theta) = w(u,\Theta)\left[\mu_t + \sum_{\nu=1}^{N} c_\nu p_\nu(u)\right]. \qquad (46)$$

Here, $w = w(u,\Theta)$ is the reference density and $\{p_\nu(u), q_\nu(u)\}$ is the biorthonormal system of polynomials, $C_t = K_t^{-1}$; K_t is the covariance matrix and c_ν is the coefficient of ellipsoidal expansion

$$c_\nu = \mu_t \mathrm{E}^{EA}\left[q_\nu(U_t)\right] = \left[q_\nu(U_\lambda)\tilde{g}_t^{EA}(\lambda,\Theta)\right]_{\lambda=0}, \qquad (47)$$

with the notation

$$u = \left(x^{\mathrm{T}} - \hat{X}_t^{\mathrm{T}}\right)C_t\left(x - \hat{X}_t\right); \quad U_t = \left(X_t^{\mathrm{T}} - \hat{X}_t^{\mathrm{T}}\right)C_t\left(X_t - \hat{X}_t\right);$$
$$U_\lambda = \left(\partial^{\mathrm{T}}/i\partial\lambda - \hat{X}_t\right)C_t\left(\partial/i\partial\lambda - \hat{X}_t\right); \qquad (48)$$

E^{EA} is the expectation for the ellipsoidal distribution (46).

According to [11], in order to compile the stochastic differential equations for the coefficients c_ν, one has to find the stochastic Ito differential of the product $q_\chi(u)\tilde{g}_t(\lambda)$ bearing in mind that u depends on the estimate $\hat{X}_t = m_t/\mu_t$ and the expectation m_t and the normalizing function μ_t obey the stochastic differential equations. Therefore, one has to replace the variables x and u and the operators $\partial/i\partial\lambda$ and U_λ, carry out differentiation, and then assume that $\lambda = 0$.

So by repeating [11], we get that the equations for m_t and μ_t with the function $\hat{\varphi}_1$ obey the formula

$$\hat{\varphi}_1 = \mathrm{E}_{\Delta^x}^{p_t}[\varphi_1], \qquad (49)$$

with regard to the notation

$$\sigma_0 = \psi\nu_0\psi^{\mathrm{T}}, \quad \sigma_1 = \psi\nu_0\psi_1^{\mathrm{T}}, \quad \sigma_2 = \psi_1\nu_0\psi_1^{\mathrm{T}} \tag{50}$$

and the equation for $\tilde{g}_t(\lambda, \Theta)$ is representable as

$$dm_t = fdt + hdY_t, \quad d\mu_t = bdY_t, \tag{51}$$

$$d\tilde{g}_t = Adt + BdY_t. \tag{52}$$

It is denoted here that

$$f = \mu_t f_0 + \sum_{\nu=1}^{N} c_\nu f_\nu, \quad h = \mu_t h_0 + \sum_{\nu=1}^{N} c_\nu h_\nu, \quad b = \mu_t b_0 + \sum_{\nu=1}^{N} c_\nu b_\nu,$$

$$f_0 = f_0(Y_t, \hat{X}_t, \Theta, t) = \mathrm{E}_{\Delta^x}^{w}[\varphi], \; f_\nu = f_\nu(Y_t, \hat{X}_t, , \Theta, t) = \mathrm{E}_{\Delta^x}^{wp_\nu}[\varphi],$$

$$h_0 = h_0(Y_t, \hat{X}_t, \Theta, t) = \mathrm{E}_{\Delta^U}^{w}\left[\sigma_1(Y_t, \Theta, t) + X\varphi_1(X, Y_t, \Theta, t)^{\mathrm{T}}\right]\sigma_2(Y_t, \Theta, t)^{-1},$$

$$h_\nu = h_\nu(Y_t, \hat{X}_t, \Theta, t) = \mathrm{E}_{\Delta^U}^{wp_\nu}\left[\sigma_1(X, Y_t, \Theta, t) + X\varphi_1(X, Y_t, \Theta, t)^{\mathrm{T}}\right]\sigma_2(Y_t, \Theta, t)^{-1},$$

$$b_0 = b_0(Y_t, \hat{X}_t, \Theta, t) = \mathrm{E}_{\Delta^U}^{w}\left[\varphi_1(X, Y_t, \Theta, t)^{\mathrm{T}}\right]\sigma_2(Y_t, \Theta, t)^{-1}, \quad b_\nu = b_\nu(Y_t, \hat{X}_t, \Theta, t)$$
$$= \mathrm{E}_{\Delta^U}^{wp_\nu}[\varphi_1(X, Y_t, \Theta, t)]\sigma_2(Y_t, \Theta, t)^{-1},$$

$$A = \mathrm{E}_{\Delta^x}^{\tilde{p}_t}\left\{i\lambda^{\mathrm{T}}\varphi(X, Y_t, \Theta, t) - \frac{1}{2}\lambda^{\mathrm{T}}\left(\psi'\nu_0\psi'^{\mathrm{T}}\right)(X, Y_t, \Theta, t)\lambda\right.$$

$$\left.\int_{R_0^q}\left[e^{i\lambda^{\mathrm{T}}\psi''(X, Y_t, \Theta, t, v)} - 1 - i\lambda^{\mathrm{T}}\psi''(X, Y_t, \Theta, t, v)\right]\nu_P(t, dv)e^{i\lambda^{\mathrm{T}}X}\right\},$$

$$B = \mathrm{E}_{\Delta^x}^{\tilde{p}_t}\left[\varphi_1(X, Y_t, \Theta, t)^{\mathrm{T}} + i\lambda^{\mathrm{T}}\left(\psi'\nu_0\psi_1'^{\mathrm{T}}\right)(X, Y_t, \Theta, t)\right]e^{i\lambda^{\mathrm{T}}X}\left(\psi_{1'}\nu_0\psi_1'^{\mathrm{T}}\right)^{-1}(X, Y_t, \Theta, t). \tag{53}$$

The equations for coefficient of MOE in (46) and (47) in virtue of [11] have the form

$$dc_\chi = \mathrm{E}_{\Delta^x}^{p^*}\{q_{\chi'}(u)\left(2\varphi^{\mathrm{T}}C_t(X - \hat{X}_t) + \mathrm{tr}[C_t\sigma_0]\right) + 2q_{\chi''}(u)\left(X^{\mathrm{T}} - \hat{X}_t^{\mathrm{T}}\right)C_t\sigma_0C_t(X - \hat{X}_t)$$
$$+\int_{R_0^q}\left[q_\chi(\bar{u}) - q_\chi(u) - 2q_{\chi'}(u)\left(X^{\mathrm{T}} - \hat{X}_t^{\mathrm{T}}\right)C_t\psi''\right]\nu_P(t, dv) - q_{\chi'}(u)\left(X^{\mathrm{T}} - \hat{X}_t^{\mathrm{T}}\right)C_t(h + \hat{X}_t b)\varphi_1/\mu_t$$
$$+q_{\chi'}(u)\mathrm{tr}\left[(h + \hat{X}_t b)\sigma_1^{\mathrm{T}}C_t\right]/\mu_t + 2q_{\chi''}(u)\mathrm{tr}\left[(h + \hat{X}_t b)\sigma_1^{\mathrm{T}}C_t(X - \hat{X}_t)\left(X^{\mathrm{T}} - \hat{X}_t^{\mathrm{T}}\right)C_t\right]/\mu_t\}dt$$
$$+\left\{\frac{1}{2n}(c_{\chi-1} + 2\chi c_\chi)\mathrm{tr}\left[\dot{C}_t K_t\right] + \frac{c_{\chi-1}}{2n}\left(\mathrm{tr}\left[C_t h\sigma_2 h^{\mathrm{T}}\right] - 2\hat{X}_t^{\mathrm{T}}C_t h\sigma_2 b^{\mathrm{T}} + \hat{X}_t^{\mathrm{T}}C_t\hat{X}_t b\sigma_2 b^{\mathrm{T}}\right)/\mu_t^2\right\}dt$$
$$+\mathrm{E}_{\Delta^x}^{p^*}\left\{\left[q_\chi(u)\varphi_1^{\mathrm{T}} + q_{\chi'}(u)\left(X^{\mathrm{T}} - \hat{X}_t^{\mathrm{T}}\right)C_t\sigma_1\right]\sigma_2^{-1}\right\}dY_t. \tag{54}$$

In addition to the notation (54), we assume that

$$\gamma_{\chi 0} = \gamma_{\chi 0}(Y_t, \hat{X}_t, \Theta, t) = E_{\Delta^x}^{w}\{q_{\chi'}(u)\left(2\varphi(X, Y_t, \Theta, t)^{\mathrm{T}} C_t(X - \hat{X}_t) + \mathrm{tr}[C_t \sigma_0(X, Y_t, \Theta, t)]\right)$$
$$+2q_{\chi''}(u)\left(X^{\mathrm{T}} - \hat{X}_t^{\mathrm{T}}\right)C_t\sigma_0(X, Y_t, \Theta, t)C_t(X - \hat{X}_t) + \int_{R_0^q}\left[q_{\chi}(\overline{u}) - q_{\chi}(u) - 2q_{\chi'}(u)\left(X^{\mathrm{T}} - \hat{X}_t^{\mathrm{T}}\right)C_t\psi''(X, Y_t, \Theta, t, v)\right] \nu_P(t, dv)\},$$

$$\gamma_{\chi\nu} = \gamma_{\chi\nu}(Y_t, \hat{X}_t, \Theta, t) = E_{\Delta^x}^{wp_\nu}\{q_{\chi'}(u)\left(2\varphi(X, Y_t, \Theta, t)^{\mathrm{T}} C_t(X - \hat{X}_t) + \mathrm{tr}[C_t \sigma_0(X, Y_t, \Theta, t)]\right)$$
$$+2q_{\chi''}(u)\left(X^{\mathrm{T}} - \hat{X}_t^{\mathrm{T}}\right)C_t\sigma_0(X, Y_t, \Theta, t)C_t(X - \hat{X}_t)$$
$$+\int_{R_0^q}\left[q_{\chi}(\overline{u}) - q_{\chi}(u) - 2q_{\chi'}(u)\left(X^{\mathrm{T}} - \hat{X}_t^{\mathrm{T}}\right)C_t\psi''(X, Y_t, \Theta, t, v)\right] \nu_P(t, dv)\},$$

$$\varepsilon_{\chi 0} = \varepsilon_{\chi 0}(Y_t, \hat{X}_t, \Theta, t) = \mathrm{E}_{\Delta^x}^{w}\{q_{\chi'}(u)\left[\sigma_1(X, Y_t, \Theta, t)^{\mathrm{T}} - \varphi_1(X, Y_t, \Theta, t)\left(X^{\mathrm{T}} - \hat{X}_t^{\mathrm{T}}\right)\right]$$
$$+2q_{\chi''}(u)\sigma_1(X, Y_t, \Theta, t)^{\mathrm{T}} C_t(X - \hat{X}_t)\left(X^{T} - \hat{X}_t^{\mathrm{T}}\right)\},$$

$$\varepsilon_{\chi\nu} = \varepsilon_{\chi\nu}(Y_t, \hat{X}_t, \Theta, t) = \mathrm{E}_{\Delta^x}^{wp_\nu}\{q_{\chi'}(u)\left[\sigma_1(X, Y_t, \Theta, t)^{\mathrm{T}} - \varphi_1(X, Y_t, \Theta, t)\left(X^{\mathrm{T}} - \hat{X}_t^{\mathrm{T}}\right)\right]$$
$$+2q_{\chi''}(u)\sigma_1(X, Y_t, \Theta, t)^{\mathrm{T}} C_t(X - \hat{X}_t)\left(X^{T} - \hat{X}_t^{\mathrm{T}}\right)\},$$

$$\eta_{\chi 0} = \eta_{\chi 0}(Y_t, \hat{X}_t, \Theta, t) = E_{\Delta^x}^{w}\left\{q_{\chi}(u)\varphi_1(X, Y_t, \Theta, t)^{\mathrm{T}} + q_{\chi'}(u)\left(X^{\mathrm{T}} - \hat{X}_t^{\mathrm{T}}\right)C_t\sigma_1(X, Y_t, \Theta, t)\right\}\sigma_2(Y_t, \Theta, t)^{-1},$$

$$\eta_{\chi\nu} = \eta_{\chi\nu}(Y_t, \hat{X}_t, \Theta, t) = E_{\Delta^x}^{wp_\nu}\left\{q_{\chi}(u)\varphi_1(X, Y_t, \Theta, t)^{\mathrm{T}} + q_{\chi'}(u)\left(X^{\mathrm{T}} - \hat{X}_t^{\mathrm{T}}\right)C_t\sigma_1(X, Y_t, \Theta, t)\right\}\sigma_2(Y_t, \Theta, t)^{-1}. \quad (55)$$

and then we can rearrange Eq. (54) in

$$dc_{\chi} = \left\{\mu_t\gamma_{\chi 0}(Y_t, \hat{X}_t, \Theta, t) + \sum_{\nu=1}^{N} c_\nu\gamma_{\chi\nu}(Y_t, \hat{X}_t, \Theta, t) + \mathrm{tr}\left[\{\mu_t(h_0(Y_t, \hat{X}_t, \Theta, t) + \hat{X}_t b_0(Y_t, \hat{X}_t, \Theta, t)\}\right.\right.$$
$$\left.+\sum_{\nu=1}^{N} c_\nu\left(h_\nu(Y_t, \hat{X}_t, \Theta, t) + \hat{X}_t b_\nu(Y_t, \hat{X}_t, \Theta, t)\right)\left\{\varepsilon_{\chi 0}(Y_t, \hat{X}_t, \Theta, t) + \sum_{\nu=1}^{N} c_\nu\varepsilon_{\chi\nu}(Y_t, \hat{X}_t, \Theta, t)/\mu_t\right\}C_t\right]$$
$$+\frac{1}{2n}(c_{\chi-1} + 2\chi c_{\chi})\mathrm{tr}[\dot{C}_t K_t] + \frac{c_{\chi-1}}{2n}\mathrm{tr}\left[C_t\left(h_0(Y_t, \hat{X}_t, \Theta, t) + \sum_{\nu=1}^{N} c_\nu h_\nu(Y_t, \hat{X}_t, t)/\mu_t\right)\sigma_2(Y_t, \Theta, t)\right.$$
$$\left.\times\left(h_0(Y_t, \hat{X}_t, \Theta, t)^{\mathrm{T}} + \sum_{\nu=1}^{N} c_\nu h_\nu(Y_t, \hat{X}_t, \Theta, t)^{\mathrm{T}}/\mu_t\right)\right] - 2\hat{X}_t^{\mathrm{T}} C_t\left(h_0(Y_t, \hat{X}_t, \Theta, t) + \sum_{\nu=1}^{N} c_\nu h_\nu(Y_t, \hat{X}_t, \Theta, t)/\mu_t\right)$$
$$\times\sigma_2(Y_t, \Theta, t)\left(b_0(Y_t, \hat{X}_t, \Theta, t)^{\mathrm{T}} + \sum_{\nu=1}^{N} c_\nu b_\nu(Y_t, \hat{X}_t, \Theta, t)^{\mathrm{T}}/\mu_t\right)$$
$$+\hat{X_t\mathrm{T}}C_t\hat{X}_t\left(b_0(Y_t, \hat{X}_t, \Theta, t) + \sum_{\nu=1}^{N} c_\nu b_\nu(Y_t, \hat{X}_t, \Theta, t)/\mu_t\right)\sigma_2(Y_t, \Theta, t)$$
$$\left.\times\left(b_0(Y_t, \hat{X}_t, \Theta, t) + \sum_{\nu=1}^{N} c_\nu b_\nu(Y_t, \hat{X}_t, \Theta, t)^{\mathrm{T}}/\mu_t\right)\right\}dt$$
$$+\left\{\mu_t\eta_{\chi 0}(Y_t, \hat{X}_t, \Theta, t) + \sum_{\nu=1}^{N} c_\nu\eta_{\chi\nu}(Y_t, \hat{X}_t, \Theta, t)\right\}dY_t \quad (\chi = 1, \ldots, N). \quad (56)$$

The modified ellipsoidal suboptimal filter (MESOF) is defined by Eqs. (51), (52), and (56) and the relation $\hat{X}_t = m_t/\mu_t$ under the initial conditions

$$m(t_0) = \mathrm{E}[X_0|Y_0], \quad \mu(t_0) = 1, \quad c_{\chi}(t_0) = c_{\chi 0} \quad (\chi = 1, \ldots, N), \quad (57)$$

($c_{\chi 0}$ $(\chi = 1, \ldots, N)$ are the coefficients of the expansion (46) of the probability density $\tilde{p}_{t_0}(x) = p_0(x|Y_0)$ of the vector X_0 relative to Y_0).

Upon solution of Eqs. (51), (52), (56), and (57), the rms optimal estimate of the state vector and the covariance matrix of filtration error in MESOF obey the following approximate formulae:

$$\hat{X}_t = m_t/\mu_t;$$
$$R_t = \mathrm{E}^{w}_{\Delta^x}\left[\left(X - \frac{m_t}{\mu_t}\right)\left(X^{\mathrm{T}} - \frac{m_t^{\mathrm{T}}}{\mu_t}\right)\right] + \sum_{\nu=1}^{N}\frac{c_\nu}{\mu_t}\mathrm{E}^{wp_\nu}_{\Delta^x}\left[\left(X - \frac{m_t}{\mu_t}\right)\left(X^{\mathrm{T}} - \frac{m_t^{\mathrm{T}}}{\mu_t}\right)\right]. \quad (58)$$

Note that the order of the obtained MESOF, especially under high dimension n of the system state vector, is much lower than the order of other conditionally optimal filters. It is the case at allowing for the moments of up to the 10th order. Then, already for $n>3$ and $N = 5$, we have $n + N + 1 \leq n(n+3)/2$. We conclude that for $n>3$ and $N = 5$, MECOF has a lower order than the filters of the method of normal approximation, generalized second-order Kalman-Bucy filters, and Gaussian filter. Thus, the following theorems underlie the algorithm of modified ellipsoidal conditionally optimal filtration.

Statement 10. Under the conditions of Statement 8, if there is MECOF, then it is defined by Eqs. (51), (52), and (56) under the conditions (57) and (58).

Statement 11. Under the conditions of Statement 9, if there is MESOF, then it is defined by the equations of Statement 10 under the conditions (45).

The aforementioned methods of MESOF construction offer a basic possibility of getting a filter close to the optimal-in-estimate one with any degree of accuracy. The higher the EA coefficient, the maximal order of the allowed for moments, the higher accuracy of approximation of the optimal estimate. However, the number of equations defining the parameters of the a posteriori one-dimensional ellipsoidal distribution grows rapidly with the number of allowed for parameters. At that, the information about the analytical nature of the problem becomes pivotal.

For approximate analysis of the filtration equations by following [11] and allowing for random nature of the parameters Θ, we come to the following equations for the first-order sensitivity functions [11]:

$$\begin{aligned} d\nabla^\Theta \hat{X}_s &= \nabla^\Theta A^{\hat{X}_s}dt + \nabla^\Theta B^{\hat{X}_s}dY_t, \quad \nabla^\Theta B^{\hat{X}_s}(t_0) = 0,\\ d\nabla^\Theta R_{sq} &= \nabla^\Theta A^{R_{sq}}dt + \nabla^\Theta B^{R_{sq}}dY_t, \quad \nabla^\Theta R_{sq}(t_0) = 0,\\ d\nabla^\Theta c_\kappa &= \nabla^\Theta A^{c_\kappa}dt + \nabla^\Theta B^{c_\kappa}dY_t, \quad \nabla^\Theta c_\kappa(t_0) = 0. \end{aligned} \quad (59)$$

Here the procedure of taking the derivatives is carried out over all input variables, and the coefficients of sensitivity are calculated for $\Theta = m^\Theta$. It is assumed at that the variance is small as compared with their expectations. Obviously, at differentiation with respect to Θ $\left(\nabla^\Theta = \partial/\partial\Theta\right)$, the order of the equations grows in proportion to the number of derivatives. The equations for the elements of the matrices of the second sensitivity functions are made up in a similar manner.

To estimate the MESOF (MECOF) performance, we follow [5, 8] and introduce for the Gaussian Θ with the expectation m^Θ and covariance matrix K^Θ the conditional loss function admitting quadratic approximation, the factor $\varepsilon = \varepsilon_2^{1/4}$, as well as

$$\rho^{\hat{X}_s} = \rho^{\hat{X}_s}(\Theta) = \rho(m^\Theta) + \sum_{ii=1}^{n^\Theta}\rho_i'(m^\Theta)\Theta_s^0 + \sum\sum_{i,j=1}\rho_{ij}''(m^\Theta)\Theta_i^0\Theta_j^0. \quad (60)$$

It is denoted here

$$\varepsilon_2 = \mathrm{E}^{EA}\left[\rho(\Theta)^2\right] - \rho\left(m^{\Theta}\right)^2,$$
$$\mathrm{E}^{EA}\left[\rho(\Theta)^2\right] = \rho\left(m^{\Theta}\right)^2 + \rho'\left(m^{\Theta}\right)^{\mathrm{T}} K^{\Theta} \rho'\left(m^{\Theta}\right) + 2\rho\left(m^{\Theta}\right)\mathrm{tr}\left[\rho''\left(m^{\Theta}\right)K^{\Theta}\right]$$
$$+\left\{\mathrm{tr}\left[\rho''\left(m^{\Theta}\right)K^{\Theta}\right]\right\}^2 + 2\mathrm{tr}\left[\rho''\left(m^{\Theta}\right)K^{\Theta}\right]^2. \tag{61}$$

At that, in (61) the functions ρ' and ρ'' are determined through certain formulas on the basis of the first and second sensitivity functions. Therefore, we come to the following result.

Statement 12. Estimation of MESOF (MECOF) performance under the conditions of Statements 10 and 11 relies on Eqs. (59)–(61) under the corresponding derivatives in the right sides of Eq. (59).

8. New types of continuous MECOF

Based on Statements 10 and 11 in [18], continuous MECOF were described. We consider the problem of continuous conditionally optimal filtration for the general case of Eqs. (34) and (35) where it is desired to determine the optimal estimate $\hat{X}_t$ of process X_t at the instant $t>t_0$ from the results of observation of this process until the instant t, that is, over the interval $[t_0, t)$, in the class of permissible $\hat{X}_t = AZ_t$ estimates and with a stochastic differential equation given by

$$dZ_t = [\alpha_t \xi(Y_t, Z_t, \Theta, t) + \gamma_t]dt + \beta_t \eta(Y_t, Z_t, \Theta, t)dY_t \tag{62}$$

under the given vector and matrix structural functions ξ and η and every possible time functions $\alpha_t, \beta_t, \gamma_t$ (α_t and β_t are matrices and γ_t is a vector). The criterion for minimal rms error of the estimate Z_t is used as the optimality criterion. It is common knowledge that selection of the class of permissible filters defined by the structural functions ξ and η in Eq. (62) is the greatest challenge in practice of using the COF theory [1, 3, 11]. In principle they can be defined arbitrarily. One can select ξ and η at will so that the class of permissible filters contained an arbitrarily defined COF. In this case, COF is in practice more precise than the given COF. At the same time, by selecting a finite segment of some basis in the corresponding Hilbertian space L_2 as components of the vector function ξ and elements of the matrix function η, one can obtain an approximation with any degree of precision to the unknown optimal functions ξ and η. This technique of selecting the functions ξ and η on the basis of the equations of the theory of suboptimal filtration seems to be the most rational one. At that, the COF equations obtained from the equation for the nonnormalized a posteriori characteristic function open up new possibilities.

To use the equations obtained from nonnormalized equations for the a posteriori distribution, one needs to change the formulation of the COF problems [3, 11] so as to use the equation for the factor μ_t. For that, we take advantage of the following equations to determine the class of permissible continuous MECOF (62):

$$d\mu_t = \rho_t \chi(Y_t, Z_t, \Theta, t)dY_t, \tag{63}$$
$$\hat{X}_t = AZ_t/\mu_t, \tag{64}$$

where $\chi(Y_t, Z_t, \Theta, t)$ is a certain given structural matrix function and ρ_t is the row matrix of coefficients depending on t and subject to rms optimization along with the coefficients α_t, β_t, and γ_t in the filter Eq. (62).

Relying on the results of the last section and generalizing [7], one can specify the following types of the permissible MECOF:

1. This type of permissible MECOF can be obtained by assuming $Z_t = m_t, A = I_n$ and determining the functions ξ, η, χ in Eqs. (62) and (63) and obeying Eq. (51) which gives rise to the following expressions for the structural functions:

$$\xi = \xi(Y_t, Z_t, \Theta, t) = \left[\mu_t f_0(Y_t, Z_t/\mu_t, \Theta, t)^{\mathrm{T}} f_1\left(Y_t, Z_t/\mu_t, \Theta, t\right)^{\mathrm{T}} \dots f_N\left(Y_t, Z_t/\mu_t, \Theta, t\right)^{\mathrm{T}}\right]^{\mathrm{T}}; \tag{65}$$

$$\begin{aligned} \eta &= \eta(Y_t, Z_t, \Theta, t) \\ &= \left[\mu_t h_0(Y_t, Z_t/\mu_t, \Theta, t)^{\mathrm{T}} h_1\left(Y_t, Z_t/\mu_t, \Theta, t\right)^{\mathrm{T}} \dots h_N\left(Y_t, Z_t/\mu_t, \Theta, t\right)^{\mathrm{T}}\right]^{\mathrm{T}}; \end{aligned} \tag{66}$$

$$\begin{aligned} \chi &= \chi(Y_t, Z_t, \Theta, t) \\ &= \left[\mu_t b_0(Y_t, Z_t/\mu_t, \Theta, t)^{\mathrm{T}} b_1\left(Y_t, Z_t/\mu_t, \Theta, t\right)^{\mathrm{T}} \dots b_N\left(Y_t, Z_t/\mu_t, \Theta, t\right)^{\mathrm{T}}\right]^{\mathrm{T}}. \end{aligned} \tag{67}$$

At that, the order of MECOF defined by Eqs. (62) and (63) is equal to $n+1$. This type of MECOF may be designed for Z_t being constant Z_0 and $A = I_\ell (\ell < n)$.

2. To obtain a wider class of permissible MECOF, rearrange Eq. (56) in

$$\begin{aligned} dc_\chi = &\left\{ F_{\chi 0}(Y_t, Z_t, \Theta, t) + \sum_{\nu=1}^{N} c_\nu F_{\chi\nu}(Y_t, Z_t, \Theta, t) + \sum_{\lambda, \nu=1}^{N} c_\lambda c_\nu F_{\chi\lambda\nu}(Y_t, Z_t, \Theta, t) \right. \\ &\left. + c_{\chi-1} \sum_{\lambda, \nu=1}^{N} c_\lambda c_\nu F_{\chi\lambda\nu'}(Y_t, Z_t, \Theta, t) \right\} dt \\ &+ \left\{ \mu_t \eta_{\chi 0}(Y_t, Z_t, \Theta, t) + \sum_{\nu=1}^{N} c_\nu \eta_{\chi\nu}(Y_t, Z_t, \Theta, t) \right\} dY_t \end{aligned} \tag{68}$$

with the following notations:

$$F_{\chi 0}(Y_t, Z_t, \Theta, t) = \mu_t \gamma_{\chi 0}(Y_t, Z_t, \Theta, t) + \mu_t \mathrm{tr}\left[(h_0(Y_t, Z_t, \Theta, t) Z_t b_0(Y_t, Z_t, \Theta, t))\varepsilon_{\chi 0}(Y_t, Z_t, \Theta, t)\right];$$

$$F_{\chi\nu}(Y_t, Z_t, \Theta, t) = \gamma_{\chi 0}(Y_t, Z_t, \Theta, t) + \mathrm{tr}[(h_\nu(Y_t, Z_t, \Theta, t) + Z_t b_\nu(Y_t, Z_t, \Theta, t))\varepsilon_{\chi 0}(Y_t, Z_t, \Theta, t) + (h_0(Y_t, Z_t, \Theta, t)$$

$$+ Z_t b_0(Y_t, Z_t, \Theta, t))\varepsilon_{\chi\nu}(Y_t, Z_t, \Theta, t)] + \frac{1}{2n}\delta_{\chi-1,\nu}\{\mathrm{tr}\left[b_\nu K_t + C_t h_0(Y_t, Z_t, \Theta, t)\sigma_2(Y_t, \Theta, t) h_0(Y_t, Z_t, \Theta, t)^{\mathrm{T}}\right]$$

$$-2Z_t^{\mathrm{T}} C_t h_0(Y_t, Z_t, \Theta, t)\sigma_2(Y_t, \Theta, t) b_0(Y_t, Z_t, \Theta, t)^{\mathrm{T}} + Z_t^{\mathrm{T}} C_t Z_t b_0(Y_t, Z_t, \Theta, t)\sigma_2(Y_t, \Theta, t) b_0(Y_t, Z_t, \Theta, t)^{\mathrm{T}}\} + \frac{1}{n}\chi\delta_{\chi\nu}\mathrm{tr}\left[\dot{C}_t K_t\right];$$

$$F_{\chi\lambda\nu} = \mathrm{tr}\left[C_t(h_\lambda(Y_t, Z_t, \Theta, t) + Z_t b_\lambda(Y_t, Z_t, \Theta, t))\varepsilon_{\chi\nu}(Y_t, Z_t, \Theta, t)\right]/\mu_t$$

$$+\frac{1}{n}\delta_{\chi-1,\lambda}\{\mathrm{tr}\left[C_t h_0(Y_t, Z_t, \Theta, t)\sigma_2(Y_t, \Theta, t) h_\nu(Y_t, Z_t, \Theta, t)^{\mathrm{T}}\right] - Z_t^{\mathrm{T}} C_t(h_0(Y_t, Z_t, t)\sigma_2(Y_t, t) b_\nu(Y_t, Z_t\Theta, , t)^{\mathrm{T}}$$

$$+h_\nu(Y_t, Z_t, \Theta, t)\sigma_2(Y_t, \Theta, t) b_0(Y_t, Z_t, \Theta, t)^{\mathrm{T}} + Z_t^{\mathrm{T}} C_t Z_t b_0(Y_t, Z_t, \Theta, t)\sigma_2(Y_t, \Theta, t) b_\nu(Y_t, Z_t, \Theta, t)^{\mathrm{T}}\}/\mu_t;$$

$$\begin{aligned} F_{\chi\lambda\nu'} = \frac{1}{2n} &\left\{ \mathrm{tr}\left[C_t(h_\lambda(Y_t, Z_t, \Theta, t) + \sigma_2(Y_t, \Theta, t) h_\nu(Y_t, Z_t, \Theta, t)^{\mathrm{T}}\right] \right. \\ &- 2Z_t^{\mathrm{T}} C_t h_\lambda(Y_t, Z_t, \Theta, t)\sigma_2(Y_t, \Theta, t) b_\nu(Y_t, Z_t, \Theta, t)^{\mathrm{T}} \\ &\left. + Z_t^{\mathrm{T}} C_t Z_t b_\lambda(Y_t, Z_t, \Theta, t)\sigma_2(Y_t, t) b_\nu(Y_t, Z_t, \Theta, t)^{\mathrm{T}} \right\} / \mu_t^2. \end{aligned} \tag{69}$$

By taking as the basis for the type of permissible MECOF Eqs. (51), (52), and (68), one has to regard all components of the vector Z_t as all components of the vector m_t and coefficients $c_1, \dots, c_N$ so that $Z_t = \left[m_t^T c_1 \dots c_N\right]^T$. At that, the order of all permissible filters is equal to n + N + 1. Putting $Z_t = Z_0$ and $A = I_l \quad (l < n)$, one gets the corresponding MECOF.

3. The widest class of permissible filters providing MECOF of the maximal reachable accuracy can be obtained if one takes the function ξ in (62) as the vector with all components of the vector functions $\mu f_{\chi 0}, c_\nu f_{\chi\nu} (\chi, \nu = 1, \dots, N)$ in Eqs. (65) and (66) all addends involved in the scalar functions $F_{\chi 0}, c_\nu F_{\chi\nu}, c_\lambda c_\nu F_{\chi\lambda\nu}$ $(\chi, \lambda, \nu = 1, \dots, N)$ $c_{\chi-1} c_\lambda c_\nu F_{\chi\lambda\nu'}$, $(\chi - 1, \lambda, \nu = 1, \dots, N)$ in (68) and as the function η in (62), the matrix whose rows are the row matrices $\mu_t h_{\chi 0}, c_\nu h_{\chi\nu} (\chi, \nu = 1, \dots, N)$ in (68) and all row matrices $\mu\eta_{\chi 0}, c_\nu \eta_{\chi\nu} (\chi, \nu = 1, \dots, N)$ in (69). As for the function χ in Eq. (63), it is determined through (67) as in the case of the simplest types of permissible filters. The so-determined class of permissible filters has ECOF defined by Eqs. (51), (52), and (69), at that ECOF is more precise than ESOF. We notice that this class of permissible filters can give rise to an overcomplicated ECOF because of high dimension of the structural vector function ξ. So we distinguish the following new type of permissible filters.

4. Components of the vector function ξ are all components of the vector functions $\mu_t f_{\chi 0}, c_\nu f_{\chi\nu} (\chi, \nu = 1, \dots, N)$ and all scalar functions $F_{\chi 0}, c_\nu F_{\chi\nu}, c_\lambda c_\nu F_{\chi\lambda\nu}$ $(\chi, \lambda, \nu = 1, \dots, N)$, $c_{\chi-1} c_\lambda c_\nu F_{\chi\lambda\nu'}$ $(\chi - 1, \lambda, \nu = 1, \dots, N)$ without decomposing them into individual addends. This class of permissible filters also includes ECOF (51), (52), and (69).

To determine the coefficients α_t, β_t, and γ_t of the equation MECOF (62), one needs to know the joint one-dimensional distribution of the random processes X_t and $\hat{X}_t$. It is determined by solving the problem of analysis of the system obeying the stochastic differential Eqs. (62) and (63). As always in the theory of conditionally optimal filtration, all complex calculations required to determine the optimal coefficients of the MECOF Eq. (62) or (63) are based only on the a priori data and therefore can be carried out in advance at designing MECOF. At that, the accuracy of filtration can be established for each permissible MECOF. The process of filtration itself comes to solving the differential equation, which enables one to carry out real-time filtration.

Consequently, we arrive to the following results.

Statement 13. Under the conditions of Statement 8, the MECOF equations like (62) and (63) coincide with the equations of continuous MECOF where the structural functions belong to the four aforementioned types.

Statement 14. The rms MECOF of the order $n_x + 1$ coinciding with MECOF is defined for CStS (34) and (35), Eqs. (62)–(64), and the structural functions of the first class.

Statement 15. The rms of MECOF of the order $n_x + N + 1$ coinciding with MECOF obeys for the CStS (34) and (35), Eqs. (62)–(64), and the structural functions of Statement 14.

Statement 16. If accuracy of MECOF determined according to Statement 14 is insufficient, then the functions of the Statement 15 can be used as structural ones.

Statement 17. The relations of Statement 12 underlie the estimate of quality of MECOF under the conditions of Statements 13–15, provided that there are corresponding derivatives in the right sides of the equations.

Example 3. The presented MECOF for linear CStS coincide with Kalman-Bucy filter [2–4, 11].

Example 4. MECOF for linear CStS with parametric noises coincide with linear Pugachev conditionally optimal filter.

Finally let us consider quasilinear CStS (36) and (37), reducible to the following differential one:

$$\dot{X}_t = \varphi(X_t, \Theta, t) + \psi(t, \Theta)V_1^{EL}, \tag{70}$$

$$\dot{Y}_t = \varphi_1(X_t, \Theta, t) + V_2^{EL} \tag{71}$$

where V_1 and V_2 are non-Gaussian white noises. In this case using ELM and Kalman-Bucy filters with parameters depending on m_t^x, K_t^x and c_{1t}^x, we get the following interconnected set of equations:

$$\dot{m}_1^x = \varphi_{00} \quad m_{t_0}^x = m_0^x, \quad \dot{m}_1^y = \varphi_{10} \quad m_{t_0}^y = m_0^y, \tag{72}$$

$$\dot{X}_t^0 = \varphi_{01}X_t^0 + \psi(t, \Theta)V_1, \quad \dot{Y}_t^0 = \varphi_{11}X_t^0 + V_2^{EL}, \quad Y_{t_0}^0 = Y_0^0, \tag{73}$$

$$\dot{K}_t^x = \varphi_{11}K_t^x + K_t^x\varphi_{11}^T + \psi(t, \Theta)G_1^{EL}(t, \Theta)\psi(t, \Theta)^T, \quad K_{t_0}^x = K_0^x, \tag{74}$$

$$\begin{aligned} &\dot{\hat{X}}_t = \varphi_{00} - \varphi_{01}m_t^x + \varphi_{01}\hat{X}_t + R_tG_2^{EL}(t, \Theta)^{-1}\left[\dot{Y}_t - \varphi_{11}\hat{X}_t - \varphi_{10} + \varphi_{11}m_t^x\right), \\ &\hat{X}_0 = M^{EL}\hat{X}_{t_0}, \end{aligned} \tag{75}$$

$$\begin{aligned} &\dot{R}_t = \varphi_{01}R_t - R_t\varphi_{01} - R_t\varphi_{11}G_2^{EL}(t, \Theta)^{-1}\varphi_{11}R_t + \psi(t)G_1^{EL}(t, \Theta)\psi(t)^T, \\ &R_0 = M^{EL}\left[\left(X_0 - \hat{X}_0\right)\left(X_0 - \hat{X}_0\right)^T\right]. \end{aligned} \tag{76}$$

Here the following notations are used:

$$\begin{aligned} &m_t^x = M^{EL}X_t, \quad m_t^y = M^{EL}Y_t \quad \text{and} \quad K_t^x = M^{EL}\left[X_t^0\, X_t^{0T}\right], \\ &R_t = M^{EL}\left[\left(X_t - \hat{X}_t\right)\left(X_t - \hat{X}_t\right)^T\right] \end{aligned} \tag{77}$$

being the mathematical expectations, state, and error covariance matrices

$$\begin{aligned} &\varphi_{00} = \varphi_{00}\left(m_t^x, K_t^x, c_{1t}^x, t, \Theta\right), \quad \varphi_{10} = \varphi_{10}\left(m_t^x, K_t^x, c_{1t}^x, t, \Theta\right), \\ &\varphi_{01} = \varphi_{01}\left(m_t^x, K_t^x, c_{1t}^x, t, \Theta\right) = \frac{\partial \varphi_{01}}{\partial m_t^x}, \quad \varphi_{11} = \varphi_{11}\left(m_t^x, K_t^x, c_{1t}^x, t, \Theta\right) = \frac{\partial \varphi_{10}}{\partial m_t^x} \end{aligned} \tag{78}$$

being ELM ecoefficiencies, G_i^{EL} $(i = 1, 2)$ are intensities of normal EL equivalent white noises. So Eqs. (72)–(78) define the corresponding Statement 18.

9. Ellipsoidal Pugachev conditionally optimal continuous control

The idea of conditionally optimal control (COC) was suggested by Pugachev (IFAC Workshop on Differential Games, Russia, Sochi, 1980) and developed [13]. The COC essence is in the search of optimal control among all permissible controls (as in classical control theory) but in the restricted class of permissible controls. These controls are computed in online regime. At practice the permissible continuous class of controls may be defined by the set of ordinary differential equations of the given structure.

So let us consider the following Ito equations:

$$dX = \varphi(X, Y, U, t)dt + \psi_1(X, Y, U, t)dW_0 + \int_{R_0^q} \psi_2(X, Y, U, t, v)P^0(dt, dv), \quad (79)$$

$$dY = \varphi'(X, Y, U, t)dt + \psi_1'(X, Y, U, t)dW_0 + \int_{R_0^q} \psi_2'(X, Y, U, t, v)P^0(dt, dv). \quad (80)$$

Here X is the nonobservable state vector; Y is the observable vector; $U \in \mathcal{D}$ is the control vector; W_0 being the Wiener StP, $P^0(A, B)$ being the independent of W_0 centered Poisson measure; φ, ψ_1, ψ_2 and $\varphi', \psi_1', \psi_2'$ being the known functions. Integration is realized in R^q space with the deleted origin. Initial conditions X_0 and Y_0 do not depend on X and Y. Functions φ, ψ_1, ψ_2 in (79) as a rule do not depend on Y, but depend on U components that are governed by Eq. (79). Functions $\varphi', \psi_1', \psi_2'$ in Eq. (80) depend on U components that govern observation.

The class of the admissible controls is defined by the equations

$$dU = [\alpha\xi(Y, U, t) + \gamma]dt + \beta\eta(Y, U, t)dY \quad (81)$$

without restrictions and with restrictions

$$dU = [\alpha\xi(Y, U, t) + \gamma]dt + \beta\eta(Y, U, t)dY - \max\left\{0, n(U)^T[\alpha\xi(Y, U, t) + \gamma]dt + n(U)^T\beta\eta(Y, U, t)dY\right\}n(U)\mathbf{1}_{\partial\mathcal{D}}(U). \quad (82)$$

Here $n(U)$ is the unit vector of external normal for boundary $\partial\mathcal{D}$ in point U; $1_{\partial\mathcal{D}}(U)$ is the set indicator.

Conditionally optimal criteria is taken in the form of mathematical expectation of some functional depending on $X_{t_0}^t = \{X(\tau) : \tau \in [t_0, t]\}$ and $U_{t_0}^t = \{U(\tau) : \tau \in [t_0, t]\}$:

$$\rho = \mathrm{E}\ell\left(X_{t_0}^t, U_{t_0}^t, t\right), \quad (83)$$

where E is the mathematical expectation and ℓ is the loss function at the given realizations $x_{t_0}^t, u_{t_0}^t$ of $X_{t_0}^t, U_{t_0}^t$.

So according to Pugachev we define COC as the control realized by minimization (83) by choosing α, β, γ and by satisfying (82) at every time moment and at a given α, β, γ for all preceding time moments. For the loss function (83) depending on X and U at the same time, moment t is necessary to compute ellipsoidal one-dimensional distribution of X and Y in Eqs. (79), (80), and (82) using EAM (ELM). This problem is analogous to COF and MCOF design (Section 8).

For high accuracy and high availability CStS especially functioning in real-time regime, software tools "StS-Analysis," "StS-Filtering," and "StS-Control" based on NAM, EAM, and ELM were developed for scientists, engineers, and students of Russian Technical Universities.

These tools were implemented for solving safety problems for system engineering [19].

In [18, 20] theoretical propositions of new probabilistic methodology of analysis, modeling, estimation, and control in stochastic organizational-technical-economical systems (OTES) based on stochastic CALS informational technologies (IT) are considered. Stochastic integrated logistic support (ILS) of OTES modeling life cycle (LC), stochastic optimal of current state estimation in stochastic media defined by internal and external noises including specially organized OTES-NS (noise support), and stochastic OTES optimal control according to social-technical--economical-support criteria in real time by informational-analytical tools (IAT) of global type are presented. Possibilities spectrum may be broaden by solving problems of OTES-CALS integration into existing markets of finances, goods, and services. Analytical modeling, parametric optimization and optimal stochastic processes regulation illustrate some technologies and IAT given plans. Methodological support based on EAM gives the opportunity to study infrequent probabilistic events necessary for deep CStS safety analysis.

10. Conclusion

Modern continuous high accuracy and availability control stochastic systems (CStS) are described by multidimensional differential linear, linear with parametric noises, and nonlinear stochastic equations. Right-hand parts of these equations also depend on stochastic factors being random variables defining the dispersion in engineering systems parameters. Analysis and synthesis CStS needs computation of non-Gaussian probability distributions of multidimensional stochastic processes. The known analytical parametrization modeling methods demand the automatic composing and the integration of big amount interconnected equations.

Two methods of analysis and analytical modeling of multidimensional non-Gaussian CStS were worked out: ellipsoidal approximation method (EAM) and ellipsoidal linearization method (ELM). In this case one achieves cardinal reduction the amount of distribution parameters.

Necessary information about ellipsoidal approximation methods is given and illustrated. Some important remarks for engineers concerning EAM accuracy are given. It is important to note that all complex calculations are performed on design stage. Algorithms for composition of EAM (ELM) equation are presented. Application to problems of shock and vibroprotection are considered.

For statistical CStS offline and online analysis approximate methods based on EAM (ELM) for a posteriori distributions are developed. In this case one has twice reduction of equation amount. Special bank of approximate suboptimal and Pugachev conditionally optimal filters for typical identification and calibration problems based on the normalized and nonnormalized was designed and implemented.

In theoretical propositions of new probabilistic methodology of analysis, modeling, estimation, and control in stochastic OTES based on stochastic CALS information technologies (IT) are considered. Stochastic integrated logistic support (ILS) of OTES modeling life cycle, stochastic optimal of current state estimation in stochastic media defined by internal and external noises including specially organized OTES-NS (noise support), and stochastic OTES optimal control according to social-technical-economical-support criteria in real time by informational-analytical tools (IAT) of global type are presented. Possibility spectrum may be broaden by solving problems of OTES-CALS integration into existing markets of finances, goods, and services. Methodological support based on EAM (ELM) gives the opportunity to study infrequent probabilistic events necessary for deep CStS safety analysis.

Acknowledgements

The authors would like to thank Russian Academy of Science for supporting the work presented in this chapter. Authors are much obliged to Mrs. Irina Sinitsyna and Mrs. Helen Fedotova for translation and manuscript preparation.

Author details

Igor N. Sinitsyn*, Vladimir I. Sinitsyn and Edward R. Korepanov
Federal Research Center "Computer Science and Control of Russian Academy of Sciences", Moscow, Russia

*Address all correspondence to: sinitsin@dol.ru

References

[1] Pugachev VS, Sinitsyn IN. Stochastic Differential Systems. Analysis and Filtering. Chichester: John Wiley & Sons; 1987

[2] Pugachev VS, Sinitsyn IN. Theory of Stochastic Systems. Moscow: Logos; 2000 (in Russian)

[3] Pugachev VS, Sinitsyn IN. Stochastic Systems. Theory and Applications. Singapore: World Scientific; 2001

[4] Sinitsyn IN, Sinitsyn VI. Lectures of Normal and Ellipsoidal Approximation of Distributions in Stochastic Systems. Moscow: Torus Press; 2013 (in Russian)

[5] Pugachev VS, Sinitsyn IN. Lectures on Functional Analysis and Applications. Singapore: World Scientific; 1999

[6] Pugachev VS. Probability Theory and Mathematical Statistics for Engineers. Oxford: Pergamon Press; 1984

[7] Sinitsyn IN. Suboptimal orthogonal filters for nonlinear stochastic systems on manifolds. Informatika i Ee Primeneniya. 2016;**10**(1):34-44

[8] Sinitsyn IN. Normal and orthogonal suboptimal filters for nonlinear stochastic systems on manifolds. Systems and Means of Informatics. 2016;**26**(1):199-226

[9] Sinitsyn IN, Sinitsyn VI, Korepanov ER. Ellipsoidal suboptimal filters for nonlinear stochastic systems on manifolds. Informatika i Ee Primeneniya. 2016;**10**(2):24-35

[10] Sinitsyn IN, Sinitsyn VI, Korepanov ER. Modified ellipsoidal suboptimal filters for nonlinear stochastic systems on manifolds. Systems and Means of Informatics. 2016;**26**(2):79-97

[11] Sinitsyn IN. Fil'lry Kalmana i Pugacheva (Kalman and Pugachev Filters). 2nd ed. Moscow, Russia: Logos; 2007

[12] Sinitsyn IN, Sinitsyn VI, Korepanov ER. Modificated ellipsoidal conditionally optimal filters for nonlinear stochastic systems on manifolds. Informatika i Ee Primeneniya. 2017;**11**(2):101-111

[13] Pugachev VS. Conditionally optimal control. In: 8th World Congress of IFAC Proceedings, Vol. 1. Pergamon Press; 1991. pp. 373-376

[14] Sinitsyn IN, Sergeev IV. Methodological support of measurement, inspection, and testing of the computing equipment under shock actions. In: Trudy konferentsii "Tekhnicheskie i programmnye sredstva sistem upravleniya, kontrolya i izmereniya" (Proc. Conf. "Hardware and Software Facilities of Systems of Control, Inspection, and Measurement" (UKI-2010)). Moscow: IPU RAN; 2010. pp. 127-136

[15] Sinitsyn IN, Sinitsyn VI, Sergeev IV, Korepanov ER. Methods of ellipsoidal filtration in nonlinear stochastic systems on manifolds. Automation and Remote Control. 2018;**79**(1):118-128

[16] Zakai M. On the optimal filtering of diffusion processes. Zeitschrift für Wahrscheinlichkeitstheorie und Verwandte Gebiete. 1969;**11**:230-243

[17] Wonham M. Some application of stochastic differential equations to optimal nonlinear filtering. Journal of the Society for Industrial and Applied Mathematics, Series A: Control. 1965;**2**: 347-369

[18] Sinitsyn IN, Shalamov AS. Lectures on Theory of Integrated Logistic

Support Systems. 2nd ed. Moscow, Russia: Torus Press; 2019 (in Russian)

[19] Kostogryzov A. Probabilistic Modeling in System Engineering. Rijeka: IntechOpen. 2018. DOI: 10.5772/intechopen.71396

[20] Sinitsyn IN, Shalamov AS. Probabilistic modeling, estimation and control for CALS organization-technical-economic systems. In: Probability, Combinatorics and Control. Rijeka: IntechOpen; 2019 (in print)

Section 2

Autonomous Vehicles

Chapter 2

Reconfigurable Minimum-Time Autonomous Marine Vehicle Guidance in Variable Sea Currents

Kangsoo Kim

Abstract

In this chapter, we present an approach of reconfigurable minimum-time guidance of autonomous marine vehicles moving in variable sea currents. Our approach aims at suboptimality in the minimum-time travel between two points within a sea area, compensating for environmental uncertainties. Real-time reactive revisions of ongoing guidance followed by tracking controls are the key features of our reconfigurable approach. By its reconfigurable nature, our approach achieves suboptimality rather than optimality. As the basic tool for achieving minimum-time travel, a globally working numerical procedure deriving the solution of an optimal heading guidance law is presented. The developed solution procedure derives optimal reference headings that achieve minimum-time travel of a marine vehicle in any deterministic sea currents including uncertainties, whether stationary or time varying. Pursuing suboptimality, our approach is robust to environmental uncertainties compared to others seeking rigorous optimality. As well as minimizes the traveling time, our suboptimal approach works as a fail-safe or fault-tolerable strategy for its optimal counterpart, under the condition of environmental uncertainties. The efficacy of our approach is validated by simulated vehicle routings in variable sea currents.

Keywords: guidance, minimum-time, marine vehicle, sea current, suboptimal, environmental uncertainty

1. Introduction

It is well known that the sea environment contains several kinds of flows, which possibly interacts with the motion of surface or submerged vehicles. Among these, sea or ocean currents are the most significant ones, directly affecting traveling speed, power consumption, and thus the endurance and range of a vehicle. Suppose that a marine vehicle is to travel to a given destination starting from a point in the region of flow disturbance. Then, it is quite natural that the traveling time of the vehicle should change according to the selection of a specific path. In case the power consumption of a vehicle is controlled to be constant throughout the travel, the total energy consumption is directly proportional to the traveling time.

Recently, autonomous marine vehicles (AMVs) are playing important roles in diverse applications, such as oceanographic survey, marine patrol, undersea oil/gas production, and various military applications [1]. Relying on onboard energy storage as the main energy source, the endurance and moving range of an AMV are limited by its power consumption and its capacity of energy storage. Therefore, it can be said that the reduced traveling time of an AMV enhances vehicle safety and mission effectiveness [2].

Considerable research works have been done on the guidance or path planning for a mobile vehicle through varied fluid environments. Though aiming at the same objectives, the most notable difference between the guidance and the path planning of a vehicle is the consideration of its dynamical constraints. While, in general, dynamical constraints in vehicle motion are incorporated into the formulation of vehicle guidance problems [3, 4], they are ignored in most path planning problems [5, 6]. This allows great flexibility in the target path generation, enabling the use of combinatorial optimization techniques in path planning approaches. Dynamic programming (DP) might be one of the most classical and popular techniques for combinatorial optimization. In [6], the problem of minimal-time vessel routing in a region of deterministic wave environment is treated on the basis of the dynamic programming approach. In this problem, sea region is subdivided into several subregions of different sea states. The optimal path is derived by determining the sequence of subregions to be visited, which minimizes the traveling time to a given destination. Aside from the difficulty in establishing a practically available numerical procedure adjoining the formulation, the significant solution dependency on the regional subdivision is a critical issue in this approach. Some recent researches reported the application of a generic algorithm (GA) to path planning for an underwater vehicle in a variable ocean [5]. Major advantages of the GA over dynamic programming are reduced computational complexity and time, although it is susceptible to local minima, however. Also, one of its significant drawbacks is a strong constraint in generating the optimal path. In a path planning application on the basis of GA, a user-defined primary coordinate should strictly maintain a monotonic increase in the optimal path [5]. This is such a strong constraint that makes it impossible to generate the optimal path containing interim backward intervals. Minimum-time guidance of a mobile vehicle in a fluid environment of arbitrary flow field is a strongly nonlinear optimization problem, quite difficult to solve numerically as well as analytically. One of the recent approaches to treating this sort of problems is cell mapping [3]. Though it is known to be especially adequate for strongly nonlinear problems, computational demand of cell mapping for obtaining a stable solution is enormous.

Path finding or guidance algorithms can be classified into two categories according to the instant when its solution is generated. While a pregenerative one derives an unchangeable solution prior to a mission, a reactive algorithm allows revised solution during the mission [5, 7]. In this research, as a reactive strategy for achieving minimum-time travel in varied sea current environments, we propose an approach of suboptimal guidance. In our problem, minimum-time travel of a vehicle is attempted on the basis of the optimal guidance law presented by Bryson and Ho [8]. The solution of this guidance law is a time sequence of the optimal headings. In an actual field application for the minimum-time travel, obtained optimal headings are tracked by a vehicle as the reference in its heading control. Compact as it is, the optimal guidance law is derived without considering any specific dynamic constraint, like many other path planning approaches. In our suboptimal strategy, we compensate for this drawback by incorporating reactive revisions of optimal reference heading. Once there happens a failure in tracking current optimal reference attributed to the ignorance of limitations in vehicle dynamics, onboard autopilot reroutes the vehicle by reapplying the optimal guidance law.

In addition to the dynamic constraints, there are several unfavorable environmental factors that might be fatal in achieving the proposed optimal vehicle routing.

Examples of such factors are uncertainties in sea environments, severe sensor noises, or temporally faulty actuators [9, 10]. As a fail-safe or fault-tolerable strategy, our suboptimal approach can compensate for the failure in ongoing minimum-time travel due to any of the abovementioned factors. The suboptimal guidance does not achieve rigorous optimality. However, it achieves a near-optimality realized by the utmost in-situ actions as possible, which is useful and important in a practical sense.

Though provides superior adaptiveness, robustness, and more flexibility, the reactive approach in marine vehicle guidance incurs a heavy computational cost in its onboard implementation [3, 5, 10]. In this research, we present a practical solution procedure of highly reduced computational cost required for implementing our minimum-time guidance in a suboptimal manner. This is a simple procedure applicable to any sea current whether stationary or time-varying, provided that its distribution at a specified instant is deterministic. Robust global convergence is another advantage of our procedure. On the basis of the minimum principle [8, 11], our procedure realizes an efficient search space reduction, enabling optimal solution search in a global manner. Due to this algorithmic nature, our numerical procedure bears crucially lower possibility of taking local minima than other search algorithms primarily relying on initial guesses.

As mentioned previously, deterministic sea current is the prerequisite for implementing our suboptimal and optimal strategies. It is noted that, however, in many cases, it is not easy to obtain a predescribed current distribution of the sea region of interest. One of the simplest ways to build up the database of sea current distribution is direct measurement. Many governmental, public, or private institutions related to maritime affairs provide tabulated surface current distributions which are obtained by field measurements [12, 13]. The availability of these data is more or less restrictive because there are many sea regions for which the current distribution data are not built up or treated as confidential. As another source of sea current information, numerical estimation models are playing an important role. By assimilating the field measurement into them, some recent numerical models provide both forecasts and nowcasts of ocean fields with sufficiently accurate meso-scale resolution [14]. In this research, we employ two kinds of sea current data generated in totally different ways; the measurement-based stationary current distribution in Northwest Pacific, near Japan, and the sequential tidal current distribution in Tokyo Bay obtained by a numerical forecasting model.

Unlike path-planning approaches, our approach leads to simulation-based resultant optimal trajectory rather than optimal reference path. In our optimal guidance problem, not the position but the heading of a vehicle is employed as the design variable to be optimized. And, since we consider the dynamics of a specific vehicle as a constraint, optimal trajectory is to be generated by the simulated vehicle routing following the optimal reference heading. It is noted here that while the term of path is used as a route or track between one place and another, the trajectory means a curved path that an object describes on the basis of its kinematic scheme in this research [15].

2. Minimum-time guidance

2.1 Problem statement

The problem of minimum-time guidance for AMVs moving in flow fields is described in this section. Consider a marine vehicle traveling through a sea region of flows such as sea currents, whose properties are the function of space, or both space and time. The vehicle is to travel to a predetermined destination starting from the initial position at the initial time t_0. Then, it is easily anticipated that the traveling time of the vehicle varies depending on its traveling path (**Figure 1**). Furthermore, it is also anticipated that an ingenious traveling path can minimize the traveling time to the destination. In this research, we refer to the optimal path as the path of minimum traveling time to the destination. As a minimum-time problem, our problem merely takes the traveling time as the performance index. In other words, in the general form used in optimal control, the integrand of performance index takes the value of 1.

$$J = \int_{t_0}^{t_f} L d\tau = \int_{t_0}^{t_f} d\tau = t_f - t_0 \tag{1}$$

In Eq. (1), J is the performance index, and t_0 and t_f are the initial and final time of the travel.

2.2 Formulation

As mentioned previously, we employ the vehicle heading as the output in our optimal guidance problem. Here, it is noted that we adopt the so-called GNC (Guidance, Navigation, and Control) system based on the hierarchical control architecture consisting of two control layers. That is, the high-level control for guidance and navigation, and the low-level control for pure tracking purpose (**Figure 2**). The optimal heading derived by solving the optimal guidance law is used as the reference output for low-level heading tracking control. In this research, we use the optimal heading guidance law presented by Bryson and Ho [8]. In deriving the optimal guidance law, two sets of coordinate systems are used: the inertial (earth-fixed) coordinate system o-xy and the body fixed coordinate system o'-$x'y'$ (**Figure 3**).

As the marine vehicle used in our problem, we employ an autonomous underwater vehicle (AUV) "r2D4" as described by Kim and Ura [10]. In **Figure 3**, actuator inputs and kinematic variables are described. ψ is the yaw displacement of the vehicle. While δ_{pr} denotes the main thruster axis deflection, δ_{el} and δ_{er} are the deflections of elevators on left and right sides, respectively.

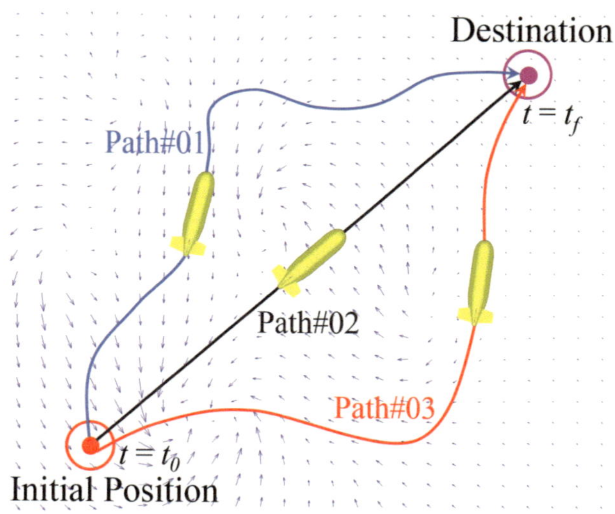

Figure 1.
Dependence of traveling time on traveling path.

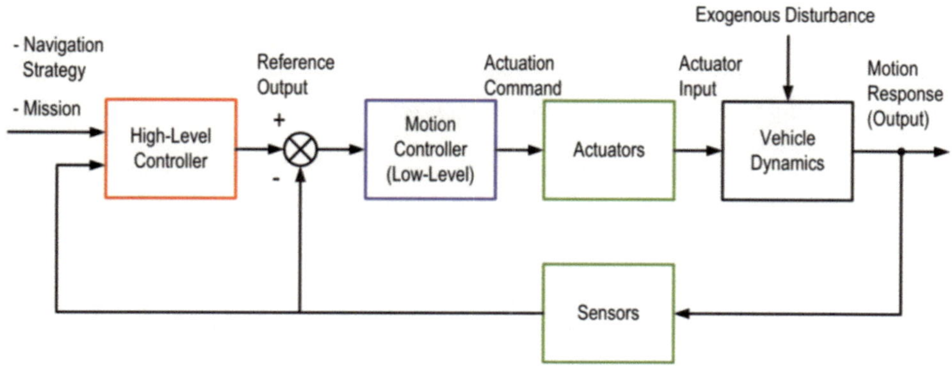

Figure 2.
Two-layer hierarchical control architecture for an AMV.

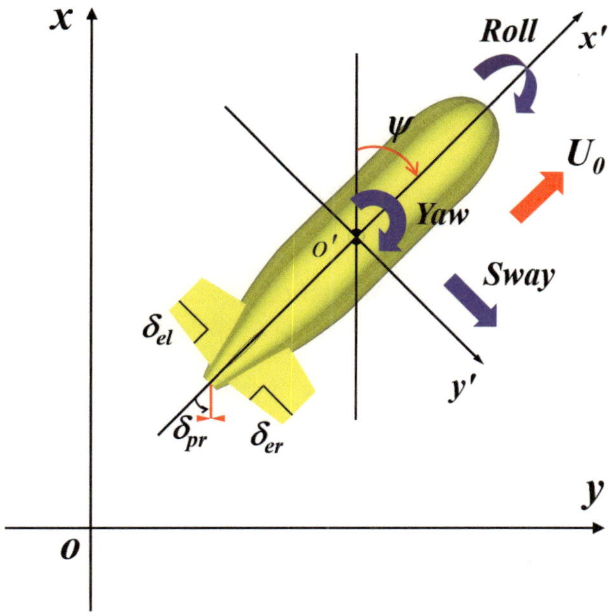

Figure 3.
Coordinate systems for optimal guidance problem formulation.

In this research, we approximate that the direction of the vehicle's advance velocity coincides with the x'-axis. It is arguable in the rigorous definition since there certainly occurs sideslip during a turning motion of an underactuated vehicle. However, as mentioned by Lewis et al. [16], the hydrodynamic sideslip induced by a low-speed, slender vehicle is bounded within a sufficiently small range, justifying our approximation. Since the distribution of a sea current is considered to be deterministic in our research, current velocity is described as a function of the position and time. Therefore, on the assumption that the advance velocity of a vehicle and the current velocity are superimposable, the resultant vehicle velocity is expressed as follows:

$$u = \dot{x} = U_0 cos\,\psi + u_c(x, y, t) \tag{2}$$

$$v = \dot{y} = U_0 sin\,\psi + v_c(x, y, t) \tag{3}$$

where u and v are the components of the vehicle velocity relative to the inertial frame, U_0 is the advance speed of the vehicle in still water, and u_c and v_c are the components of current velocity at a given position and time. It is noted that we assume U_0 is the constant throughout a travel, which implies the operating condition of steady cruise.

Eq. (4) shows the minimum-time guidance law originally presented by Bryson and Ho [8]. Detailed procedures deriving Eq. (4) are well explained by Kim and Ura [10]. It is noted here that if only deterministic, there is no restriction on the type of the sea current in Eq. (4). That is, not only stationary but also time-varying sea current can be applied to Eq. (4). This leads to one of the most powerful aspects of our approach over many other path planning approaches based on combinatorial optimization.

$$\dot{\psi} = \sin^2\psi \frac{\partial v_c}{\partial x} + \left(\frac{\partial u_c}{\partial x} - \frac{\partial v_c}{\partial y} \right) \sin\psi \cos\psi - \cos^2\psi \frac{\partial u_c}{\partial y} \quad (4)$$

The optimal guidance law shown above is a nonlinear ordinary differential equation of unknown vehicle heading. The solution of optimal guidance law is used as the optimal reference heading, by tracking which a vehicle achieves the minimum-time travel to the destination, leaving the trail of optimal trajectory.

3. Numerical solution procedure

Eq. (4) is a nonlinear ordinary differential equation (ODE) for an unspecified vehicle heading $\psi(t)$. If the functions $u_c(x,y,t)$ and $v_c(x,y,t)$ describing current velocity distribution are differentiable and deterministic, the solution of Eq. (4) seems to be attainable with an initial value of $\psi(t)$, in terms of an appropriate numerical solution algorithm such as Runge-Kutta. However, in practice, with an arbitrary initial heading a vehicle following the guidance law Eq. (4) does not reach the destination, as depicted in **Figure 4**.

More precisely, consisting of a part of the solution, the initial vehicle heading is not arbitrary but is to be assigned correctly. This is because Eq. (4) is derived from

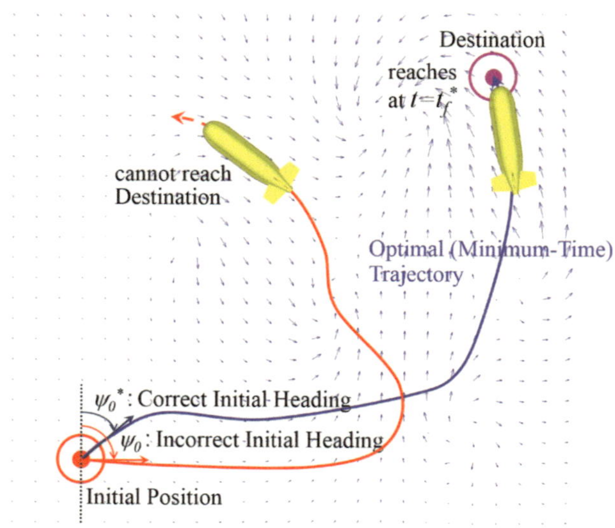

Figure 4.
Solution convergence affected by initial heading.

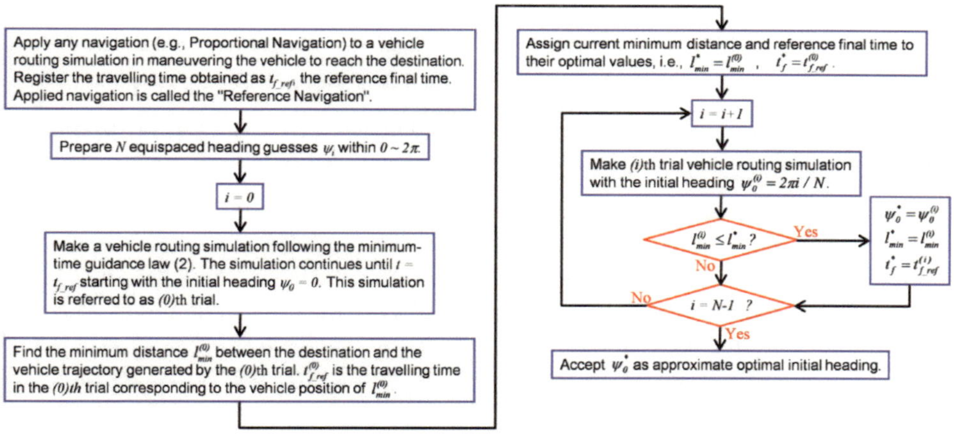

Figure 5.
Schematic of the numerical solution procedure AREN.

the Euler-Lagrange equation, which is a typical example of the two-point boundary value problem, characterized by split boundary conditions in states and costates [8, 11]. To obtain the solution of a two-point boundary value problem, an iterative solution procedure is usually required. The most famous and commonly used numerical procedures for such purpose are the shooting and relaxation methods [17]. However, direct applications of these methods to our problem have significant difficulties. In applying shooting method to a two-point boundary problem in time domain, governing ODEs with proper initial guesses should be integrated until reaching the upper limit of the boundary. However, as noticeable from its name, i.e., the minimum-time guidance, our problem is a so-called free boundary one, having unspecified upper limit in time domain. In treating a free boundary problem by relaxation method, on the other hand, the independent variable should be transformed into a new one defined between 0 and 1. Here, we can anticipate an intrinsic serious difficulty in determining the stepsize in free boundary problems. Properness of temporal grid distribution ensuring convergence is initially unknown, and to know, it is extremely difficult before the end time-marching computation. Moreover, strong initial guess dependency of the solution is another serious concern in applying the relaxation method to our problem. Inappropriate initial guess possibly leads to local optimality or divergence [17].

As a new approach for deriving the numerical solution of the optimal guidance law Eq. (4), we presented a search procedure, which determines correct initial heading. Being named AREN (Arbitrary REference Navigation), our procedure works globally on the basis of the minimum principle. **Figure 5** summarizes the algorithmic scheme of our solution procedure.

Note that in **Figure 5** and hereafter, an asterisked variable denotes the one corresponding to the optimal solution. In applying AREN, we first have to make a vehicle routing simulation in which the vehicle travels to the destination following an arbitrary guidance. It is noted here that the traveling time must be registered at the final stage of this simulation. We call the traveling time the reference final time and use it as the key criterion in seeking optimal initial heading. The navigation applied to the simulation is called reference navigation, which is arbitrary if only the vehicle's arrival at the destination is assured. Therefore, simple one such as proportional navigation (PN) based on the line-of-sight (LOS) guidance is frequently used as the reference navigation. To find the correct initial heading, the interval of 0–2π is divided by equally spaced N-1 subintervals, as represented by:

$$\psi_0^{(i)} = i\Delta\psi \quad for \quad i=0,1,\ldots N\text{-}1 \tag{5}$$

where $\psi_0^{(i)}$ is *(i)*th initial heading guess, and $\Delta\psi$ is the increment of the guess. Next, by applying an initial heading guess $\psi_0^{(i)}$ to Eq. (4), we solve Eq. (4) in time domain. This produces a simulated vehicle routing starting from $\psi_0^{(i)}$. The routing having been produced here is called the *(i)*th trial adjoining to $\psi_0^{(i)}$. Once the vehicle passes through the destination by the *(i)*th trial, it is regarded as a possible optimal routing since the correct initial heading incorporated into the optimal guidance law lets a vehicle reach the destination. Therefore, *N* trials are the candidates for the simulated minimum-time routing. In practice, however, discretization error in the optimal initial heading causes the residual in the optimal trajectory, making the optimal solution identified in an approximate manner. For the vehicle trajectory generated by a trial, we define the "minimum distance" as the shortest distance between the destination and the trajectory. In **Figure 6**, $l_{min}^{(k-1)}$, $l_{min}^{(k)}$, and $l_{min}^{(k+1)}$ are the minimum distances corresponding to *(k-1)*th, *(k)*th, and *(k + 1)*th trials, respectively.

When the minimum distance of *(k)*th trial is smaller than any other one, satisfying:

$$l_{min}^{(k)} \le l_{min}^{(i)} \; for \quad i=0,1,\ldots N\text{-}1 \tag{6}$$

we choose the *(k)*th trial as the optimal routing because the vehicle approaches the destination marking the smallest deviation. In determining the optimal routing among the trials, however, there still remains a serious drawback. We have no idea how long we have to continue a trial not to miss the true minimum distance of the trial. We settle this problem by exploiting the result of reference navigation. The reference navigation is apparently a nonoptimal one based on an arbitrary guidance only assuring the arrival at the destination. Therefore, the reference final time t_{f_ref} must be larger or equal to that of the optimal routing as follows:

$$0 < t_f^* \le t_{f\_ref} \tag{7}$$

where t_f^* represents the traveling time of the optimal routing. It should be noted here that by the minimum principle [8, 11], we can set up a sufficient condition for

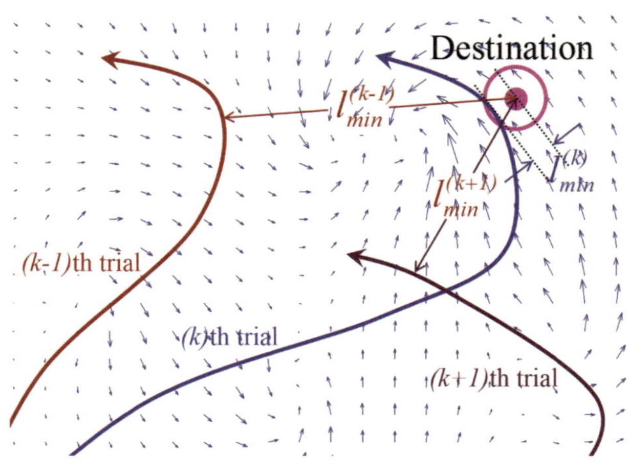

Figure 6.
Minimum distances of trials.

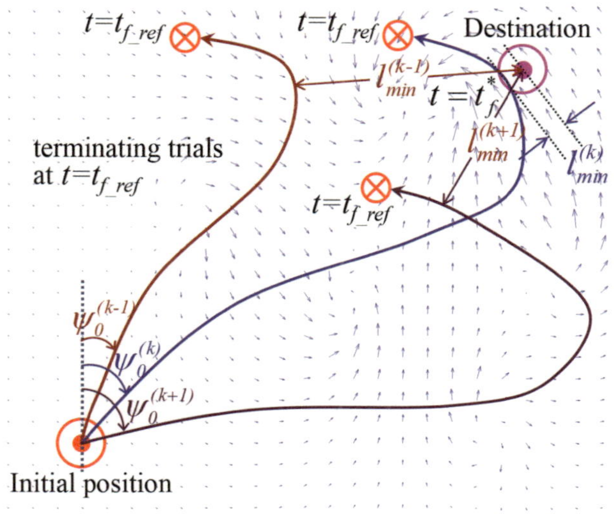

Figure 7.
Determining optimal routing among trials.

seeking the optimal solution. By the minimum principle, once a trial has started with an initial heading sufficiently close to the optimal value, the vehicle obviously passes by the vicinity of the destination at the traveling time smaller than t_{f_ref}. In other words, the reference final time qualifies as the upper limit of the necessary simulation time of any trial, which assures the convergence to the vicinity of the destination in case the trial is near optimal. In **Figure 7**, *(k)*th trial is selected as the optimal routing among all trials terminated at t_{f_ref} since $l_{min}^{(k)}$ is the smallest minimum distance.

The minimum distance of the optimal routing is to be interpreted as the residual error in the converged solution. Therefore, it can be said that the smaller the minimum distance is, the better the convergence is. When $l_{min}^{(k)}$ is still unacceptably large though the *(k)*th trial has been accepted as the optimal routing, the initial heading interval of $\psi_0^{(k-1)} \sim \psi_0^{(k+1)}$ is subdivided, and the trials are repeated starting from these subdivisions pursuing finer convergence.

4. Dynamic constraint

As mentioned previously, we adopt GNC system based on the hierarchical control architecture consisting of two control layers. In the high-level control layer,

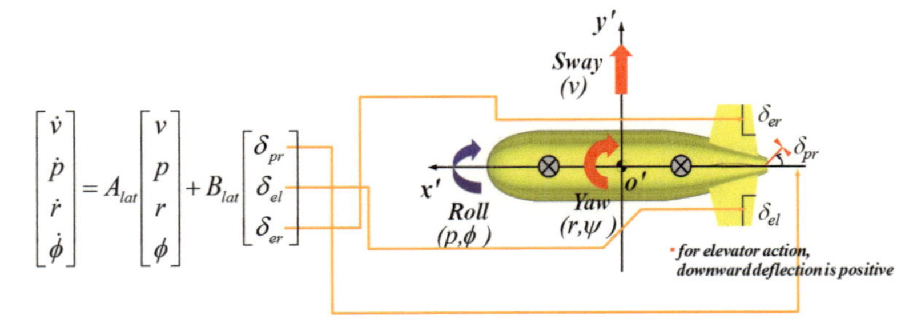

Figure 8.
Graphical description of the equation of motion for lateral dynamics.

i.e., the optimal guidance, the guidance law derived irrespective of specific vehicle dynamics is used. In the low-level control layer, however, vehicle dynamics is implemented as an implicit constraint. When conducting the trials explained in previous section, closed-loop dynamics of a specific vehicle is used. Therefore, vehicle trajectories generated by the trials are feasible ones subject to the dynamic constraint of a specific vehicle. Eq. (8) is the state-space model of the lateral dynamics of r2D4 describing its sway, roll, and yaw responses. In **Figure 8**, kinematic variables and actuations appearing in Eqs. (8)–(10) are described graphically. By solving Eq. (8) in time domain, velocities and attitudes of the vehicle are obtained.

$$\begin{bmatrix} \dot{v} \\ \dot{p} \\ \dot{r} \\ \dot{\phi} \end{bmatrix} = A_{lat} \begin{bmatrix} v \\ p \\ r \\ \phi \end{bmatrix} + B_{lat} \begin{bmatrix} \delta_{pr} \\ \delta_{el} \\ \delta_{er} \end{bmatrix} \tag{8}$$

where

$$A_{lat} = \begin{bmatrix} -0.5266 & 0.0029 & -0.7754 & -0.0052 \\ -3.5308 & -4.7898 & 7.9422 & -10.5536 \\ -0.1472 & -0.0334 & -0.5742 & -0.0722 \\ 0.0 & 1.0 & -0.0129 & 0.0 \end{bmatrix} \tag{9}$$

$$B_{lat} = \begin{bmatrix} -0.0335 & -0.0011 & 0.0011 \\ 0.0244 & -2.1704 & 2.1704 \\ 0.0486 & -0.0149 & 0.0149 \\ 0.0 & 0.0 & 0.0 \end{bmatrix} \tag{10}$$

5. Minimum-time guidance in stationary current flows

5.1 Reference navigation

As mentioned previously, in order to practice the numerical procedure AREN, a vehicle routing simulation by the reference navigation has to be conducted beforehand. Among several strategies for mobile vehicle navigation, the simplest one ensuring arrival at the destination might be the PN based on LOS guidance. In all of our optimal guidance examples presented in this paper, we employ PN as the reference navigation.

5.2 Linear shearing flow

The first example of the optimal guidance in this paper is the minimum-time routing in a current disturbance of linear shearing flow, taken from Bryson and Ho [8]. The current velocity in this problem is described by:

$$u_c(x, y) = 0 \tag{11}$$

$$v_c(x, y) = -U_c x/h \tag{12}$$

In Eq. (12), U_c and h are constants whose numerical values are set to be 1.54 m/s and 100 m, respectively. In this example, the vehicle is to travel to the destination located at the origin, starting from the initial position at (−186 m, 366 m). As an operating condition, the vehicle is assumed to maintain its thrust power constant throughout its travel, producing a constant advance speed of 1.54 m/s. This is an important operating condition applied to all examples presented thereafter. With the current distribution given as Eqs. (11) and (12), we can derive the analytic optimal guidance law in a closed form, shown as follows:

$$\frac{x}{h} = csc\psi - csc\psi_f \tag{13}$$

$$\frac{y}{h} = \frac{1}{2}\left[csc\psi_f(cot\psi - cot\psi_f) + cot\psi(csc\psi_f - csc\psi) + log\frac{csc\psi_f - cot\psi_f}{csc\psi - cot\psi}\right] \tag{14}$$

where ψ_f represents the final vehicle heading taken at the destination. The analytic optimal guidance law shown above is similar to that found by Bryson and Ho [8] but has some differences due to the switched x and y.

Vehicle trajectories are shown in **Figure 9**. In **Figure 10**, vehicle headings obtained by the routings conducted by PN, that is, the reference navigation, and by optimal guidance are shown.

During the vehicle routing by PN, significant adverse drift happens at the initial stage. This is because the speed of current flow exceeds the advance speed of the vehicle in the region $|x| > 100$ m, as noticeable from Eqs. (11) and (12). On the other hand, optimal guidance detours the vehicle across the upper half plane of the flow region on purpose, taking advantage of the strong current flowing to favorable direction. This leads to the dramatic decrease in traveling time. The traveling times by PN and optimal guidance are 353.8 and 739.2 s, respectively, implying a 52% reduction in the optimal guidance. As seen in **Figure 9**, the optimal trajectory

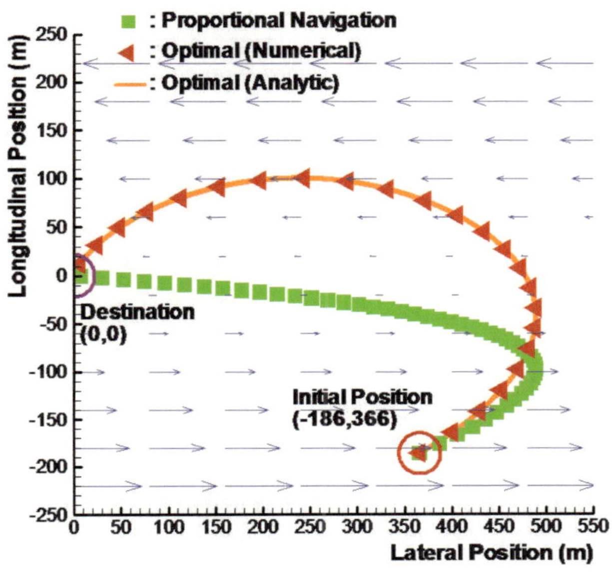

Figure 9.
Vehicle trajectories in a stationary linear shearing flow.

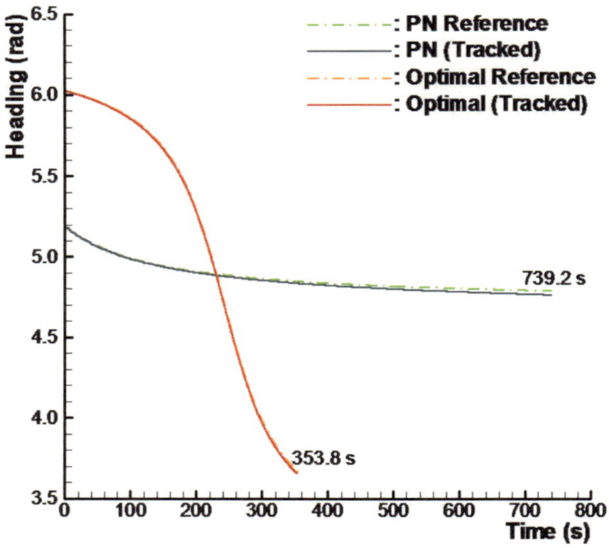

Figure 10.
Vehicle headings during travels in a stationary linear shearing flow.

obtained by the numerical solution shows extremely good agreement with the analytic one.

As a criterion for evaluating the reference tracking performance in our two-layer control architecture, we employ Normalized Root-Mean Square Error (NRMSE) fit defined as follows:

$$fit = 1 - \frac{\left\| \xi_{ref} - \xi \right\|}{\left\| \xi_{ref} - \bar{\xi} \right\|} \tag{15}$$

where $\boldsymbol{\xi}_{ref}$ and $\boldsymbol{\xi}$ are the vectors of output reference and output, and $\bar{\xi}$ represents the mean value of $\boldsymbol{\xi}$. The value of NRMSE fit varies between $-\infty$ to 1, implying full decorrelation to perfect fit between the output reference and the output. In this simulation, NMRES fit between the optimal reference heading and the actually tracked one has marked 0.993. This means highly good heading tracking result, which is also found in **Figure 10**. The numerical solution approximates the analytic solution with extremely high accuracy (**Figure 9**), validating AREN as an effective numerical procedure for the optimal guidance law Eq. (4). As shown in this example, our approach based on the numerical procedure AREN and two-layer control architecture works properly achieving minimum-time AMV routing in a given sea current field.

5.3 Sea current in Northwest Pacific near Japan

In response to the successful result obtained from the benchmark example shown in previous section, we apply our optimal guidance to the minimum-time routing problem in an actual sea region of stationary sea current. The sea region selected is located in the Northwest Pacific Ocean near Japan. The daily updated sea current data of this region are available from [18] presented by the Japan Meteorological Agency. The most notable environmental characteristic in this sea region is the current field dominated by Kuroshio. The Kuroshio is a strong western

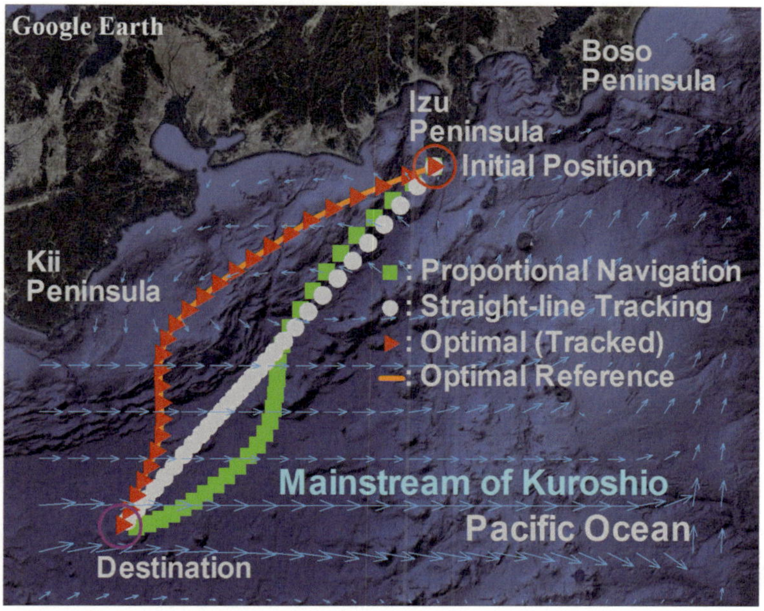

Figure 11.
Vehicle trajectories in a sea current in Northwest Pacific near Japan.

boundary current flowing northeastward along the coast of Japan. In the sea current data from [18], current velocity is defined only on the grid nodes covering the region. As noticeable from Eq. (4), however, in order to derive the optimal heading reference, current velocity and its gradient at every vehicle position have to be available. In the previous example, their exact values are easily obtained by analytic formulae. In this example, however, since they are defined only on the grid nodes, current velocity and its gradient are estimated by interpolating the predefined values on grid nodes surrounding the present vehicle position. In applying the current velocity interpolation, the grid node nearest the present vehicle position is identified first. Then, the current velocity at the present vehicle position is estimated by 2-D bi-quadratic interpolation utilizing the values on the nearest node and

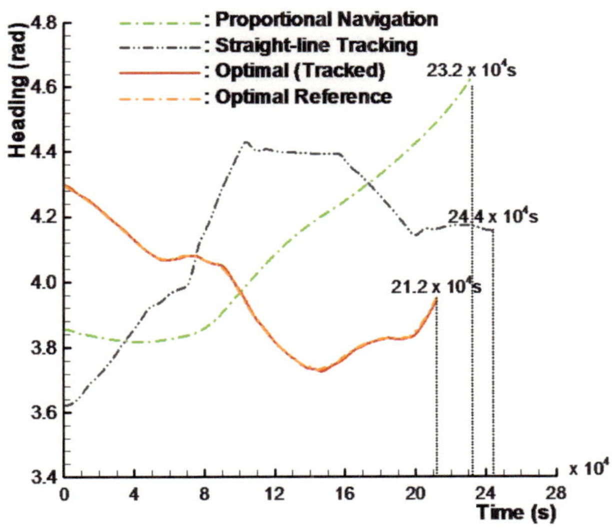

Figure 12.
Time sequences of vehicle headings during the travel.

eight nodes surrounding current vehicle position. Gradients of current velocities are obtained by the same manner. Since the velocity gradients are not provided from the database, however, prior to the interpolation, we calculate their nodal values by finite difference approximation.

Figure 11 shows the vehicle trajectories obtained by the guidance of three different objectives already explained in the previous example. Time sequences of the vehicle headings corresponding to the vehicle trajectories shown in **Figure 11** are depicted in **Figure 12**.

As shown in the figure, like the preceding examples in which the current velocities and their gradients are analytically available anywhere in the region, optimal reference trajectory has successfully been derived by interpolation-based current velocities and gradients. Moreover, subject to its dynamic constraint, the vehicle tracks the optimal reference trajectory with a negligibly small deviation, resulting in the NMRES fit to be 0.986. This demonstrates the validity of our optimal guidance strategy in any actual sea currents, if only their distribution is deterministic.

6. Suboptimal strategy

6.1 Environmental uncertainty

In the following example, we apply our optimal guidance strategy to a vehicle routing in the same sea region shown in the preceding example. The only thing different from the preceding example is that we consider uncertainty in our sea current data in this example. The uncertainty components in sea current velocities are expressed as additive white Gaussian noise (AWGN). Taking the sea current velocities in the Northwest Pacific Ocean used beforehand as the mean values, the on-site current velocity including uncertainty is given by:

$$u_{cs}(x,y) = u_c(x,y) + e_u(\sigma) \tag{16}$$

$$v_{cs}(x,y) = v_c(x,y) + e_v(\sigma) \tag{17}$$

where u_{cs} and v_{cs} are the components of onsite current velocity, u_c and v_c are the components of deterministic current velocity taken from the database, and $e_u(\sigma)$ and $e_v(\sigma)$ are the AWGNs with standard deviation σ. As the parameter for specifying the value of σ, we introduce the regional mean current speed U_{cm} defined as follows:

$$U_{cm} = \frac{\sum_{i=1}^{N} \sqrt{u_{ci}^2 + v_{ci}^2}}{N} \tag{18}$$

where i represents the index covering all grid nodes on which the database-based current velocities are defined.

Vehicle trajectories by optimal vehicle routings conducted on two different level uncertainties are shown in **Figure 13**. When the level uncertainty is such that $\sigma = 2U_{cm}$, optimal heading reference derived without considering any uncertainty still seems acceptable. As a result, though slightly deviating from the destination, the final position of the vehicle remains in the vicinity of the destination. When the level of uncertainty increases up to $\sigma = 4U_{cm}$, however, the vehicle following the

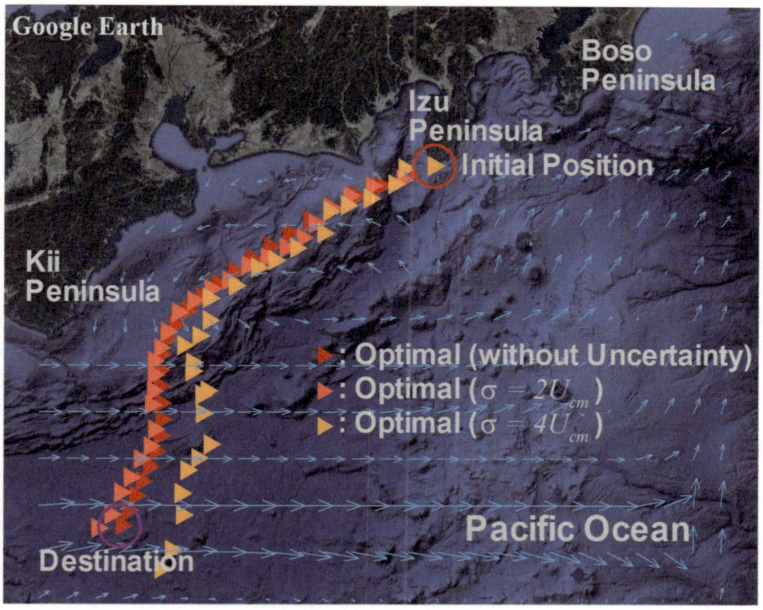

Figure 13.
Vehicle trajectories in Northwest Pacific near Japan. The sea current velocities in this example include uncertainties modeled by AWGN.

optimal reference heading can no longer approach the destination, which means the failure in accomplishing the minimum-time travel to the destination.

6.2 Suboptimal guidance

The suboptimal guidance proposed in this research is a fail-safe or fault-tolerable strategy toward robust field implementation of our optimal guidance strategy. The optimal reference heading obtained by our approach is the one derived without considering the dynamics of a specific vehicle. This means the optimal trajectory may not be realized by a specific vehicle. Hence, we note that the dynamic constraint is one possible source of the failure in putting our approach into practice for an actual field application. Another significant source of the failure is the environmental uncertainty, as already shown above. It is easily expected that as a vehicle progresses following the optimal heading in the sea region of environmental uncertainty, due to the interaction with the current flow different from that was used in deriving the optimal heading, its actual trajectory deviates away from the optimal reference trajectory, and eventually, it might fail in reaching the destination. The basic idea of our suboptimal approach is rather simple. Let d_1 denotes the deviation distance between the present vehicle position and the preassigned one on the optimal reference trajectory obtained by AREN. When d_1 exceeds a prescribed acceptable limit d_a, the high-level controller in autopilot is activated to reroute the vehicle by reapplying AREN. This rerouting is repeated whenever d_1 exceeds a predefined acceptable limit. The resulting vehicle routing is not rigorously optimal, since it includes past nonoptimal travels. However, Bellman's principle of optimality [8, 11] states that it is evidently the best strategy we can take under the condition we are faced with. We, therefore, call this approach the suboptimal guidance. **Figure 14** depicts the schematic of our suboptimal guidance explained thus far.

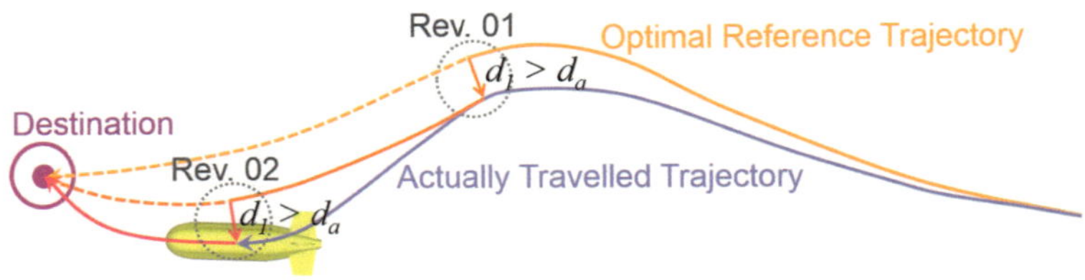

Figure 14.
Schematic of the vehicle routing by suboptimal guidance.

6.3 Suboptimal vehicle routing in a stationary sea current

Here we show another example of vehicle routing in the Northwest Pacific Ocean containing the environmental uncertainty of $\sigma = 4U_{cm}$. Vehicle trajectories are shown in **Figure 15**.

In this example, a vehicle does not merely track the pregenerated optimal heading reference throughout but regenerates and follows new ones whenever necessary. In other words, the vehicle follows the optimal heading references revised repeatedly on the basis of our suboptimal strategy. In this example, we set d_a, the acceptable limit of position deviation, to be 12,500 m. As noted in the figure, optimal vehicle routing is revised five times, making the vehicle to reach the destination, at last. Repeating the vehicle rerouting five times, our suboptimal guidance has succeeded in taking the vehicle to the destination. Traveling times of vehicle routings are summarized in **Table 1**. It is noted here that, while the results of vehicle routing by PN, straight-line tracking, and optimal guidance are deterministic, the result by suboptimal guidance is not since it is derived using the model including environmental uncertainties. Therefore, care should be taken in interpreting the suboptimal result. The result of suboptimal vehicle routing is event-dependent, so that it differs in every event. In **Table 1**, we find that the traveling time of suboptimal vehicle routing is 208372.0 s. Notably, it is even shorter than that of the

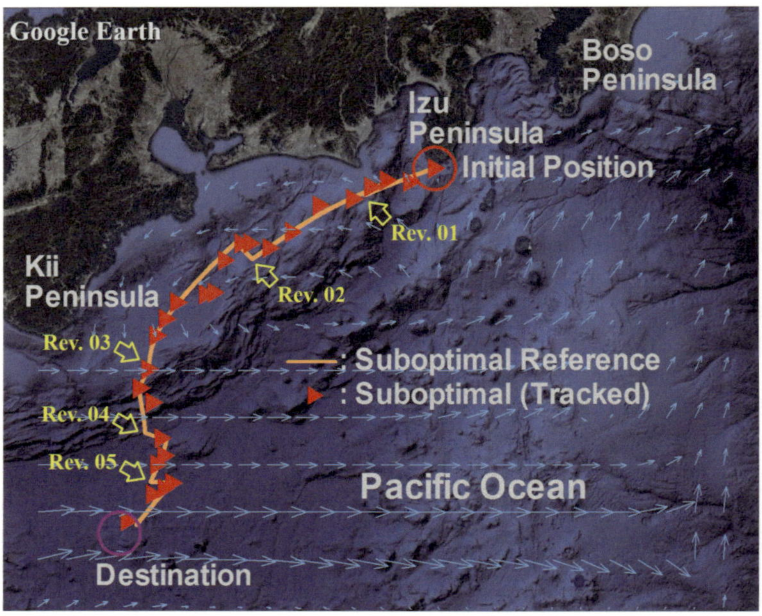

Figure 15.
Vehicle trajectories in NW Pacific generated by suboptimal routing.

			Traveling time
Guidance	Nonoptimal	PN	232198.0 s
		Straight-line tracking	244312.0 s
	Optimal		212006.5 s
	Suboptimal		208372.0 s

Table 1.
Traveling times of vehicle routings in NW Pacific.

optimal routing in the sea current without uncertainty. This implies that in total, the uncertainties have affected the travel in a favorable manner, which may not be the case in other events, however.

6.4 Suboptimal vehicle routing in a time-varying tidal flow

The last example presented in this chapter is an underwater vehicle routing in Tokyo Bay. In this example, we consider the mission of minimum-time homing to a destination. Due to its narrow entrance and shallow depth, sea currents in Tokyo Bay are hardly affected by the outer ocean currents such as Kuroshio. Instead, like many other littoral zones, currents in Tokyo Bay are dominated by the tidal flow. In this research, we use the time-varying sea current distribution data in Tokyo Bay, generated by a numerical tidal flow simulation model by Kitazawa et al. [19]. As was the case in the previous examples, we first conduct optimal vehicle routing without considering environmental uncertainties. In **Figure 16**, vehicle trajectories generated by the routings of different guidance strategies are shown. As found in the figure, our approach of optimal guidance successfully accomplishes the minimum-time homing mission even the sea current is time varying. Note that the current distribution shown in the figure is the one taken at the traveling time of straight-line tracking, which makes the last arrival at the destination.

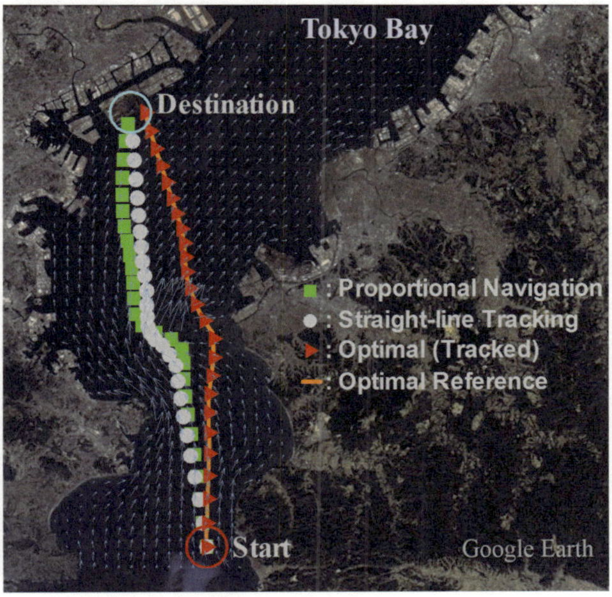

Figure 16.
Vehicle trajectories in the tidal flow in Tokyo Bay.

Next, on the basis of our suboptimal strategy, we conduct the vehicle routing simulation, incorporating uncertainties into the time-varying tidal flow in Tokyo Bay. The standard deviation of the uncertainties is set to be $4U_{cm}$, as was the case in the preceding examples. **Figure 17(a)–(f)** shows sequential vehicle trajectories created by our suboptimal vehicle routing. The traveling time of the suboptimal

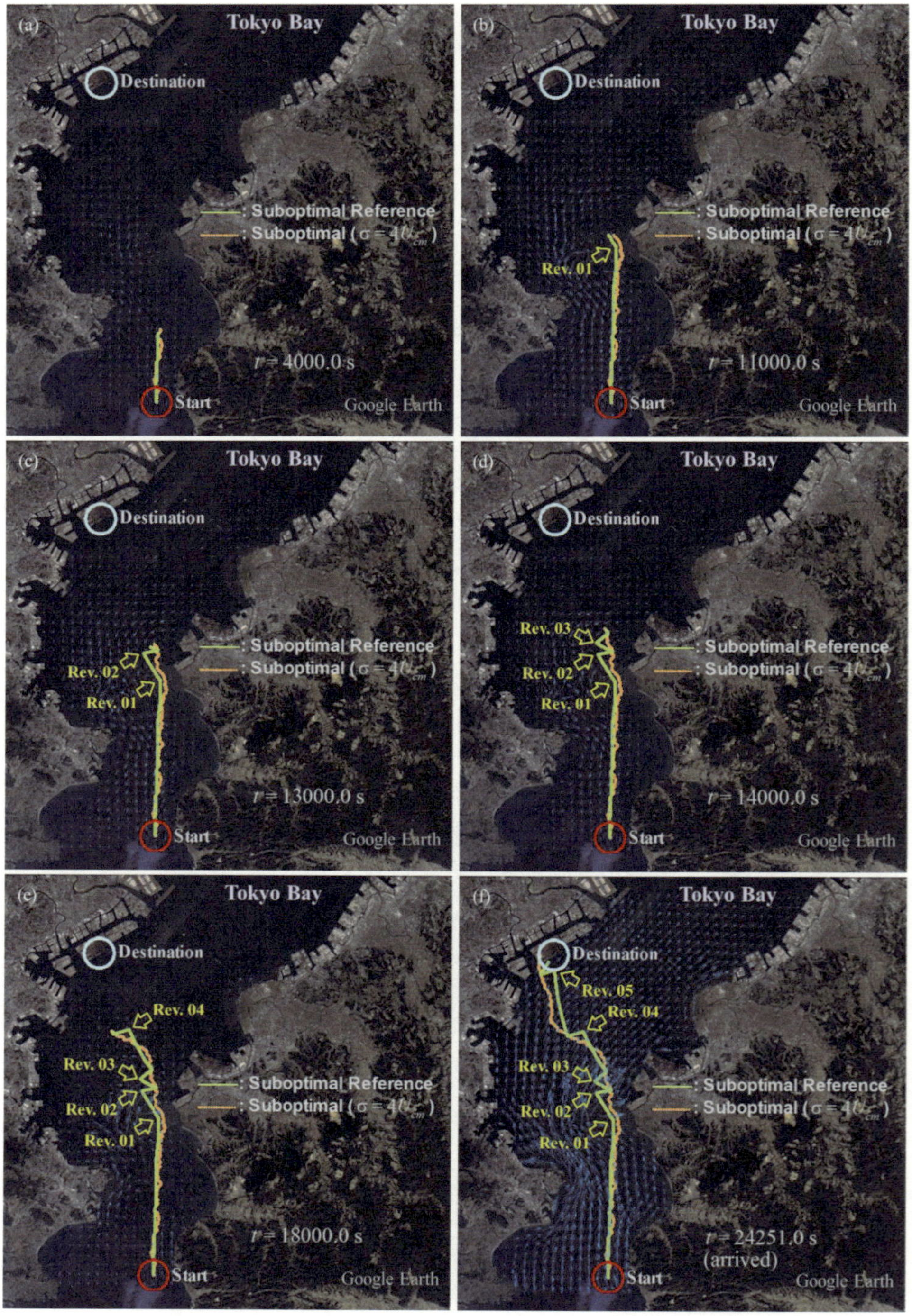

Figure 17.
Sequential vehicle trajectories in Tokyo Bay generated by suboptimal routing (a) t = 4000 s (b) t = 11000 s (c) t = 13000 s (d) t = 14000 s (e) t = 18000 s (f) t =24251 s.

			Traveling time
Guidance	Nonoptimal	PN	25848.0 s
		Straight-line tracking	27744.5 s
	Optimal		24002.0 s
	Suboptimal		24251.0 s

Table 2.
Traveling times of vehicle routings in Tokyo Bay.

vehicle routing and those of the optimal, PN, and straight-line tracking obtained in the previous example are summarized in **Table 2**.

In this example, we set d_a, the acceptable limit of deviation distance, to be 1800 m. As seen in the figures, the vehicle has successfully accomplished its homing mission by the suboptimal routing, repeating five revised travels. In view of the results obtained by this example, we find that our suboptimal approach works effectively even in a time-varying environment including uncertainties.

7. Conclusion

In this chapter, a systematic procedure for obtaining the numerical solution of the optimal guidance law to achieve the minimum-time routing in a region of sea current has been presented. The optimal heading is obtained as the solution of the optimal guidance law, which is fed to the heading control system as the reference.

Reduced computational cost is one of the outstanding features of the proposed procedure. While linearly proportional to the area of a search region in DP, the computational time in our procedure exhibits square root dependence. Moreover, unlike the other path finding algorithms such as DP or GA, when applied to a time-varying environment, our procedure does not increase the search space, resulting in the same computational cost as required in the time-invariant ones.

The performance of the optimal guidance has strong dependency on the current distribution. While an extremely simple configuration, such as uniform flow, hardly allows navigation time reduction by the optimal guidance, a multi-directional complicated flow distribution enhances the potential efficacy of the optimal guidance.

As a fail-safe or fault-tolerable strategy in optimal guidance, the concept of suboptimal guidance has been proposed. The fact that there actually are several possible actions lessening the chance of optimality emphasizes the practical importance of our suboptimal strategy.

We have not considered the problem of unknown or nondeterministic currents. Our approach cannot be applied to an entirely unknown environment. For a sea region with partially or coarsely defined current flow, however, an estimated distribution can be built by means of interpolation and extrapolation. As has already been shown in the optimal and suboptimal vehicle routing examples in actual sea regions, spatiotemporal interpolation of the current velocity successfully derives the converged solution.

Acknowledgements

The author would like to thank Prof. D. Kitazawa of IIS, the University of Tokyo for providing simulated tidal flow data of Tokyo Bay. Also, the author is particularly grateful to T. Ura, an emeritus professor of the University of Tokyo, for his guidance and support concerning this research work.

Author details

Kangsoo Kim
National Maritime Research Institute, National Institute of Maritime, Port, and Aviation Technology, Tokyo, Japan

*Address all correspondence to: kim@m.mpat.go.jp

References

[1] Nicholson JW, Healey AJ. The present state of autonomous underwater vehicle applications and technologies. Marine Technology Society Journal. 2008;**42**(1):44-51

[2] Kim K, Ura T. Applied model-based analysis and synthesis for the dynamics, guidance, and control of an autonomous undersea vehicle. Mathematical Problems in Engineering. 2010;**2010**: 149385. DOI: 10.1155/2010/149385

[3] Crespo LG, Sun JQ. Optimal control of target tracking with state constraints via cell mapping. Journal of Guidance, Control, and Dynamics. 2001;**24**(5): 1029-1031

[4] Zhao Y, Bryson AE. Optimal paths through downbursts. Journal of Guidance, Control, and Dynamics. 1990;**13**(5):813-818

[5] Alvarez A, Caiti A, Onken R. Evolutionary path planning for autonomous underwater vehicles in a variable ocean. IEEE Journal of Oceanic Engineering. 2004;**29**(2):418-429

[6] Papadakis NA, Perakis AN. Deterministic minimal time vessel routing. Operations Research. 1990; **38**(3):426-438

[7] Kamon I, Rivlin E. Sensory-based motion planning with global proofs. IEEE Transactions on Robotics and Automation. 1997;**13**(6):814-822

[8] Bryson AE, Ho YC. Applied Optimal Control. 1st ed. New York: Taylor & Francis; 1975

[9] Burken JJ. Two reconfigurable flight-control design methods: Robust servomechanism and control allocation. Journal of Guidance, Control, and Dynamics. 2001;**24**(3):482-493

[10] Kim K, Ura T. Optimal guidance for AUV navigation within undersea area of current disturbances. Advanced Robotics. 2009;**23**(5):601-628

[11] Lewis FL, Syrmos VL. Optimal Control. 2nd ed. New York: John Wiley & Sons, Inc.; 1995

[12] McCormick H, editor. Reed's Nautical Almanac: North American West Coast 2008. Boston: Thomas Reed Publications; 2008

[13] National Ocean Service. Tidal Current Tables 2003, Pacific Coast of North America and Asia. North America and Asia, Camden: International Marine Publishing; 2002

[14] Robinson AR. Forecasting and simulating coastal ocean processes and variabilities with the Harvard Ocean Prediction System. In: Mooers CNK, editor. Coastal Ocean Prediction. Vol. 56. Washington: American Geophysical Union; 1999. pp. 77-99. DOI: 10.1029/CE056

[15] Cambridge Dictionary. Available from: https://dictionary.cambridge.org/ [Accessed: 10 December 2019]

[16] Lewis EV, editor. Principles of Naval Architecture Vol. III, Motions in Waves and Controllability. Jersey: The Society of Naval Architects and Marine Engineers; 1989

[17] Press WH, Flannery BP, Teukolsky SA, Vetterling WT. Numerical Recipes in Fortran. Cambridge: Cambridge University Press; 1992

[18] Sea Current Data by Japan Meteorological Agency. Available from: https://www.data.jma.go.jp/kaiyou/data/db/kaikyo/daily/current_HQ.html?areano=2 [Accessed: 22 November 2019]

[19] Kitazawa D et al. Predictions of coastal ecosystem in Tokyo Bay by

pelagic-benthic coupled model. In: Proceedings of 20th International Conference on Offshore Mechanics and Arctic Engineering (OMAE '01), 2001; Rio de Janeiro: ASME; 2001. No. OSU-5035

Chapter 3

Knowledge-Based Controller Optimised with Particle Swarm Optimisation for Adaptive Path Tracking Control of an Autonomous Heavy Vehicle

Noor Hafizah Amer, Khisbullah Hudha, Hairi Zamzuri, Vimal Rau Aparow, Amar Faiz Zainal Abidin, Zulkiffli Abd Kadir and Muhamad Murrad

Abstract

This chapter discusses the development of an adaptive path tracking controller equipped with a knowledge-based supervisory algorithm for an autonomous heavy vehicle. The controller was developed based on a geometric/kinematic controller, the Stanley controller. One of the mostly known issues with any geometric/kinematic controller is that a properly tuned controller may not be valid in a different operating region than the one it was being tuned/optimised on. Therefore, this study proposes an adaptive algorithm to automatically choose an optimal controller parameter depending on the manoeuvring and vehicle conditions. An optimal knowledge database is developed for an adaptive algorithm to automatically obtain the parameter values based on the vehicle speed, v, and heading error, ϕ. Several simulations are carried out with different trajectories and speeds to evaluate the effectiveness of the controller against its predecessors, namely, Stanley and the non-adaptive modified Stanley (Mod St) controllers. The simulated steering actions are then compared against human driver's experimental data along the predefined paths. It was shown that the proposed adaptive algorithm managed to guide the heavy vehicle successfully and adapt to various trajectories with different vehicle speeds while recording lateral error improvement of up to 82% compared to the original Stanley controller.

Keywords: heavy vehicle, autonomous trajectory tracking, path tracking, Stanley controller, particle swarm optimisation

1. Introduction

This study proposed a new adaptive steering control strategy for trajectory tracking controller of a heavy vehicle. The controller aims to automatically steer the vehicle along the desired trajectory and adapt to various speeds and trajectories during varied manoeuvrings. A path tracking controller is a controller module that is developed to provide electronic actuation to the vehicle system while navigating the vehicle automatically. An effective controller needs to be developed to ensure a functional steering module while autonomously navigating through various paths. Commonly, there are several types of trajectory tracking controllers as reviewed previously [1, 2]. One of the most common types of controllers is kinematic controllers such as Pure Pursuit and Follow-the-Carrot due to the simplicity and stability it can provide. This type of controllers relies on the kinematic properties of the vehicles such as speed and acceleration, as well as the travelled distance for the controller feedback. Compared to other dynamic controllers that require the kinetic properties of the vehicle such as torques, moments, and forces, geometric and kinematic properties are relatively easier to measure.

In geometric/kinematic controllers, one of the most established controllers is the Stanley controller as published in [3, 4]. The controller was developed by the Stanford University racing team on their autonomous vehicle (Stanley) in winning the DARPA Challenge in 2005 [5]. Compared to other geometric/kinematic controllers, the Stanley controller does not include a lookahead distance in its formulation. This enables the controller to be robust enough without depending on what lies ahead. However, the Stanley controller poses different problems, which is common in most of the geometric/kinematic controllers. It was found that the performance of this controller on any given trajectory depends on how well the parameter-tuning process was. A finely tuned Stanley controller will be effective only on the driving conditions that it was tuned for. However, it needs to be retuned to work with another road course and speed range. Previous studies have been discussing this issue and stated the same conclusion for most of geometric/kinematic controllers [1, 3, 6].

Therefore, an improvement is proposed to include an adaptive algorithm that will adjust the controller's parameters based on the driving conditions. Adaptive controllers for trajectory tracking controller have been proposed in numerous studies recently [7–10] to improve the adaptability and stability of the controller under varying conditions. These controllers are designed to cater robustness in a specific area, such as for slippery roads [7], unknown slip conditions [9], and unknown skidding conditions [10]. While the adaptiveness of these controllers in the designated area was proven, respectively, it may not be as effective when dealing with multiple types of disturbances other than the ones it was designed for. For example, an adaptive controller designed to cater various skidding conditions may not be able to cater unknown yaw disturbance. Also, most of the studies for an autonomous vehicle are using the linear vehicle model to develop the control structure where most of the nonlinearity in vehicle motions is neglected such as frictions and aerodynamic effects. In addition, some adaptive algorithms consist of algorithms that require high computational capability due to its associated complexity. Therefore, the adaptive controller proposed in this study is aimed to solve these issues by (1) considering a nonlinear vehicle model containing most of the nonlinearity of a vehicle motion in the controller development phase; (2) using adaptive inputs as vehicle speed and heading error, which both are directly dependent on the sharpness of turns and vehicle slips; and (3) using a simple geometric/kinematic controller as the basic controller to be modified.

Overall, the main contribution of this work is on the development of a knowledge-based algorithm using the adaptive mechanism of path tracking controller to accommodate the varying trajectories and vehicle's speed setting.

Two inputs are considered in triggering the adaptive algorithm, which are the heading error, ϕ, and vehicle speed, v. The range of speed region catered in this study is based on the limitations and the expected operating region of the autonomous heavy vehicle, which is up to 72 km/h. Variations in road course trajectory are observed in terms of the instantaneous difference between vehicle heading and trajectory direction, ϕ with a range between 0 and 75 deg. in both directions. This may as well cater sharp turns. The basic controller is modified to increase its sensitivity to disturbance. Then, by optimising the controller parameters for different disturbance input combinations, a knowledge database is developed. With this, a set of optimum parameters can be chosen depending on the instantaneous speed and trajectory experienced by the vehicle. An algorithm has been developed to carry out the selection process. The controller's performance is then evaluated on six different trajectories and four random speed values to evaluate its effectiveness against the basic Stanley controller and the modified controller without an adaptive algorithm. Results show a promising prospect for the proposed controller.

This chapter starts with Introduction section that covers a brief background of the study, followed by the modelling of a nonlinear seven-degree-of-freedom (7DoF) vehicle model used to simulate the vehicle's behaviour. Next, the proposed adaptive controller is explained in Section 3, beginning with the basic controller's and the adaptive algorithm development. The simulation and experiment procedures in evaluating the controllers including the development of road courses used are presented in Section 4, with the findings discussed in Section 5. Conclusions for this work are presented in the final section.

2. Modelling of a 7DoF heavy vehicle

In this study, a full vehicle model is developed with the aim to simulate the vehicle's behaviour in lateral direction. The model, developed based on a heavy vehicle, namely, a High Mobility Multipurpose Wheeled Vehicle (HMMWV), is closely related to an armoured vehicle as shown in **Figure 1**. It consists of several

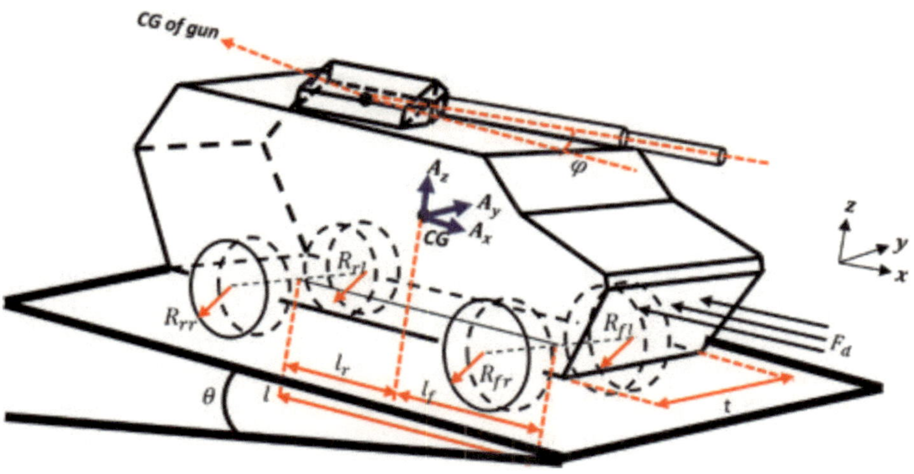

Figure 1.
Heavy vehicle model [11–13].

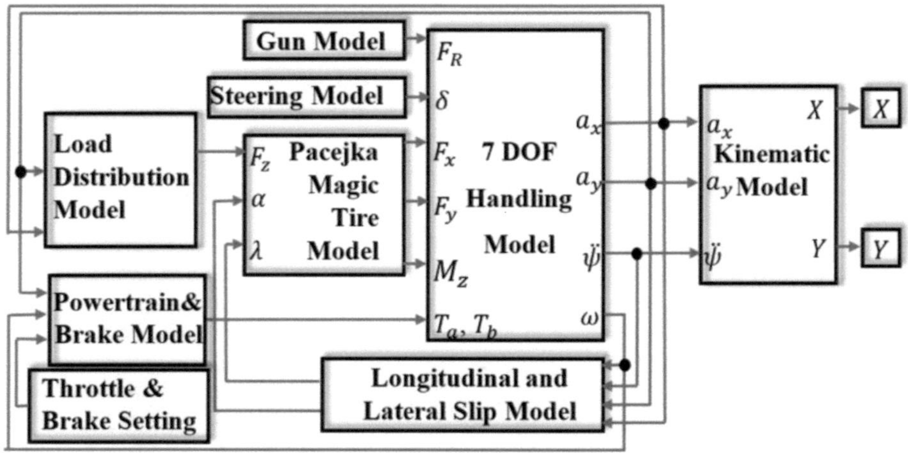

Figure 2.
Configuration of the heavy vehicle model [12, 14, 15].

subsystems in order to simulate the different elements in the vehicle system that contribute to the overall vehicle performance in lateral directions: a 7DoF handling model to simulate the vehicle dynamic responses at its centre of gravity during manoeuvrings; a tyre model in order to simulate all four tyres' behaviour from the road surface interactions; a slip model to calculate the generated slips in lateral and longitudinal directions during driving manoeuvrings, which will be an important input to the tyre model; a load distribution model to estimate the static and dynamic load transfers in longitudinal and lateral directions, which are significant during accelerating, braking, and/or cornering; an engine model; and lastly, a kinematic model to evaluate the vehicle's position relative to the local and global coordinates. The configuration of the overall vehicle model and its respective subsystem is shown in **Figure 2**.

Few assumptions and simplifications are made in developing this model. Firstly, this model focuses mainly on the vehicle responses in lateral and longitudinal directions. Responses and disturbances in vertical direction are considered less significant by assuming an ideal suspension system between vehicles' sprung body, and the vehicle is assumed to travel on even and smooth roads. With these two assumptions, the ride model involving suspension forces and road disturbances is not considered in this model. Next, the vehicle is modelled as a rigid body with concentrated sprung mass at the centre of gravity and four wheels that are connected to each of the vehicle four corners. The vehicle is moving on a level, unbanked, and uninclined road, and therefore, the weight is acting along the z-axis. The vehicle body is represented in three-dimensional x-y-z planes, which is allowed to displace in lateral and longitudinal directions, as well as rotating about the z-axis (yaw). Each of the connected wheels is modelled as a rigid body, which is allowed to rotate about its rotational axis, which is parallel with the vehicle's lateral axis. Only the two front wheels are allowed to steer, which is equipped with an active Pitman arm steering system. Modelling of this system is also considered in this chapter. In calculating the tyre responses, the vertical load on each tyre is calculated using the vehicle's load transfer model by considering the amount of load transfer during manoeuvrings. The load transfer is based on weight distribution in lateral direction during cornering and longitudinal direction during acceleration/braking. To ensure that the model generate response as close as possible to a real condition, rolling and air resistance are considered in this model. In terms of the steering system, the armoured vehicle is equipped with a Pitman arm steering system where the output steering values were saturated at ±10 deg. This is due to the consideration of the system's limitation as well as the controller's stability region that has been tested as described previously [3, 14].

The previous work from the authors has demonstrated the derivations and verifications of this model [12, 15]. Also, several previous studies have been evaluating its controller's performance on the same model [11, 14, 16]. **Figure 1** also shows the symbols and vehicle parameters used in developing the model. Perhaps it is worth noting a common confusion between the fixed global coordinate axes (X, Y, and Z), which are commonly associated with the Earth's longitude, latitude, and altitude, respectively, and the moving local coordinate axes (x, y, and z). This model has been validated by verifying the simulated vehicle responses against the HMMWV responses from a CarSim software as described in detail by Aparow et al. [12]. The verification was done using the standard manoeuvring procedures, namely, Slalom tests, double lane change, and step steer manoeuvrings.

3. Knowledge-based adaptive controller development for a path tracking system

3.1 Basic structure of trajectory tracking controller

In this study, the adaptive controller was based on the previously developed controller from the previous publication by Amer et al. [14], namely, the modified Stanley (Mod St) controller. Equation (1) shows the original Stanley controller from Hoffmann et al. [3]. The same publication has extended the controller to include the yaw rate compensation as shown in Eq. (2). A modification has been carried out to these controllers to increase their sensitivity towards the disturbance input as proposed previously in [14]. The Mod St controller used in this work is shown in Eq. (3), where the main parameters are shown in **Figure 3**. Here, e is the lateral error of the vehicle (m), measured between the vehicle and the perpendicular point on path; ϕ is the heading error (rad), which is the difference between the vehicle's and the path's instantaneous directions; v is the vehicle speed (m/s); $\dot{\psi}$ is the instantaneous yaw rate of the vehicle (rad/s); and $\dot{\psi}_{traj}$ is the trajectory yaw rate (rad/s). Meanwhile, k_ϕ, k_1, k, and k_ψ are the tuneable gains. In this study, the original Stanley controller from Eq. (1) is used as the benchmark, while the Mod St controller from Eq. (3) is used as the base controller for the adaptive control structure.

$$\delta = \phi + \tan^{-1}\left(\frac{ke(t)}{v(t)}\right) \tag{1}$$

$$\delta(t) = \phi + \arctan\left(\frac{ke(t)}{1+v(t)}\right) + k_\psi\left(\dot{\psi} - \dot{\psi}_{traj}\right) \tag{2}$$

$$\delta(t) = k_\phi \phi + k_1 \arctan\left(\frac{ke(t)}{1+v(t)}\right) + k_\psi\left(\dot{\psi} - \dot{\psi}_{traj}\right) \tag{3}$$

Six trajectories are chosen for this study in order to portray the different kinds of road courses, shown in **Figure 4**. Each was named based on the trajectory shape,

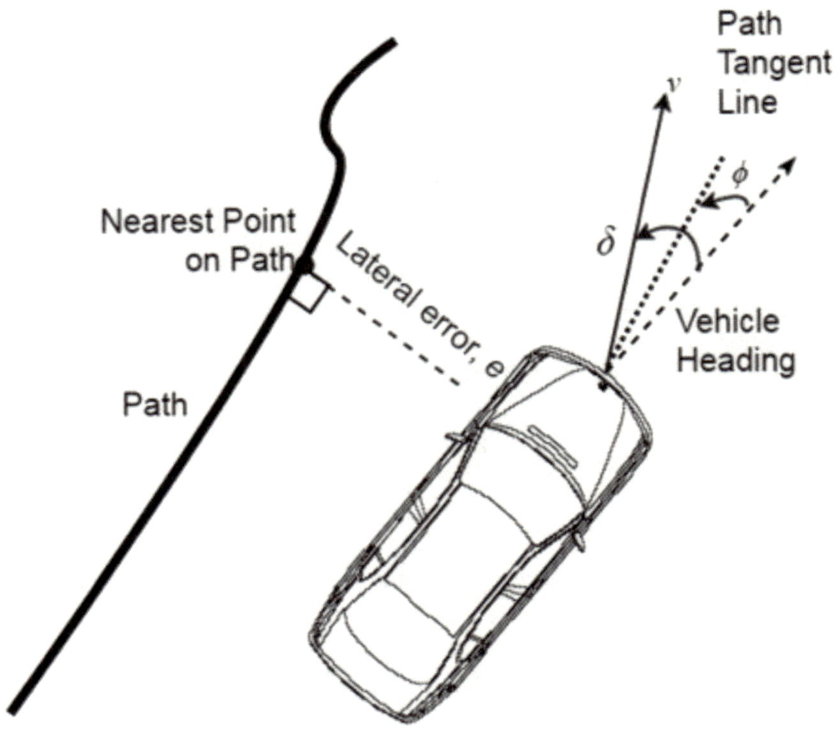

Figure 3.
Main parameters for the Stanley controller and its variants.

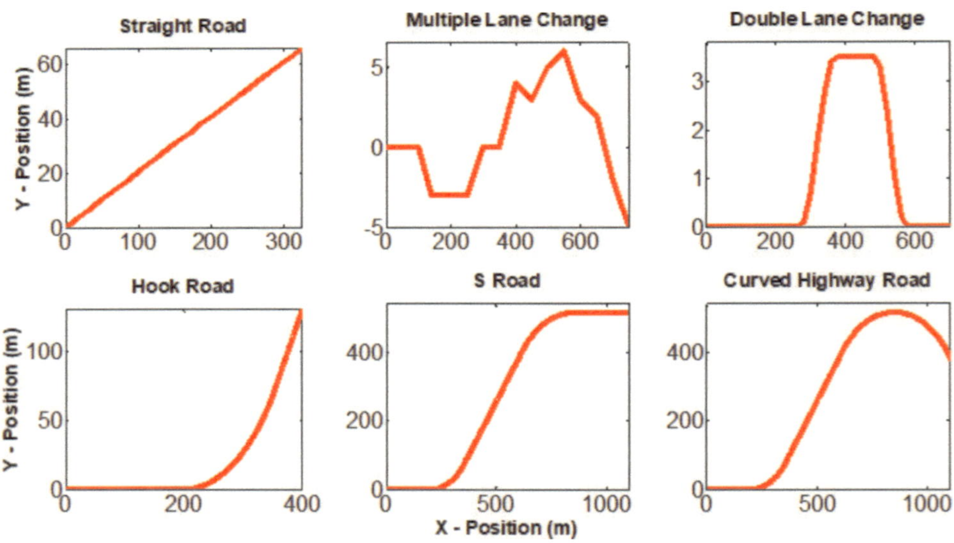

Figure 4.
Road courses for controller testing.

namely, "straight road", "multiple lane change", "double lane change", "curve", "S", and "hook". Each trajectory is defined as a set of points with *X-Y* coordinates. Therefore, lateral error is obtained based on the current *X* position for the vehicle, such that $e = Y_{path}(X_{vehicle}) - Y_{vehicle}(X_{vehicle})$.

Further analysis on the application of controller on a heavy vehicle yields an important issue in tuning the parameters as discussed previously by Snider [6] and Shan et al. [17]. A perfectly tuned controller will work properly within a certain range of vehicle speed, as well as a certain type of trajectory and road courses. Also, the more parameters included in the controller, the more sensitive it is to changes, which will make it less robust. These effects can be shown in **Figure 5** where all the three variants of the Stanley controller were simulated on the heavy vehicle model with different speeds on S road. Each of the controllers was properly tuned for 6 m/s speed on the particular trajectory using particle swarm optimisation (PSO) algorithm with the same approach as presented in [13–15]. **Table 1** shows the parameter values for each controller upon optimisations for each trajectory at 6 m/s speed, which was used throughout the testing for all speed values. From the figures, it can be seen that the properly tuned controller is valid only for a certain range of speed and the number of parameters is one of the main factors that affect the robustness of the controller. The controller with the least parameters showed better robustness and performance in varying speeds. However, significantly large error (0.4–0.8 m) can be noticed. This is agreeable to the finding by Wallace et al. [18], which concludes that it is a conflicting factor in tuning a geometric controller between stability and robustness. Therefore, this study proposes an adaptive controller with the ability to supervise the selection of an optimum set of controller parameters based on the speed and trajectory experienced by the vehicle. With this, the controller can be used regardless of the manoeuvring conditions without the need to be re-tuned.

With the original Stanley controller modified into the Mod St controller in Eq. (3) to increase its sensitivity, an adaptive algorithm is proposed to automatically tune the four parameters in Mod St, namely, k_ϕ, k_1, k, and k_ψ. An adaptive algorithm using a three-dimensional control surface is used. Since it was known that the main factors affecting the validity of any set of controller parameters are both the vehicle speed and type of trajectory, two inputs are chosen for the algorithm, which are the heading error, ϕ, and vehicle speed, v. Each of the controller parameters is optimised for each input combination to form a knowledge database. With this, the adaptive algorithm is developed and converted into separate control surfaces for each parameter where the adaptive algorithm will be developed to automatically

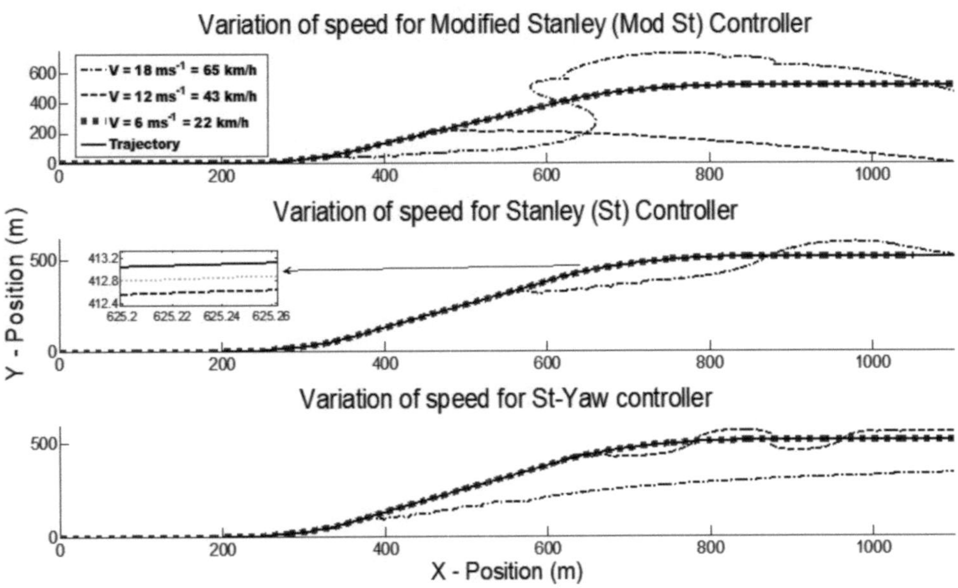

Figure 5.
Effect of varying speeds on the S road for the Mod St controller [Eq. (3)], Stanley controller [Eq. (1)], and Stanley controller with yaw compensation [Eq. (2)].

	St	Mod-St			
	k	k_1	k	k_ψ	k_ϕ
Straight	10	10	10	-2.964	0.7719
Multiple Lane Change	10	10	10	0.3046	1.892
Double Lane Change	10	10	9.689	0.0901	0.819
Hook	10	10	10	0.0423	-0.058
S	10	10	9.757	0.0642	0.0199
Curve	10	10	10	0.1762	-0.0051

Table 1.
Controller parameters for St and Mod St controllers for each trajectory.

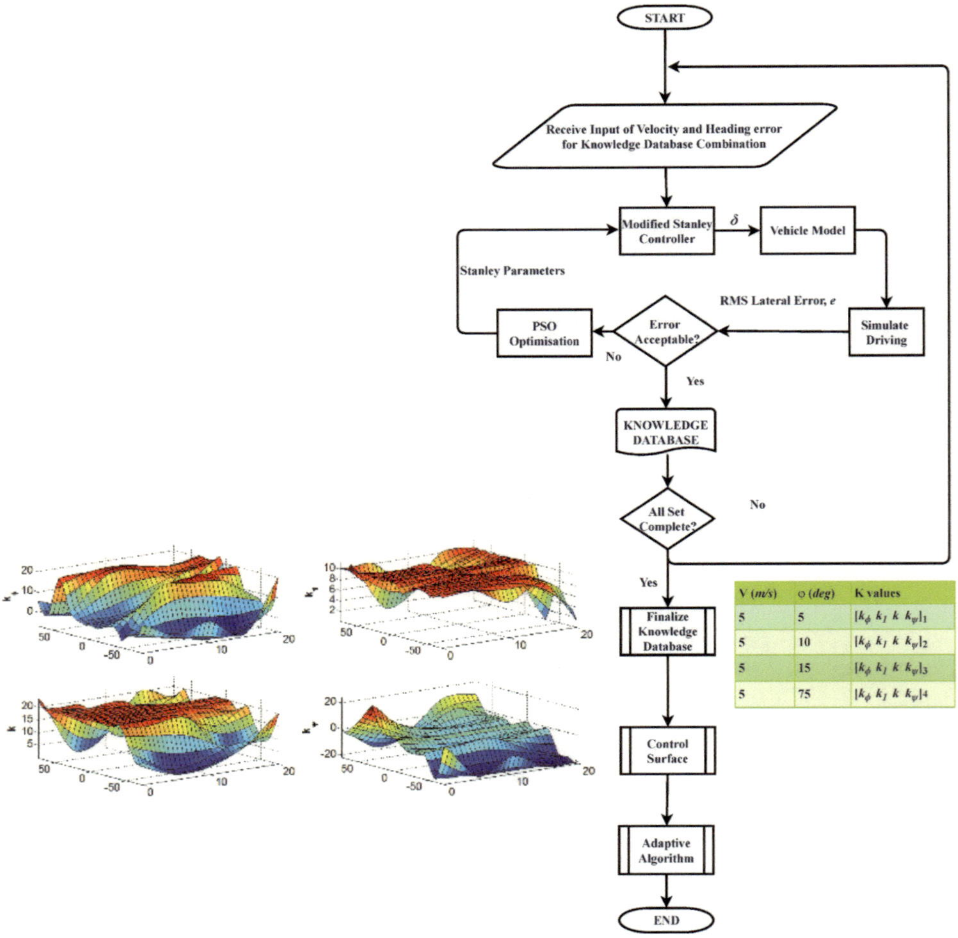

Figure 6.
Procedure in developing an adaptive modified Stanley controller.

choose parameter values from the surfaces. The overall procedure can be illustrated in **Figure 6**.

3.2 Development of the adaptive controller

An adaptive controller with the ability to supervise the selection of an optimum set of controller parameters depending on the heading error (ϕ) and vehicle speed

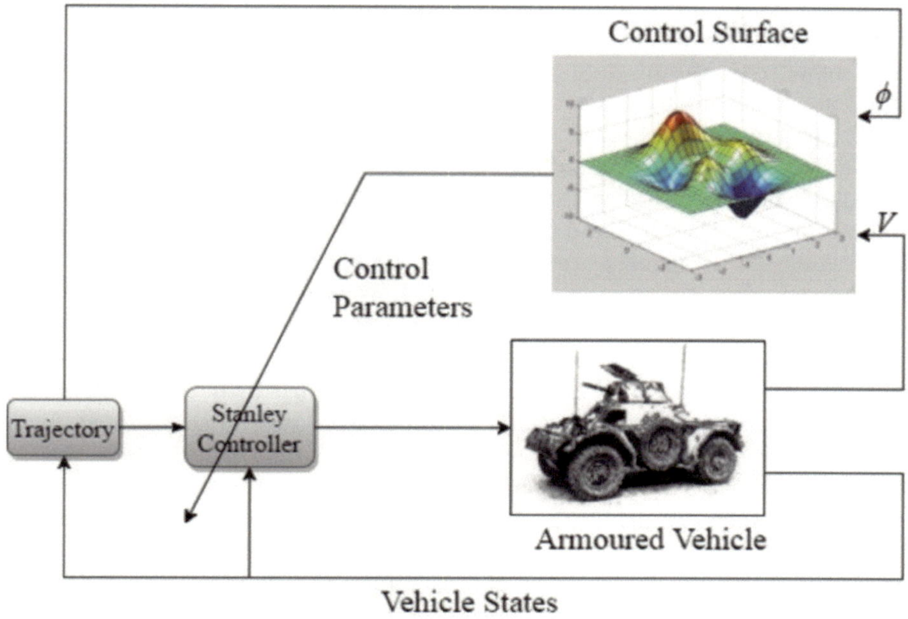

Figure 7.
Structure of the adaptive Stanley controller.

v (m/s)	1	5	10	15	20									
ϕ (deg)	1	5	15	30	45	60	75	-1	-5	-15	-30	-45	-60	-75

Table 2.
Interval for knowledge database for each input.

(v) is proposed in this study. These two variables are chosen since they represent the vehicle's instantaneous speed and the state of trajectory heading with respect to vehicle's heading at a particular time. By choosing this as the input for the adaptive mechanism, and optimising the controller at different values of these two states, the controller will be able to react to variable speed and trajectory it operates on. The overall structure for the knowledge-based adaptive Stanley controller is shown in **Figure 7**.

3.2.1 Building knowledge database using particle swarm optimisation

In this study, a knowledge database is built to provide artificial insights for the controller to choose a suitable set of parameter values for the controller based on the vehicle's driving conditions, namely, speed (v) and heading error (ϕ). The database should have a complete set of parameters (k_ϕ, k_1, k, and k_ψ) that correspond to each range of v and ϕ.

Table 2 shows the interval values for input variables that were chosen for the knowledge database. In choosing the interval values, several sensitivity analyses were carried out to test the suitability of an optimised parameters set in varying speeds. It was found that each optimised parameters will yield satisfactory results within ±5 m/s and ± 15 deg. Therefore, these intervals are chosen. A minimum of 1 m/s (≈ 3.6 km/h) and a maximum of 20 m/s (≈ 72 km/h) were chosen based on the common and safe operating speed for the vehicle. A minimum of 1 deg. and a maximum of 75 deg. were chosen for ϕ values, which correspond to the minimum and maximum values of the cornering angle for the vehicle.

A metaheuristic optimisation algorithm, namely, PSO method, is used in building the knowledge database, as outlined in Amer et al. [13, 15] to optimise the four controller parameters for each of the v and ϕ combinations. Each of the four controller parameters is optimised for each of the v and ϕ combinations. PSO is one of the established optimisation algorithms based on a natural phenomenon. First introduced in 1995 [19], it is one of the many optimisation algorithms based on swarm intelligence available as research tools. A comprehensive study on PSO algorithm and other swarm intelligence algorithms with their associated optimisation performance can be found in [20]. The study has found that PSO has better performance in terms of solution consistencies and convergence than the other swarm algorithms. Zhang [21] has also stated that PSO has better convergence in achieving global optimum with better accuracy than conventional methods. Also, with a lesser number of algorithm parameters to tweak than genetic algorithm (GA), PSO is always preferable. Overall, it was found that PSO converged faster and the study concluded that PSO provides better accuracy with a fast convergence for a standard computing capability. The PSO performance has also been studied in various applications for controller parameter tuning such as [22]. Here, PSO was used to tune a model predictive controller, and the optimisation performance was compared against ant colony optimisation (ACO) and gravitational search algorithm (GSA). It was shown that PSO was more superior than the others with up to 90% quicker convergence than ACO and GSA. The PSO-optimised controller also performed better, showing that PSO managed to find better optimum points than its counterparts.

PSO algorithm simulates the behaviour of swarmed particles, which move within the swarm towards an intended position. Each particle will be assigned random positions within the swarm, and its next movement (position and velocity) will be determined based on its relative position to an optimum position. Optimum position will be the position of any particle with an optimum fitness value. Every particle will have its own fitness value and the memory of its own best position, *pbest*, as well as the overall swarm best position, *gbest*, based on the optimum fitness value. The particle's next position will be determined by considering these memories. This will guarantee that the particles will not be moving too quick towards the new position that can turn out to be another local optimum. In other words, any of the i_{th} particle within the swarm will have position coordinates, x_{id}, in each of the d_{th} dimensions and move with velocity v^t. Its next position, $x_{id}^{(t+1)}$, will be determined based on its next velocity $v^{(t+1)}$ shown in Eqs. (4) and (5), respectively. These processes will be iterated over several cycles, and finally, the swarm will converge to an optimum position:

$$x(t+1)_{id} = x(t)_{id} + v(t+1) \tag{4}$$

$$v^{(t+1)} = iw \times v^{(t)} + c \times rand(0,1) \times \left(p_{best}{}^{(t)} - x^{(t)}{}_{id}\right) + s \times rand(0,1) \times \left(g_{best}{}^{(t)} - x^{(t)}{}_{id}\right) \tag{5}$$

In this study, the optimisation process of Stanley controllers using PSO can be shown in **Figure 8**. Referring back to **Figure 6**, the upper part of the chart shows the optimisation process of each input combination using PSO, while **Table 3** tabulates the main parameters used in PSO algorithm. In addition, since lateral error is one of the most significant performance indices for a path tracking controller, fitness function for each particle will be evaluated by the lateral error between the vehicle

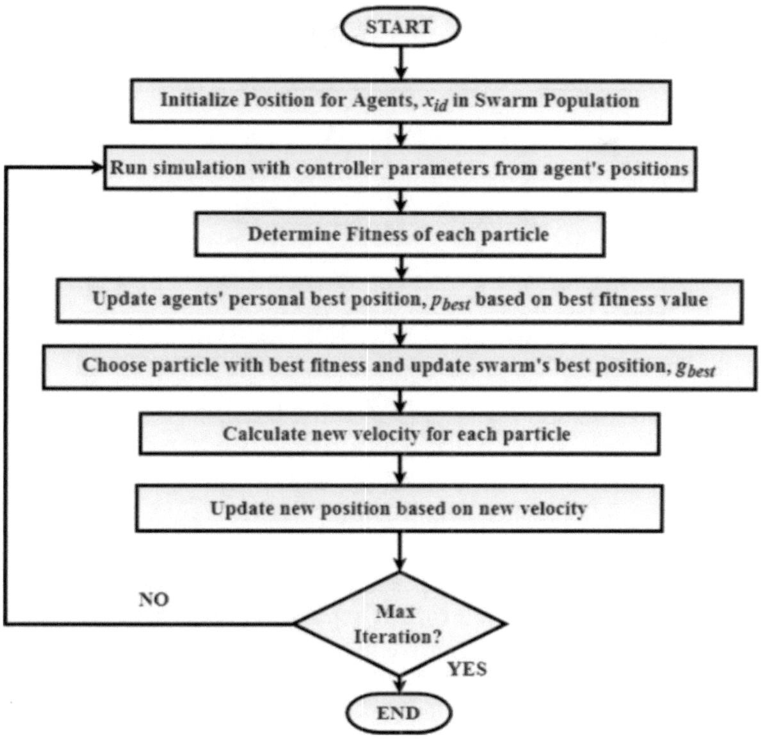

Figure 8.
Procedure for PSO.

Parameter	Values
Social Coefficient, s	1.42
Cognitive Coefficient, c	1.42
Inertial Weight, iw	0.9
No. of Dimensions, N_d	4 (k_1, k_ϕ, k, and k_ψ)
Upper Bound Limit	[10;10;10;10]
Lower Bound Limit	[-10; -10; -10; -10]
No. of Particles, N_p	150
No. of Iterations, N_i	20

Table 3.
PSO parameters used in building a knowledge database [13, 15].

and trajectory (e), which resulted from the parameter values for each particle's position. Lateral error is quantified by the root mean square (RMS) value, as shown in Eq. (6):

$$\text{Fitness Function, } f(k_f, k_1, k, k_y) = \sqrt{\frac{\sum[e(t)]^2}{n}} \tag{6}$$

Upon completion, all sets of parameters are compiled and integrated into a knowledge database consisting all parameters for each combination of ϕ and v

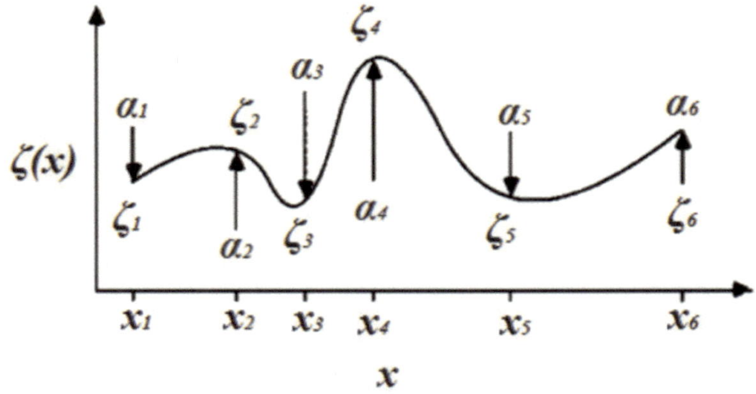

Figure 9.
Wrapping an imaginary spline around data points (xi, ζi) with point forces (αi) acting at each point.

values. The database is then used to form separate control surfaces for each corresponding parameter. For the proposed basic controller in Eq. (3), there are four parameters, which correspond to four separate control surfaces formed from the knowledge database. For the adaptive method, the controller parameters are adjusted by an algebraic algorithm that will interpolate a suitable set of parameter values based on the developed knowledge database. The interpolation will be on a set of control surfaces that map the optimum set of parameters according to its respective ϕ and v values.

3.2.2 Generating control surface from the knowledge database

In this study, the developed knowledge database in Section 3.2.1 is transformed into a set of control surfaces (Φ^k) that correspond to each of the four controller parameters k_ϕ, k_1, k, and k_ψ. The transformation of the database to control surfaces was carried out using the concept named biharmonic spline [23], originally used to deduce geographical contour from a set of altitude data of the topography. It worked by assuming an imaginary spline across a one-dimensional data point and applying a point force (α_j) to each of the data points with coordinates (x_j, ζ_j). This can be shown in **Figure 9**.

Using the biharmonic spline method, the line in **Figure 9** satisfies a biharmonic partial differential equation (PDE) with $N = 6$ shown in Eq. (7). The solution of this PDE is satisfied with a point-force Green's function that is centred at each point as shown in Eq. (8). Solving for α_j in Eq. (9) will yield the strength of each point force (α_j):

$$\frac{d^4\zeta}{dx^4} = \sum_{j=1}^{N} 6\alpha_j \delta\left(x - x_j\right) \tag{7}$$

$$\zeta(x) = \sum_{j=1}^{N} \alpha_j \left|x - x_j\right|^3 \tag{8}$$

$$\zeta(x_i) = \sum_{j=1}^{N} \alpha_j \left|x_i - x_j\right|^3 \tag{9}$$

Extending this method to a three-dimensional space, one can assume spreading an imaginary surface over the points instead of a spline. In this study, the biharmonic interpolation concept is applied on the knowledge database, which consists of data that are scattered in a three-dimensional space. The control surface produced needs to pass through each of the data in the database using this concept

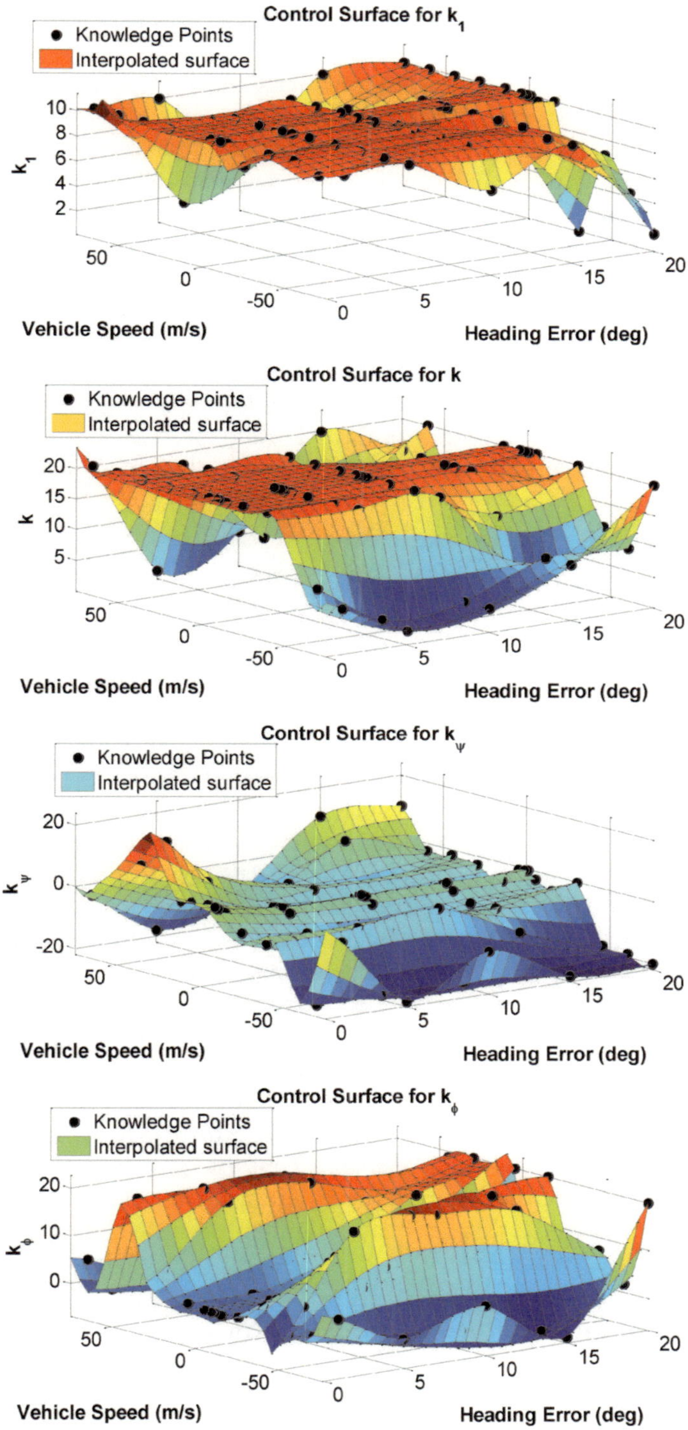

Figure 10.
Control surface for each controller parameter.

by wrapping an imaginary thin lamina around all the data points. The database is separated into four sets of data, which will generate four control surfaces, which corresponds to each controller parameter as shown in **Figure 10**. Each surface shows the interpolated surface relating to the control parameter values to its corresponding vehicle speed and heading error.

3.2.3 Using the control surface to generate an optimum set of parameters

Using the control surface generated previously, the interpolation algorithm will choose corresponding parameter values within the surface based on instantaneous heading error and vehicle's speed. This will adaptively tune the modified Stanley controller, adjusting it to various conditions during manoeuvring regardless of any path and vehicle speed. In cases where the v and ϕ values are outside of the database range from **Table 2** and surfaces from **Figure 10**, no extrapolation method will be used. Instead, parameter values at surface boundaries will be considered. This is to avoid the controller to enter an unstable controller region and ensure that the algorithm will continuously tune the parameters even when the v and ϕ values are outside of the range in the knowledge database. By denoting the kth control surface as $\Phi^k(v, \phi)$ where k = 1, 2, 3, 4 and K_k corresponds to the controller parameters k_ϕ, k_1, k, and k_ψ, the adaptive algorithm to automatically tune the controller parameters has been developed as a set of controller commands as shown in Eq. (10):

$$K_k = \begin{cases} \phi^k(\phi, V) & \text{if } \phi_{\min} \le \phi \le \phi_{\max},\ \text{AND } v_{\min} \le v \le v_{\max} \\ \phi^k(\phi, V_{\min/\max}) & \text{if } \phi_{\min} \le \phi \le \phi_{\max},\ \text{AND } v \le v_{\min} \text{ OR } v \ge v_{\max} \\ \phi^k(\phi_{\min/\max}, V) & \text{if } \phi \le \phi_{\min} \text{ OR } \phi \ge \phi_{\max},\ \text{AND } v_{\min} \le v \le v_{\max} \\ \phi^k(\phi_{\min/\max}, V_{\min/\max}) & \text{if } \phi \le \phi_{\min} \text{ OR } \phi \ge \phi_{\max},\ \text{AND } v \le v_{\min} \text{ OR } v \ge v_{\max} \end{cases} \tag{10}$$

4. Simulation and experimental procedures

The effectiveness of the developed controller is evaluated through simulations within the MATLAB/Simulink using ODE solver settings, namely, Heun ODE2 and a fixed-step size of 0.001 s. The controller is implemented on the validated vehicle model from Section 2 that was set to enter each of the trajectories in **Figure 4** with an initial velocity, zero brake, and throttle settings to keep the vehicle at a constant speed of 6 m/s. The adaptive controller is compared against the two of its predecessors, namely, the Mod St from Eq. (3) and the original Stanley controller as shown in Eq. (1). Each of these controllers has been optimised using PSO with similar procedures as described in Section 3.2.1 and simulated using the same simulation settings. In evaluating the controller's performance under varying input disturbances, a set of simulations were carried out using a different trajectory as shown in **Figure 4** as well as different speed settings. The vehicle was set to travel with different speed values that are chosen randomly within and outside of the range of the knowledge database.

4.1 Experimental validation against the human driver

The adaptive path tracking controller developed in this study is validated experimentally against the human-driven responses when navigating similar paths. For this, an instrumented heavy vehicle was driven, and the driver's steering input was recorded. The same manoeuvring was simulated using the developed controller, and the resulted steering input from the controller was compared against the human's driving input, to observe the validity of the automatic steering input from the controller.

(a) (b)

Figure 11.
Road course used in the experiment: (a) single lane change and (b) double lane change.

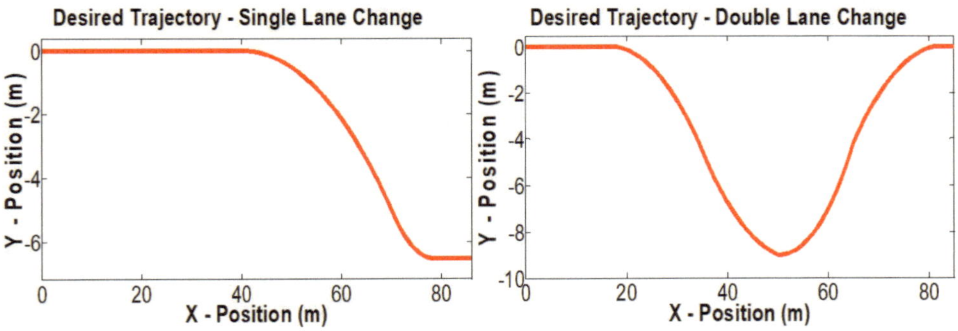

(a) (b)

Figure 12.
Road course for simulations taken from real dimensions for (a) single lane change and (b) double lane change.

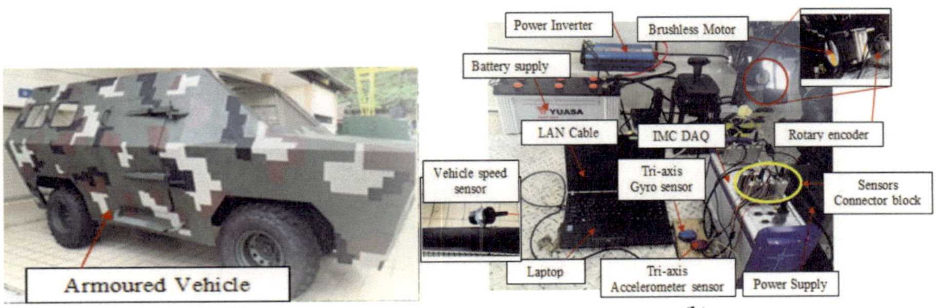

(a) (b)

Figure 13.
(a) Prototype of heavy vehicle and (b) sensor configuration in driver's cockpit.

In the experiment, two types of manoeuvrings were conducted, namely, single and double lane changes. The dimensions of the desired road were set on a level, open space as shown in **Figure 11** for both single and double lane change manoeuvrings, based on ISO-3888 [24]. These dimensions were converted into a series of coordinate points for simulation stages as shown in **Figure 12**. As one can see, the dimensions are quite short due to the limited space available. For this, only 20 km/h speed was chosen to ensure good manoeuvrings.

The manoeuvring tests are carried out with an instrumented prototype of a heavy vehicle as shown in **Figure 13**. The prototype was installed with various sensors and actuators as shown. However, for this experiment, the main sensors used are the accelerometer, gyro sensor, and rotary encoder. The rotary encoder will record the driver's steering input from steering column rotation, and from this, the wheel angle data can be deduced. Meanwhile, position data is acquired by a built-in GPS sensor in IMC DAQ that is attached to the roof of the vehicle to ensure good data transmission. Should the GPS fail, one can use the recorded acceleration data from the accelerometer, as well as vehicle speed data from the speed sensor to calculate the vehicle position in local coordinates.

From this experiment, automated manoeuvrings from the controller will be compared against the actual driving inputs from the human driver. Two observations can be deduced due to the nature of the short trajectories. First, the controller's ability in mimicking human driver behaviour can be evaluated. Second, the path tracking performance of the controller in guiding the heavy vehicle along short roads and sharp cornering can be evaluated also. The findings will be detailed in the next section.

5. Results and discussions

In evaluating the performance of the proposed adaptive controller, trajectory tracking performance was observed. Two aspects were compared, namely, (a) the vehicle trajectories and (b) the lateral error while navigating the trajectories with each respected controller. The performance of the proposed knowledge-based adaptive controller was compared against its two predecessors, which are the original St controller from Eq. (1) and Mod St from Eq. (3) as explained in Section 3.1. All simulation results for the six trajectories are shown in **Figure 14** for the straight road, **Figure 15** for the Multiple Lane Change Road, **Figure 16** for the Double Lane Change Road, **Figure 17** for the Hook Road, **Figure 18** for the S Road, and **Figure 19** for the Curved Highway Road.

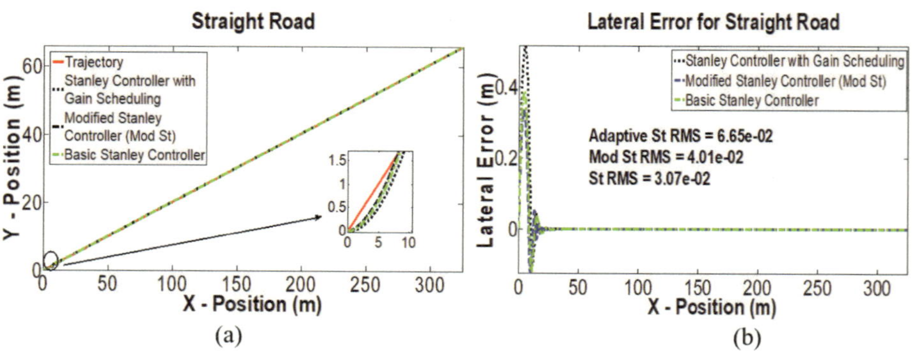

Figure 14.
Trajectory tracking performance for straight road: (a) vehicle trajectories, (b) lateral error, e.

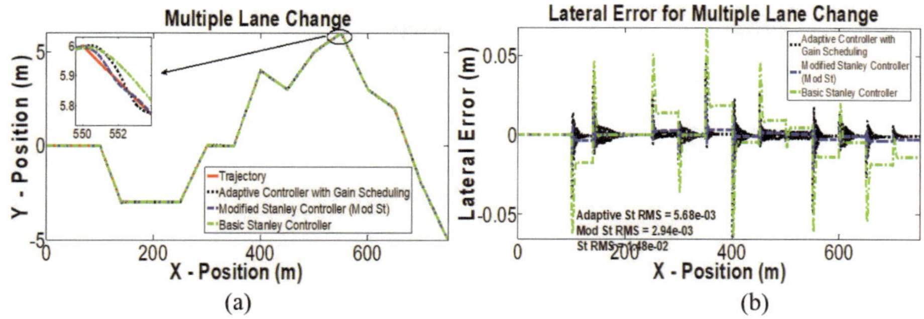

Figure 15.
Trajectory tracking performance for curved highway road: (a) vehicle trajectories, (b) lateral error, e.

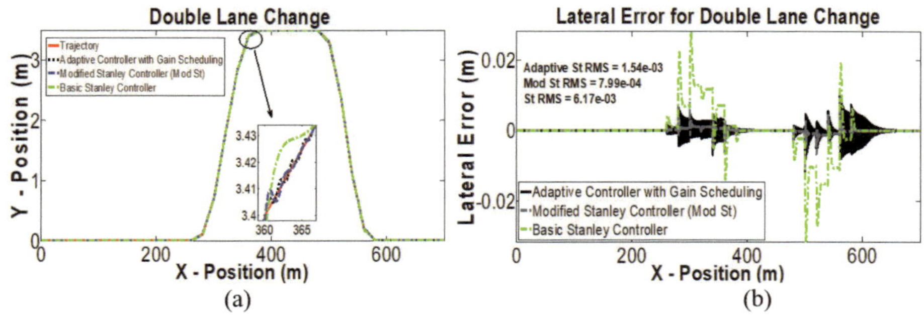

Figure 16.
Trajectory tracking performance for multiple lane change road: (a) vehicle trajectories, (b) lateral error, e.

It can be seen that the proposed controller managed to guide the heavy vehicle along the desired trajectories successfully. Overall, the adaptive controller performed significantly better than the St controller. The proposed controller managed to guide the vehicle with better precision closer to the path, which can be indicated further by the lateral error graphs in **(b)**. These improvements are mainly caused by the fact that the adaptive controller adopts Mod St steering command from Eq. (3) as the base controller to be automatically tuned. This controller considers the yaw rate error feedback, which can improve the overall trajectory tracking performance. Also, it has more controller parameters, which made the controller more sensitive to tuning, which, in turn, improve the tracking performance upon parameter selection. This also explains the exceptional performance by the Mod St controller shown in the graphs. For shorter and tight manoeuvring road

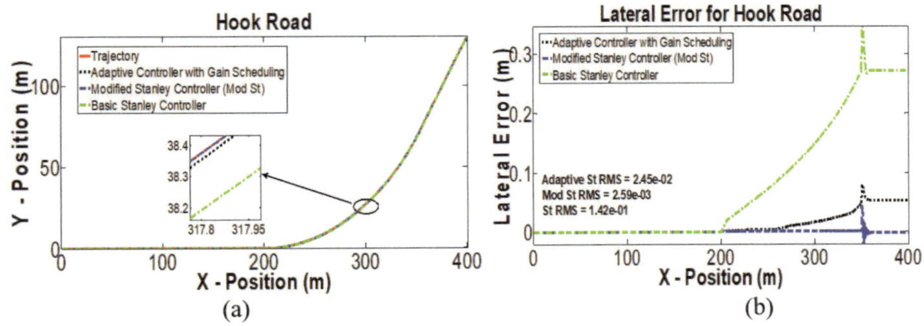

Figure 17.
Trajectory tracking performance for double lane change road: (a) vehicle trajectories, (b) lateral error, e.

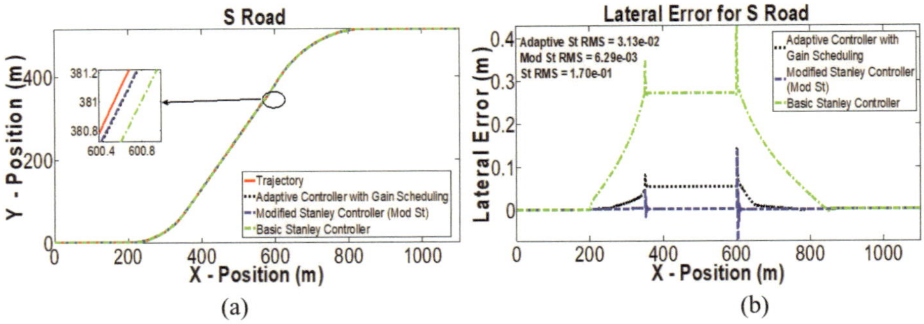

Figure 18.
Trajectory tracking performance for hook road: (a) vehicle trajectories, (b) lateral error, e.

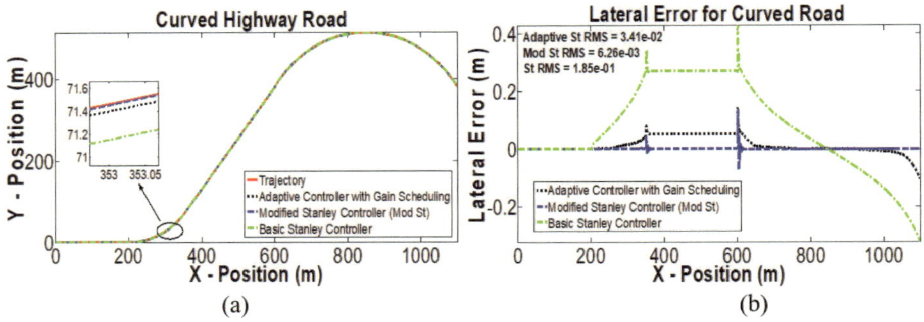

Figure 19.
Trajectory tracking performance for S road: (a) vehicle trajectories, (b) lateral error, e.

courses, namely, straight, multiple lane change, and lane change, shown in **Figures 14–16** respectively, the vehicle was manoeuvred successfully along the intended road courses with significantly better lateral error than the original Stanley controller. However, unwanted oscillations can be observed in **Figures 16** and **17** due to rapid cornering that exists in these courses. This can be minimised by preparing smoother curvature for the vehicle to follow [6].

However, one might notice the inferior performance shown by the proposed controller when compared against the Mod St controller. As stated before, this is the base controller where the adaptive algorithm was built on. In the simulation, the Mod St controller was tuned specifically for each trajectory using a metaheuristic optimisation algorithm, namely, PSO. The procedures are explained by Amer et al. [15]. Since the controller was specifically tuned for each trajectory and the 6 m/s speed, the controller parameter has been chosen to optimise the vehicle performance for each of the roads. This explains the fact that the base controller performed better than the adaptive controller, which was automatically tuned by the adaptive algorithm. However, despite the inferior performance compared to the Mod St controller, the adaptive controller still managed to guide the vehicle with a satisfactory performance. Looking at the RMS values for lateral error, the adaptive controller recorded a lateral error of 0.00154–0.0341 m across all the six trajectories, which are well below the average lateral error of 0.1 m recorded by the Stanley vehicle in the original publication [3]. Therefore, it can be concluded that the proposed adaptive controller performed well in navigating various trajectories. Overall comparison results for the RMS values on lateral error between the evaluated controllers are listed in **Table 4**.

The response of the controller under various vehicle speeds was studied next for hook, S, and curved highway road, as shown in **Figure 20**. These roads were chosen

	Adaptive	Mod St	St	% Diff w.r.t. Mod St	% Diff w.r.t. St
Straight	3.11E-02	5.43E-02	8.85E-02	-74.60	-64.86
Multiple Lane Change	5.68E-03	2.94E-03	1.48E-02	48.24	-61.62
Double Lane Change	1.54E-03	7.99E-04	6.17E-03	48.12	-75.04
Hook	2.45E-02	2.59E-03	1.42E-01	89.43	-82.75
S	3.13E-02	6.29E-03	1.70E-01	79.90	-81.59
Curved	3.41E-02	6.26E-03	1.85E-01	81.64	-81.57

Table 4.
Comparison of RMS values for lateral error between the controllers.

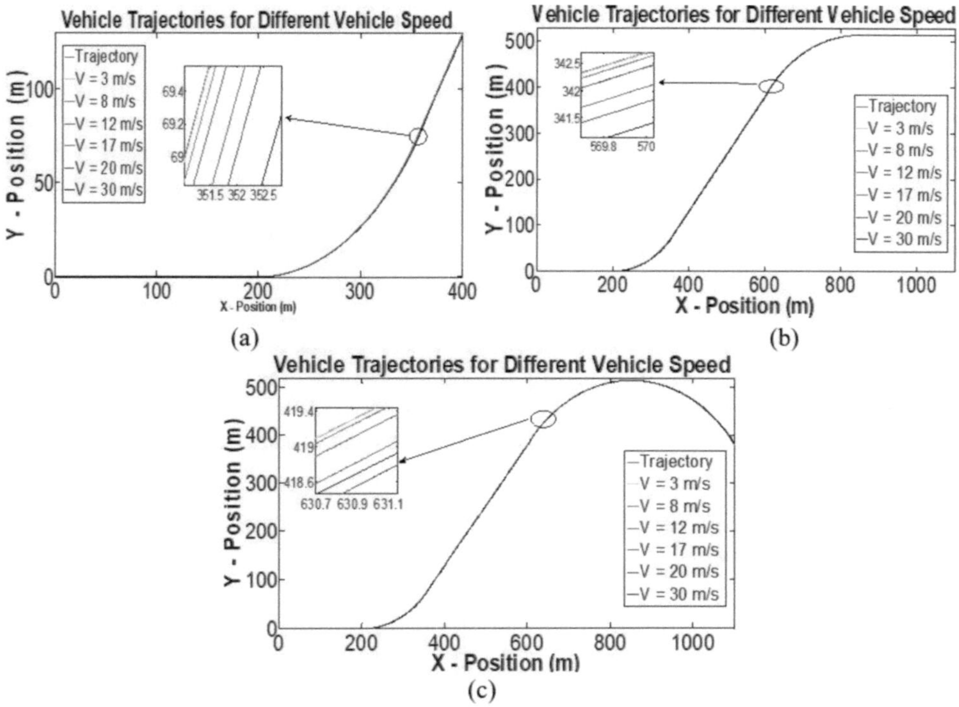

Figure 20.
Effect of varying speeds for (a) hook road, (b) S road, and (c) curved highway road.

due to its suitability for a high-speed testing. Straight, multiple lane change, and double lane change roads have shorter courses and extreme manoeuvrings, which are not suitable for a high-speed testing. In each analysis, six constant speed values were chosen to evaluate the controller, namely, 3, 8, 12, 17, 20, and 30 m/s. Based on knowledge database range and intervals listed in **Table 2**, 20 and 30 m/s were chosen to observe the controller's behaviour on the knowledge database boundary and outside the boundary. Other values were chosen randomly to observe the controller's performance with vehicle speeds well within the knowledge database boundary. From the figures, one can see that the proposed adaptive controller managed to steer the heavy vehicle well along the desired trajectory. However, as the vehicle speeds increase, larger error was observed since the vehicle is moving further than the intended trajectory. This is understandable since an increasing speed means that the vehicle can be diverted faster. Nevertheless, the controller still managed to bring the vehicle back to its intended direction with lateral error of well within 1 m as shown in **Table 5**.

Road Course	Vehicle Constant Speed (m/s)					
	3	*8*	*12*	*17*	*20*	*30*
Hook	0.0089	0.0511	0.1448	0.3290	0.4721	0.7547
S	0.0100	0.0625	0.1728	0.3976	0.5706	0.8488
Curve	0.0098	0.0676	0.1870	0.4337	0.6163	0.9322

Table 5.
RMS values of lateral error for various vehicle speeds with the adaptive controller.

5.1 Validation of steering angle against the human driver

For this analysis, three responses were observed, namely, vehicle trajectories, lateral error, and vehicle wheel angle throughout the manoeuvrings, which are single lane change and double lane change from **Figure 11**, and described in detail in Section 4.1. The actual data from the instrumented vehicle are compared against the simulation results with the adaptive controller using the same trajectory and road courses in **Figure 12** and constant vehicle speed of 20 km/h. **Figure 21** shows the comparison between experimental results by the human driver and simulation results by the adaptive controller for the single lane change manoeuvring. Meanwhile, **Figure 22** shows the same comparison for the double lane change manoeuvring.

In both figures, graphs in (*a*) show the vehicle trajectories, which indicate that the proposed controller managed to automatically provide correctional steering input in guiding the vehicle closer to the desired trajectories. There is a noticeable

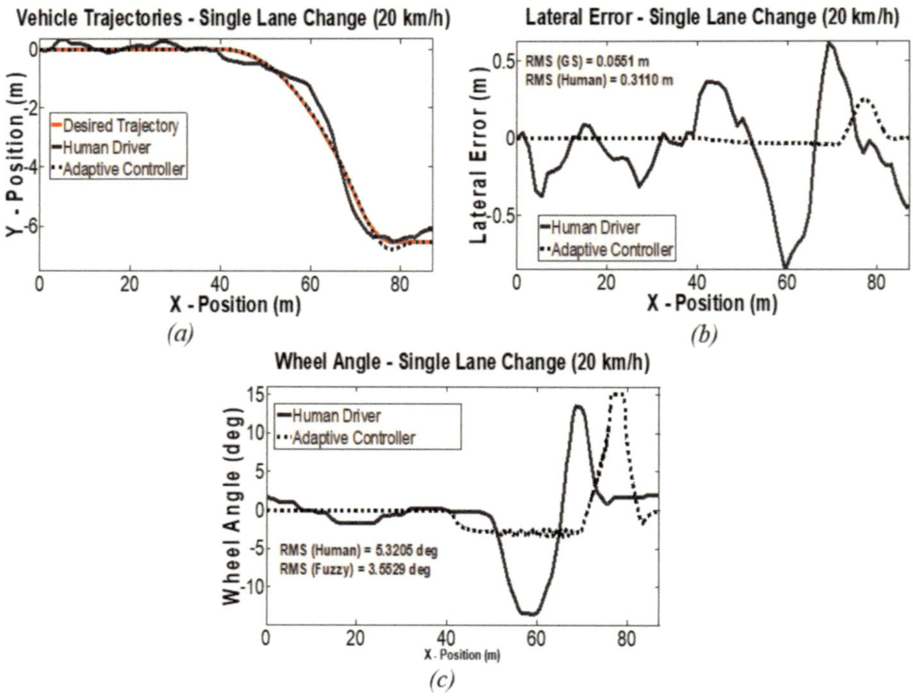

Figure 21.
Comparison between experimental results and simulations for single lane change road: (a) vehicle trajectory, (b) lateral error w.r.t. trajectory, (c) steering input required.

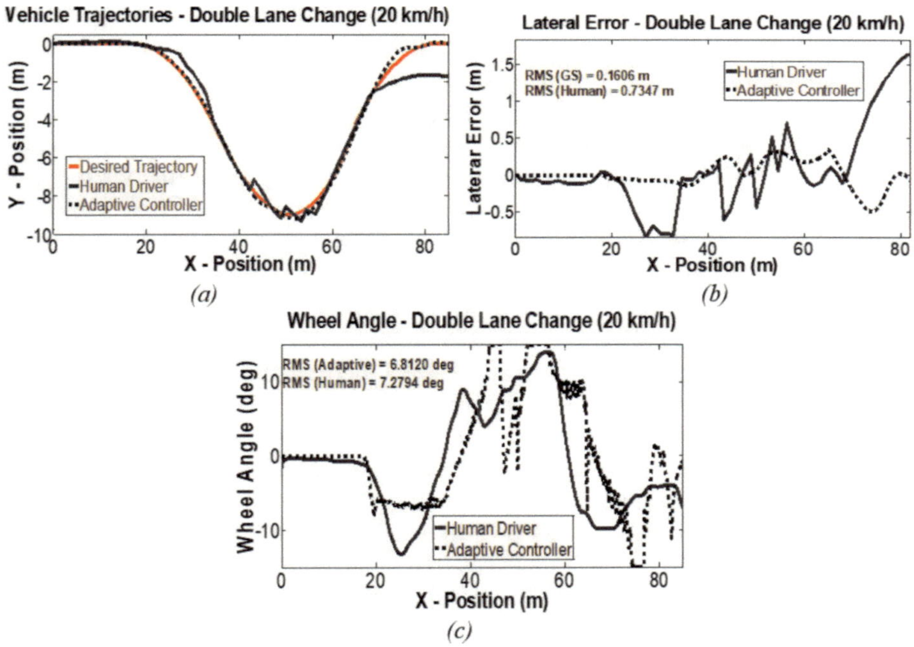

Figure 22.
Comparison between experimental results and simulations for double lane change road: (a) vehicle trajectory, (b) lateral error w.r.t. trajectory, (c) steering input required.

overshoot in vehicle trajectory from simulation (X = 80 m) for both manoeuvrings, which can be attributed to the speed of the vehicle. In the experiment, it was hard to keep the constant speed, and human instinct has caused the driver to slow down the vehicle while navigating sharp corners, whereby the speed was kept constant throughout the simulations. This has caused dissimilarities between the two results. Also, the road courses used here are very short, which may account to the inabilities of the controller, as well as the human driver, to perfectly track the trajectories. Nevertheless, the controller did well in steering the vehicle. This can be further studied through the lateral error results shown in graphs (b). The adaptive controller recorded better lateral error 82 and 78% less RMS values for single lane change and double lane change manoeuvrings, respectively.

Looking at the wheel angle from graphs (c), one can observe the correctional steering input provided by the human driver in the experiment, as well as the inputs from the adaptive controller in simulation. Compared against the human driver for the same manoeuvrings, the controller provides the steering input to the wheel with the same trend and input shape as the human driver, but with a faster response. Having a fast controller is always advantageous for any unaccountable delays and uncertainties that may happen in real implementation. Also, the controller is able to adapt to various speeds and trajectories while mimicking the human driver actions, which was proven from the results presented here.

6. Conclusions

In this study, an adaptive controller for an autonomous heavy vehicle is presented. The controller was developed based on an established Stanley controller that was modified to increase its sensitivity to the parameter changes. An adaptive algorithm was constructed to automatically tune the controller parameters based on the instantaneous vehicle speeds, v, and heading error, ϕ, between the vehicle and road course trajectory. In constructing the adaptive algorithm, a knowledge database was built by optimising a set of parameters for the modified Stanley controller that corresponds to various combinations of v and ϕ using particle swarm optimisation.

The developed controller was applied on a validated 7DOF, nonlinear heavy vehicle model. Six trajectories were chosen representing long and short courses as well as courses with large and very small turning curvatures. With these, a series of simulations were carried out, and the performance of the proposed controller was compared against the basic original Stanley and also the modified Stanley controllers.

From the simulations, it was found that the proposed adaptive controller performed well in guiding the vehicle along all trajectories. It recorded significantly lower lateral error RMS between 61 and 82% improvement when compared against the original Stanley controller. However, 48–89% increases in lateral error RMS were observed when comparing the controller against the modified Stanley controller. This can be explained by the fact that the Mod St controller was optimised specifically for the respective courses and, therefore, performed exceptionally well on the roads.

In terms of the controller's ability in adapting to various trajectories and vehicle speeds, the controller was tested with various vehicle speeds within and outside of the knowledge database region. It was capable of navigating the vehicle smoothly regardless of the vehicle speed, with understandably larger error in maximum speed. However, the overall lateral error was well kept below 1 m. While the maximum testing speed was 30 m/s, the autonomous heavy vehicle will be operating with speeds much lower than that. Previous studies and implementations of autonomous passenger's vehicles have been recorded to operate with about 10 m/s and lower [4, 25, 26]. In this study, the proposed adaptive controller has managed to perform well without depending on any planner to provide a smooth trajectory. This ensures the applicability of the proposed controller to be operated on a heavy vehicle on various trajectories and vehicle speed values.

Acknowledgements

The authors would like to thank the Malaysian Ministry of Education through research grants LRGS (LRGS/B-U/2013/UPNM/DEFENCE and SECURITY-P1).

Author details

Noor Hafizah Amer[1,2*], Khisbullah Hudha[1], Hairi Zamzuri[2], Vimal Rau Aparow[3], Amar Faiz Zainal Abidin[4], Zulkiffli Abd Kadir[1] and Muhamad Murrad[1]

1 Faculty of Engineering, Universiti Pertahanan Nasional Malaysia, Kuala Lumpur, Malaysia

2 Malaysia-Japan International Institute of Technology, Universiti Teknologi Malaysia, Kuala Lumpur, Malaysia

3 Faculty of Engineering, The University of Nottingham Malaysia, Selangor, Malaysia

4 Faculty of Engineering Technology, Universiti Teknikal Malaysia Melaka, Melaka, Malaysia

*Address all correspondence to: noorhafizah@upnm.edu.my

References

[1] Paden B, Cap M, Yong SZ, Yershov D, Frazzoli E. A survey of motion planning and control techniques for self-driving urban vehicles. IEEE Transactions on Intelligent Vehicles. 2016;**1**(1):33-55

[2] Amer NH, Zamzuri H, Hudha K, Kadir ZA. Modelling and control strategies in path tracking control for autonomous ground vehicles: A review of state of the art and challenges. Journal of Intelligent and Robotic Systems. 2017;**86**(2):225-254

[3] Hoffmann GM, Tomlin CJ, Montemerlo D, Thrun S. Autonomous automobile trajectory tracking for off-road driving: Controller design, experimental validation and racing. In: American Control Conference, New York, USA. Canada, USA: IEEE; 2007. pp. 2296-2301

[4] Thrun S et al. Stanley: The robot that won the DARPA Grand Challenge. Journal of Field Robotics. 2006;**23**(9): 661-692

[5] Buehler M, Iagnemma K, Singh S. The 2005 Darpa Grand Challenge: The Great Robot Race. Berlin Heidelberg: Springer Science+Business Media, Springer-Verlag; 2007

[6] Snider JM. Automatic Steering Methods for Autonomous Automobile Path Tracking. Pittsburgh, Pennsylvania, USA: Robotics Institute, Carnegie Mellon University, Technical Report CMU-RI-TR-90-17; 2009

[7] Lucet E, Lenain R, Grand C. Dynamic path tracking control of a vehicle on slippery terrain. Control Engineering Practice. 2015;**42**:60-73

[8] Hima S, Lusseti B, Vanholme B, Glaser S, Mammar S. Trajectory tracking for highly automated passenger vehicles. In: International Federation of Automatic Control (IFAC) World Congress, Vol. 44, No. 1, Milan, Italy. Netherlands: Elsevier; 2011. pp. 12958-12963

[9] Pratama PS et al. Adaptive backstepping control design for trajectory tracking of automatic guided vehicles. In: Duy HV, Dao TT, Zelinka I, Choi H-S, Chadli M, editors. AETA 2015: Recent Advances in Electrical Engineering and Related Sciences. Cham: Springer International Publishing; 2016. pp. 589-602

[10] Li X, Wang Z, Zhu J, Chen Q. Adaptive tracking control for wheeled mobile robots with unknown skidding. In: IEEE Conference on Control Applications (CCA). Sydney, NSW, Australia: IEEE; 2015. pp. 1674-1679

[11] Kadir ZA, Mazlan SA, Zamzuri H, Hudha K, Amer NH. Adaptive fuzzy-PI control for active front steering system of armoured vehicles: Outer loop control design for firing on the move system. Strojniski Vestnik-Journal of Mechanical Engineering. 2015;**61**(3): 187-195

[12] Aparow VR, Hudha K, Ahmad MMHM, Jamaluddin H. Development and verification of 9-DOF armored vehicle model in the lateral and longitudinal directions. Jurnal Teknologi. 2016;**78**(6):117-137

[13] Amer NH, Zamzuri H, Hudha K, Aparow VR, Kadir ZA, Abidin AFZ. Modelling and trajectory following of an armoured vehicle. In: 2016 SICE International Symposium on Control Systems (ISCS), Nagoya, Japan. Canada, USA: IEEE; 2016. pp. 1-6

[14] Amer NH, Zamzuri H, Hudha K, Aparow VR, Kadir ZA, Abidin AFZ. Path tracking controller of an autonomous armoured vehicle using modified Stanley controller optimized

with particle swarm optimization. Journal of the Brazilian Society of Mechanical Sciences and Engineering. 2018;**40**(2):104

[15] Amer NH, Hudha K, Zamzuri H, Aparow VR, Kadir ZA, Abidin AFZ. Hardware-in-the-loop simulation of trajectory following control for a light armoured vehicle optimised with particle swarm optimisation. International Journal of Heavy Vehicle Systems. 2019;**26**:663-691

[16] Amer NH. Adaptive Stanley Controller with Fuzzy Supervisory System for Trajectory Tracking Control of an Armoured Vehicle. Malaysia: Malaysia-Japan International Institute of Technology Universiti Teknologi Malaysia; 2018

[17] Shan Y, Yang W, Chen C, Zhou J, Zheng L, Li B. CF-pursuit: A pursuit method with a clothoid fitting and a fuzzy controller for autonomous vehicles. International Journal of Advanced Robotic Systems. 2015; **12**(134):1-13

[18] Wallace R, Stentz A, Thorpe CE, Maravec H, Whittaker W, Kanade T. First results in robot road-following. In: Proceedings of the 9th International Joint Conference on Artificial Intelligence, Vol. 2, Los Angeles, California, USA. USA: Morgan Kaufmann Publishers Inc; 1985. pp. 1089-1095

[19] Eberhart RC, Kennedy J. A new optimizer using particle swarm theory. In: Proceedings of the Sixth International Symposium on Micro Machine and Human Science, Vol. 1, New York, USA. Canada, USA: IEEE; 1995. pp. 39-43

[20] Ab Wahab MN, Nefti-Meziani S, Atyabi A. A comprehensive review of swarm optimization algorithms. PLOS One. 2015;**10**(5):e0122827

[21] Zhang K. A minimum zone method for evaluating straightness errors using PSO algorithm. In: Advanced Intelligent Computing Theories and Applications. With Aspects of Theoretical and Methodological Issues. Berlin, Heidelberg: Springer; 2008. pp. 308-314. DOI: 10.1007/978-3-540-87442-3_39

[22] Merabti H, Belarbi K, Bouchemal B. Nonlinear predictive control of a mobile robot: A solution using metaheuristics. Journal of the Chinese Institute of Engineers. 2016;**39**(3):282-290

[23] Sandwell DT. Biharmonic spline interpolation of GEOS-3 and SEASAT altimeter data. Geophysical Research Letters. 1987;**14**(2):139-142

[24] International Organization for Standardization. Passenger Cars - Test Track for a Severe Lane-Change Manoeuvre. ISO-3888. 1999. Available from: https://www.iso.org/standard/67973.html

[25] Buehler M, Iagnemma K, Singh S. The DARPA Urban Challenge: Autonomous Vehicles in City Traffic. Berlin Heidelberg: Springer-Verlag; 2009

[26] Wang X, Fu M, Ma H, Yang Y. Lateral control of autonomous vehicles based on fuzzy logic. Control Engineering Practice. 2015;**34**:1-17

Chapter 4

Integral Backstepping Controller for UAVs Team Formation

Wesam M. Jasim and Dongbing Gu

Abstract

In this chapter, two controllers are investigated for stabilisation, path tracking and leader-follower team formation. The first controller is a PD2 implemented for attitude stability. The second controller is an Integral Backstepping IBS control algorithm presented for the path tracking and leader-follower team formation problems of quadrotors. This nonlinear control technique divide the control into two loops, the inner loop is for the attitude stabilisation and the outer loop is for the position control. The dynamic model of a quadrotor is represented based on Euler angles representation and includes some modelled aerodynamical effects as a nonlinear part. The IBS controller is designed for the translational part to track the desired trajectory and to track the leader quadrotor by the followers. Stability analysis is achieved via a suitable Lyapunov function. The external disturbance and model parameters uncertainty are considered in the simulation tests. The proposed controllers yielded good results in terms of Root Mean Square Error RMSE values, time-consumption, disturbance rejection and model parameter uncertainties change coverage.

Keywords: integral backstepping, adaptive controller, Euler angles, UAVs quadrotors, team formation

1. Introduction

In recent years, research on the control of Unmanned Aerial Vehicles (UAVs) has been growing due to its simplicity in design and low cost. Quadrotor helicopters have several advantages over fixed-wing air crafts, such as taking off and landing vertically in a limited space and hovering easily over fixed or dynamic targets, which gives them efficiency in applications that fixed-wing air crafts cannot do, in addition to being safer [1–3]. Based on its structure the UAV offers the power of sensing and computing in many applications. Quadrotor UAVs can be used to perform several tasks in the applications of dangerous areas for a manned aircraft in a high level of accuracy. They can be utilised in different applications, such as inspection of power lines, oil platforms, search and rescue operations, and surveillance [4–6]. Increasing the applications of quadrotors encourages the growth in their technologies and raises the requirements on autonomous control protocols. Moreover, using swarm robotics has advantages over individual robots in that they perform their tasks faster with high accuracy and use a minimum number of sensors by distributing them to the robots [7]. Researchers are focusing on the design and implementation of many types of controllers to control the take-off, landing and hovering of individual quadrotor UAVs with some applications which require the creation of a trajectory and tracking in three dimensions, benefiting from the wide developments in sensors.

Research in the field of control of individual and multi-robot quadrotor team formation is still facing some challenges. Challenges of individual quadrotor control come from the complexity of modelling its dynamic system because of its complex structure and the design issue. The dynamic model equations present four input forces with six output states, which mean that the system is in under-actuated range [5, 7]. Further challenges of multi-robot control come from evaluating the control architecture and communication network limitations.

The formation problem of quadrotors has had a vast area of interesting research in the past few years. Researchers have been motivated to contribute to this field of research by the development of materials, sensors and electronics used in designing quadrotors, which consequently has an effect on minimising their size, weight and cost. Working as a team of quadrotors has many benefits over using a single quadrotor in several applications.

Team formation control includes many problems to be addressed, including communication loss, delay between the robots or packet drop problems [8–11]. Simultaneous localization and mapping is another problem in team formation control, in which the vehicle builds up its maps and estimates its location precisely at the same time; this problem has also been addressed in [12–14]. The third problem is the collision and obstacle avoidance, which includes avoiding collisions with both other robots and static or moving unknown obstacles while flying to their destination and maintaining their positions. Solutions to this problem have been handled by [12, 15]. Now, team formation control adopts a combination of some functions; the first is to perform the mission between two points, the second is to preserve the comparative positions of the robots over the formation and maintain the shape consequently, the third is to avoid obstacles and the forth is to divide the formation. In this chapter, we focus on designing only a control law for the leader-follower team formation problem with collision avoidance between team members by maintaining the distance between the leader and the follower.

In the leader-follower approach, at least one vehicle performs as a leader and the other robots are followers. The leader vehicle tracks a predefined path, whereas the followers maintain a certain distance with the leader and among themselves to obtain the desired shape. Each robot has its own controller and the robots keep the desired relative distance between themselves. However, two types of control architecture may be used to control the vehicle: one loop control scheme and two loop control scheme. If a two loop control scheme is used to control each vehicle, the outer loop is used for position control and its x and y output is the desired roll and pitch angles. These desired angles with the desired yaw angle are used to calculate the vehicle torques; in other words, they stabilise the quadrotor angles. This type of control is built according to time scale separation, where the attitude dynamics are much faster than translation dynamics. In the one loop control scheme, on the other hand, separation of the vehicle dynamics to attitude and translation is not considered. In this case, the position tracking error is used directly to calculate the vehicle torques to achieve its path tracking. According to these definitions, leader-follower team formation requires attitude stabilisation and path tracking to be achieved.

Abundant literature exists on the subject of attitude stabilisation, path tracking and leader-follower team formation control. Several control techniques have been demonstrated to control a group of quadrotors varying between the linear PID, PD or LQR controllers to more complex nonlinear controllers as neural networks and BS controllers. These controllers achieved good results and some of them guaranteed the performance, such as the LQR controller, and some of them guaranteed their stability. The performance of an individual quadrotor or a group of quadrotors in formation control is often affected by external disturbances such as payload changes (or mass changes), wind disturbance, inaccurate model parameters, etc. Therefore, the IBS controller was proposed to reject the effect of disturbances and handle the change in model parameter uncertainties. On the other hand, improving the control performance is another aspect.

Dynamic model representation of the quadrotors is a major demand for designing these controllers. In this chapter, Euler angles technique was used to represent the quadrotors.

2. Dynamic model

In this section, Euler angles are used to represent the quadrotor dynamical model. External disturbances and model parameter uncertainties change are considered as well. An IBS controller is derived and tested in simulation. The stability analysis is obtained via a selected Lyapunov function. The full quadrotor dynamic model including the gyroscope effects $G(\omega)$ is

$$\begin{cases} \dot{\mathbf{p}} = \mathbf{v} \\ \dot{\mathbf{v}} = -g\mathbf{e} + \dfrac{f}{m} R_\theta \mathbf{e} \\ \dot{R}_\theta = R_\theta S(\omega) \\ J\dot{\omega} = -S(\omega)J\omega - G(\omega) + \tau_E \end{cases} \tag{1}$$

and the rotational matrix R_θ from the inertial frame to the body frame is

$$R_\theta = \begin{bmatrix} c\psi c\theta & c\psi s\theta s\varphi - s\psi c\varphi & c\psi s\theta c\varphi + s\psi s\varphi \\ s\psi c\theta & s\psi s\theta s\varphi + c\psi c\varphi & s\psi s\theta c\varphi - c\psi s\varphi \\ -s\theta & c\theta s\varphi & c\theta c\varphi \end{bmatrix}. \tag{2}$$

where m is the quadrotor mass, $\omega = \left[\omega_x, \omega_y, \omega_z\right]^T$ is the angular velocity in the body frame, J is the 3×3 diagonal matrix representing three inertial moments in the body frame, τ_E is the torque vector applied on the quadrotor, $\mathbf{v} = \left[v_x, v_y, v_z\right]^T$ is the linear velocity, $\mathbf{p} = [x, y, z]^T$ is the position vector, S is the skew-symmetric cross product matrix, and the vector $\mathbf{e} = [0, 0, 1]^T$.

Assuming that φ, θ, ω_x, ω_y and ω_z are very small, $\zeta = [\varphi, \theta, \psi]^T$, $\eta = \dot{\zeta} = \left[\dot{\varphi}, \dot{\theta}, \dot{\psi}\right]^T = \left[\omega_x, \omega_y, \omega_z\right]^T$ and $\dot{\eta} = \left[\ddot{\varphi}, \ddot{\theta}, \ddot{\psi}\right]^T = \left[\dot{\omega}_x, \dot{\omega}_y, \dot{\omega}_z\right]^T$, then the attitude control part of Eq. (1) can be written as:

$$\begin{cases} f = mg \\ \ddot{\varphi} = \dot{\theta}\dot{\psi}\dfrac{J_y - J_z}{J_x} + \dfrac{J_r}{J_x}\dot{\theta}\Omega + \dfrac{\tau_\varphi}{J_x} \\ \ddot{\theta} = \dot{\varphi}\dot{\psi}\dfrac{J_z - J_x}{J_y} - \dfrac{J_r}{J_y}\dot{\varphi}\Omega + \dfrac{\tau_\theta}{J_y} \\ \ddot{\psi} = \dot{\varphi}\dot{\theta}\dfrac{J_x - J_y}{J_z} + \dfrac{\tau_\psi}{J_z} \end{cases} \tag{3}$$

3. Quadrotors formation problem

The full dynamic model based on Euler angles (1) of a quadrotor can be written as:

$$\begin{cases} \dot{\mathbf{p}}_i = \mathbf{v}_i \\ \dot{\mathbf{v}}_i = -g\mathbf{e} + \dfrac{f_i}{m_i} R_{i\theta}\mathbf{e} \\ \dot{\zeta}_i = \eta_i \\ J_i\dot{\eta}_i = S(\eta_i)J_i\eta_i + G(\eta_i) - \tau_{iE} \end{cases} \tag{4}$$

where i is L for the leader and F for the follower.

The leader-follower formation control problem to be solved in this chapter is a distributed control scheme of one leader and one follower. The leader control problem is formulated as a trajectory tracking, and the follower control problem is also formulated as a tracking problem, but with a different tracking target.

The follower keeps its yaw angle the same as the leader when it maintains the formation pattern. It moves to a desired position $\mathbf{p}_{Fd}$, which is determined by a desired distance d, a desired incidence angle ρ, and a desired bearing angle σ. A new frame F' is defined by the translation of the leader frame L to the frame with the desired follower position $\mathbf{p}_{Fd}$ as the origin. As shown in **Figure 1**, the desired incidence angle is measured between the desired distance d and the $x-y$ plane in the new frame F', and the desired bearing angle is measured between the x axis and the projection of the d in $x-y$ plane in the new frame F'. The desired position $\mathbf{p}_{Fd}$ is

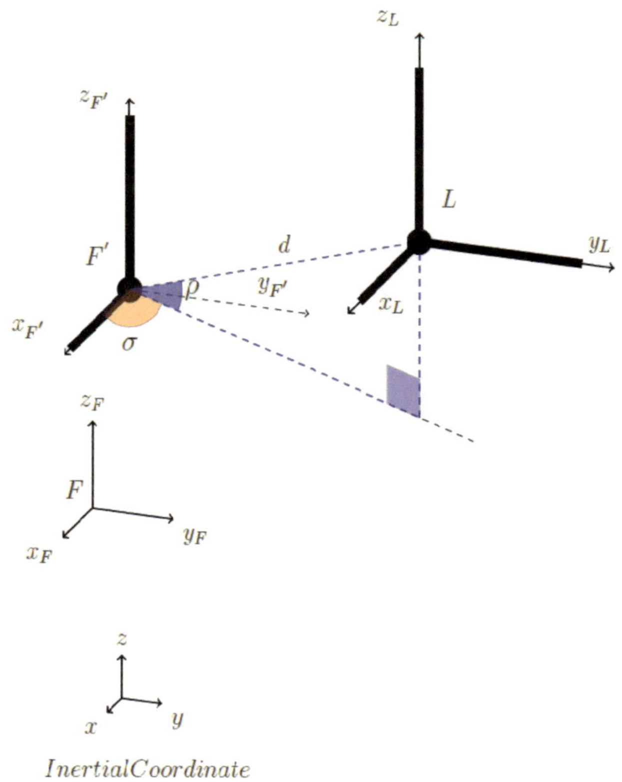

Figure 1.
Body frames in formation.

$$\mathbf{p}_{Fd} = \mathbf{p}_L - R_{Lq}^T d \begin{bmatrix} \cos\rho\cos\sigma \\ \cos\rho\sin\sigma \\ \sin\rho \end{bmatrix}. \tag{5}$$

Assume both the leader and the follower are able to obtain their own pose information and the follower is able to obtain the leader's pose information via wireless communication. The design goal of the controllers is to find the state feedback control law for the thrust and torque inputs for both the leader and the follower. The leader-follower formation control problem is solved if both conditions (6) and (7) are satisfied.

$$\begin{cases} \lim_{t\to\infty}(\mathbf{p}_{Fd} - \mathbf{p}_F) = 0 \\ \lim_{t\to\infty}(\psi_L - \psi_F) = 0 \end{cases} \tag{6}$$

and

$$\begin{cases} \lim_{t\to\infty}(\mathbf{p}_{Ld} - \mathbf{p}_L) = 0 \\ \lim_{t\to\infty}(\psi_{Ld} - \psi_L) = 0 \end{cases} \tag{7}$$

The communication among the robots is assumed to be available. The position $\mathbf{p}_L$, yaw angle ψ_L of the leader L and its first and second derivatives $\dot{\psi}_L$ and $\ddot{\psi}_L$ are assumed to be available and measurable. The linear velocity of the leader L and its derivatives $\mathbf{v}_L$ and $\dot{\mathbf{v}}_L$ are assumed bounded and available for the follower.

4. Formation IBS controllers

Integral backstepping control is one of the popular control approaches for both individual and multiple quadrotors. In this section, the integral backstepping control is applied for the individual quadrotor path tracking and leader-follower formation problems. This nonlinear control technique divide the control into two loops, the inner loop is for the attitude stabilisation and the outer loop is for the position control as shown in **Figure 2**.

In this case, the leader and the follower desired roll and pitch angles are assumed to be $\theta_{Ld} = \theta_{Fd} = 0$ and $\varphi_{Ld} = \varphi_{Fd}$. The dynamic model of a quadrotor is

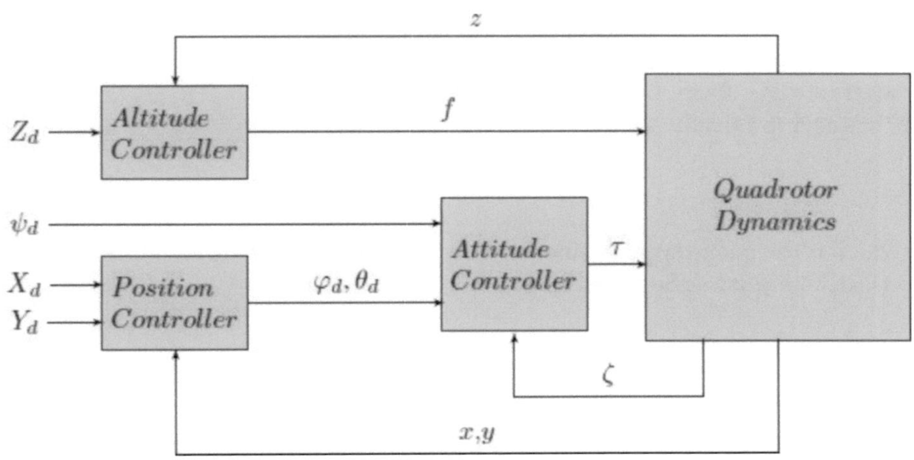

Figure 2.
Two-loop control block diagram.

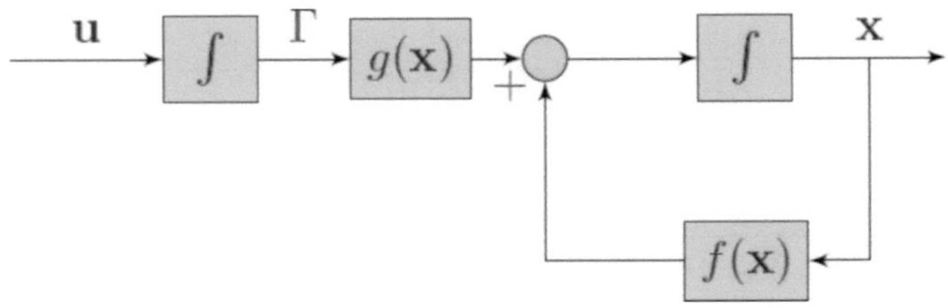

Figure 3.
Initial system.

represented based on Euler angles representation and includes some modelled aerodynamical effects as a nonlinear part. The IBS controller is designed for the translational part to track the desired trajectory. Stability analysis is achieved via a suitable Lyapunov function. The external disturbance and model parameters uncertainty are considered in the simulation tests in all circumstances.

4.1 Backstepping control concept

Backstepping is a recursive design mechanism to asymptotically stabilise a controller for the following system [16]:

$$\begin{cases} \dot{\mathbf{x}} = f(\mathbf{x}) + g(\mathbf{x})\Gamma \\ \dot{\Gamma} = \mathbf{u} \end{cases}. \tag{8}$$

This system is described as an initial system in **Figure 3**, where $\mathbf{x} \in \mathbb{R}^n$ and $\Gamma \in \mathbb{R}$ are the system state and $\mathbf{u} \in \mathbb{R}$ is the control input. $f, g : \mathcal{D} \to \mathbb{R}^n$ are assumed to be smooth and $f(0) = 0$. A stabilising state feedback control law $\Gamma = \Phi(\mathbf{x})$, assuming $\Phi(0) = 0$, exists, in addition to a Lyapunov function $V_1 : \mathcal{D} \to \mathbb{R}^+$ such that

$$\dot{V}_1(\mathbf{x}) = \frac{\partial V_1}{\partial \mathbf{x}}[f(\mathbf{x}) + g(\mathbf{x})\Phi(\mathbf{x})] \leq -V_\varepsilon(\mathbf{x}), \forall \mathbf{x} \in \mathcal{D} \tag{9}$$

where $V_\varepsilon(\mathbf{x}) : \mathcal{D} \to \mathbb{R}^+$ is a positive semidefinite function. Now, the following algebraic manipulation is required: by adding and subtracting the term $g(\mathbf{x})\Phi(\mathbf{x})$ to/from the subsystem (8) we can have the following system:

$$\dot{\mathbf{x}} = f(\mathbf{x}) + g(\mathbf{x})\Phi(\mathbf{x}) + g(\mathbf{x})s \tag{10}$$

where $s = \Gamma - \Phi(\mathbf{x})$, by this construction, when $s \to 0$, $\dot{\mathbf{x}} = f(\mathbf{x}) + g(\mathbf{x})\Phi(\mathbf{x})$ which is asymptotically stable. The derivative of s is

$$\dot{s} = \dot{\Gamma} - \dot{\Phi}(\mathbf{x}) = \mathbf{u} - \dot{\Phi}(\mathbf{x}) = v \tag{11}$$

which is the backstepping, since $\Phi(\mathbf{x})$ is stepped back by differentiation as described in **Figure 4**. So we have

$$\begin{aligned} \dot{\mathbf{x}} &= f(\mathbf{x}) + g(\mathbf{x})\Phi(\mathbf{x}) + g(\mathbf{x})s \\ \dot{s} &= v \end{aligned} \tag{12}$$

This system is equivalent to the initial system (8), where $\dot{\Phi} = \frac{\partial \Phi}{\partial \mathbf{x}}\dot{\mathbf{x}} = \frac{\partial \Phi}{\partial \mathbf{x}}[f(\mathbf{x}) + g(\mathbf{x})\Gamma]$. The next step is to stabilise the system (12), and the following Lyapunov function is considered:

$$V(\mathbf{x}, s) = V_1(\mathbf{x}) + \frac{1}{2}s^2. \tag{13}$$

Then

$$\begin{aligned}\dot{V} &= \frac{\partial V_1}{\partial \mathbf{x}}[f(\mathbf{x}) + g(\mathbf{x})\Phi(\mathbf{x}) + g(\mathbf{x})s] + sv \\ &\leq -V_\varepsilon(\mathbf{x}) + \left[\frac{\partial V_1}{\partial \mathbf{x}}g(\mathbf{x}) + v\right]s.\end{aligned} \tag{14}$$

Let

$$\begin{cases} v = -\dfrac{\partial V_1}{\partial \mathbf{x}}g(\mathbf{x}) - \varepsilon s \\ \varepsilon > 0 \end{cases}. \tag{15}$$

Then

$$\dot{V} \leq -V_\varepsilon(\mathbf{x}) - \varepsilon s^2 < 0. \tag{16}$$

This signifies that the origin $(\mathbf{x} = 0, s = 0)$ is asymptotically stable. Since , then the origin $\mathbf{x} = 0$ and $\Gamma = 0$ is also asymptotically stable. In the next step an integral part is added to the BS controller to eliminate the steady state error which occurred in the simulation results and is called IBS.

4.2 Follower integral backstepping controller

The IBS controller for the follower is to track the leader and maintain a desired distance between them with desired incidence and bearing angles.

In this subsection, we derive the IBS controller for the follower. Let us use the follower translational part (17):

$$\ddot{\mathbf{p}}_F = f(\mathbf{p}_F) + g(\mathbf{p}_F)f_F \tag{17}$$

where

$$f(\mathbf{p}_F) = [0 \;\; 0 \;\; -g]^T \tag{18}$$

$$g(\mathbf{p}_F) = \left[u_{Fx}/m_F \quad u_{Fy}/m_F \quad c\theta_F c\varphi_F/m_F\right]^T \tag{19}$$

where

$$u_{Fx} = (c\psi_F s\theta_F c\varphi_F + s\psi_F s\varphi_F) \tag{20}$$

$$u_{Fy} = (s\psi_F s\theta_F c\varphi_F - c\psi_F s\varphi_F). \tag{21}$$

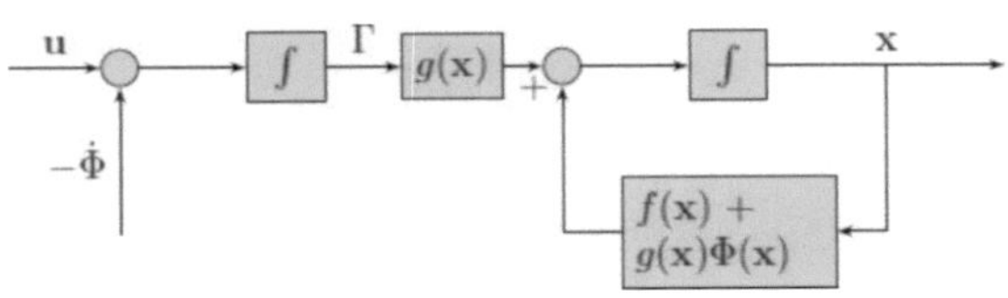

Figure 4.
Backstepping system.

Then the position tracking error between the leader and the follower can be calculated as:

$$\tilde{\mathbf{p}}_F = \mathbf{p}_{Fd} - \mathbf{p}_F = \mathbf{p}_L - R_L^T d \begin{bmatrix} \cos\rho\cos\sigma \\ \cos\rho\sin\sigma \\ \sin\rho \end{bmatrix} - \mathbf{p}_F \tag{22}$$

and its derivative is

$$\dot{\tilde{\mathbf{p}}}_F = \dot{\mathbf{p}}_{Fd} - \dot{\mathbf{p}}_F = \dot{\mathbf{p}}_{Fd} - \mathbf{v}_F \tag{23}$$

where $\mathbf{v}_F$ is a virtual control, and its desirable value can be described as:

$$\mathbf{v}_F^d = \dot{\mathbf{p}}_{Fd} + b_F\tilde{\mathbf{p}}_F + k_F\overline{\mathbf{p}}_F \tag{24}$$

where the integration of the follower position error is added to minimise the steady-state error.

Now, consider the linear velocity error between the leader and the follower as:

$$\tilde{\mathbf{v}}_F = \mathbf{v}_F^d - \dot{\mathbf{p}}_F. \tag{25}$$

By substituting (24) into (25) we obtain

$$\tilde{\mathbf{v}}_F = \dot{\mathbf{p}}_{Fd} + b_F\tilde{\mathbf{p}}_F + k_F\overline{\mathbf{p}}_F - \dot{\mathbf{p}}_F \tag{26}$$

and its time derivative becomes

$$\dot{\tilde{\mathbf{v}}}_F = \ddot{\mathbf{p}}_{Fd} + b_F\dot{\tilde{\mathbf{p}}}_F + k_F\tilde{\mathbf{p}}_F - \ddot{\mathbf{p}}_F. \tag{27}$$

Then from (24) and (25) we can rewrite (23) in terms of the linear velocity error as:

$$\dot{\tilde{\mathbf{p}}}_F = \tilde{\mathbf{v}}_F - b_F\tilde{\mathbf{p}}_F - k_F\overline{\mathbf{p}}_F. \tag{28}$$

By substituting (17) and (28) into (27), the time derivative of the linear velocity error can be rewritten as:

$$\dot{\tilde{\mathbf{v}}}_F = \ddot{\mathbf{p}}_{Fd} + b_F\tilde{\mathbf{v}}_F - b_F^2\dot{\tilde{\mathbf{p}}}_F - b_F k_F\overline{\mathbf{p}}_F + k_F\tilde{\mathbf{p}}_F - f(\mathbf{p}_F) - g(\mathbf{p}_F)f_F. \tag{29}$$

The desirable time derivative of the linear velocity error is supposed to be

$$\dot{\tilde{\mathbf{v}}}_F = -c_F\tilde{\mathbf{v}}_F - \tilde{\mathbf{p}}_F. \tag{30}$$

Now, the total thrust f_F, the longitudinal u_{Fx} and the lateral u_{Fy} motion control can be found by subtracting (29) from (30) as follows:

$$\begin{aligned} f_F = (g + \dot{v}_{Lz} + (1 - b_{Fz}^2 + k_{Fz})\tilde{z}_F + (b_{Fz} + c_{Fz})\tilde{v}_{Fz} - b_{Fz}k_{Fz}\overline{z}_F \\ -d(R_{\theta 31}\cos\rho\cos\sigma + R_{\theta 32}\cos\rho\sin\sigma + R_{\theta 33}\sin\rho))\frac{m_F}{c\theta_F c\varphi_F} \end{aligned} \tag{31}$$

$$\begin{aligned} u_{Fx} = (\dot{v}_{Lx} + (1 - b_{Fx}^2 + k_{Fx})\tilde{x}_F + (b_{Fx} + c_{Fx})\tilde{v}_{Fx} - b_{Fx}k_{Fx}\overline{x}_F \\ -d(R_{\theta 11}\cos\rho\cos\sigma + R_{\theta 12}\cos\rho\sin\sigma + R_{\theta 13}\sin\rho))\frac{m_F}{f_F} \end{aligned} \tag{32}$$

$$u_{Fy} = (\dot{v}_{Ly} + \left(1 - b_{Fy}^2 + k_{Fy}\right)\tilde{y}_F + (b_{Fy} + c_{Fy})\tilde{v}_{Fy} - b_{Fy}k_{Fy}\overline{y}_F \\ -d(R_{\theta 21}\cos\rho\cos\sigma + R_{\theta 22}\cos\rho\sin\sigma + R_{\theta 23}\sin\rho))\frac{m_F}{f_F}. \tag{33}$$

For the attitude stability, the following nonlinear PD^2 controller (34) proposed in [17] was implemented and tested in simulation for both the leader and the follower:

$$\tau_E = \omega \times J\omega + G(\tilde{\omega}) - (\mu_3 + \mu_2\mu_1)\tilde{\mathbf{q}} - \mu_1 J\dot{\tilde{\mathbf{q}}} - \mu_2\tilde{\omega}. \tag{34}$$

where μ_1, μ_2 and μ_3 are constants.

Next, we show the stability of the follower's translational part.

4.3 Follower controller stability analysis

The following candidate Lyapunov function is chosen for the stability analysis for the follower's translational part with the IBS controller:

$$V = \frac{1}{2}\left(\tilde{\mathbf{p}}_F^T\tilde{\mathbf{p}}_F + \tilde{\mathbf{v}}_F^T\tilde{\mathbf{v}}_F + k_F\overline{\mathbf{p}}_F^T\overline{\mathbf{p}}_F\right) \tag{35}$$

and its time derivative is

$$\dot{V} = \tilde{\mathbf{p}}_F^T\dot{\tilde{\mathbf{p}}}_F + \tilde{\mathbf{v}}_F^T\dot{\tilde{\mathbf{v}}}_F + k_F\overline{\mathbf{p}}_F^T\dot{\overline{\mathbf{p}}}_F. \tag{36}$$

By substituting $\dot{\overline{\mathbf{p}}}_F = \tilde{\mathbf{p}}_F$ and Eqs. (28) and (30) into (36), Eq. (36) becomes

$$\dot{V} = -b_F\tilde{\mathbf{p}}_F^T\tilde{\mathbf{p}}_F - c_F\tilde{\mathbf{v}}_F^T\tilde{\mathbf{v}}_F \leq 0. \tag{37}$$

Finally, (37) is less than zero provided b_F and c_F are positive diagonal matrices, i.e. $\dot{V} < 0$, $\forall\left(\tilde{\mathbf{p}}_F, \tilde{\mathbf{v}}_F\right) \neq 0$ and $\dot{V}(0) = 0$. It can be concluded from the positive definition of V and applying LaSalle theorem that a global asymptotic stability is guaranteed. This leads us to conclude that $\lim_{t\to\infty}\tilde{\mathbf{p}}_F = 0$ and $\lim_{t\to\infty}\tilde{\mathbf{v}}_F = 0$, which meets the position condition of (6).

4.4 Leader IBS controller

The leader is to track a desired trajectory $\mathbf{p}_{Ld}$. Its IBS controller is developed by following the procedure described for the follower quadrotor.

The result is that the total force and horizontal position control laws f_L, u_{Lx} and u_{Ly} can be written using Euler angles dynamic model representation as:

$$f_L = \left(\ddot{z}_{Ld} + g + \left(1 - {b_{Lz}}^2 + k_{Lz}\right)\tilde{z}_L + (b_{Lz} + c_{Lz})\tilde{v}_{Lz} - b_{Lz}k_{Lz}\overline{z}_L\right)\frac{m_L}{c_{\theta L}c_{\varphi L}} \tag{38}$$

$$u_{Lx} = \left(\ddot{x}_{Ld} + \left(1 - {b_{Lx}}^2 + k_{Lx}\right)\tilde{x}_L + (b_{Lx} + c_{Lx})\tilde{v}_{Lx} - b_{Lx}k_{Lx}\overline{x}_L\right)\frac{m_L}{f_L} \tag{39}$$

$$u_{Ly} = \left(\ddot{y}_{Ld} + \left(1 - {b_{Ly}}^2 + k_{Ly}\right)\tilde{y}_L + (b_{Ly} + c_{Ly})\tilde{v}_{Ly} - b_{Ly}k_{Ly}\overline{y}_L\right)\frac{m_L}{f_L}. \tag{40}$$

The torque vector applied to the leader quadrotor $\tau_{LE} \in \mathbb{R}^3$ is a nonlinear PD^2 controller (34). These leader controllers are used for path tracking tests.

5. Simulations

In order to determine the efficiency of the proposed controller, a MATLAB quadrotor simulator is used to test it numerically. The design parameters of the quadrotor used in the simulator are listed in **Table 1**. Two paths were presented in the simulation to show the performance of using the proposed controller with four different circumstances for quadrotors team formation. The first desired path to be tracked by the leader was.

$$\begin{cases} x_{Ld} = 2\cos(t\pi/80); & y_{Ld} = 2\sin(t\pi/80) \\ z_{Ld} = 1 + 0.1t; & \psi_{Ld} = \pi/6 \end{cases}. \tag{41}$$

The IBS controllers were tested in simulation to track a desired path by the leader and maintain the desired distance, desired incidence angle and desired bearing angle between them for the follower. The parameters chosen for both paths were $b_L = diag(180, 0.34, 0.34)$, $c_L = diag(0.7, 0.02, 0.02)$, $k_L = diag(0.0516, 0.0081, 0.0081)$, $b_F = diag(12, 0.7, 0.7)$, $c_F = diag(1.4, 0.02, 0.02)$ and $k_F = diag(0.01, 0.001, 0.001)$.

The leader initial positions were $\left[x_L, y_L, z_L\right]^T = [2, 0, 0]^T$ metres and the initial angles were $[\varphi_L, \theta_L, \psi_L]^T = [0, 0, 0]^T$ radian. Then the follower followed the leader and maintained the desired distance between them $d = 2$ metres, the desired incidence and bearing angles $\rho = -\pi/6$, $\sigma = \pi/6$ radian, respectively. The follower initial positions were $\left[x_F, y_F, z_F\right]^T = [0.5, 0, 0]^T$ metres and the initial angles were $[\varphi_F, \theta_F, \psi_F]^T = [0, 0, 0]^T$ radian.

The second desired path to be tracked by the leader was

$$\begin{cases} x_{Ld} = 4\cos(t\pi/40); & y_{Ld} = 4\sin(t\pi/40) \\ z_{Ld} = 1 + 0.1t; & \psi_{Ld} = \pi/6 \end{cases}. \tag{42}$$

The leader initial positions were $\left[x_L, y_L, z_L\right]^T = [4, 0, 0]^T$ metres and the initial angles were $[\varphi_L, \theta_L, \psi_L]^T = [0, 0, 0]^T$ radian. Then the follower followed the leader and maintained the desired distance between them $d = 3$ metres, the desired incidence and bearing angles $\rho = 0$, $\sigma = \pi/6$ radian, respectively. The follower initial positions were $\left[x_F, y_F, z_F\right]^T = [1.4, -1.5, 0]^T$ metres and the initial angles were $[\varphi_F, \theta_F, \psi_F]^T = [0, 0, 0]^T$ radian.

The four circumstances included: (17) no disturbance, (32) force disturbance $d_{v_{ix}} = -2$ Nm during $10 \le t \ge 10.25$ seconds, $d_{v_{iz}} = 2$ Nm during

Symbol	Definition	Value	Units
J_x	Roll Inertia	4.4×10^{-3}	$kg.m^2$
J_y	Pitch Inertia	4.4×10^{-3}	$kg.m^2$
J_z	Yaw Inertia	8.8×10^{-3}	$kg.m^2$
m	Mass	0.5	kg
g	Gravity	9.81	m/s^2
l	Arm Length	0.17	m
J_r	Rotor Inertia	4.4×10^{-5}	$kg.m^2$

Table 1.
Quadrotor parameters.

$20 \leq t \geq 20.25$ seconds, $d_{v_{iy}} = 2$ Nm during $30 \leq t \geq 30.25$ seconds in the first path, $d_{v_{ix}} = -0.5$ Nm during $20 \leq t \geq 20.25$ seconds, $d_{v_{iz}} = 0.5$ Nm during $60 \leq t \geq 60.25$ seconds, $d_{v_{iy}} = 0.5$ Nm during $100 \leq t \geq 100.25$ seconds in the second path, and the attitude part for the leader and the follower is disturbed using (43), applied at the same time for both the leader and the follower, (33) +30% model parameter uncertainty, and (44) −30% model parameter uncertainty.

$$\mathbf{d} = 0.01 + 0.01 \sin(0.024\pi t) + 0.05 \sin(1.32\pi t) \qquad (43)$$

Figures 5 and **6** indicate the response of the IBS controller while the leader was tracking the first and second desired path, respectively. **Figure 7** shows the distance

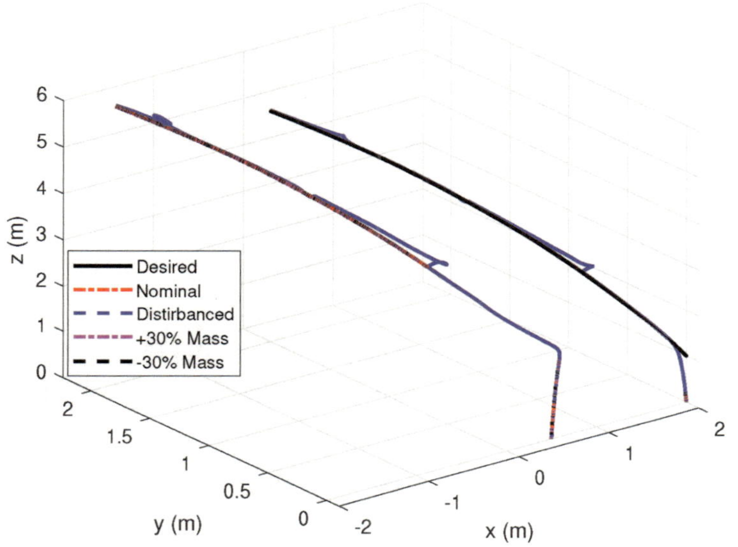

Figure 5.
Leader-follower formation in first path.

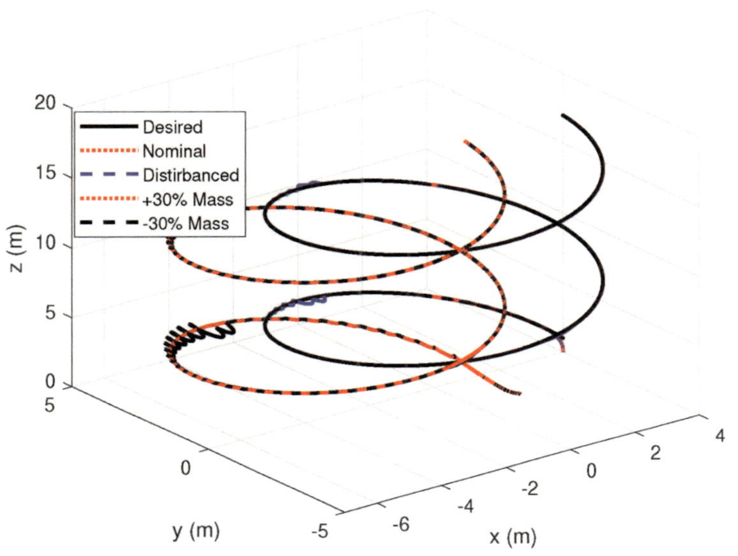

Figure 6.
Leader-follower formation in second path.

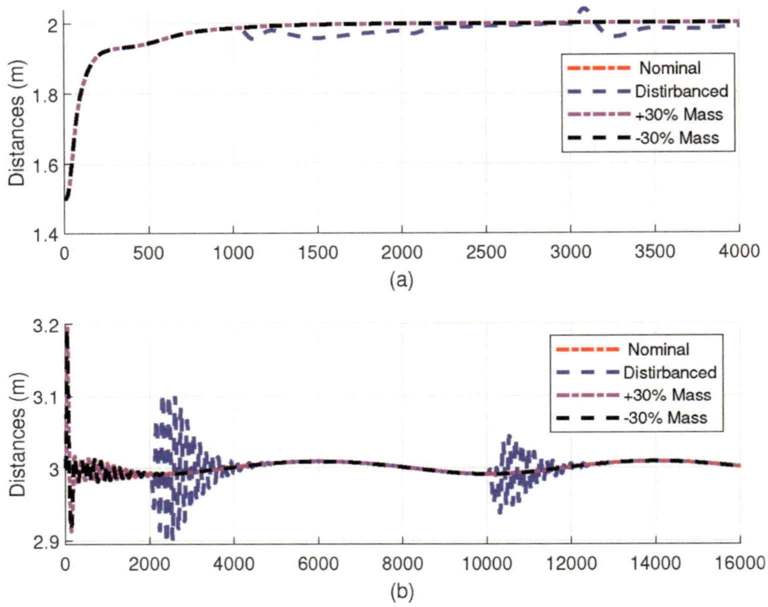

Figure 7.
The distance between the leader and the follower in (a) the first path, (b) the second path.

between the leader and the follower via the two paths, and **Figures 8–11** illustrate the yaw angles' behaviour for the leader and the follower via the two paths respectively.

It can be noticed from these figures that not only the overshoot but also the error in distance between the leader and the follower was low. It was also rejecting the disturbances in the two paths.

Table 2 demonstrates the RMSE values of the two paths positions and yaw angle. It is clear that the RMSE values of the IBS controller were almost the same

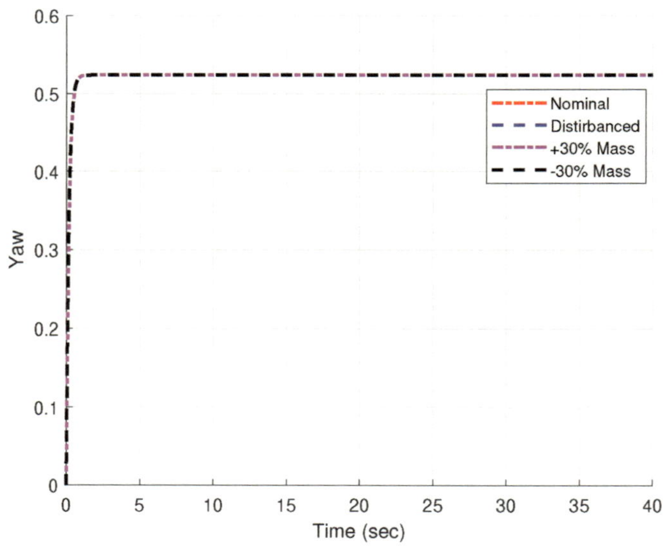

Figure 8.
Leader yaw angle in first path.

when using the IBS controller in normal conditions and with ±30% model parameter uncertainty in both paths, while they significantly increased with the disturbance. It can be seen that the IBS controller was able to track the desired trajectories with small position tracking errors in less than 3 s and it could reject the disturbances and cover the change in model parameter uncertainties.

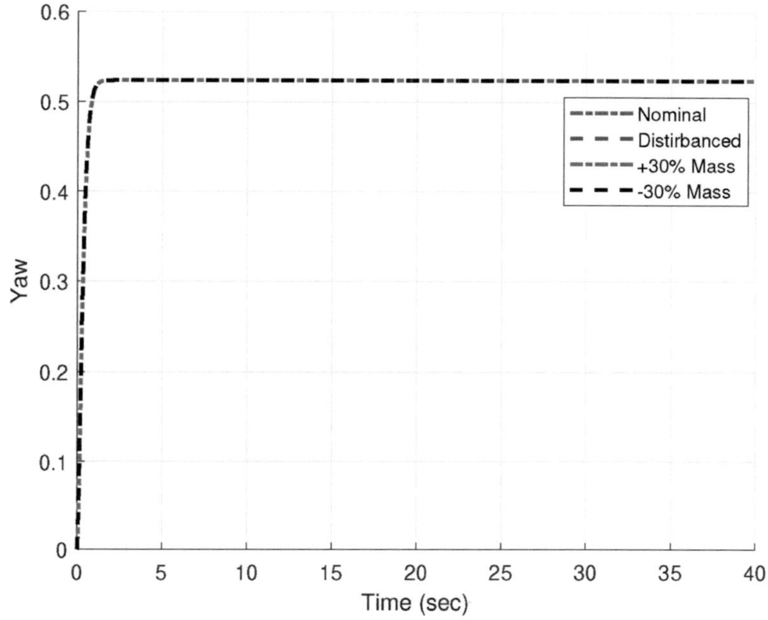

Figure 9.
Follower yaw angle in first path.

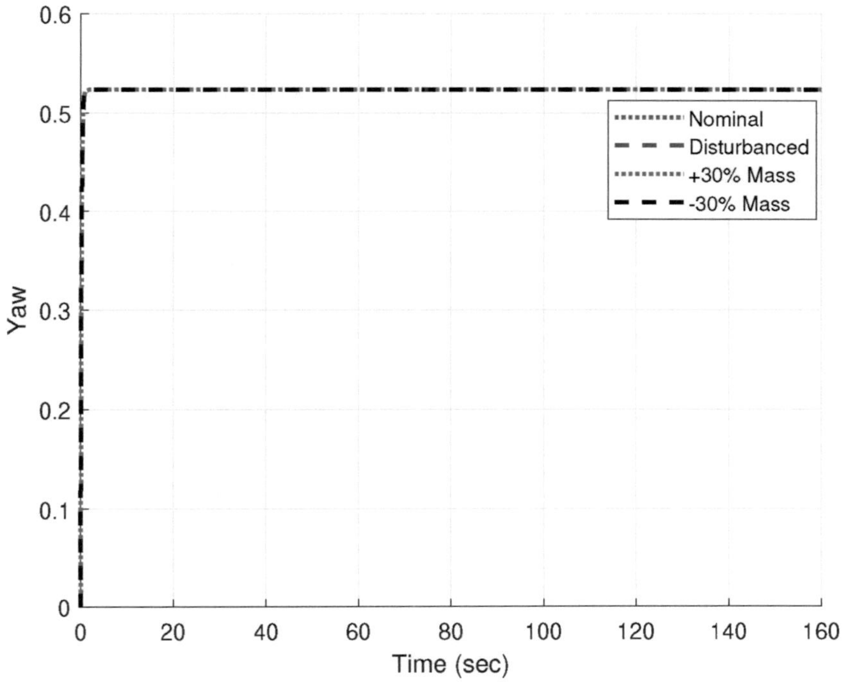

Figure 10.
Leader yaw angle in second path.

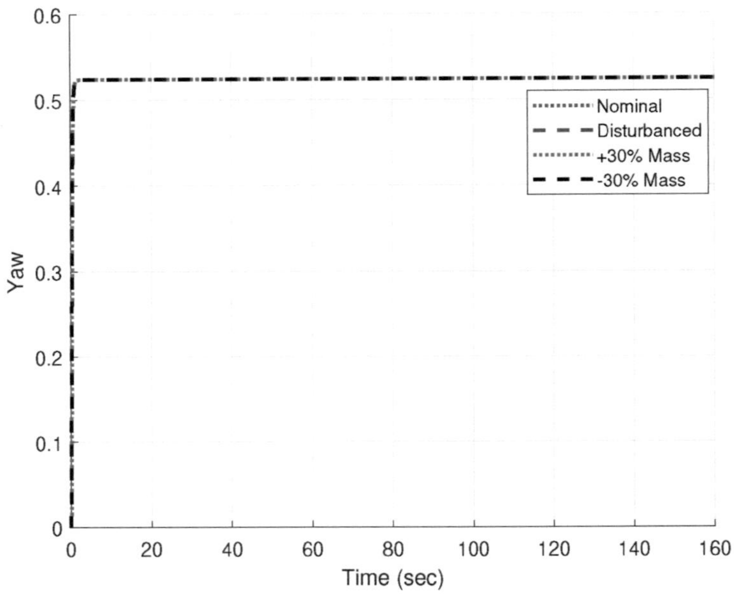

Figure 11.
Follower yaw angle in second path.

	Path 1				Path 2			
RMSE	$x\ (m)$	$y\ (m)$	$z\ (m)$	ψ (deg.)	x	$y\ (m)$	$z\ (m)$	ψ (deg.)
IBS	0.0005	0.0040	0.0936	0.0004	0.0030	0.0248	0.0936	0
IBS + **d**	0.0112	0.0772	0.0943	0.0004	0.0804	0.1018	0.0910	$7e^{-6}$
IBS + 30%	0.0005	0.0040	0.0936	0.0004	0.0030	0.0248	0.0936	0
IBS − 30%	0.0005	0.0040	0.0936	0.0004	0.0030	0.0248	0.0936	0

Table 2.
Position and ψ RMSE values for the two paths.

In conclusion, it is obvious that the proposed IBS controller maintained the distance between the leader and the follower and keep them in the desired formation.

6. Discussions

This chapter presented an IBS controller derived based on the Backstepping controller for quadrotor UAVs leader-follower team formation problem. Two loops control scheme was used in simulation to find the total thrust and torques. A PD^2 controller was used for attitude part control, while the IBS controller was used to control the translation part of the quadrotors. The dynamic model of the quadrotor was derived based on Euler angles and the effect of the external disturbance and the model parameter uncertainties are also considered.

It is well-known that IBS control is a methodical approach to build the Lyapunov function ahead with the control input design. Thus by the cancellation of the indefinite error terms, the stability of the derivative of the Lyapunov function can be secured. Although the stability of the Lyapunov function is guaranteed, this does not guarantee the performance of the system. In this work, a suitable Lyapunov function was used to derive the controller stability conditions.

The simulation results prove that the performance by using the IBS controller had significantly small errors. It is also obvious that using the IBS controller led to a smooth and fast performance with small overshoot. Moreover, the response of using the proposed controller in rejecting the external disturbances was fast enough.

As a result, the proposed IBS controller indeed produced good control performance in all circumstances.

Author details

Wesam M. Jasim[1]* and Dongbing Gu[2]

1 College of Computer Science and Information Technology, University of Anbar, Ramadi, Iraq

2 School of Computer Science and Electronic Engineering, University of Essex, Colchester, UK

*Address all correspondence to: co.wesam.jasim@uoanbar.edu.iq

References

[1] Min BC, Cho CH, Choi KM, Kim DH. Development of a micro quad-rotor UAV for monitoring an indoor environment. In: Advances in Robotics. Vol. 5744. Stuttgart, Germany: Holtzbrinck Publishing Group and BC Partners; 2009. pp. 262-271

[2] Srikanth MB, Dydek ZT, Annaswamy AM, Lavretsky E. A robust environment for simulation and testing of adaptive control for mini-UAVs. In: American Control Conference ACC '09. St. Louis, MO, USA: Hyatt Regency Riverfront; 2009. pp. 5398-5403

[3] Erginer B, Altug E. Modeling and PD control of a quadrotor vtol vehicle. In: Proceeding of the 2007 IEEE Intelligent Vehicles Symposium. Istanbul, Turkey; 2007. pp. 894-899

[4] Jaimes A, Kota S, Gomez J. An approach to surveillance an area using swarm of fixed wing and quad-rotor unmanned aerial vehicles UAVs. In: IEEE International Conference on System of Systems Engineering SoSE '08. Singapore; 2008. pp. 1-6

[5] Bouabdallah S, Siegwart R. Full control of a quadrotor. In: Proceeding of the 2007 IEEE/RSJ International Conference on Intelligent Robots and Systems. San Diego, CA, USA; 2007. pp. 153-158

[6] Curiac DI, Volosencu C. Path planning algorithm based on Arnold cat map for surveillance UAVs. Defence Science Journal. 2015;**65**:483-488

[7] Nathan PT, Almurib HAF, Kumar TN. A review of autonomous multi-agent quad-rotor control techniques and applications. In: 2011 4th International Conference on Mechatronics (ICOM). Kuala Lumpur, Malaysia; 2011. pp. 1-7

[8] Pilz U, Popov A, Werner H. Robust controller design for formation flight of quad-rotor helicopters. In: Joint 48th IEEE Conference on Decision and Control and 28th Chinese Control Conference. Shanghai, P.R. China; 2009. pp. 8322-8327

[9] Guerrero JA, Castillo P, Challal Y. Quadrotors formation control a wireless medium access aware approach. Journal of Intelligent and Robotic Systems. 2013;**4**:221-231

[10] Dong X, Yu B, Shi Z, Zhong Y. Time-varying formation control for unmanned aerial vehicles: Theories and applications. IEEE Transactions on Control Systems Technology. 2015;**23**: 340-348

[11] Turpin M, Michael N, Kumar V. Capt: Concurrent assignment and planning of trajectories for multiple robots. The International Journal of Robotics Research. 2014;**33**(1):98-112

[12] Roy N, Newman P, Srinivasa S. Towards a Swarm of Agile Micro Quadrotors. USA, Canada: MIT Press; 2013. p. 504

[13] Nägeli T, Conte C, Domahidi A, Morari M, Hilliges O. Environment-independent formation flight for micro aerial vehicles. In: 2014 IEEE/RSJ International Conference on Intelligent Robots and Systems. 2014. pp. 1141-1146

[14] Saska M, Kasl Z, Preucil L. Motion planing and control of formations of micro aerial vehicles. In: Preprints of the 19th World Congress, the International Federation of Automatic Control IFAC. Cape Town, South Africa; 2014. pp. 1228-1233

[15] Eskandarpour A, Majd VJ. Cooperative formation control of quadrotors with obstacle avoidance and self collisions based on a hierarchical mpc approach. In: 2014 Second RSI/ISM

International Conference on Robotics and Mechatronics (ICRoM). 2014. pp. 351-356

[16] Marquez HJ. Nonlinear Control Systems. New Jersey: Hoboken; 2003

[17] Tayebi A, McGilvray S. Attitude stabilization of a VTOL quadrotor aircraft. IEEE Transactions on Control Systems Technology. 2006;**14**(3): 562-571

Section 3

Mechatronics

Chapter 5

An Implementable and Stabilizing Model Predictive Control Strategy for Inverted Pendulum-Like Behaved Systems

Odilon S.L. de Abreu, Márcio A.F. Martins and Leizer Schnitman

Abstract

In control theory, the inverted pendulum is a class of dynamic systems widely used as a benchmarking for evaluating several control strategies. Such a system is characterized by an underactuated behavior. It is also nonlinear and presents open-loop unstable and integrating modes. These dynamic features make the control more difficult, mainly when the controller synthesis seeks to include constraints and the guarantee of stability of the closed-loop system. This chapter presents a stabilizing model predictive control (MPC) strategy for inverted pendulum-like behaved systems. It has an offset-free control law based on an only optimization problem (one-layer control formulation), and the Lyapunov stability of the closed-loop system is achieved by adopting an infinite prediction horizon. The controller feasibility is also assured by imposing a suitable set of slacked terminal constraints associated with the unstable and integrating states of the system. The effectiveness of the implementable and stabilizing MPC controller is experimentally demonstrated in a commercial-didactic rotary inverted pendulum prototype, considering both cases of stabilization of the pendulum in the upright position and the output tracking of the rotary arm angle.

Keywords: rotary inverted pendulum, model predictive control, nonlinear system, Lyapunov stability, feasible-optimization problem

1. Motivation

Dynamic inverted pendulum-featured apparatuses are widespread in systems and control theory. These represent a class of nonlinear and underactuated electromechanical systems, which, in turn, are composed of open-loop unstable and integrating modes. The scale-up of inverted pendulum-based conceptual sketches in practical mechanisms and real applications has been in progress, of which one can cite stabilization of rocket launch, robot balance, and segway-like means of transportation, among others [1, 2].

In control theory, inverted pendulum-type systems have enabled extensive studies concerning controller architectures ranging from proportional-integrative-

IntechOpen

derivative (PID)-like classical strategies to more advanced technique ones, such as optimal and adaptive strategies. In the middle of the advanced control strategies, the so-called class of model predictive control (MPC) strategies has been preferred by systematically handling system constraints. In fact, any MPC algorithm makes explicit use of a system model to predict its outputs and to obtain an optimal control law that minimizes the prediction error and control efforts [3–5]. Since it requires an online solution to its associated optimization problem, it first became popular in applications of slow dynamic systems, such as those in petroleum refineries and petrochemical industries [6]. However, with advances in hardware development and optimization techniques, MPC applications have been extended to fast dynamic systems, including inverted pendulum-like mechatronic systems [7–9].

On the other hand, when one seeks to control open-loop unstable systems such as inverted pendulum-like behaved ones, the guarantee of stability associated with control laws plays a crucial role concerning practical implementation purposes. In particular, the synthesis of stabilizing MPC laws deals essentially with terminal state constraints. One of the most heavily studied stability approach is one based on a dual-mode framework, which is composed of two distinct control modes: in the first mode, a conventional MPC law forces the system states to converge to a certain invariant set at the end of the finite prediction horizon, while in the second control mode, a local state-feedback controller takes over and drives the state to the desired operating point within this set [10]. This approach, however, requires the computation of the invariant set parameters, which is obtained from an offline numerical procedure. Although this set can be obtained offline with standard algorithms, undesired convergence and numerical issues may appear as the system dimension increases. Furthermore, the control horizon should be large enough such that the system states at the end of this horizon lie in the invariant set; otherwise the resulting optimization problem becomes infeasible, thus compromising both feasibility and control performance.

Another way to guarantee the closed-loop system stability of MPC controllers is to adopt an infinite prediction horizon, the so-called infinite-horizon model predictive control (IHMPC). Since the infinite-horizon problems cannot be directly handled by an optimization algorithm, the realization of IHMPC controllers is obtained from the combination between a terminal cost term and terminal equality constraints [11]. The terminal cost, associated with open-loop stable modes of the system, is calculated through the solution of the Lyapunov equation, whereas the terminal constraints are necessary to limit the objective function when the system is composed of integrating and unstable modes. However, stability proof is only achieved if the constrained optimization problem is feasible. The feasibility is also a critical issue of this approach, particularly because the domain of attraction of the controller becomes quite reduced by virtue of the associated hard constraints. Although there is already a rich theory in this field, the applications are heavily limited to theoretical works [12]. Among the methods developed to circumvent this issue so far, the approach based on slacked terminal constraints seems to be more adequate for practical implementation purposes, with recent applications reported in the literature, one implementation in a crude oil distillation [13] and the other in an inverted pendulum mechatronic-like fast dynamic system, namely, customized engine control unit [14].

This class of controllers formulates optimization problems that are always feasible through the suitable inclusion of slack variables in the control laws, without compromising their convergence and stabilizing properties. Also, these IHMPC controllers make use of the customized state-space models, obtained from an analytical expression of the step response of the system. Because of this, their formulations have been gradually developed over time. Odloak [15] focused on systems with open-loop stable poles, his work being extended to contemplate simple integrating poles as well in [6], commonly found in systems of the process industry. Then, Santoro and Odloak [16] encompassed time delays to the formulation proposed in [6]. For open-loop stable and unstable time-delay processes, Martins and Odloak [12] synthesized their IHMPC controller. More recently, the master's dissertation [17] included integrating poles in the last work formulation [12], such that its associated IHMPC controller can be directly applied to rotary inverted pendulum-behaved systems, the case understudy of this chapter. The implementation in a real system of the feasible-optimization problem-based stabilizing MPC controller proposed in [17] has not yet been documented in the literature. This gap will be filled in the present work.

2. System description

The objective of the MPC controller to be explored in this work is to stabilize the pendulum rod in the upright position while it leads the rotary arm angle to the desired positions. To this end, the rotary inverted pendulum used here will be a commercial-didactic prototype manufactured by Quanser. This prototype is installed in the Control Laboratory of the Center for Technological Training in Industrial Automation (CTAI) at the Federal University of Bahia (UFBA). **Figure 1** illustrates the features of such a system [18].

The rotary inverted pendulum prototype consists of a servomotor system, whose voltage V_m applied to it is responsible for generating torque in the rotary arm of angle (θ). The long pendulum rod is connected to the end of the rotary arm, and its angle, α, is zero when it is upright in the vertical position (cf. **Figure 1**).

The governing mathematical model of this system can be obtained by the Euler-Lagrange formalism, resulting in the following well-known equations [18]:

$$\ddot{\theta} = \frac{-bcsin(\alpha)(\dot{\alpha})^2 + bdsin(\alpha)\cos(\alpha) - ce(\dot{\theta}) + cfV_m}{ac - b^2\cos^2(\alpha)},$$
$$\ddot{\alpha} = \frac{adsin(\alpha) - b^2\sin(\alpha)\cos(\alpha)(\dot{\alpha}^2) - becos(\alpha)(\dot{\theta}) + bfcos(\alpha)V_m}{ac - b^2\cos^2(\alpha)}, \quad (1)$$

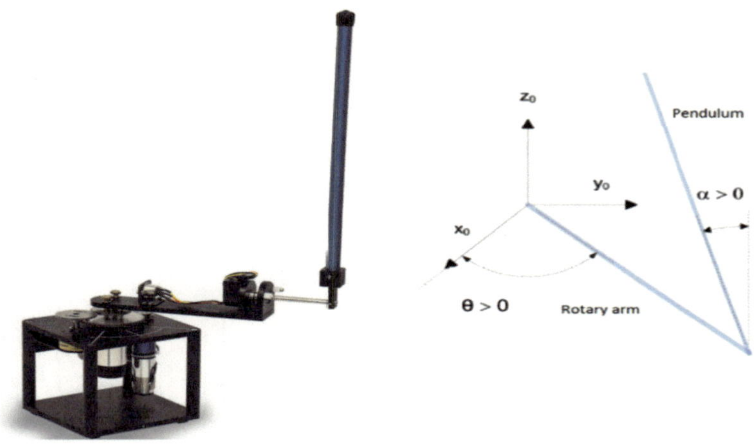

Figure 1.
Rotary inverted pendulum prototype manufactured by Quanser (left) and the schematic diagram (right).

where $\ddot{\theta}$, $\dot{\theta}$, $\ddot{\alpha}$, and $\dot{\alpha}$ represent angular accelerations and velocities associated with rotary arm angle and inverted pendulum angle, respectively. In addition, the parameters (a, b, c, d, e, f) are constants related to the physical dimensions of the various components that make up the inverted pendulum prototype. Information about the modeling and physical parameters of the system can be referred to [18].

Since one of the control objectives aims to the stabilization of the pendulum in the upright position, it is quite adequate to assume that α will suffer small variations, which implies that $\sin(\alpha) \approx \alpha$, $\cos(\alpha) \approx 0$, and $\dot{\alpha}^2 \approx 0$. Then, after some algebraic manipulations in Eq. (1), applying the Laplace transform as well, one turns out to be the following transfer function matrix ($G(s)$):

$$\begin{bmatrix} \theta(s) \\ \alpha(s) \end{bmatrix} = \underbrace{\begin{bmatrix} \dfrac{fcs^2 - fd}{s\left((ac - b^2)s^3 + ecs^2 - ads - ed\right)} \\ \dfrac{bfs}{(ac - b^2)s^3 + ecs^2 - ads - ed} \end{bmatrix}}_{G(s)} V_m(s). \tag{2}$$

This model representation of the system in terms of transfer functions is useful to obtain the state-space formulation to be used in the stabilizing MPC control law, as will be shown in the next section.

3. Stabilizing MPC formulation

The stabilizing MPC control law used in this work seeks to solve an infinite-horizon optimization problem, such that its objective function is composed of the following terms:

$$\begin{aligned} J_k = {} & \sum_{j=1}^{\infty} \left\| \boldsymbol{y}(k+j|k) - \boldsymbol{y}_{sp} - \boldsymbol{\delta}_{y,k} - \Psi_{un} \mathbf{F}_{un}{}^{(j-m)} \boldsymbol{\delta}_{un,k} - (j-m)\Delta t \delta_{i,k} \right\|_{\mathbf{Q}_y}^2 \\ & + \sum_{j=0}^{m-1} \| \Delta \boldsymbol{u}(k+j|k) \|_{\mathbf{R}}^2 + \| \boldsymbol{\delta}_{y,k} \|_{\mathbf{S}_y}^2 + \| \boldsymbol{\delta}_{un,k} \|_{\mathbf{S}_{un}}^2 + \| \boldsymbol{\delta}_{i,k} \|_{\mathbf{S}_i}^2, \end{aligned} \tag{3}$$

where m is the control horizon, $\Delta \boldsymbol{u}(k+j|k) \in \mathbb{R}^{nu}$ is the vector of input moves at time step $k+j$, $\mathbf{Q}_y \in \mathbb{R}^{ny \times ny}$ is a positive-definite weighting matrix of the controlled outputs, $\mathbf{R} \in \mathbb{R}^{nu \times nu}$ is a positive-definite weighting matrix of the input moves, $\boldsymbol{y}_{sp} \in \mathbb{R}^{ny}$ is the vector of references of the controlled variables, and $\boldsymbol{y}(k+j|k) \in \mathbb{R}^{ny}$ is the vector of the predicted outputs at time step $k+j$ computed at time step k, considering a state-space model obtained from an analytical expression of the step response of the system described as in Eq. (2), namely:

$$\begin{bmatrix} \boldsymbol{x}^s(k+1) \\ \boldsymbol{x}^{st}(k+1) \\ \boldsymbol{x}^{un}(k+1) \\ \boldsymbol{x}^i(k+1) \end{bmatrix} = \underbrace{\begin{bmatrix} \mathbf{I}_{ny} & \mathbf{0}_{ny \times nd} & \mathbf{0}_{ny \times nun} & \Delta t \mathbf{I}_{ny} \\ \mathbf{0}_{nd \times ny} & \mathbf{F}_{st} & \mathbf{0}_{nd \times nun} & \mathbf{0}_{nd \times ny} \\ \mathbf{0}_{nun \times ny} & \mathbf{0}_{nun \times nd} & \mathbf{F}_{un} & \mathbf{0}_{nun \times ny} \\ \mathbf{0}_{ny} & \mathbf{0}_{ny \times nd} & \mathbf{0}_{ny \times nun} & \mathbf{I}_{ny} \end{bmatrix}}_{A} \begin{bmatrix} \boldsymbol{x}^s(k) \\ \boldsymbol{x}^{st}(k) \\ \boldsymbol{x}^{un}(k) \\ \boldsymbol{x}^i(k) \end{bmatrix} + \underbrace{\begin{bmatrix} \mathbf{B}_s \\ \mathbf{B}_{st} \\ \mathbf{B}_{un} \\ \mathbf{B}_i \end{bmatrix}}_{B} \Delta \boldsymbol{u}(k), \tag{4}$$

$$y(k) = \underbrace{\begin{bmatrix} \mathbf{I}_{ny} & \Psi_{st} & \Psi_{un} & \mathbf{0}_{ny \times nu} \end{bmatrix}}_{C} \begin{bmatrix} x^s(k) \\ x^{st}(k) \\ x^{un}(k) \\ x^i(k) \end{bmatrix}. \tag{5}$$

In the state-space model defined in the pair of Eqs. (4) and (5), $x^s(k) \in \mathbb{R}^{ny}$ are the artificial integrating states introduced by the incremental form of inputs, $x^{st}(k) \in \mathbb{C}^{nst}$ are the stable states of the system, $x^{un}(k) \in \mathbb{C}^{nun}$ are the unstable states of the system, and $x^i(k) \in \mathbb{R}^{ny}$ are the true integrating states of the system. $\mathbf{I}_n$ and $\mathbf{0}_n$ are identity and null matrices of $n \times n$ dimension, respectively. The remaining matrices ($\mathbf{F}_{st}$, $\mathbf{F}_{un}$, $\mathbf{B}_s$, $\mathbf{B}_{st}$, $\mathbf{B}_{un}$, $\mathbf{B}_i$, $\Psi_{\mathbf{st}}$ e $\Psi_{\mathbf{un}}$) are obtained from step-response coefficients of the transfer function matrix of the system, and the details can be referred to [19, 20].

In the objective function, there are also $\delta_{y,k}$, $\delta_{un,k}$, and $\delta_{i,k}$ that are slack variables introduced into the control law so as to provide additional degrees of freedom to the resulting optimization problem, thus assuring the feasibility of the controller. These slack variables are weighted by positive defined matrices $\mathbf{S_y} \in \mathbb{R}^{ny \times ny}$, $\mathbf{S}_{un} \in \mathbb{R}^{nun \times nun}$, and $\mathbf{S}_i \in \mathbb{R}^{ni \times ni}$, respectively. In fact, the set of slack variables adopted in the problem formulation is responsible for softening, when necessary, terminal constraints that are imposed to limit the infinite-horizon objective function, owing to the existence of open-loop unstable and integrating modes.

It should be kept in mind that the objective function defined in Eq. (3) can be rewritten as follows:

$$\begin{aligned} J_k = & \sum_{j=1}^{m} \left\| y(k+j|k) - y_{sp} - \delta_{y,k} - \Psi_{un}\mathbf{F}_{un}{}^{(j-m)}\delta_{un,k} - (j-m)\Delta t\delta_{i,k} \right\|_{\mathbf{Q}_y}^2 \\ & + \sum_{j=1}^{\infty} \left\| y(k+m+j|k) - y_{sp} - \delta_{y,k} - \Psi_{un}(\mathbf{F}_{un})^j\delta_{un,k} - j\Delta t\delta_{i,k} \right\|_{\mathbf{Q}_y}^2 \\ & + \sum_{j=0}^{m-1} \| \Delta u(k+j|k) \|_{\mathbf{R}}^2 + \| \delta_{y,k} \|_{\mathbf{S}_y}^2 + \| \delta_{un,k} \|_{\mathbf{S}_{un}}^2 + \| \delta_{i,k} \|_{\mathbf{S}_i}^2. \end{aligned} \tag{6}$$

Then, with the aid of the state-space model used to carry out the prediction of the system, it is possible to demonstrate that the objective function becomes:

$$\begin{aligned} J_k = & \sum_{j=1}^{m} \left\| y(k+j|k) - y_{sp} - \delta_{y,k} - \Psi_{un}\mathbf{F}_{un}{}^{(j-m)}\delta_{un,k} - (j-m)\Delta t\delta_{i,k} \right\|_{\mathbf{Q}_y}^2 \\ & + \sum_{j=1}^{\infty} \left\| \begin{matrix} x^s(k+m|k) + j\Delta t x^i(k+m|k) + \Psi_{st}(\mathbf{F}_{st})^j x^{st}(k+m|k) + \\ \Psi_{un}(\mathbf{F}_{un})^j x^{un}(k+m|k) - y_{sp} - \delta_{y,k} - \Psi_{un}(\mathbf{F}_{un})^j\delta_{un,k} - j\Delta t\delta_{i,k} \end{matrix} \right\|_{\mathbf{Q}_y}^2 \\ & + \sum_{j=0}^{m-1} \| \Delta u(k+j|k) \|_{\mathbf{R}}^2 + \| \delta_{y,k} \|_{\mathbf{S}_y}^2 + \| \delta_{un,k} \|_{\mathbf{S}_{un}}^2 + \| \delta_{i,k} \|_{\mathbf{S}_i}^2. \end{aligned} \tag{7}$$

It is worth emphasizing that if constraints are not imposed at the end of the control horizon, the objective function value will increase unbounded. To this end, the following terminal constraints are imposed on the optimization control problem:

$$x^s(k+m|k) - y_{sp} - \delta_{y,k} = \mathbf{0}, \tag{8}$$

$$x^{un}(k+m|k) - \delta_{un,k} = \mathbf{0}, \tag{9}$$

$$x^i(k+m|k) - \delta_{i,k} = \mathbf{0}. \tag{10}$$

Furthermore, the term associated with stable modes of the system comprises a convergent series, giving rise to the so-called terminal cost, namely:

$$\sum_{j=1}^{\infty} \| \Psi_{st}(\mathbf{F}_{st})^j x^{st}(k+m|k) \|^2_{\mathbf{Q}_y} = \| x^{st}(k+m|k) \|^2_{\overline{\mathbf{Q}}}, \tag{11}$$

where $\overline{\mathbf{Q}}$ is the terminal weighting matrix obtained from the solution to the Lyapunov equation of the system. In symbols:

$$\overline{\mathbf{Q}} = (\mathbf{F}_{st})^{\top}(\Psi_{st})^{\top}\mathbf{Q}_y\Psi_{st}\mathbf{F}_{st} + (\mathbf{F}^{st})^{\top}\overline{\mathbf{Q}}(\mathbf{F}_{st}). \tag{12}$$

Therefore, the feasible-optimization problem-based stabilizing MPC control law is summarized as follows:

Problem 1.

$$\min_{\Delta u_k,\, \delta_{y,k},\, \delta_{un,k},\, \delta_{i,k}} J_k = \sum_{j=1}^{m} \left\| y(k+j|k) - y_{sp} - \delta_{y,k} - \Psi_{un}\mathbf{F}_{un}{}^{(j-m)}\delta_{un,k} - (j-m)\Delta t \delta_{i,k} \right\|^2_{\mathbf{Q}_y}$$

$$+\| x^{st}(k+m|k) \|^2_{\overline{\mathbf{Q}}} + \sum_{j=0}^{m-1} \| \Delta u(k+j|k) \|^2_{\mathbf{R}} + \| \delta_{y,k} \|^2_{\mathbf{S}_y} + \| \delta_{un,k} \|^2_{\mathbf{S}_{un}} + \| \delta_{i,k} \|^2_{\mathbf{S}_i},$$

subject to Eqs. (8), (9), and (10), and

$$\Delta u(k+j|k) \in \mathbb{U}, \quad j = 0, \ldots, m-1, \tag{13}$$

$$\mathbb{U} = \left\{ \begin{array}{c} -\Delta u_{max} \le \Delta u(k+j|k) \le \Delta u_{max} \\ \Delta u(k+j|k) = 0, j \ge m \\ u_{min} \le u(k-1) + \sum_{i=0}^{j} \Delta u(k+i|k) \le u_{max} \end{array} \right\}, \tag{14}$$

where $\Delta u_k = \left[\Delta u(k|k)^{\top} \cdots \Delta u(k+m-1|k)^{\top} \right]^{\top}$ is the sequence of control moves along the control horizon.

Remark 1. The slack variables play a remarkable role with respect to the feasibility of the control formulation, i.e., the control law of **Problem 1** will always provide a feasible solution, either the nominal case (linear model) or plant-model mismatch, an object under study of this work.

Remark 2. The weighting matrices $\mathbf{S}_y$, $\mathbf{S}_{un}$, and $\mathbf{S}_i$ (additional tuning parameters when compared to conventional MPC strategies) should be carefully selected. For instance, the values of $\mathbf{S}_y$ should be chosen sufficiently large, e.g., orders of magnitude larger than $\mathbf{Q}_y$ ($\ge 10^3\mathbf{Q}_y$), to guarantee that the solution of the slacked optimization problem will only use the slack vector when the terminal constraints need to be softened. While $\delta_{un,k}$ and $\delta_{i,k}$ do not need to be minimized a priori, by issues of achieving the closed-loop stability as fast as possible, one seeks their minimization weighted by large enough values of positive-definite $\mathbf{S}_{un}$ and $\mathbf{S}_{in}$ ($\ge 10^2\mathbf{Q}_y$) in order to enforce them to zero in a finite number of steps.

Remark 3. From the stability point of view, the master's dissertation [17] demonstrates the conditions necessary to prove that the objective function behaves as a Lyapunov function, thus assuring that the control actions obtained from the solution of **Problem 1** drive the system asymptotically to the reference value (desired steady state), if it is reachable; otherwise, the system will converge to an equilibrium point (reachable steady state) lying at a minimum distance from the desired steady state.

4. Results and discussion

This section is devoted to present the implementation results of the feasible-optimization problem-based stabilizing MPC controller (**Problem 1**) in the rotary inverted pendulum prototype described in Section 2. The ultimate goal of the controller is to maintain the pendulum rod in the upright position after it has been swung up to this position by the energy-based swing-up control scheme embedded in the system. In addition, the IHMPC controller is simultaneously designed to track the desired positions to be configured for rotary arm angles. In Quanser apparatus, an unconstrained linear-quadratic regulator (LQR) controller makes up the control system, besides the swing-up control strategy. The existing LQR strategy will be replaced by the IHMPC controller, and this scheme is depicted in **Figure 2**.

The architecture used for this real-time implementation of the IHMPC controller is summarized in **Figure 3**. From this figure, it is possible to note the information exchange among software-hardware-equipment mechanisms of the prototype. The control law is solved at each sampling time on the computer i7-8550H with 1.80GHz processor and 16GB of RAM, using Matlab script and *Quarc real-time control* toolbox. The software-hardware interface is done via USB communication through the Q8-USB acquisition board. This acquisition system acts, in turn, as an interface between the digital part of the system (controller) and the analogic one that is composed of the amplifier (VoltPAQ-X1).

For the experimental results presented as follows, we consider a scenario of square wave-type tracking on the rotary arm, while the controller must maintain the pendulum rod around the upright vertical position, even in the existing unmeasured disturbance scenarios. The constraints associated with the control signal and control actions (decision variables) are those established in **Table 1**.

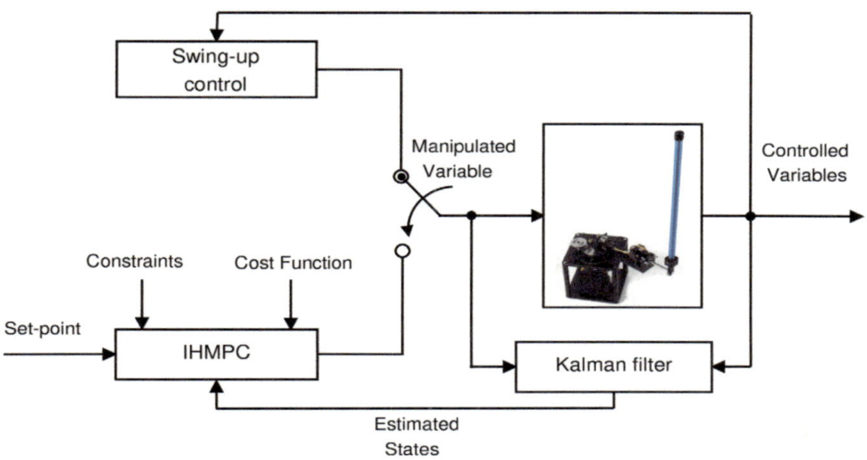

Figure 2.
Schematic representation of the application of the IHMPC controller in the rotary inverted pendulum prototype.

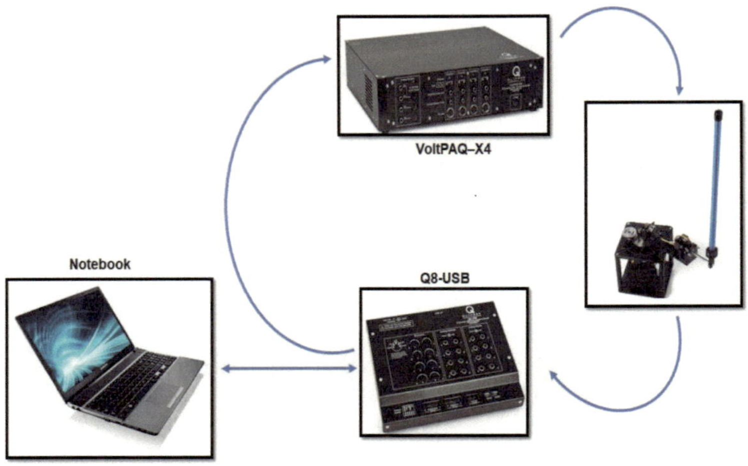

Figure 3.
The architecture used in communication among software-hardware-equipment mechanisms.

Variables	Minimum value (V)	Maximum value (V)
Control signal	−12	12
Control actions	−1	1

Table 1.
Constraints on system inputs.

Note that there is a strict condition of ± 1 V on the control actions. Also, the IHMPC tuning parameters considered were sampling period Δt = 2 milliseconds, $\mathbf{Q}_y = diag(1, 6 \times 10^2)$, $\mathbf{R} = 9.8 \times 10^{-2}$, $\mathbf{S}_y = diag(10^5, 10^4)$, $\mathbf{S}_{un} = diag(10^2, 6 \times 10^4)$, and $\mathbf{S}_i = 10^2$. The state estimator used here was the Kalman filter, whose tuning parameters associated with process noise and measurement noise were the following covariance matrices $\mathbf{Q}_{Kalman} = \mathbf{I}_{9\times9}$ and $\mathbf{R}_{Kalman} = 2.4 \times 10^{-6}\mathbf{I}_{2\times2}$, respectively. Finally, a control horizon of $m = 9$ has been adopted as an appropriate value to attain the desired control performance, which was chosen from a sensitivity analysis, as will also be shown here.

The closed-loop system results are depicted in **Figures 4** and **5**. From **Figure 4**, one can see that after about 5.5 seconds, the time necessary that the swing-up control acted to lead the pendulum rod to its upright position, the IHMPC controller takes over and performs quite well both tasks associated with the rotary arm angle tracking and the stabilization of the pendulum rod within an acceptable range lying at about $\pm 2°$. It is also noteworthy that after the execution of the square-wave trajectory on the rotary arm angle, the controller had a great performance concerning impulse-like external disturbances inserted in the pendulum rod since the controlled variables are momentarily moved away from their set points, but soon they are brought back to their original positions.

Even though the stability of the IHMPC controller is only related to the nominal case (linear model), it proved to be very sufficient in a realistic plant-model mismatch scenario, including nonlinearities existing in the rotary inverted pendulum apparatus, such as dead zone, friction, backlash, hysteresis, and so on. This model uncertainty scenario was responsible for non-prohibitive oscillations, within a practical implementation purpose, on the constrained control signal (cf. **Figure 5**), which were reflected in the controlled outputs.

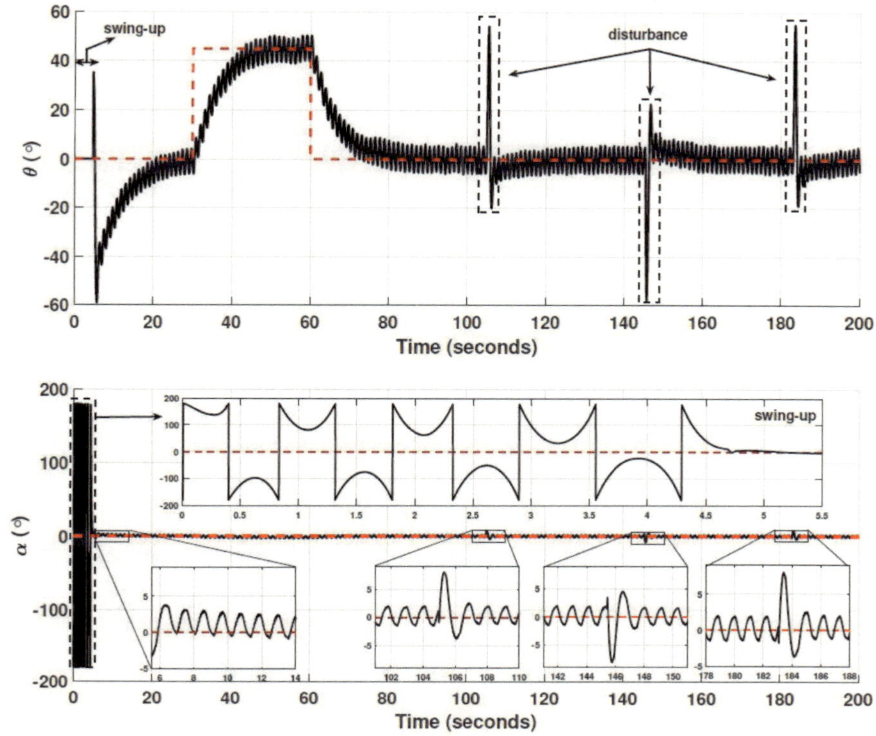

Figure 4.
Controlled variables: rotary arm angle (θ) and pendulum rod angle (α).

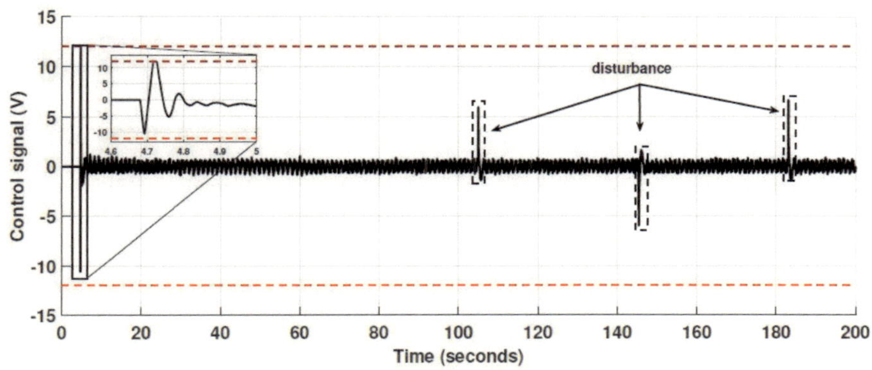

Figure 5.
Behavior of the control signal (tension applied to the servomotor).

Furthermore, it is worth mentioning that fulfilling a tighter constraint ($\pm 1V$) by a conventional stabilizing MPC controller, e.g., [11, 21], could result in an unfeasibility scenario; however, since IHMPC controller used here is based on a feasible-optimization formulation, its control law always will provide a feasible solution while the system is controllable, thus becoming it implementable in practice. The IHMPC controller uses its additional degrees of freedom (slack variables), when necessary, in order to comply with the terminal constraints. **Figure 6** illustrates the use of the slack variables in the control problem. It is observed that the controller makes use of these variables immediately after a perturbation in the system occurred namely, set-point changes and unmeasured disturbance entrance, situations in which it can be hard to comply with non-slacked terminal constraints.

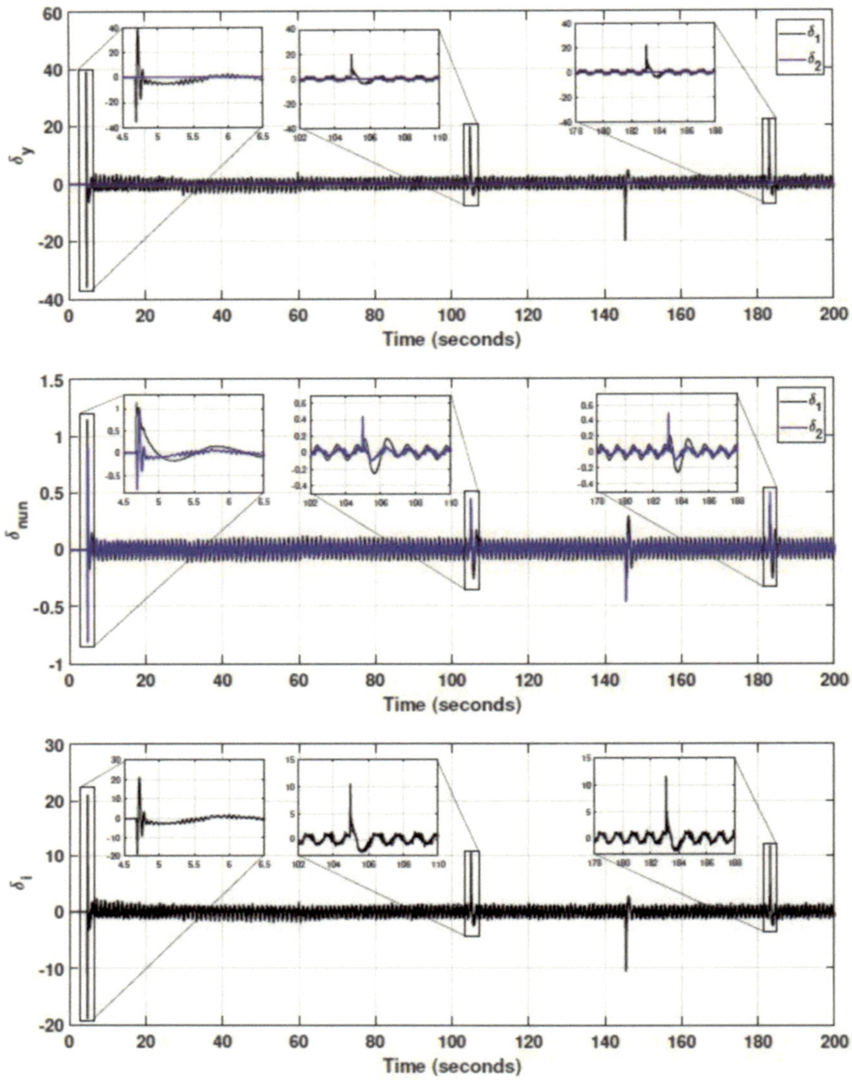

Figure 6.
Behavior of the slack variables associated with the feasible-optimization formulation of the IHMPC controller.

However, as the system goes to an acceptable cyclic steady state around its set point, due to the noise degree intrinsic to the system, the slack variables converge to the origin very fast and systematically.

On the other hand, in order to obtain a satisfactory performance as in the results presented earlier, an effort with respect to the controller tuning was necessary. This tuning task is easier in the stabilizing MPC controllers than the conventional finite-horizon MPC ones [13], as demonstrated in what follows. In this case, it was sufficient to handle only the control horizon. In inverted pendulum-like fast dynamic systems, when one applies a more aggressive control policy, i.e., small control horizon, it can cause undesired overshoots, while adopting large control horizons, it cannot have time sufficient to act with control action properly, thus bringing unnecessary oscillations or even causing the instability of the closed-loop system.

To work around this trade-off, we proceeded with sensitivity analysis on the control horizon, keeping the same remaining tuning parameters shown in the

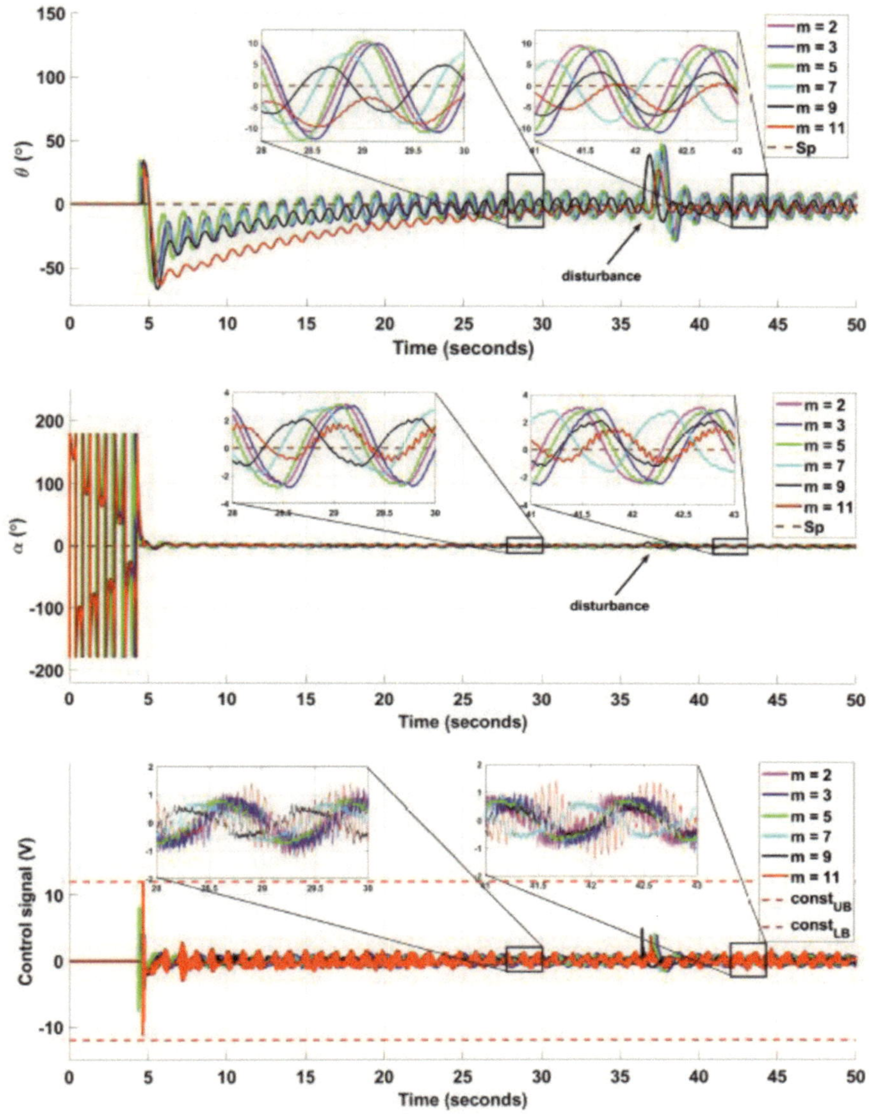

Figure 7.
Sensitivity analysis on the control horizon of the closed-loop system.

preceding experimental results. **Figure 7** summarizes the aforementioned analysis. Note that as the control horizon increases, the oscillations decrease until the control horizon $m = 9$. However, a value greater than $m = 9$ makes the closed-loop system go back to having undesired and larger oscillations, thus jeopardizing the use of energy associated with the control signal. Therefore, the use of the IHMPC controller enabled a simple analysis concerning only one tuning parameter, which yielded an appropriate value to meet the desired control performance in the real case.

5. Conclusions

In this chapter, we have investigated the application of an implementable and stabilizing model predictive control model strategy in a commercial-didactic rotary inverted pendulum apparatus, hitherto unexplored in the literature. Although the guarantee of stability of the controller is devoted to the nominal case (linear model), its formulation based on a feasible-optimization problem allows it to be used in any plant-model mismatch scenario in practice, such as one series of nonlinearities existing in the real system, namely, dead zone, friction, and backlash, among other unmodelled dynamics. The experimental results showed the effectiveness and robustness of the controller in the aforementioned plant-model mismatch setting by performing quite well its task in the rotary arm angle tracking and stabilization of the pendulum rod around the upright position as well as in the optimum use of energy associated with control efforts.

A simple tuning procedure was adopted by virtue of using a stabilizing MPC controller, which allowed us to handle only one tuning parameter through sensitivity analysis on the control horizon. The value found was quite adequate to attain the control objectives in terms of the trade-off existing between the performance on the controlled variables and the use of energy related to the control signal.

The future direction for this research is to further extend this controller to guarantee the stability of the nonlinear case, including the energy-based swing-up control schemes.

Acknowledgements

The authors would like to thank the Brazilian research agencies CAPES, CNPq, and FAPESB for their financial support.

Author details

Odilon S.L. de Abreu, Márcio A.F. Martins* and Leizer Schnitman
Programa de Pós-Graduação em Mecatrônica, Escola Politécnica, Universidade Federal da Bahia, Salvador, BA, Brazil

*Address all correspondence to: marciomartins@ufba.br

References

[1] Ogata K. Modern Control Engineering. 5th ed. Upper Saddle River: Prentice Hall; 2010

[2] Nise NS. Linear Systems Theory. 7th ed. Hoboken: Wiley; 2015

[3] Rossiter JA. Model-Based Predictive Control: A Practical Approach. Boca Raton: CRC Press; 2003

[4] Camacho EF, Bordons C, Alba CB. Model Predictive Control. London: Springer; 2004

[5] Maciejowski JM. Predictive Control: With Constraints. Harlow: Pearson Education; 2002

[6] Carrapiço OL, Odloak D. A stable model predictive control for integrating processes. Computers and Chemical Engineering. 2005;**29**(5):1089-1099

[7] Askari M, Mohamed HAF, Moghavvemi M, Yang SS. Model predictive control of an inverted pendulum. In: International Conference for Technical Postgraduates, TECHPOS. 2009. DOI: 10.1109/TECHPOS.2009.5412069

[8] Mohamed HAF. Hard constraints explicit model predictive control of an inverted pendulum. 1st International Conference on Energy, Power and Control (EPC-IQ). 2010;**6**(1):28-32

[9] Peyrl H, Zanarini A, Besselmann T, Liu J, Boéchat M-A. Parallel implementations of the fast gradient method for high-speed mpc. Control Engineering Practice. 2014;**33**:22-34

[10] Mayne DQ, Rawlings JB, Rao CV, Scokaert POM. Constrained model predictive control: Stability and optimality. Automatica. 2000;**36**(6):789-814

[11] Rawlings JB, Muske KR. The stability of constrained receding horizon control. IEEE Transactions on Automatic Control. 1993;**38**(10):1512-1516

[12] Martins MAF, Odloak D. A robustly stabilizing model predictive control strategy of stable and unstable processes. Automatica. 2016;**67**(5):132-143

[13] Martin PA, Zanin AC, Odloak D. Integrating real-time optimization and model predictive control of a crude distillation unit. Brazilian Journal of Chemical Engineering. 2019;**36**(3):1205-1222

[14] Brugnolli MM, Angelico BA, Laganá AAM. Predictive adaptive cruise control using a customized ecu. IEEE Access. 2019;7:55305-55317

[15] Odloak D. Extended robust model predictive control. AIChE Journal. 2004;**50**(8):1824-1099

[16] Santoro BF, Odloak D. Closed-loop stable model predictive control of integrating systems with dead time. Journal of Process Control. 2012;**22**(7):1209-1218

[17] Abreu OSL. Stabilizing model predictive control applied to open-loop unstable and integrating systems (in Portuguese) [Master dissertation]. Federal University of Bahia; 2018

[18] Quanser. Inverted Pendulum Experiment for Matlab/Simulink Users; 2011

[19] Yamashita AS, Santoro BF, Martins MAF, Odloak D. Application of an infinite horizon mpc to a nonlinear open-loop unstable reactor system. In: 8th International Conference on Integrated Modeling and Analysis in Applied Control and Automation, IMAACA. 2015. pp. 22-28

[20] Qin SJ, Badgwell TA. A survey of industrial model predictive control technology. Control Engineering Practice. 2003;**11**(7):733-764

[21] Muske KR, Rawlings JB. Model predictive control with linear models. AIChE Journal. 1993;**39**(2):262-287. ISSN 0001-1541

Chapter 6

Stewart-Gough Platform: Design and Construction with a Digital PID Controller Implementation

Flabio Dario Mirelez-Delgado, José Ronaldo Díaz-Paredes and Miguel Abraham Gallardo-Carreón

Abstract

This document presents the design of a digital PID control for a Stewart-Gough platform, delimited by six degrees of freedom (DoF) that allow the final effector to have displacement in the XYZ axes and rotation with warpage, pitch, and yaw restrictions. It includes the study and resolution of the direct and inverse kinematics of the platform, as well as the workspace described by the final effector and its corresponding simulation of movements and joints to study extreme points and possible singularities. From the definition of characteristics, the CAD design generated from the generalized mathematical model of the public domain, and the general selection of materials for the construction of the functional prototype, a study of applied forces is generated to observe the points with stress concentrators, the safety factor, and possible deformations. The estimation of the sampling period for the selection of the microcontroller and an approximate definition of the response time are also considered. The development of this prototype and its documentation are proposed as didactic material for the study, design, and control of parallel mechanisms.

Keywords: Stewart-Gough platform, PID control, parallel robots, inverse kinematics, CAD design

1. Introduction

When talking about industrial robotics, one of the first associated thoughts is about the serial robots that are highly applicable to this field; however, parallel robots also have a great importance and range of applications in this environment. Among its most important characteristics, its high load/power ratio is distinguished. In the case of the Stewart-Gough platform, this characteristic lies in its possibility of distributing the load in approximately six equal parts, whereby the total load capacity of the parallel robot approximates six times the load capacity of each actuator individually.

The Stewart platform is a parallel type manipulator that involves a configuration of six degrees of freedom, and each degree of freedom corresponds to an actuator. The six actuators can be linear or angular and join the bottom of the robot, which has no mobility, called a base with the part that does have mobility, the final effector. Due to this configuration, the Stewart-Gough platform has high rigidity, which can be translated into greater precision.

The final effector is the most interesting part of the robot, since its position and even orientation are the characteristics that determine if the robot is able to meet the precision necessary to implement and satisfy a need, whether industrial, educational, research, etc. It can have different applications, for example, orientation of satellites, flight simulators, and shakers (or also called agitators that are part of the chemistry laboratory instruments), among others.

This document explains the design, construction, and implementation of a discrete PID control to a Stewart-Gough platform.

2. Preliminaries

2.1 Control

To enter into definition of what control is, we begin by explaining what a process is, which is referred to as a set of equipment or devices attached or implemented to perform operations that help fulfill a task [1]. To enforce this task, a series of additional devices are needed that regulate the process in general, which are called a control system as a whole [1].

Ogata [2] mentions four methods (belonging to the classical control or also called conventional control) for the design of a controller. These are the following:

- Root place
- Frequency response
- PID controllers
- Modified PID controllers

In general, a control system needs a mathematical model that describes its behavior when receiving inputs [2]. For the Stewart platform, it can be determined by inverse kinematics [3].

The PID controller is the most common form of feedback, was an essential element of the first governors, and became a model or standard tool when process control emerged in the 1940s [4]. Consider a control loop of an input and an output. The block diagram [5] is shown in **Figure 1**.

Of the PID control family, there are members that involve the three actions: proportional (P), integral (I), and derivative (D) [5].

There are two control design techniques that are analog and digital, and for this they work with continuous time and discrete time systems, respectively. Many of the systems are described by differential equations, and analog control design techniques have become popular. Also most of the controllers are made of digital systems [6]. You can also discretize analog controllers to obtain digital controllers. Next, characteristics of control systems are mentioned, analog and digital [7]:

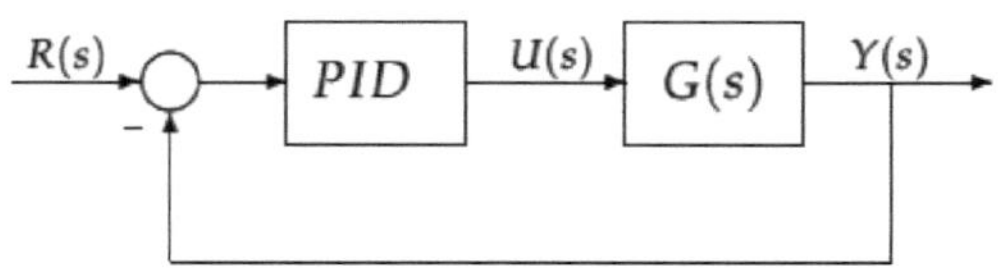

Figure 1.
SISO system block diagram.

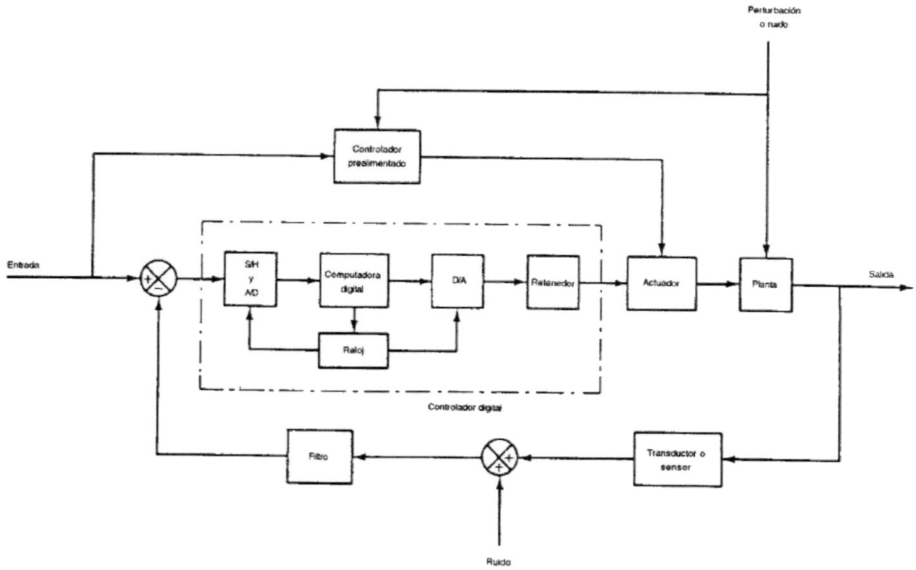

Figure 2.
Block diagram of a digital control system.

- Analog
    - Noise is a problem in all signal transmissions.
- Digital
    - They have the ability to reject noise.
    - It is easy to implement and modify digital controllers simply by changing the coefficient values.

A digital or discrete time system is a dynamic system in which one or more variables may vary only at certain times called sampling indicated by kT (k = 0,1,2, ..., n) [8].

The block diagram of a digital control system is shown in **Figure 2** [9].

3. Parallel robots

The term parallel robot, also known as closed-chain robot or parallel manipulator, basically consists of a fixed base attached to an end effector or mobile base which can perform its movement based on the movement of the actuators that are located in such a way that they close the kinematic chain between both bases [9]. The principle of operation initially arose with James E. Gwinnett, who designed a platform for the entertainment industry in the 1928 patent application [10]. However, it was not until 1940 when W. Pollard designed the first industrial sparallel robot for spray painting processes [11] (**Figures 3** and **4**).

After the first advances of the Stewart platforms applied to industrial processes, some companies began to make variants of these platforms, such as the case of the Redifon company that was asked by Lufthansa German Airlines to produce a simulator of flight, with the initial model designed for its then new Boeing 727 fleet. This simulator model included three axes, which gave the mechanism the mobility

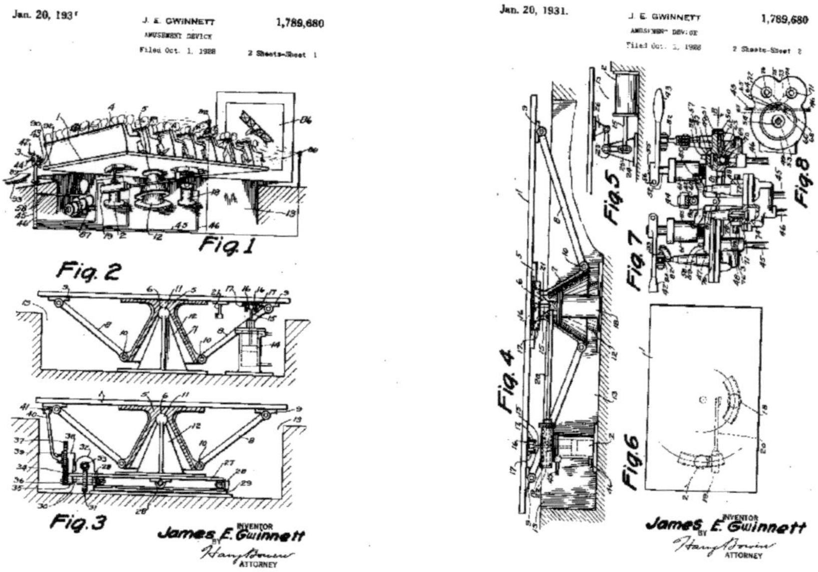

Figure 3.
James E. Gwinnett patent (industrial entertainment platform).

needed to recreate the behavior of the aircraft. Redifon is currently in service since its commissioning in 1949, when it began with the production of flight simulators, trainers, and the development of new techniques [12] (**Figure 5**).

4. Inverse kinematics

Inverse kinematics is a mathematical modeling of a manipulator, either serial (open kinematic chain) or parallel (closed kinematic chain). Such modeling requires as input parameters the position of the final effector in order to calculate the angles that exist between links and thus determine the position of each actuator in the reference *XYZ* coordinate space [13]. It should be noted that the final effector is the part of interest of any robot.

The fixed coordinates (F_{xyz}) are placed in the center of the fixed base, and the other mobile coordinate system (M_{uvw}) is positioned in the center of the mobile platform (**Figure 6**).

Points F_i and M_j are the points of the joints between one actuator end with the fixed base and the other end of the actuator with the movable base, respectively. The separation angles between points F_1 and F_2, F_3 and F_4, and F_5 and F_6 are denoted by θ_b. Similarly, the angle of separation between points M_1 and M_2, M_3 and M_4, and M_5 and M_6 is denoted by θ_p. To locate the links or points F_i, use Eq. (1):

$$F_i = \begin{bmatrix} F_{Xi} \\ F_Y \\ F_{Zi} \end{bmatrix} = \begin{bmatrix} r_b \cos(\mu_i) \\ r_b \mathrm{sen}(\mu_i) \\ 0 \end{bmatrix} \tag{1}$$

And for points M_j:

$$M_j = \begin{bmatrix} M_{Uj} \\ M_{Vj} \\ M_{Wj} \end{bmatrix} = \begin{bmatrix} r_p \cos(\lambda_j) \\ r_p \mathrm{sen}(\lambda_j) \\ 0 \end{bmatrix} \tag{2}$$

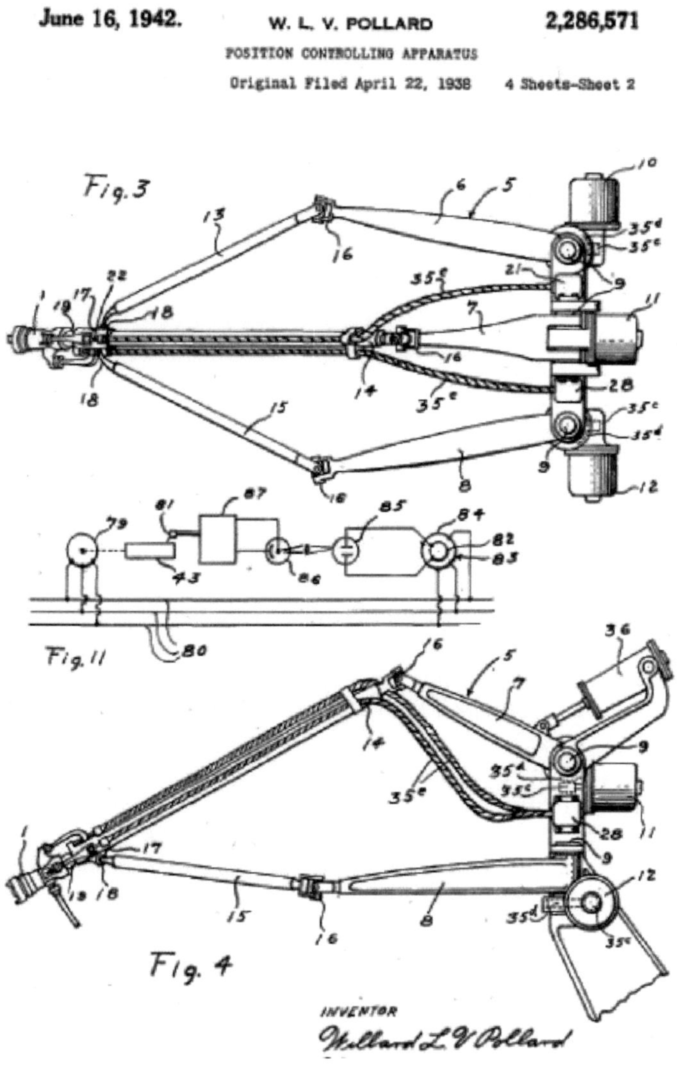

Figure 4.
First parallel robot designed by W. Pollard.

Figure 5.
Flight simulator developed by Redifon for the Boeing 747 model.

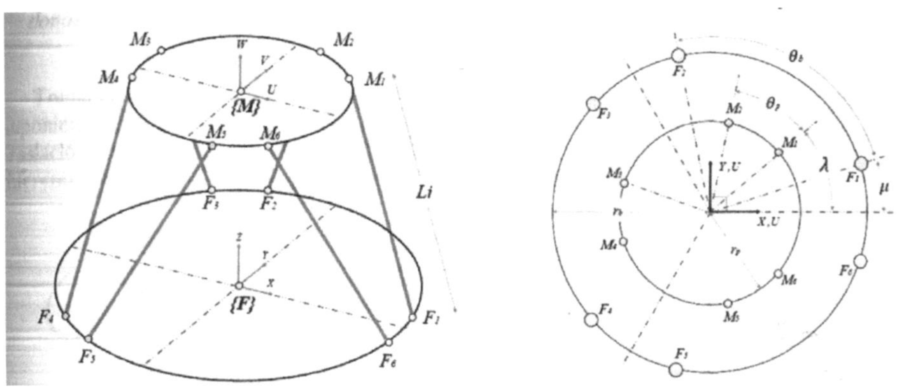

Figure 6.
Isometric and top view of the platform with coordinate systems and actuator junction points.

5. Direct kinematics

As mentioned in the previous chapter, reverse kinematics needs the position of the final effector as input parameters to calculate the angles between links of the robot. So, the direct kinematics is the counterpart of inverse kinematics, since its initial data are the angles established between each actuator and what is calculated is the position of the final effector. A very common method of using is Denavit-Hartenberg, explained by Barrientos, which follows a 16-step algorithm to get a single homogeneous transformation matrix from the premultiplication of submatrices of homogeneous transformation. The steps of this methodology are the following [14]:

DH1. Number the links starting with 1 (first mobile link in the chain) and ending with n (last mobile link). The fixed base of the robot will be numbered as link 0.

DH2. Number each joint starting with 1 (the one corresponding to the first degree of freedom) and ending in n.

DH3. Locate the axis of each joint. If this is rotary, the axis will be its own axis of rotation. Yes it is prismatic, it will be the axis along which the displacement occurs.

DH4. To i from 0 to $n-1$, place the axis z_i on the axis of the joint $i+1$.

DH5. Place the origin of the base system $\{S_0\}$ at any point on the z_0 axis. The axes x_0 and y_0 are they will place so that they form a dextrogyre system with z_0.

DH6. For i from 1 to $n-1$, place the system $\{S_0\}$ (in solidarity with the link) at the intersection of the axis z_i with the normal line common to z_{i-1} and z_i. If both axes were cut, it would be located $\{S_i\}$ at the point of cut. If they were parallel S_0, it would be located in the joint $i-1$.

DH7. Place x_i on the normal line common to z_{i-1} and z_i.

DH8. Position y_i so that it forms a dextrogyre system with z_i and x_i.

DH9. Position the system $\{S_n\}$ at the end of the robot so that z_n matches the direction of z_{n-1} and x_n be normal to z_{n-1} and z_n.

DH10. Get θ_i as the angle to be rotated around z_{n-1} so that x_{n-1} and $\{S_i\}$ remain parallel.

DH11. Get d_i as the distance measured along z_{i-1}, which should be displaced $\{S_i\}$ so that x_{i-1} and x_i were aligned.

DH12. Obtain α_i as the distance measured along x_i (which would now match x_{i-1}), so that the new $\{S_{i-1}\}$ would have to be moved so that its origin coincided with $\{S_i\}$.

DH13. Get α_i as the angle that should be turned around x_i (which would now match x_{i-1} so that the new $\{S_{i-1}\}$ totally coincided with $\{S_i\}$.

DH14. Obtain the transformation matrices ${}^{i-1}A_i$ defined as:

$$ {}^{i-1}A_i = \begin{bmatrix} c\theta_i & -c\alpha_i s\theta_i & s\alpha_i s\theta_i & a_i c\theta_i \\ s\theta_i & c\alpha_i c\theta_i & -s\alpha_i c\theta_i & a_i s\theta_i \\ 0 & s\alpha_i & c\alpha_i & d_i \\ 0 & 0 & 0 & 1 \end{bmatrix} \tag{3} $$

6. Microcontroller selection

One of the main parameters to be defined within the control system implementation is the sampling period T, which is a design element that allows different components to be selected among the most important ones such as the microcontroller. There are several methods for obtaining it; however, for this project the selection has been considered using the method described in [8] which selects the period based on the commitment between the following factors:

1. The calculation time of the processor
2. Numerical precision in the implementation
3. Loss of information in the sampling
4. Response to disturbances

In general, the sampling period must be selected in compromise between a range of time that avoids the deterioration of the quality of the control that can produce a high value of T and the amount of calculations necessary to execute the control algorithm with small values that can produce information loss and frequency overlap (aliasing).

In these, three cases are considered, which can be monitored based on the available elements, the ease of calculation, and the nature of the project.

- **Take the bandwidth of the system:** This considers the system as a closed loop system in which each of the elements and their respective frequency are raised. From this, the bandwidth of the system that will serve as a reference for the base frequency of sampling is determined.

- **Establishment time required for the transient response:** This method can be performed by simulation or experimental, since it considers obtaining a response time based on the reach of 63.3% of the final value in the transient response. In [16] an oscilloscope was used to measure the response curve and the Tao time by applying a pulse signal and a mechanism adapted to the actuators. In turn, a great advantage of this method is to observe the response curve and the actual efficiency of the actuators with respect to the data provided by the manufacturer.

- **Select the highest frequency component:** This is a method that allows an estimation of the sampling period based on the system component that requires sampling more frequently. Because currently the plant calculations have not been performed, nor have the components been obtained for obtaining transient response, this is considered the most successful method for estimating the sampling period.

Once the method is defined, the component that requires the highest working frequency is selected. For the selection, the actuators are directly discarded, because they only require a PWM pulse that varies depending on the desired position that does not exceed 1.25 ms, which is why the components to be considered will be the gyroscope sensors and the resistive panel that will measure the external forces applied to the final effector and the external movements applied to the platform in inclined planes.

Later, the selected sensors and their detailed characteristics will be described; however, in this section, only the basic principle of their operation will be seen, which will follow up the methodology described above to obtain the sampling time.

The sensor placed at the top of the final effector and responsible for measuring the movements of moving objects will be a five-wire resistive panel (can be found in 4, 5, or 8 wires), which consists of a partially conductive layer (resistive) that is applied evenly to the panel. Conductive bus bars are protected with a silver paint through the opposite edges of the panel. Rigid and flexible panels are mounted with bus bars perpendicular to each other as shown in **Figure 7**. The sensor measurement is done by applying a force on the ITO layer which generates a voltage gradient across one of the layers and measuring the tension in the other layer. This gradient is normally produced by grounding a bus bar and applying +5 V on the other bar, which produces a voltage gradient on the axes that cross the panel; this applies only to one layer, the rigid layer, while the other layer is the sensitive one in both measurements.

The measurement is done sequentially in which the events are presented with the voltage gradient on the x-axis and the voltage in the sensor layer is measured; followed by this, the voltage gradient is switched to the y-axis, and another measurement is taken of the same sensor layer. To determine the sampling rate of the resistive panel, the use that will be given must be considered, for example, if it is to be used as a human interface where the contact will be a finger, the necessary resolution is not very high. Being a 6-inch-long panel and the contact resolution is 2 in (average finger size), then the analog-to-digital converter (ADC) must represent 30 points. With 5 bits, 32 points can be represented, so an ADC converter of 8 to 12 bits is sufficient in this case and a sampling rate of 10 samples per second is sufficient. In the case of wanting to make a more accurate representation in the panel using pens or objects of greater precision, the touch screen should have a resolution of at least 320 points. This gives a converter of at least 9 bits (29 = 512 levels).

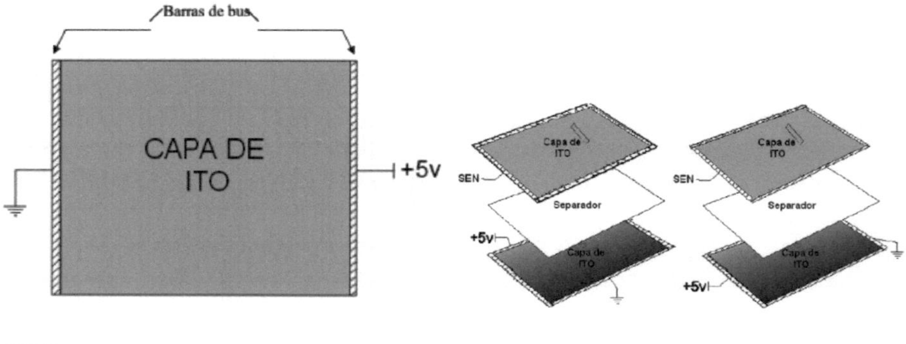

Figure 7.
Configuration of the layers of a resistive panel.

The sampling rate of the touch screen controller in this case must be higher so as not to lose information on the movements being captured; therefore, a speed of 100 samples per second would suffice. For the actual position in movements on the inclined plane, the MPU-6050 model is contemplated within the gyroscope selection with something important to point out, and among its main features is the precise capture of fast and slow movement; it has a scale range of 250/500/1000/2000 °/s. That is why, by requiring a greater sampling range per second than the resistive screen, this gyroscope will be selected as the component with the highest working frequency and which will be used to monitor the sampling time selection process. To measure the movements of the final effector, it is not considered necessary to work at the maximum capacity of the gyroscope because the maximum angle of action in warping and pitching will be ± 20 according to [22], since the configuration of the angular actuators. Considering that the system performs an extreme compensation movement of 40 by programming a minimum sampling of 250, the movements should be carried out with a speed of 1/250 (0.004) s, which is approximate to the speed of the angular actuator that varies between 0.0028 and 0.002 s/g (characteristic that will be seen later), which is why it is considered that the sensor at its maximum power would be wasting most of the sampling because the actuators would not be able to compensate for the said signal (**Figure 8**).

Despite having sufficient sampling time, the case in which the sensor configured at its maximum capacity of 2000/s will be used (in which case the movements made on it should be 0.0005 seconds, this value is higher to the tests to which it will be subjected) to have an extra safety factor in the selection of the sampling; this leads to the next step, which mentions the cases in which overlaps (aliasing) or loss of information with a too large period may occur (**Figures 9** and **10**). To prevent these problems from large sampling periods, the Nyquist theorem [22] is applied, which dictates that the sampling frequency Ws, defined as $2\pi/T$ where T is the sampling period, is greater than $2W_c$ (frequency to be sampled), that is:

$$W_s > 2W_c$$

From this, it is mentioned in [8] that it should be considered as a general rule that the system must be sampled between 8 and 12 times during a cycle of the dampened frequency of the transient response if the system is under-absorbed or between 8 and 12 times during the time of establishment of the response if the system is overdamped. In which case, the minimum of this factor will be taken

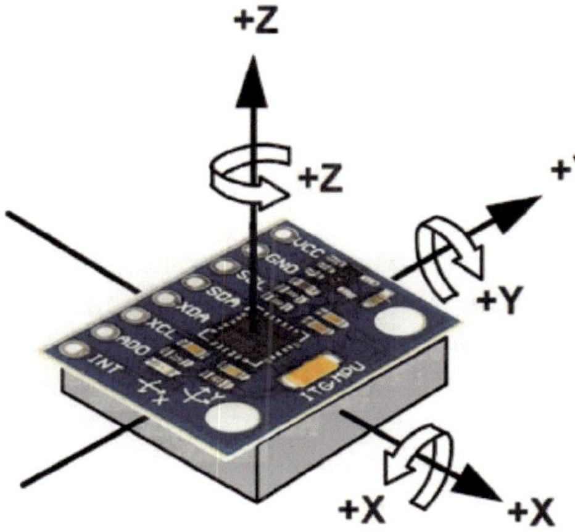

Figure 8.
Gyroscope sensor and accelerometer MPU-6050.

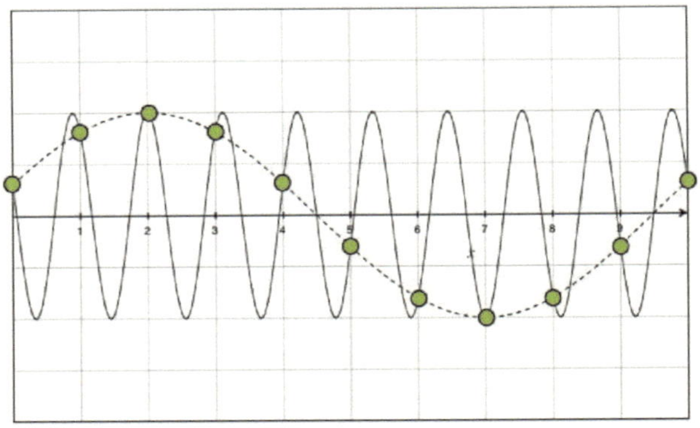

Figure 9.
Signal overlap (aliasing).

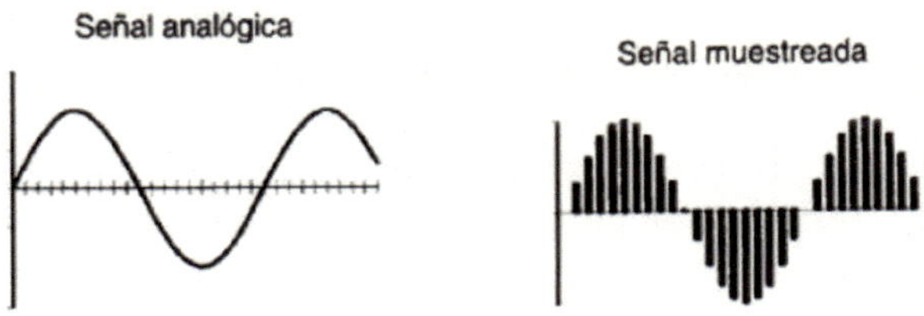

Figure 10.
Signal sampled by Nyquist theorem.

because it is desired to obtain a system with an under-absorbed or over-absorbed response.

Once the gyroscope is selected as the component with the highest frequency within the system, the largest measuring range provided by this device is taken as a reference, and as already mentioned, the sensor has a capacity to measure 2000/s and deliver it. A digital signal by I2C protocol, however, although this sampling configuration will not be required when incorporated into the system (since it exceeds the value of the speed performed by the actuators, in addition to requiring an ADC with higher resolution than even would be exceeded because the workspace of the platform is limited and does not exceed 100) is considered as the base sampling frequency because it is considering the critical case in which you want to obtain the highest frequency required for. The processor fulfills this task.

It is important to note that $W_s = 16$ kHz will be the selected working frequency that requires at least the processor to operate optimally in the sampling of the gyroscope signal; this is because later considerations will be made (of which currently are ignored to simplify calculations) such as the incorporation of resistive screen sampling in the ADC, the working frequency of the actuators individually, and in the kinematic chain, as well as may or may not affect the size of the program along the lines of code because when considering the implementation of a conventional microcontroller, the programming must be done in series and not in parallel.

Once an estimated time is defined for the sampling period, the microcontroller options (or microprocessor with minimum system) that meet these requirements are searched. Likewise, some of the main characteristics that they present and are required by other system components are compared, such as PWM peripherals for the six actuators, the I2C communication ports for the gyroscope, the ADC channels for the resistive panel, as well as its resolution and an extra feature that would be the type of assembly due to the didactic purpose of the platform, since when pretending

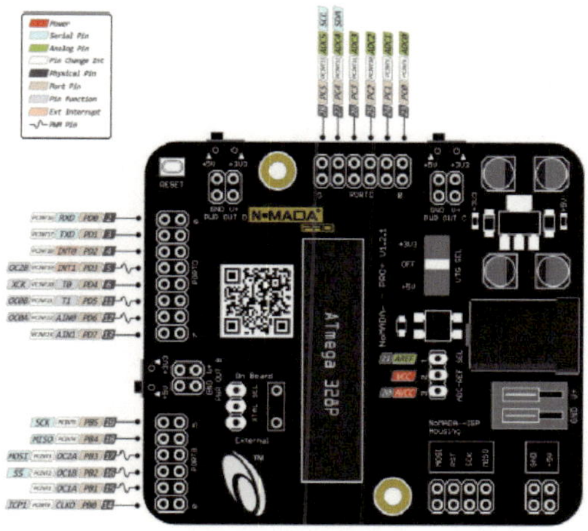

Figure 11.
ATmega328P microprocessor peripherals incorporated in NomadaPro + development card.

to be manipulated by students of the institution, against any problem that may arise, the components are easy to replace.

It is currently considered the best candidate for the ATMega328P microprocessor, either with a minimum system implemented or using the nomadic development card Pro + that incorporates it, and being a THT type assembly is easy replacement, as well as being economical and accessible. The microprocessor incorporates six ADC channels, of which one will be destined for the resistive panel and the other two directly coincide with the SCL and SDA channels through which I2C communication is performed. In addition, six PWM channels (one channel for each actuator) that are distributed in ports B and D are required for actuator control (**Figure 11**).

7. Mathematic model

The Stewart platform mechanism, until the mid-1980s, mostly maintained the design of the triangular platform connected by spherical joints to three linear actuators adjustable in length. This configuration linked the base (whose design can vary between a triangular and hexagonal shape) with the final effector (which maintains its triangular shape) by means of the actuators coinciding two by two at the vertices of the final effector with a total of six different points contained in the base plane. These configurations are denoted as 3-3 Stewart platform and 6-3 Stewart platforms [23], respectively, as can be seen in **Figure 12**.

Considering a point of coincidence in spherical joints in the triangular platform, it restricted the mobility in the manipulator, producing overstrain and a reduced workspace; from this, the configuration is obtained whose base retains the regular or semi-regular hexagonal shape and the final effector is generated from an equilateral triangle with the trimmed vertices known as 6-6 Stewart platform. The said design allows some parameters and components to be varied according to the needs and/or specifications, such as the actuators that can go from being linear to angular by means of a crank-crank type conditioning mechanism. Likewise, the type of joint can be modified, in which as is the case, spherical and non-universal joints will be used. Once the different configurations of the structure of the Stewart-Gough platform are present, the configuration 6-6 Stewart platform was selected, which allowed a better manipulation of the final effector, which avoids as much as possible

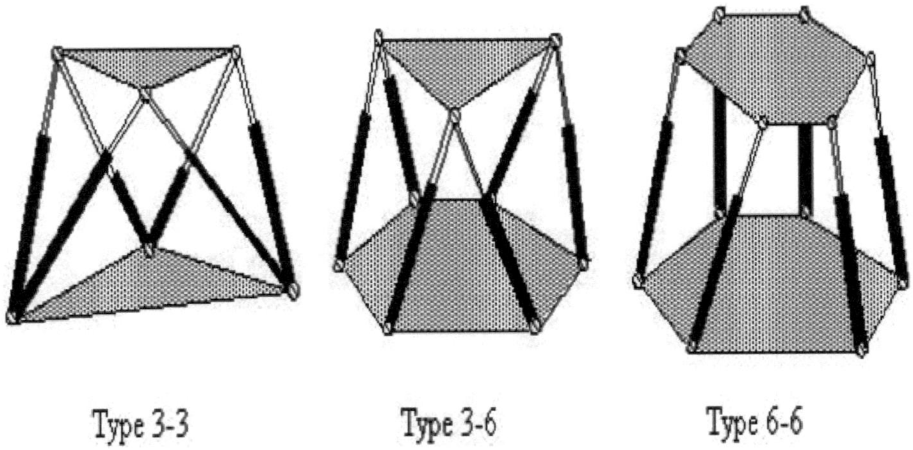

Figure 12.
Different configurations of the Stewart-Gough platform [23].

the efforts by singularities, and given the application, a large workspace is not necessary as it is proposed to carry out movements that are not so abrupt for the stabilization of objects that move on the surface, application of external forces, or movements on the inclined plane.

From this, it is observed that there is a wide variety of mechanical configurations for this type of manipulator with the 6-6 Stewart platform kinematic chain according to the previous selection. This kinematic structure considers a variant known as 6-SPS that is observed in **Figure 13**, where 6 refers to the number of degrees of freedom of the robot and SPS comes from the English spherical-prismatic-spherical referring to the leg from point to point, where the joints at the ends are prismatic and the link between them is a prismatic joint [23]. Once this configuration is selected that allows spherical joints to be placed on both sides of each actuator, a modification is made to the actuators such that instead of a prismatic joint between the joints relative to the

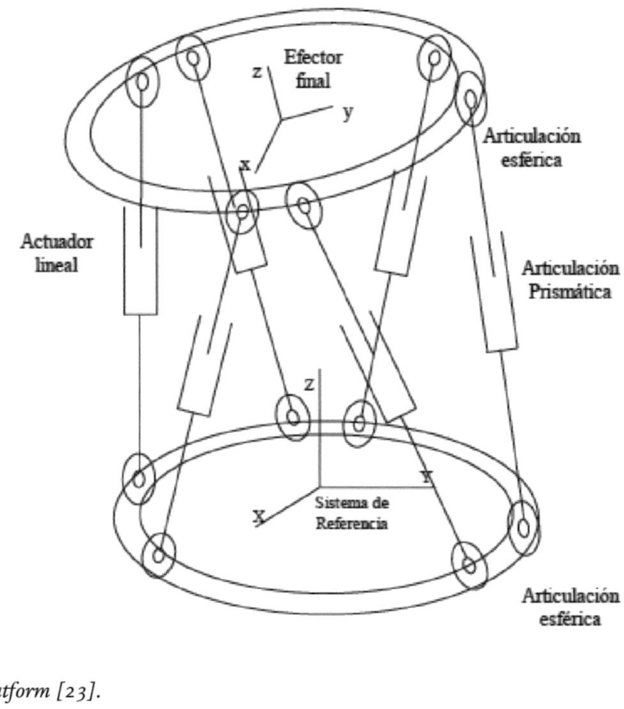

Figure 13.
Stewart 6-SPS platform [23].

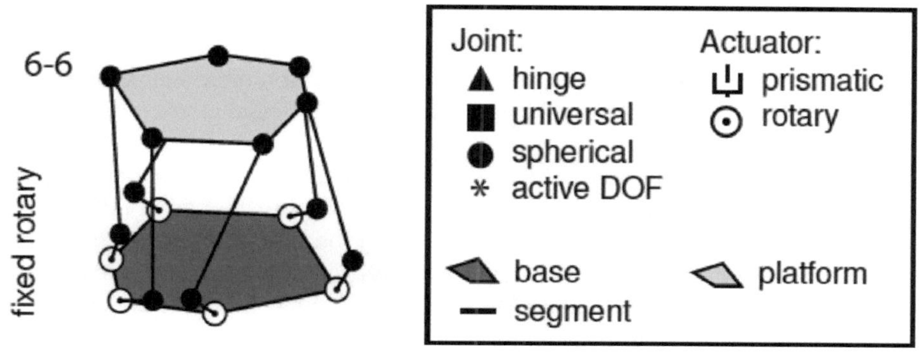

Figure 14.
Stewart 6-SPS platform with angular actuators [18].

linear actuator, it is replaced by a link element between both unions and the extension movement is generated by a configuration with angular actuators, as will be seen later in the left, thus remaining a configuration like the one presented in **Figure 14**. Using the Grubler criterion [13], the number of degrees of freedom on a 6-SPS type platform can be determined using the equation:

$$m = \lambda(n - j - 1) + \sum_{i=1}^{j} f_i - I_f \tag{4}$$

where: $m =$ degrees of freedom of the system.

$\lambda =$ degrees of freedom of the space where the mechanism is, $\lambda = 3$, two-dimensional and $\lambda = 6$ for the spatial case.

$n =$ number of fixed links of the mechanism, including the base and the final effector.

$j =$ number of joints in the mechanism.

$f_i =$ relative degrees of movement per board.

$I_f =$ number of passive degrees of freedom of the mechanism.

Substituting Eq. (5), you have:

$$m = 6(14 - 18 - 1) + \sum_{i=1}^{12} 3 + \sum_{i=1}^{6} 1 - 6 = 6 \tag{5}$$

The workspace, also known as the field of action, is the area or spatial volume that the robot can describe when it reaches extreme points. This described volume is determined by the dimensions, shape, and movement of the joints that make it up, as well as the degrees of freedom (depending on the configuration and type of robot), and on some occasions the applied control system can also influence [15].

Although the robot has a defined workspace, it is not confirmed that such space can be described in any orientation. There are a number of points, usually the most extreme and the closest to the origin that can be accessed only with certain configurations, and some others can be reached in any orientation. This as already mentioned depends a lot on the type of configuration of the robot.

One of the disadvantages that parallel robots have over serial ones is the limited description of the workspace; this is due to the restrictions that one joint has over another in the closed kinematic chain. In the case of the Stewart-Gough platform, you can make the proposal of a workspace described by the final effector from which the dimensions of the different elements that make up the parallel robot (base, joints, platform) are obtained, legs, actuators, and horn. On the other hand, you can obtain a supposed workspace based on the selection of components to evaluate at each point of the joints, the actuators, and links, the possible interferences that can be caused between them, as well as the possible restrictions of each meeting [16].

There are several methods to obtain the measurements of each element of the Stewart-Gough platform either by describing trajectories as they do in [19] or generating a proposal from a radius of action; this is because the said platform has a design focused on minimally invasive coronary bypass surgery. The proposal of measures can also be made based on the generation of algorithms by a method observed in [17]. Several of these methods to describe the workspace of a Stewart-Gough platform in a graphic form are based on the discretization of the Cartesian space and then evaluate the length of the links according to the actuators and thus determine that they are within the range and detect interference between actuators and/or meetings and the possible restrictions of each meeting. This method can also include simulations by implementing and evaluating the Jacobian matrix to define a movement and joint capacity [18]. Despite the variations between models of the Stewart-Gough platform, the inverse kinematics and the workspace in general have the same behavior and the same description of trajectories, as mentioned in the calculation of the inverse kinematics. The same can be used for all regardless of the type of actuators. This of course may vary slightly according to the dimensions, configuration, and/or elements used for its elaboration; however, the shadow described by the said workspace is projected with a great similarity between them. Using the methodology outlined in [19], **Figure 15** is generated, which corresponds to the definition of dimensions of the Stewart-Gough platform, making the consideration that the servomotors are located parallel to each other. For this, it is based on the definition of the base, that when the Stewart-Gough Platform was previously proposed with approximate dimensions at a volume that encloses it from 30 to 35 cm, an internal diameter coincides with the axis of the servos is established, it has a measure of 25 cm because the standard digital servomotors have a depth of approximately 5 cm (**Figures 16** and **17**).

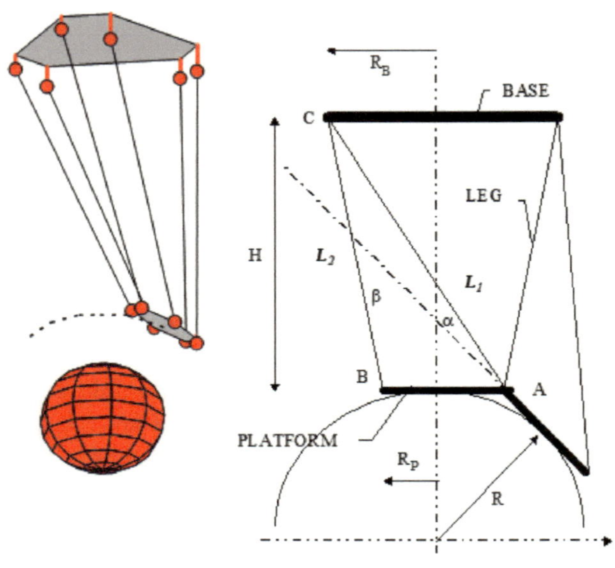

Figure 15.
Description of the trajectory of the Stewart platform for CABG surgery [19].

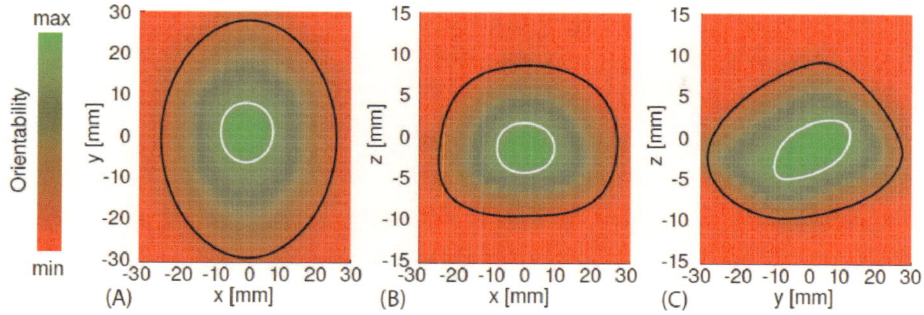

Figure 16.
Workspace described by the Stewart-Gough platform using the Jacobian matrix [18].

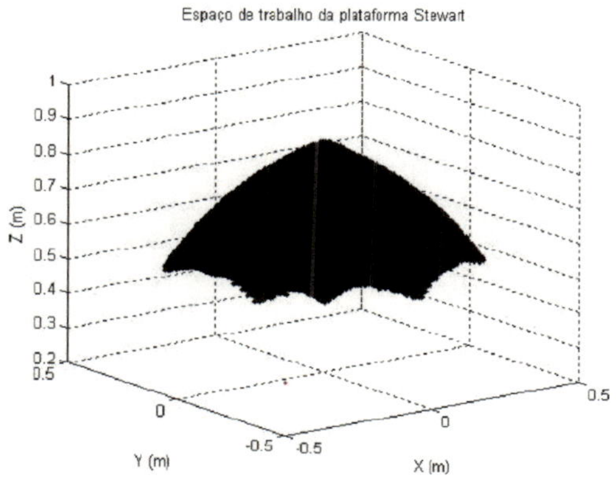

Figure 17.
Workspace of the Stewart-Gough platform in isometric view [16].

Once the base is defined, the required workspace is raised. In principle, the design of the platform is not required to describe paths with too much depth or around a radius as is done in [19]; in this case the important data to consider would be the angles of warping and pitching inclination, since they will be intended for stabilizing sliding objects on the surface of the platform.

For this situation, one of the conditions of the state space is proposed, which would be described by the yellow line in **Figure 15**, where the inclination angle is proposed of 20 present in most Stewart platforms with documented angular actuators and which is considered sufficient inclination for the stabilization of objects sliding on the surface as can be seen in [15]. It is worth mentioning that this is an extreme reach position that is physically not possible to achieve and would cause singularities, which is why it is recommended not to bring the final effector to these horizons in which the platform would lose stability and control may lose efficiency in its execution.

Finally, the possible restrictions generated by the articulation of the spherical joint are added (**Figure 15**) since according to the table in Annex F, the working angle (depending on the size of the joint) is between $\pm$ 13 and $\pm$ 18; therefore, leaving a value of 14, it is observed that the opposite angle corresponds to 76 that are assigned in the internal area of the trapezoid generated with the frontal view since in principle when placing the actuators parallel to each other, it seems to be a four-bar mechanism that can make 360 movements. However, there are restrictions generated by the other links that have to respect that angle of work of the patella.

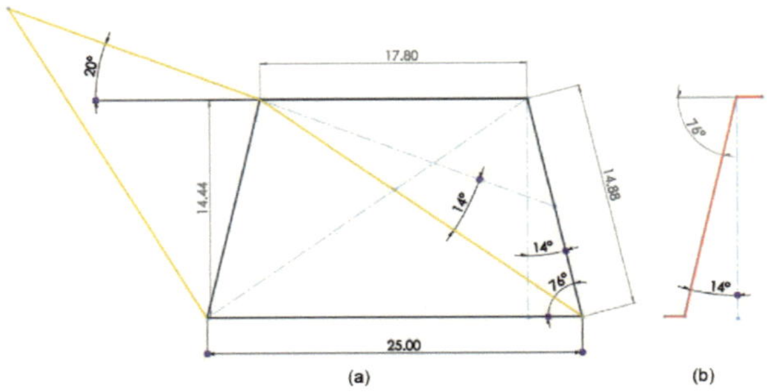

Figure 18.
Dimensions of the Stewart-Gough platform according to workspace. (a) Platform in front view with actuators parallel to each other. (b) Leg with spherical joints in side view.

This is how the values of link L = 15, the diameter of the platform Dp = 18, the diameter of the base Db = 25, and a height in the original position of 14.5 are finally obtained.

The parts that move the platform itself are called legs and are composed of a rod, threaded at each end, so that two kneecaps (upper and lower) are placed (**Figures 16–18**).

The lower ball joints connect each of the rods with their respective actuator (in this case it is the servomotor) by means of what will be referred to as horn or actuator arm. Servomotors are considered part of the "fixed base" or simply called a base [20].

To facilitate the reader's understanding, the parts just explained on the CAD model designed for this project are presented in **Figure 19**. The arms have a hole in their geometric center used to join each arm with the axis of their respective servomotor. At the end of the horn, the lower kneecaps are coupled to create a joint and join the horn with the rod (also called link), so that it allows the movement to be transmitted.

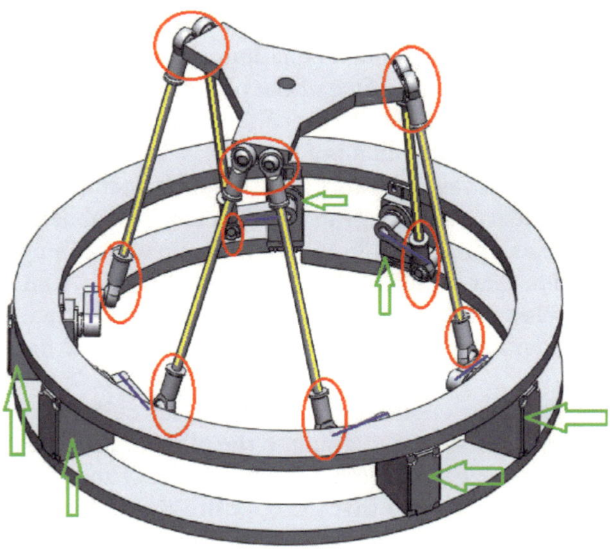

Figure 19.
Parts of the base. Ball joints (red ovals), rods (yellow lines), actuator arms or horns (blue line) and servomotors (green arrows).

Such movement is a conversion of angular movement to linear movement; this is a distinguishing feature of this project to the configuration that has the classic Stewart platform.

At the upper end of each of the rods, there is a joint to another kneecap, similar to the lower part of the rod.

On the other hand, there are a total of six other ball joints on the platform, connecting the closed kinematic chain to the platform. The platform is also often called the final effector, and this is the most interesting part of any serial or parallel robot.

Because the inverse kinematics is unique for parallel manipulators, the same inverse kinematics calculated on the Stewart platform [16] will be taken using the same views and changing some labels.

The coordinate systems $\{F\}$, of the base row, and $\{M\}$, of the final effector, will be renamed by $\{B\}$ (base) and $\{P\}$ (platform), respectively, and also the labels of the unions F_i and M_j for unions B_i and P_j, respectively. Taking into account the modifications explained in this paragraph, Eqs. (6, 7) presented in the theoretical framework will be as follows:

$$B_i = \begin{bmatrix} B_{Xi} \\ B_{ij} \\ B_{Zi} \end{bmatrix} = \begin{bmatrix} r_b \cos(\mu_i) \\ r_b \mathrm{sen}(\mu_i) \\ 0 \end{bmatrix} \quad (6)$$

$$P_i = \begin{bmatrix} P_{Ui} \\ P_{Vi} \\ P_{Wi} \end{bmatrix} = \begin{bmatrix} r_p \cos(\lambda_i) \\ r_p \mathrm{sen}(\lambda_i) \\ 0 \end{bmatrix} \quad (7)$$

Figure 20 presents the schematic diagram of the reference systems: Look at **Figure 21**, this indicates the joint positioning diagram, assuming that the *uv* plane of the reference system $\{P\}$ is parallel to the plane *xy*. Both coordinate systems ($\{P\}$ and $\{B\}$) are centered and separated by a vertical distance from the base, that is, on the z-axis (height of the platform or the final effector with respect to the coordinate system $\{B\}$): For the simulation process of external forces applied to the Stewart-Gough platform, the CAD design generated from the measurements was used, from which an estimate of the measurements for the base, platform, links, and horn of

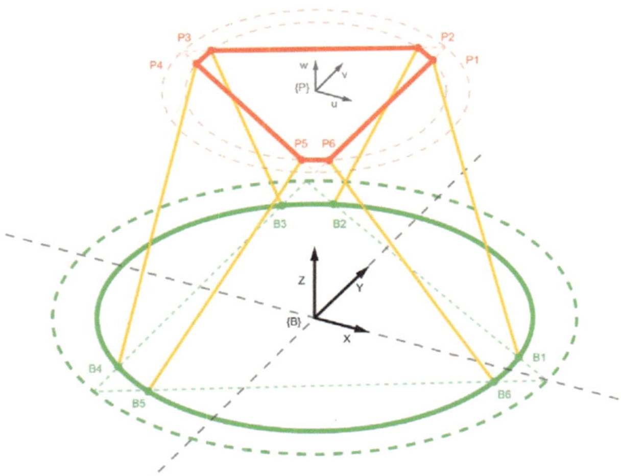

Figure 20.
Schematic diagram of new reference systems.

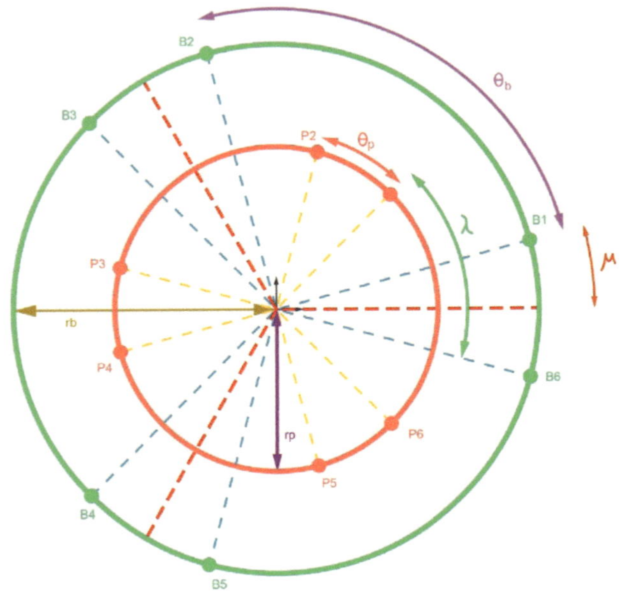

Figure 21.
Joint positioning diagram.

the servo. While a standard model was used for servomotors and spherical joints taking a CAD design from the public domain, within a platform called GRABCAD. For the generation of the design study, we worked with a CAD design software and finite element study that allowed us to observe some important aspects such as deformations, stress concentrators, and safety factor estimated for the platform in case of applying a force of overload in the final effector. Because the current work does not focus on the design of the platform as it is but of the controller for it, the pertinent calculations for a certain deformation or safety factor are not considered within the objectives. The following simulations were generated from an estimate of approximate real materials; this is because the suppliers of these products do not specify the material used.

To obtain results closer to reality, each element of the general assembly was assigned its respective material, as already mentioned, estimating the specific polymer or alloy of each metal. Starting with the bases and the final effector, in which case it was assigned an ABS plastic since, for practicality and speed, the options of laser cutting or 3D printing are currently considered, likewise, the servomotors are also considered in ABS. For the joints and links, the majority of suppliers handle metallic products for RC cars; therefore, a galvanized metal was assigned for its relative low cost and ease of machining. Finally, aluminum is mostly used for the servomotor horn, and therefore 1060 alloy aluminum is assigned.

Once each material is determined, the rest of the simulation conditions are defined starting with each restriction of the mechanism, and once all its position relationships are defined, the only geometric restriction will be the base on its lower face simulating that said platform is embedded on a regular horizontal surface. For the application of applied forces, an estimation of distributed load on the final effector is made considering that on average each actuator (depending on the manufacturer) can bear nominally 15 kg/cm of torque, and being the 3 cm servo horns, it is considered that each actuator supports a total of 5 kg at its end, for which a total force of 30 kg is applied. In the previous figure, it can be seen that within the mechanism in general, most of the effort is generated in the legs of the platform, while the largest almost imperceptible concentrator is at the junction between the leg and the servomotor horn, just where applies the highest torque due to the use of angular actuators. Despite this, it can be seen within the graph that there is no critical effort that may represent a risk of rupture (**Figure 22**).

For the following study, the deformation in each part of the platform is contemplated; note that in both the stress study and the deformation study, the elements corresponding to the base do not undergo significant changes, however, the platform and the links deform from considerable way especially the part of the final effector that can become almost completely deformed. However, it is worth mentioning that CAD design software mostly exaggerates the results obtained from the simulation; this is to clearly appreciate the effects of the applied conditions, since in a real case the materials would tend to fracture rather than deform due to its low ductility. In which case, the greatest deformation present is only 32.2 mm. From the previous points, the safety factor is studied, where you can clearly see the elements of the mechanism where it is applied without considering the elements of the base, which, as observed in previous studies, has no tendency to break due which is contemplated embedded in a fixed surface and reinforced with each other. Therefore, in the legs, which are important for the performance of the mechanism, a

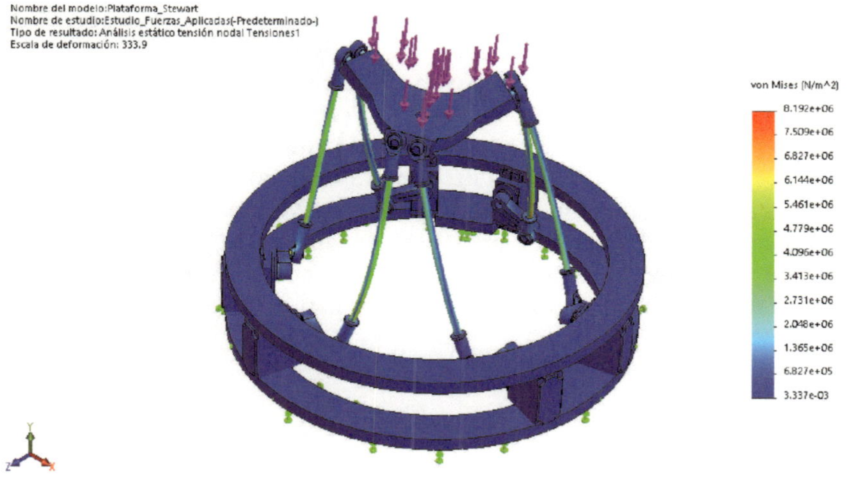

Figure 22.
von Mises stress study.

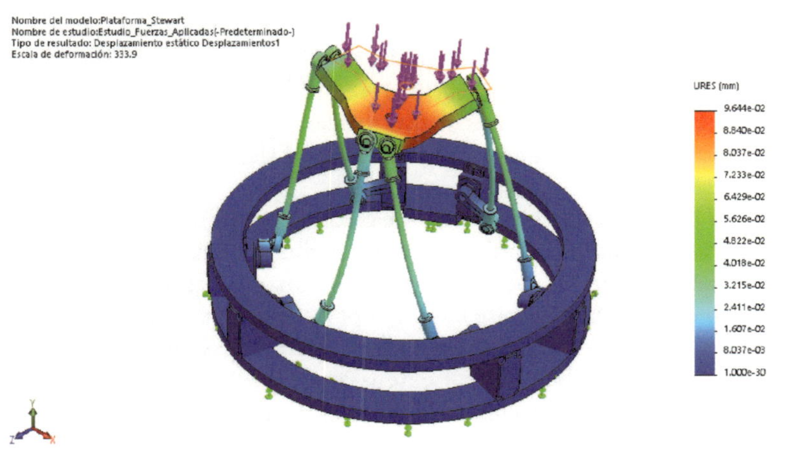

Figure 23.
Deformation study.

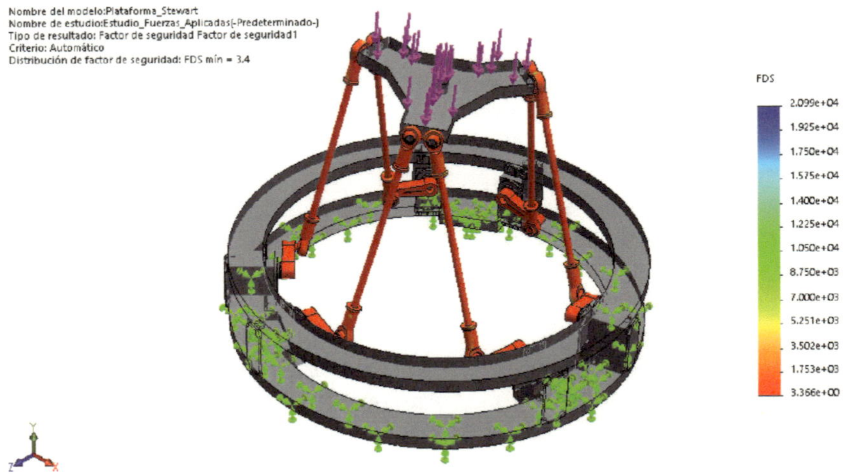

Figure 24.
Study of the safety factor.

safety factor of 3.4 is presented, which for the application that will be given to the Stewart-Gough platform is an acceptable value (**Figure 23**). Finally, the study of applied efforts is carried out in the situation where the platform has a maximum load applied to the resistive screen in one of its vertices; this is mainly contemplated (a little ahead of the selection of components) to know how much you can resist the screen to the point of rupture which in which case would be the element on the platform with greater fragility. As shown in **Figure 24**, the safety factor remains low at 0.45 and has a deformation of 1.12 cm, in which case, the CAD software deforms the part to demonstrate the displacement; however, in reality the panel would tend to fracture before having such deformation.

8. Conclusions

The present document set out the objectives and the description of the partial result; from this, a favorable product was obtained regarding the characteristics of the system and the definition of its components.

Throughout the document the different points that led to the determination or adaptation of certain elements or characteristics of the work to be carried out are observed, this (as initially mentioned in the problem statement) was subject to change as more in The theme with the purpose of bringing the system closer to the desired characteristics, is by means of this method that it was possible to determine the components that would allow the generation of the CAD design and the electronic design sketch.

One of the proposed objectives was the calculation of inverse kinematics. When investigating more about the subject, it was concluded that it is possible to use the same one as described in [1], because for parallel manipulators with the same configuration, it is unique. Despite having replaced the linear actuators with angular ones, it has no impact on the calculation of the same, since the distance of interest is the vector that describes the collinear extension between the axis of the servomotor and its respective point of coincidence on the platform (upper kneecap) as shown in **Figure 25**.

The maximum position and orientation parameters of a Stewart-Gough platform depend on the application given to this device. Since this project requires bringing a

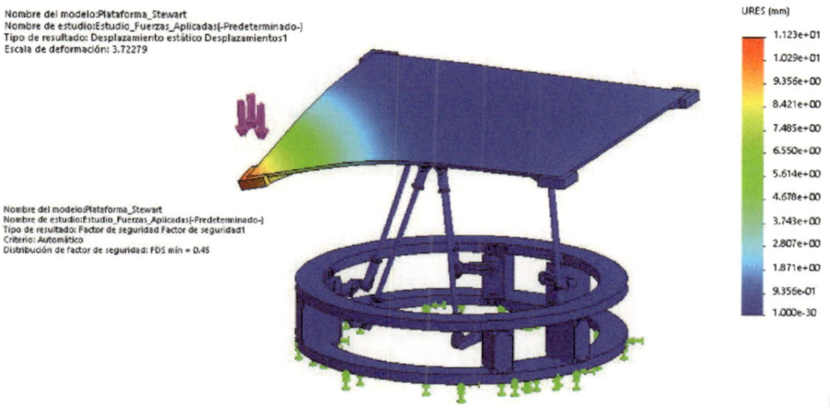

Figure 25.
Deformation and stress study of the Stewart-Gough platform applied to the resistive panel.

sliding object on the resistive panel to a desired position, it was determined that the suitable angle for this system is 20 for warping and 20 for pitching. To arrive at this characteristic, an estimation of measures was implemented based on the desired workspace, and the geometric and articular parameters were checked by simulations.

Author details

Flabio Dario Mirelez-Delgado*†, José Ronaldo Díaz-Paredes†
and Miguel Abraham Gallardo-Carreón†
Instituto Politcnico Nacional UPIIZ, Zacatecas, Mexico

*Address all correspondence to: fmirelezd@ipn.mx

† These authors contributed equally.

References

[1] Roca A. Control Automtico de Procesos Industriales. Madrid, Spain: Diaz de Santos; 2014

[2] Ogata K. Ingeniera de control moderna. Madrid: Pearson; 2010

[3] Yaovaja K. Ball Balancing on a Stewart Platform using Fuzzy Supervisory PID Visual Servo Control. 2018 5th International Conference on Advanced Informatics: Concept Theory and Applications (ICAICTA) [Internet]. IEEE; August 2018. Available from: http://dx.doi.org/10.1109/icaicta.2018.8541349

[4] Åström KJ. PID Control, de Control System Design. Santa Barbara, USA: CalTech; 2002. pp. 216-251

[5] Mazzone V. Controladores PID. de Automatizacin y Control Industrial: Quilmes, Argentina, Universidad Nacional de Quilmes; 2002. pp. 1-11

[6] Moudgalya KM. Digital Control. Wiley; 2007

[7] Jaimes LEG. Control Digital Teora y Prctica. Medelln, Colombia: Politcnico Colombiano Jaime Isaza Dadavid; 2009

[8] Ogata K. Sistemas de control en tiempo discreto. Minnesota: Prentice Hall; 1996

[9] Rodriguez DAR, Diseo de una plataforma robtica paralela de 6 DOF para asistente quirrgico en cirugas de reconstruccin craneo-facial; 2010

[10] Gwinnett JE. Amusement Device. Estados Unidos de America: Patentes; 1931

[11] Pollard W. Spray Painting Machine. Estados Unidos de America: Patentes; 1940

[12] Redifon, Another Redifon Flight Simulator Sold. Flight International, Crawley Sussex England; 1962

[13] Martnez JMR. Las diferentes formas del criterio de Kutzbach-Grbler. Facultad de Ingeniera Mecnica, Elctrica y Electrnica: Salamanca, Gto; 2015

[14] Barrientos A. Fundamentos de Robtica. New York, USA: McGraw-Hill; 2007

[15] Ximena N, Alvarez G. Estudio, diseo y construccin de una plataforma robtica didctica tipo stewart aplicada al estudio de controladores difusos. Cuenca, Ecuador: Universidad Politcnica Salesiana; 2011

[16] Acua HG, Dutra MS. Cinematica inversa y anlisis del espacio de trabajo de una plataforma. Stewart, Bucaramanga, Colombia: Universidad Autnoma de Bucaramanga; 2009

[17] Chatterjee B. Design of a Semi-Regular Stewart Platform Manipulator for a Desired Workspace. Bangalore, India: Mechanical Engineering Department, Indian Institute of Science; 2009

[18] Szufnarowski F. Stewart Platform with Fixed Rotary Actuators: A Low Cost Design Study. Bielefeld, Germany: E.E.U.U; 2018

[19] Lazatevic Z. Feasibility of a Stewart Platform with Fixed Actuators as a Platform for CABG Surgery Device. E.E. U.U: New York, USA: Columbia University Department of Bioengineering; 2003

[20] Fichter EF. A Stewart Platform-Based Manipulator: General Theory and Practical Construction. Oregon State University;

Chapter 7

Kinematics of Serial Manipulators

Ivan Virgala, Michal Kelemen and Erik Prada

Abstract

This book chapter deals with kinematic modeling of serial robot manipulators (open-chain multibody systems) with focus on forward as well as inverse kinematic model. At first, the chapter describes basic important definitions in the area of manipulators kinematics. Subsequently, the rigid body motion is presented and basic mathematical apparatus is introduced. Based on rigid body conventions, the forward kinematic model is established including one of the most used approaches in robot kinematics, namely the Denavit-Hartenberg convention. The last section of the chapter analyzes inverse kinematic modeling including analytical, geometrical, and numerical solutions. The chapter offers several examples of serial manipulators with its mathematical solution.

Keywords: algorithm, inverse kinematics, Jacobian, manipulator, optimization, redundant, robot

1. Introduction and basic definitions

In the following sections, this chapter will deal with direct and inverse kinematics of open-chain multibody systems consisting of rigid bodies. The whole problematics is analyzed from the view of robotics. Each manipulator or mechanism investigated in this chapter will be of serial kinematic structure (open chain).

Open-chain multibody systems are mechanically constructed by connecting a set of bodies, called links, by means of various types of joints. In general, the joints can be passive or active. The joints, which are moved by actuators, are active joints.

In general, from the view of robotics, there are two tasks in kinematics:

- Forward kinematics—the forward kinematics problem represents relationship between individual joints of investigated robot and end-effector.

- Inverse kinematics—the problem of inverse kinematics is as follows: given a desired configuration of end-effector of robot, find the joint angles that achieve that configuration.

Before these terms are explained and demonstrated by some study cases, we have to mention the basic definitions, necessary for the further analyses.

Degrees of freedom (DOF): is the smallest number of coordinates needed to represent the robot configuration. Thus, the number of DOF equals to the dimension of configuration space.

Joint space: Let us define all the joint variables in a vector $\mathbf{q} = [q_1, q_2, \dots, q_n]^T \in \mathbb{Q} \subset \mathbb{R}^N$. The set $\mathbb{Q}$ we call the so-called joint space and it contains all the possible values, which joint variables may acquire.

Workspace: Workspace is a subset of the Euclidean space $\mathbb{E}$, in which the robot executes its tasks. From the view of robotics, workspace is the set of all the points that mechanism may reach in Euclidean space $\mathbb{E}$ by end-effector. The workspace can be categorized as follows [1, 2]:

Maximal workspace—it is defined as locations that can be reached by end-effector at least with one orientation.

Inclusive-orientation workspace—it is defined as locations that can be reached by end-effector with at least one orientation among a range of orientations (maximal workspace is particular case).

Constant-orientation workspace—it is defined as location that can be reached by the end-effector with fixed orientation of joints.

Total-orientation workspace—it is defined as location that can be reached by the end-effector with any orientation.

Dexterity workspace—it is defined as location that can be reached by the end-effector with any orientation and without kinematic singularities.

Task space–space of positions and orientations of the end-effector frame. The workspace is a subset of task space that the end-effector frame can reach [3].

2. Rigid body motion

Rigid motion of an object is a motion that preserves distance between points [4]. Rigid body is a set of particles such that the distance between any two particles remains constant in time, regardless of any motions of the body or forces exerted on the body. If we consider $\mathbf{p}$ and $\mathbf{q}$ as two points on rigid body, while rigid body moves, $\mathbf{p}$ and $\mathbf{q}$ must satisfy $\|\mathbf{p}(t) - \mathbf{q}(t)\| = \|\mathbf{p}(0) - \mathbf{q}(0)\| = constant$, see **Figure 1**.

Let us consider an object, described as a subset O of $\mathbb{R}^3$. Then a motion of object (rigid body) is represented by mapping $f(t) : O \to \mathbb{R}^3$. This mapping describes how the points of this object move as a function of time, relative to some fixed coordinate system.

Let the inertial reference frame be $O = \{x_r, y_r, z_r\}$ and $\mathbf{i}_r, \mathbf{j}_r, \mathbf{k}_r$ represent unit vectors of the reference frame. The vector $\mathbf{p}$ can be expressed with respect to inertial reference frame $O = \{x_r, y_r, z_r\}$ by the following equation

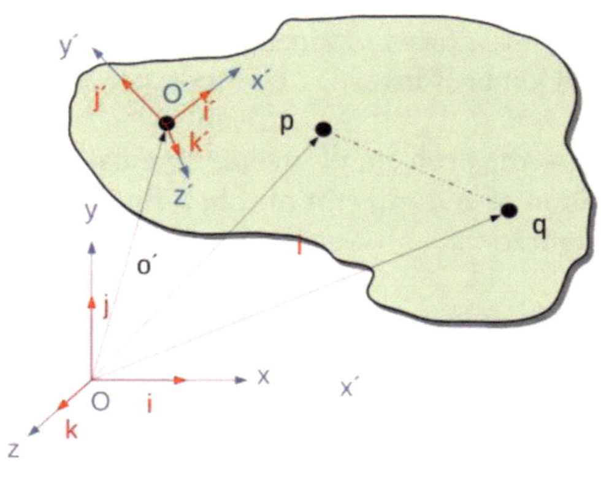

Figure 1.
Rigid body.

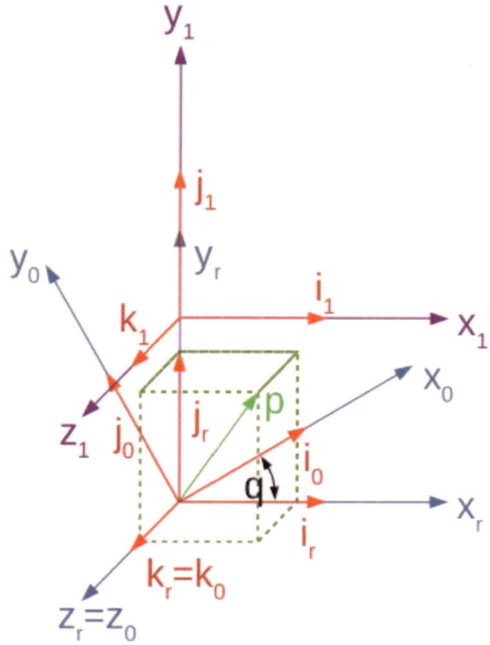

Figure 2.
Rotation of coordinate system.

$$\mathbf{p} = p_{xr}\mathbf{i}_r + p_{yr}\mathbf{j}_r + p_{zr}\mathbf{k}_r \tag{1}$$

where $\mathbf{p} = \left[p_x, p_y, p_z\right]^T \in \mathbb{R}^3$. Coordinates of vector $\mathbf{p}$ can be also expressed as its projections in directions of individual unit vectors as scalar product. In order to find the relation, the vector $\mathbf{p}$ needs to be expressed in coordinates $O_1 = \{x_1, y_1, z_1\}$

$$\begin{aligned} p_{xr} &= \mathbf{i}_r\,\mathbf{p} = \mathbf{i}_r p_{x1}\mathbf{i}_1 + \mathbf{i}_r p_{y1}\mathbf{j}_1 + \mathbf{i}_r p_{z1}\mathbf{k}_1 \\ p_{yr} &= \mathbf{j}_r\,\mathbf{p} = \mathbf{j}_r p_{x1}\mathbf{i}_1 + \mathbf{j}_r p_{y1}\mathbf{j}_1 + \mathbf{j}_r p_{z1}\mathbf{k}_1 \\ p_{zr} &= \mathbf{k}_r\,\mathbf{p} = \mathbf{k}_r p_{x1}\mathbf{i}_1 + \mathbf{k}_r p_{y1}\mathbf{j}_1 + \mathbf{k}_r p_{z1}\mathbf{k}_1 \end{aligned} \tag{2}$$

which can be rewritten in matrix form

$$\begin{bmatrix} p_{xr} \\ p_{yr} \\ p_{zr} \end{bmatrix} = \begin{bmatrix} \mathbf{i}_r\mathbf{i}_1 & \mathbf{i}_r\mathbf{j}_1 & \mathbf{i}_r\mathbf{k}_1 \\ \mathbf{j}_r\mathbf{i}_1 & \mathbf{j}_r\mathbf{j}_1 & \mathbf{j}_r\mathbf{k}_1 \\ \mathbf{k}_r\mathbf{i}_1 & \mathbf{k}_r\mathbf{j}_1 & \mathbf{k}_r\mathbf{k}_1 \end{bmatrix} \begin{bmatrix} p_{x1} \\ p_{y1} \\ p_{z1} \end{bmatrix} \tag{3}$$

that is $\mathbf{p}_b = \mathbf{R}_{r1}\mathbf{p}_1$. The meaning of this term is as follows. Coordinates of the vector $\mathbf{p}$ expressed in $O_1 = \{x_1, y_1, z_1\}$ are computed to $O = \{x_r, y_r, z_r\}$ so that they are left multiplied by transformation matrix $\mathbf{R}_{r1}$.

As can be seen in **Figure 2**, coordinate system x_0, y_0, z_0 is rotated with respect to coordinate system x_r, y_r, z_r by angle q around the axis z_r. By consideration of previous equations; and by consideration of the facts that scalar product of two perpendicular vectors equals zero, scalar product of two parallel unit vectors is one, and scalar product of concurrent unit vectors is $\cos\alpha$; and by assuming that $\cos\left(\frac{\pi}{2} \pm \alpha\right) = \mp\sin\alpha$, that is

$$\mathbf{i}^T\mathbf{i} = 1, \mathbf{j}^T\mathbf{j} = 1, \mathbf{k}^T\mathbf{k} = 1$$
$$\mathbf{i}^T\mathbf{j} = 0, \mathbf{j}^T\mathbf{k} = 0, \mathbf{k}^T\mathbf{i} = 0$$

one can obtain the following rotation matrix

$$\mathbf{R}_x = \begin{bmatrix} 1 & 0 & 0 \\ 0 & \cos\alpha & -\sin\alpha \\ 0 & \sin\alpha & \cos\alpha \end{bmatrix} \tag{4}$$

where $\mathbf{R}_x$ is a rotation matrix for rotation around the x-axis by angle α. Subsequently, rotation matrices can be also be expressed for rotation around y-axis and z-axis

$$\mathbf{R}_y = \begin{bmatrix} \cos\beta & 0 & \sin\beta \\ 0 & 1 & 0 \\ -\sin\beta & 0 & \cos\beta \end{bmatrix} \tag{5}$$

$$\mathbf{R}_z = \begin{bmatrix} \cos\gamma & -\sin\gamma & 0 \\ \sin\gamma & \cos\gamma & 0 \\ 0 & 0 & 1 \end{bmatrix} \tag{6}$$

Since rotation matrix $\mathbf{R}$ is an orthogonal matrix, for this reason

$$\mathbf{R}^T\mathbf{R} = \mathbf{I}_3 \tag{7}$$

where $\mathbf{I}_3$ is a 3×3 identity matrix. Considering the case when there is displacement of local coordinate system and at the same time also its rotation, it would be expressed as

$$\begin{bmatrix} x_r \\ y_r \\ z_r \end{bmatrix} = \mathbf{R}_{axis,angle} \begin{bmatrix} x_1 \\ y_1 \\ z_1 \end{bmatrix} + \begin{bmatrix} p_x \\ p_y \\ p_z \end{bmatrix} \tag{8}$$

Eq. (8) represents a system of three equations, which will be extended by fourth equation 1 = 0 + 0 + 0 + 1, which is

$$\begin{bmatrix} x_r \\ y_r \\ z_r \\ 1 \end{bmatrix} = \begin{bmatrix} & & & \vdots & p_x \\ & \mathbf{R}_{axis,angle} & & \vdots & p_y \\ \cdots & & \cdots & \vdots & p_z \\ 0 & & 0 & 0 & 1 \end{bmatrix} \begin{bmatrix} x_1 \\ y_1 \\ z_1 \\ 1 \end{bmatrix} \tag{9}$$

3. Forward kinematics

The forward kinematic model determines the position and orientation of the end-effector relating to base frame of the mechanism or to global coordinate system (GCS).

3.1 Open kinematic chain

We will focus on robots, which contain a set of links connected together by joints. The joints are usually revolute or prismatic or they can be more complex, such as socket joint or ball joint [5]. Within this chapter will be considered only revolute and prismatic joints, which have only a single degree-of-freedom motion. Let us consider a mechanism with N links connected together by $N-1$ joints. The i-th joint connects link $i-1$ to link i. The number of the joints starts with 1 and ends with $N-1$. The next consideration for the following mathematical model is that the first link is connected to the base fixed to inertial reference frame, while the last link is free and able to move.

The i-th joint is associated with joint variable q_i, while q_i may contain θ_i and d_i for revolute and prismatic joints, respectively. The local coordinate frame is attached to each link, so to i-th link is attached to I_i frame, $I_i = \{O_i, x_i, y_i, z_i\}$. When a mechanism performs any motion in its workspace, the coordinates of each point on i-th link are constant with respect to their coordinate frame $I_i = \{O_i, x_i, y_i, z_i\}$.

Let $\mathbf{A}_i$ be a homogeneous transformation matrix, which holds position and orientation of frame $I_i = \{O_i, x_i, y_i, z_i\}$ with respect to $I_{i-1} = \{O_{i-1}, x_{i-1}, y_{i-1}, z_{i-1}\}$. It should be noticed that values of matrix $\mathbf{A}_i$ are not constant, but they change with changing configuration of the mechanism. In general, a homogeneous transformation matrix expressing the position and orientation of $I_j = \{O_j, x_j, y_j, z_j\}$ with respect to $I_i = \{O_i, x_i, y_i, z_i\}$ is called a transformation matrix ${}^i\mathbf{T}_j$. We can also define the following matrix

$$\mathbf{H} = \begin{bmatrix} {}^0\mathbf{R}_n & {}^0\mathbf{o}_n \\ 0 & 1 \end{bmatrix} \tag{10}$$

where ${}^0\mathbf{R}_n$ is a 3×3 rotation matrix with and ${}^0\mathbf{o}_n$ is a 3×1 vector expressing position and orientation of end-effector (the last point of mechanism) with respect to inertial reference frame (base of mechanism). Eq. (10) can then be written as

$$\mathbf{H} = {}^0\mathbf{T}_n = \prod_{i=1}^{N} \mathbf{A}_i \tag{11}$$

while $\mathbf{A}_i$ equals

$$\mathbf{A}_i = \begin{bmatrix} {}^{i-1}\mathbf{R}_i & {}^{i-1}\mathbf{o}_i \\ 0 & 1 \end{bmatrix} \tag{12}$$

3.2 Denavit–Hartenberg convention

For the computation of forward kinematics for open-chain robot according to Eq. (11), a general approach was derived in order to determine the relative position and orientation of two consecutive links. This approach determines two frames attached to two links (rigid bodies) and computes the coordinate transformations between them [6].

For utilization of the Denavit-Hartenberg convention, some rules need to be observed. Let us consider **Figure 3**. Let axis i represent the axis connecting link $i-1$ and link $i+1$. In order to define link frame i, the procedure is as follows. First of all, the axis z_i and axis z_{i-1} are chosen. Next, origin O_i is located at the intersection of axis z_i with the common normal to axes z_i and z_{i-1}. By this step be get points O_i and O_i'. The common normal of these two axes is a minimum distance between them. Subsequently, the axis x_i is chosen along the common normal to axes z_{i-1} and z_i in

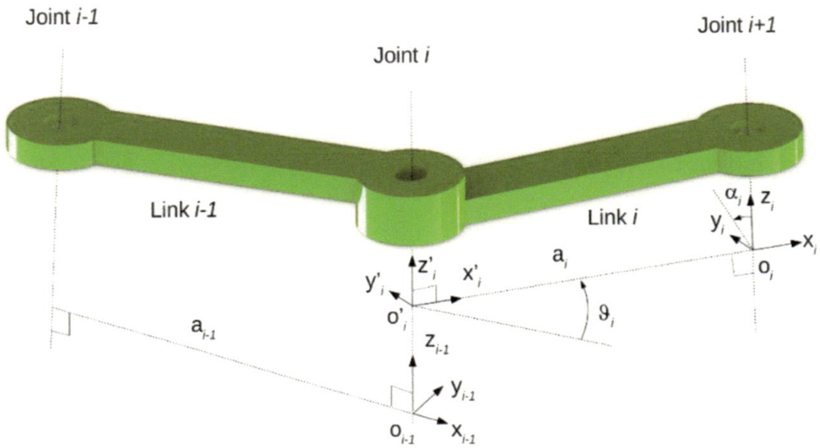

Figure 3.
Denavit-Hartenberg approach.

the direction from joint i to the joint $i+1$. In the last step, axis y_i is chosen so as to complete a right-handed frame.

After these steps, the link frames have been established and now the position and orientation of frame i with respect to frame $i-1$ can be determined by following DH parameters [7]:

- a_i: distance between the points O_i and O_i'
- d_i: distance between O_{i-1} and O_i' along the axis z_{i-1}
- α_i: angle between the axes z_{i-1} and z_i about axis x_i (positive direction—counter-clockwise rotation)
- ϑ_i: angle between axes x_i and x_{i-1} (positive direction—counter-clockwise rotation)

It should be also noted that parameters a_i and α_i are always constant, because they depend on the geometric aspect of mechanism. Considering the two other parameters d_i and ϑ_i, depending on the joint type, one is constant and other one may change as follows:

- Revolute joint: ϑ_i is the joint variable and d_i is constant
- Prismatic joint: d_i is the joint variable and ϑ_i is constant

In general, six parameters are necessary in order to describe the position and orientation of a rigid body in the 3D space. Based on previously mentioned facts, we can say about DH convention that only four parameters are required by assuming that the axis x_i intersects z_{i-1}, and that axis x_i is perpendicular to z_{i-1}.

3.2.1 Example of forward kinematics using the Denavit-Hartenberg convention

Let us consider some kind of industrial robot, namely SCARA (Selective Compliance Assembly Robot Arm) robot, which has RRP structure. Its kinematic structure is shown in **Figure 4**.

Considering the basic principles of the Denavit-Hartenberg convention introduced in the previous section, we are able to introduce D-H parameters, see **Table 1**.

Based on DH parameters, which are obvious from **Figure 4**, particular homogeneous transformation matrices can be established.

$$ {}^{0}\mathbf{A}_1 = \begin{bmatrix} 1 & 0 & 0 & 0 \\ 0 & 1 & 0 & 0 \\ 0 & 0 & 1 & d_1 \\ 0 & 0 & 0 & 1 \end{bmatrix} \begin{bmatrix} cos\left(q_1\right) & -sin\left(q_1\right) & 0 & 0 \\ sin\left(q_1\right) & cos\left(q_1\right) & 0 & 0 \\ 0 & 0 & 1 & 0 \\ 0 & 0 & 0 & 1 \end{bmatrix} \begin{bmatrix} 1 & 0 & 0 & L_1 \\ 0 & 1 & 0 & 0 \\ 0 & 0 & 1 & 0 \\ 0 & 0 & 0 & 1 \end{bmatrix} \tag{13} $$

$$ {}^{0}\mathbf{A}_1 = \begin{bmatrix} cos\left(q_1\right) & -sin\left(q_1\right) & 0 & L_1 cos\left(q_1\right) \\ sin\left(q_1\right) & cos\left(q_1\right) & 0 & L_1 sin\left(q_1\right) \\ 0 & 0 & 1 & d_1 \\ 0 & 0 & 0 & 1 \end{bmatrix} $$

$$ {}^{1}\mathbf{A}_2 = \begin{bmatrix} cos\left(q_2\right) & -sin\left(q_2\right) & 0 & 0 \\ sin\left(q_2\right) & cos\left(q_2\right) & 0 & 0 \\ 0 & 0 & 1 & 0 \\ 0 & 0 & 0 & 1 \end{bmatrix} \begin{bmatrix} 1 & 0 & 0 & L_2 \\ 0 & 1 & 0 & 0 \\ 0 & 0 & 1 & 0 \\ 0 & 0 & 0 & 1 \end{bmatrix} \tag{14} $$

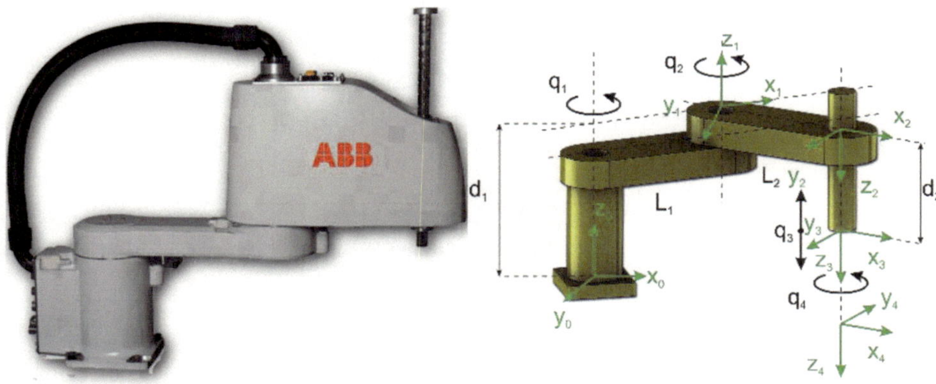

Figure 4.
SCARA robot.

Link	a_i	α_i	d_i	ϑ_i
1	L_1	0	d_1	q_1
2	L_2	0	0	q_2
3	0	π	$d_2 + q_3$	0
4	0	0	0	q_4

Table 1.
Denavit-Hartenberg parameters.

$$
{}^{1}\mathbf{A}_2 = \begin{bmatrix} cos\left(q_2\right) & -sin\left(q_2\right) & 0 & L_2 cos\left(q_2\right) \\ sin\left(q_2\right) & cos\left(q_2\right) & 0 & L_2 sin\left(q_2\right) \\ 0 & 0 & 1 & 0 \\ 0 & 0 & 0 & 1 \end{bmatrix}
$$

$$
{}^{2}\mathbf{A}_3 = \begin{bmatrix} 1 & 0 & 0 & 0 \\ 0 & cos\left(\pi\right) & -sin\left(\pi\right) & 0 \\ 0 & sin\left(\pi\right) & cos\left(\pi\right) & d_2 + q_3 \\ 0 & 0 & 0 & 1 \end{bmatrix} \tag{15}
$$

$$
{}^{3}\mathbf{A}_4 = \begin{bmatrix} cos\left(q_4\right) & -sin\left(q_4\right) & 0 & 0 \\ sin\left(q_4\right) & cos\left(q_4\right) & 0 & 0 \\ 0 & 0 & 1 & 0 \\ 0 & 0 & 0 & 1 \end{bmatrix} \tag{16}
$$

So, the final transformation matrix is

$$
{}^{0}\mathbf{T}_4 = {}^{0}\mathbf{A}_2{}^{1}\mathbf{A}_2{}^{2}\mathbf{A}_3{}^{3}\mathbf{A}_4 \tag{17}
$$

$$
{}^{0}\mathbf{T}_4 = \begin{bmatrix} & \mathbf{R} & & p_x \\ & & & p_y \\ & & & p_z \\ 0 & 0 & 0 & 1 \end{bmatrix} \tag{18}
$$

By vector ${}^{0}\mathbf{o}_4 = \left[p_x p_y p_z\right]^T$ is defined position of end-effector of SCARA manipulator with respect to its base inertial reference frame.

4. Inverse kinematics

The solution exists only if the given end-effector position and orientation are in dexterous workspace of the solved mechanism. While the forward kinematic model is expressed as

$$
\mathbf{x} = \mathbf{f}(\mathbf{q}) \tag{19}
$$

where $\mathbf{f}$ is a function defined between joint space $\mathbb{R}^n$ and workspace $\mathbb{R}^m$, which maps the joint position variables $\mathbf{q} \in \mathbb{R}^n$ to the position/orientation of the end-effector of mechanism, the inverse kinematic model is based on

$$
\mathbf{q} = \mathbf{f}^{-1}(\mathbf{x}) \tag{20}
$$

where $\mathbf{q} \in \mathbb{R}^n$ and $\mathbf{x} \in \mathbb{R}^m$. In the case of the forward kinematic model, end-effector position and orientation are computed for various kinds of mechanisms like manipulators, in a unique manner, for example, by above-mentioned transformation matrices. The inverse kinematic problem is more complex and finding the solution could be in many cases very complicated. While forward kinematics has a closed-form solution, an inverse kinematics in most cases does not have a closed-form solution. A forward kinematic model has a unique solution, while an inverse kinematic model may have multiple solutions or infinite number of solutions, especially for kinematically redundant mechanisms. In order to obtain a closed-form solution, there are two main approaches, namely algebraic approach and geometric approach.

4.1 Closed-form solution of inverse kinematics

Let us consider a two-link mechanism moving in the 2D plane, see **Figure 5**.

Considering the forward kinematic model, while the angles of joints ϑ_1 and ϑ_2 are given, the aim is to find the position of end-effector $\mathbf{x}_E = [x\,y]^T \in \mathbb{R}^m$. The forward kinematic model can be easily determined by the following equations

$$x = l_1 \cos \vartheta_1 + l_2 \cos (\vartheta_1 + \vartheta_2) \tag{21}$$

$$y = l_1 \sin \vartheta_1 + l_2 \sin (\vartheta_1 + \vartheta_2) \tag{22}$$

Now, the inverse kinematic problem is to find angles ϑ_1 and ϑ_2, while the end-effector position x and y are given by vector $\mathbf{x}_E = [x\,y]^T \in \mathbb{R}^m$.

$$c = x^2 + y^2 \tag{23}$$

$$\alpha = atan2(y, x) \tag{24}$$

$$\vartheta_2 = \pm arccos \left(\frac{x^2 + y^2 - L_1^2 - L_2^2}{2L_1 L_2} \right) \tag{25}$$

$$\beta = arccos \left(\frac{L_1^2 - L_2^2 + c^2}{2L_1 c} \right) \tag{26}$$

$$\vartheta_1 = \alpha - \beta \tag{27}$$

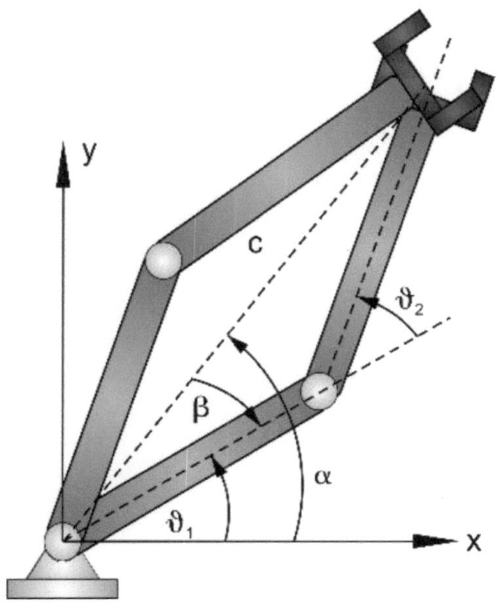

Figure 5.
Inverse kinematics solution for two-link mechanism.

As can be seen in **Figure 5**, the presented configuration of the mechanism has two solutions in inverse kinematics. These two solutions are based on signum in Eqs. (25) and (27).

5. Differential kinematics

5.1 Analytical Jacobian

In order to describe the relation between joint angles and end-effector configuration, often the relation between the joint and end-effector velocities is used. Let us consider a set of coordinates $\mathbf{x} \in \mathbb{R}^m$, their velocity is $\dot{\mathbf{x}} = d\mathbf{x}/dt \in \mathbb{R}^m$. Then, we can apply Eq. (19). Then, one obtains

$$\dot{\mathbf{x}} = \frac{\partial \mathbf{f}(\mathbf{q})}{\partial \mathbf{q}} \frac{d\mathbf{q}}{dt} = \mathbf{J}(\mathbf{q})\dot{\mathbf{q}} \tag{28}$$

$$\mathbf{J}(\mathbf{q}) = \begin{bmatrix} \frac{\partial x_1}{\partial q_1} & \cdots & \frac{\partial x_1}{\partial q_n} \\ \vdots & \ddots & \vdots \\ \frac{\partial x_m}{\partial q_1} & \cdots & \frac{\partial x_m}{\partial q_n} \end{bmatrix} \tag{29}$$

where $\mathbf{J}(\mathbf{q}) \in \mathbb{R}^{m \times n}$ is the analytical Jacobian matrix, which is very often used in kinematics and dynamics of robotic systems. Jacobian reflects differences between joint and configurations space of the investigated mechanism. In robotics, Jacobian is often used for several purposes such as for the definition of the relation between joint and configuration space, definition of the relation between forces/torques between spaces, the study of kinematic singularities, the definition of numerical solution for inverse kinematic problem, and the study of manipulability properties. We can look at Jacobian from a different perspective. Particular Jacobian columns represent the influence of i-th joint on the end-effector velocity.

The following example will demonstrate the derivation of analytical Jacobian for a three-link mechanism (**Figure 6**).

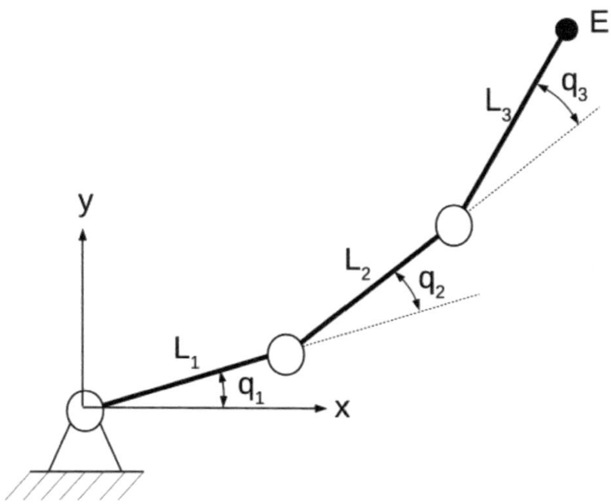

Figure 6.
Three-link mechanism.

In order to utilize Eq. (29), we need to define position of point $\mathbf{E} = \left[x_E y_E\right]^T \in \mathbb{R}^2$.

$$x_E = L_1 \cos q_1 + L_2 \cos\left(q_1 + q_2\right) + L_3 \cos\left(q_1 + q_2 + q_3\right) \tag{30}$$

$$y_E = L_1 \sin q_1 + L_2 \sin\left(q_1 + q_2\right) + L_3 \sin\left(q_1 + q_2 + q_3\right) \tag{31}$$

Assuming Eqs. (30) and (31), Jacobian according to Eq. (29) will be matrix $\mathbf{J}(\mathbf{q}) \in \mathbb{R}^{2\times 3}$

$$\mathbf{J}(\mathbf{q}) = \begin{bmatrix} \dfrac{\partial x_E}{\partial q_1} & \dfrac{\partial x_E}{\partial q_2} & \dfrac{\partial x_E}{\partial q_3} \\ \dfrac{\partial y_E}{\partial q_1} & \dfrac{\partial y_E}{\partial q_2} & \dfrac{\partial y_E}{\partial q_3} \end{bmatrix} \tag{32}$$

where the elements of the Jacobian matrix are

$$\frac{\partial x_E}{\partial q_1} = -L_1 sin\left(q_1\right) - L_2 sin\left(q_1 + q_2\right) - L_3 sin\left(q_1 + q_2 + q_3\right)$$

$$\frac{\partial x_E}{\partial q_2} = -L_2 sin\left(q_1 + q_2\right) - L_3 sin\left(q_1 + q_2 + q_3\right)$$

$$\frac{\partial x_E}{\partial q_3} = -L_3 sin\left(q_1 + q_2 + q_3\right)$$

$$\frac{\partial y_E}{\partial q_1} = L_1 cos\left(q_1\right) + L_2 cos\left(q_1 + q_2\right) + L_3 cos\left(q_1 + q_2 + q_3\right)$$

$$\frac{\partial y_E}{\partial q_2} = L_2 cos\left(q_1 + q_2\right) + L_3 cos\left(q_1 + q_2 + q_3\right)$$

$$\frac{\partial y_E}{\partial q_3} = L_3 cos\left(q_1 + q_2 + q_3\right)$$

5.2 Geometric Jacobian

Besides analytically expressing the Jacobian, we can express it by a geometric approach. To establish function $\mathbf{f}(\mathbf{q})$ in closed-form, a symbolic formalism is necessary, which could be difficult from the view of implementation. For this reason, a different way of Jacobian expression, the so-called geometric Jacobian, was developed. The geometric Jacobian can be obtained by consideration of rotational velocity vector $\boldsymbol{\omega}$. Let us consider link according to **Figure 6**. The Jacobian can be expressed as

$$\mathbf{J} = \begin{bmatrix} \mathbf{J}_v \\ \mathbf{J}_\omega \end{bmatrix} = \begin{bmatrix} \mathbf{J}_{v1} & \cdots & \mathbf{J}_{vn} \\ \mathbf{J}_{\omega 1} & \cdots & \mathbf{J}_{\omega n} \end{bmatrix} \tag{33}$$

The first term expresses the effect of $\dot{q}_1$ on linear velocity $\mathbf{v}$ and the second term expresses the effect on the rotational velocity $\boldsymbol{\omega}$. Thus,

$$\begin{aligned} \mathbf{v} &= \mathbf{J}_{v1}\dot{q}_1 + \ldots + \mathbf{J}_{vn}\dot{q}_n \\ \boldsymbol{\omega} &= \mathbf{J}_{\omega 1}\dot{q}_1 + \ldots + \mathbf{J}_{\omega n}\dot{q}_n \end{aligned} \tag{34}$$

That is, the analytical Jacobian differs from the geometrical Jacobian for the rotational part. Considering the revolute joint, the i-th column of Jacobian can be computed as

$$\begin{bmatrix} \mathbf{J}_{vi} \\ \mathbf{J}_{\omega i} \end{bmatrix} = \begin{bmatrix} {}^0\mathbf{z}_{i-1} \times \left({}^0\mathbf{p}_n - {}^0\mathbf{p}_{i-1}\right) \\ {}^0\mathbf{z}_{i-1} \end{bmatrix} \tag{35}$$

For prismatic joint, the i-th column of Jacobian can be computed as

$$\begin{bmatrix} \mathbf{J}_{vi} \\ \mathbf{J}_{\omega i} \end{bmatrix} = \begin{bmatrix} {}^0\mathbf{z}_{i-1} \\ \mathbf{0} \end{bmatrix} \tag{36}$$

where ${}^0\mathbf{p}_n$ is the end-effector position defined in transformation matrix ${}^0\mathbf{T}_n$ defined in the previous section. Next, ${}^0\mathbf{p}_{i-1}$ is the position of frame $I_{i-1} = \{O_{i-1}, x_{i-1}, y_{i-1}, z_{i-1}\}$, defined in transformation matrix ${}^0\mathbf{T}_{i-1}$. Finally, ${}^0\mathbf{z}_{i-1}$ is the third column of rotation matrix ${}^0\mathbf{R}_{i-1}$, while ${}^0\mathbf{R}_{i-1} = {}^0\mathbf{R}_1(q_1){}^1\mathbf{R}_2(q_2) \ldots {}^{i-2}\mathbf{R}_{i-1}(q_{i-1})$.

The following example will demonstrate the derivation of geometric Jacobian for a two-link mechanism (**Figure 7**).

$$\mathbf{J}(\mathbf{q}) = \begin{bmatrix} \mathbf{z}_0 \times (\mathbf{p}_2 - \mathbf{p}_0) & \mathbf{z}_1 \times (\mathbf{p}_2 - \mathbf{p}_1) \\ \mathbf{z}_0 & \mathbf{z}_1 \end{bmatrix} \tag{37}$$

where $\mathbf{p}_0 = \begin{bmatrix} 0 \\ 0 \\ 0 \end{bmatrix}$, $\mathbf{p}_1 = \begin{bmatrix} L_1 \cos q_1 \\ L_1 \sin q_1 \\ 0 \end{bmatrix}$, $\mathbf{p}_2 = \begin{bmatrix} L_1 \cos q_1 + L_2 \cos(q_1 + q_2) \\ L_1 \sin q_1 + L_2 \sin(q_1 + q_2) \\ 0 \end{bmatrix}$ and rotational axes are $\mathbf{z}_0 = \mathbf{z}_1 = \begin{bmatrix} 0 \\ 0 \\ 1 \end{bmatrix}$. By solving Eq. (37), we get

$$\mathbf{J}(\mathbf{q}) = \begin{bmatrix} -L_1 \sin q_1 - L_2 \sin(q_1 + q_2) & -L_2 \sin(q_1 + q_2) \\ L_1 \cos q_1 + L_2 \cos(q_1 + q_2) & L_2 \cos(q_1 + q_2) \\ 0 & 0 \\ 0 & 0 \\ 0 & 0 \\ 1 & 1 \end{bmatrix} \tag{38}$$

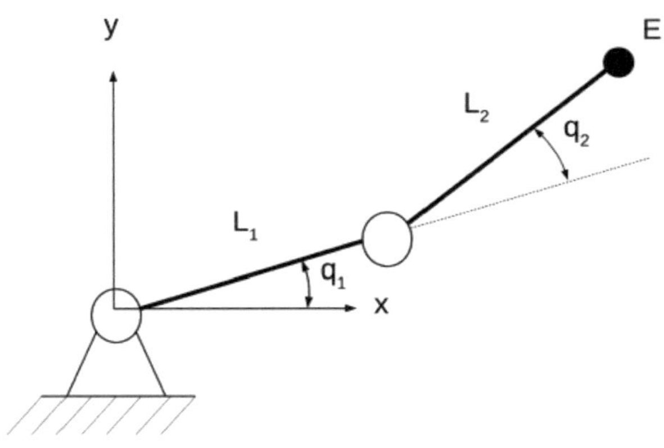

Figure 7.
Two-link mechanism.

Now, the Jacobian is a 6×2 matrix and its maximum rank is 2. That is, at most two components of angular/linear end-effector velocity can be independently assigned. In this case, when orientation is not required, only first two rows are considered.

5.3 Kinematically redundant manipulators

The next approach to inverse kinematic solution we want to focus on is numerical solutions. Nevertheless, many times, it is hard to find a closed-form solution for inverse kinematics; the basic kinematics of industrial robots have developed an approach to solve it. The problems arise with nonconventional kinematic structures, especially with kinematically redundant manipulators. A kinematically redundant manipulator has more number of DOFs than is absolutely necessary to perform the desired task. For example, conventional industrial robot has usually six DOFs, by which it is able to reach any point in its workspace. By adding an additional DOF, this robot becomes kinematically redundant due to this additional DOF.

A numerical solution is usually used when a closed-form solution for $\mathbf{q}$ does not exist or is difficult to find. In this section, we will focus on kinematically redundant mechanisms. Considering the dimension of joint space n and dimension of task space m, for kinematically redundant mechanisms $n > m$. The level of redundancy can be expressed by $r = n - m$. Kinematic redundancy is used for many tasks such as kinematic singularities avoidance, obstacle avoidance, joint limits avoidance, increasing the manipulability in specified directions, minimizing the energy consumption, minimum of motion torques, optimizing execution time, etc. As can be seen, kinematic redundancy allows many optimization tasks to be solved. On the other hand, kinematic redundancy brings some disadvantages as well; for example, a greater structural complexity of construction caused by many of DOFs (mechanical, actuators, sensors), which have an influence on final cost of this kind of mechanism. Next field of potential disadvantages is the field of control, due to complicated algorithms for inverse kinematic computation or motion control. From this reason redundant manipulators could be difficult in real-time control.

There are many approaches within numerical solution of inverse kinematics, which are still in focus of research. Most approaches deal with Jacobian matrix in order to find a linear approximation to the inverse kinematic problem. Among the most used of them are damped least squares (DLSs), Jacobian transpose, and damped least squares with singular value decomposition (SVD) [8, 9].

Another kind of approach is the approach based on Newton methods [6]. The aim of these algorithms is to find the final configuration of joints with focus on minimization problem. For this reason, the final motion of robot is smooth. This family of methods includes methods such as Powell's method [10] or Broyden, Fletcher, Goldfarb and Shanno (BFGS).

A very well-known and used method for inverse kinematics of kinematically redundant mechanisms is the so-called cyclic coordinate descent (CCD) algorithm [11]. The CCD method is a very simple and at the same time a very strong method. It is a heuristic iterative method with low computational cost per iteration. Next very know heuristic iterative method is FABRIK (Forward And Backward Reaching Inverse Kinematics) [12, 13]. The FABRIK method minimizes the system error by adjusting each joint angle one at a time.

Most of inverse kinematics numerical methods could be divided into two classes: linearization algorithms and minimization algorithms. Concerning the linearization algorithms, the idea is piecewise linearization of nonlinear inverse kinematic problem, which is based on the Jacobian matrix. An example of this kind of method is the Jacobian transpose method. In minimization algorithms, the idea is to formulate some cost function, which will be minimized, for example cyclic coordinate descent algorithm. Besides the mentioned methods, there are many other such as pseudoinverse methods, such as the Levenberg–Marquardt damped least squares methods, quasi-Newton and conjugate gradient methods, neural network and artificial intelligence methods.

The basic technique is based on Eq. (39)

$$\dot{\mathbf{x}} = \mathbf{J}(\mathbf{q})\dot{\mathbf{q}} \tag{39}$$

The above relation can be inverted to so-called Jacobian control method

$$\dot{\mathbf{q}} = \mathbf{J}^{\dagger}(\mathbf{q})\dot{\mathbf{x}} \tag{40}$$

which leads to joint velocity vector $\mathbf{q}$ with minimum norm. The term $\mathbf{J}^{\dagger}$ represents the Moore-Penrose pseudoinverse given by $\mathbf{J}^{\dagger} = \mathbf{J}^{T}\left(\mathbf{J}\mathbf{J}^{T}\right)^{-1}$ for kinematically redundant mechanisms where $m < n$ $\left(\mathbf{J}\mathbf{J}^{\dagger} = \mathbf{I}\right)$ or by $\mathbf{J}^{\dagger} = \left(\mathbf{J}^{T}\mathbf{J}\right)^{-1}\mathbf{J}^{T}$ for $m > n$ $\left(\mathbf{J}^{\dagger}\mathbf{J} = \mathbf{I}\right)$.

A very common method on which the solution is based is the Newton-Rhaphson method. The Newton-Rhaphson method is a root-finding algorithm that produces approximations of the roots of a real-valued function. The method starts with differentiable function f defined for a real variable x, derivative of function f', and initial guess x_0 for a root of function f. If these assumptions are satisfied and initial guess is close, then

$$x_1 = x_0 - \frac{f(x_0)}{f'(x_0)} \tag{41}$$

Talking about numerical motion optimization of kinematically redundant mechanisms, there are two approaches. The first approach deals with local methods, which are solved typically online, and the second one with global methods, which require quantity of computation. For this reason, the global methods are computed usually offline.

One of the commonly used local methods is the family of null-space methods. This method uses the extension of Eq. (40) and gives

$$\dot{\mathbf{q}} = \mathbf{J}^{\dagger}(\mathbf{q})\dot{\mathbf{x}} + \left(\mathbf{I} - \mathbf{J}^{\dagger}\mathbf{J}\right)\dot{\mathbf{q}}_0 \tag{42}$$

where $\dot{\mathbf{q}}_0 \in \mathbb{R}^n$ is an arbitrary joint space velocity vector, chosen according to desired behavior; so it is chosen for optimization purposes. Next, $\mathbf{I} \in \mathbb{R}^{\boldsymbol{n}\times\boldsymbol{n}}$ is the identity matrix. The term $\left(\mathbf{I} - \mathbf{J}^{\dagger}\mathbf{J}\right)$ represents the orthogonal projection matrix in the null space of $\mathbf{J}$, $\mathbf{J}^{\dagger}(\mathbf{q})\dot{\mathbf{x}}$ is orthogonal to $\left(\mathbf{I} - \mathbf{J}^{\dagger}\mathbf{J}\right)\dot{\mathbf{q}}_0$. For this reason, $\mathbf{q} = 0$. Physically, this term corresponds to self-motion, where the combined joints motion generates no motion in the task space (no motion of end-effector). So, the term $\left(\mathbf{I} - \mathbf{J}^{\dagger}\mathbf{J}\right)$ is symmetric and idempotent $\left(\left(\mathbf{I} - \mathbf{J}^{\dagger}\mathbf{J}\right)^2 = \left(\mathbf{I} - \mathbf{J}^{\dagger}\mathbf{J}\right)\right)$. Also, it ensures $\left(\mathbf{I} - \mathbf{J}^{\dagger}\mathbf{J}\right)^{\dagger} = \left(\mathbf{I} - \mathbf{J}^{\dagger}\mathbf{J}\right)$. The inverse kinematic solution expressed by Eq. (42) is equivalent to solving a quadratic programming (QP) problem based on $\mathbf{H}(\dot{\mathbf{q}}) = \min_{\dot{\mathbf{q}}} \frac{1}{2}\left(\dot{\mathbf{q}} - \dot{\mathbf{q}}_0\right)^{T}\mathbf{W}\left(\dot{\mathbf{q}} - \dot{\mathbf{q}}_0\right)$ subjected to $\dot{\mathbf{x}} = \mathbf{J}(\mathbf{q})\dot{\mathbf{q}}$.

Now the question is, how the vector $\dot{\mathbf{q}}_0$ can be chosen. One of the basic ways to choose it is the so-called projected gradient method $\dot{\mathbf{q}}_0 = \nabla_q\mathbf{H}(\mathbf{q})$. Supposing that $\dot{\mathbf{x}} = 0$, that is, the mechanism performs only self-motion, it can be written $\dot{\mathbf{q}} = \left(\mathbf{I} - \mathbf{J}^{\dagger}\mathbf{J}\right)\nabla_q\mathbf{H}$; so, $\left(\mathbf{I} - \mathbf{J}^{\dagger}\mathbf{J}\right)\nabla_q\mathbf{H} = 0$ is a necessary condition of constrained optimality. Based on these facts, an objective function can be chosen for some optimization of motion:

- Manipulability—maximize the distance from kinematic singularities

$$\mathbf{H(q)} = \sqrt{det\left[\mathbf{J(q)J}^T(\mathbf{q})\right]} \tag{43}$$

- Joint limit avoidance—minimize the distance from the middle of joints range

$$\mathbf{H(q)} = \frac{1}{2}\sum_{i=1}^{n}\left(\frac{q_i - q_i^-}{q_{M,i} - q_{m,i}}\right)^2 \tag{44}$$

where $q_i \in \left[q_{m,i}, q_{M,i}\right]$ and $q_i^- = \frac{q_{M,i} - q_{m,i}}{2}$.

- Obstacle avoidance—maximize the minimum distance to obstacle

$$\mathbf{H(q)} = \min_{\substack{\mathbf{a}\in robot \\ \mathbf{b}\in obstacle}} \|\mathbf{a(q)} - \mathbf{b}\|^2 \tag{45}$$

where $\mathbf{a(q)}$ represents points on investigated mechanism and $\mathbf{b}$ represents points on the obstacle.

Example: Let us consider a planar six-link robot connected by six revolute joints. All links have the same length $L = 100\ mm$. The purpose of the simulation is path tracking (circle form) described by matrix $\mathbf{X}_{path} = \begin{bmatrix}\mathbf{x}_d \\ \mathbf{y}_d\end{bmatrix} \in \mathbb{R}^{m\times p}$, where p is the number of geometric points of the desired path, while $\mathbf{x}_d = -2.5L + L\cos\varphi$, $\mathbf{y}_d = L\sin\varphi$, $\varphi \in \{0, \dots, 2\pi\}$ assuming the step of φ increase to be 0.2. That is, $\mathbf{x}_d = \left[x_1, \dots, x_p\right] \in \mathbb{R}^p$ and $\mathbf{y}_d = \left[y_1, \dots, y_p\right] \in \mathbb{R}^p$. From the matrix $\mathbf{X}_{path}$ will be in each path point determined the desired point $\mathbf{e}_p = \left[x_p y_p\right]^T \in \mathbb{R}^m$. Since, there is consideration of planar task with focus on end-effector position, the task space is $m = 2$. The expected solution assumes only primary solution without any secondary tasks. For inverse kinematic solution, damped least squares method, which avoids many of pseudoinverse method's problems, will be used. This method is also known as the Levenberg–Marquardt method, which arises from cost function $\|\mathbf{J}\Delta\mathbf{q} - \Delta\mathbf{x}\|^2 + \lambda^2\|\Delta\mathbf{q}\|^2$ where $\lambda \in \mathbb{R}$ is a non-zero scalar constant. By minimizing this term, one obtains

$$\Delta\mathbf{q} = \mathbf{J}^T\left(\mathbf{JJ}^T + \lambda^2\mathbf{I}\right)^{-\mathbf{1}}\Delta\mathbf{x} \tag{46}$$

where $\mathbf{I}$ $\mathbb{R}^{m\times m}$ is the identity matrix. The simulation will work according to the following algorithm.

Algorithm: Inverse kinematic model for 6-link manipulator.

1: Set desired path for end-effector by matrix $\mathbf{X}_{path}$ and parameters such as L, λ.
2: **FOR** step = $1 \rightarrow p$.
3: Set the desired position of end-effector from $\mathbf{X}_{path} \Longrightarrow \mathbf{e}_p = \left[x_p y_p\right]^T \in \mathbb{R}^m$
4: **WHILE** $\mathbf{x}_{actual} \neq \mathbf{e}_p$
5: Compute Jacobian
6: Compute $\mathbf{x}_{actual} = \left[x_{act} y_{act}\right]^T \in \mathbb{R}^m$
7: Compute $\Delta\mathbf{q} = \mathbf{q} + \mathbf{J}^T\left(\mathbf{JJ}^T + \lambda^2\mathbf{I}\right)^{-\mathbf{1}}\Delta\mathbf{x}$

8: $\mathbf{q} = \Delta\mathbf{q}$
9: **END WHILE**
10: END FOR

The results of the simulation can be seen in **Figures 8** and **9**. **Figure 8** presents the motion of planar robot. The aim is to tracking of path with circle shape (green color) by end-effector of the robot. The robot has a fixed frame in point [0,0]. The initial position of all joints is $\mathbf{q} = [0\ 0\ 0\ 0\ 0\ 0]^T$.

During the simulation, no restriction such as joint limit was considered. The simulation was done with a tolerance ±5 mm. **Figure 9** presents the variation of individual robot joints during the path tracking.

Example: Let us consider a planar 20-link robot connected with revolute joints. The links have the same length $L = 16.75\ mm$. The aim is to move the end-effector from its initial position to the end position by tracking the desired path. We will consider two cases. The first one considers free robot environment, the second one considers obstacles in the robot environment. The second solution will also consider kinematic singularities avoidance task and joint limit avoidance task.

Now, let us consider the cost function dealing with all mentioned secondary tasks [10].

$$\mathbf{H} = \|\mathbf{J}\dot{\mathbf{q}} - \dot{\mathbf{x}}\|^2 + \|\mathbf{J}_c\dot{\mathbf{q}} - \mathbf{x}_c\|^2 + \|\mathbf{J}_L\dot{\mathbf{q}} - \mathbf{x}_L\|^2 + \|\rho\dot{\mathbf{q}}\|^2 \tag{47}$$

After mathematical adjustment, we get the final formula for kinematic control

$$\dot{\mathbf{q}} = \left(\mathbf{J}^T\mathbf{W}\mathbf{J} + \mathbf{J}_c^T\mathbf{W}_c\mathbf{J}_c + \mathbf{J}_L^T\mathbf{W}_L\mathbf{J}_L + \mathbf{W}_s\right)^{-1}\left(\mathbf{J}^T\mathbf{W}\dot{\mathbf{x}}\right) \tag{48}$$

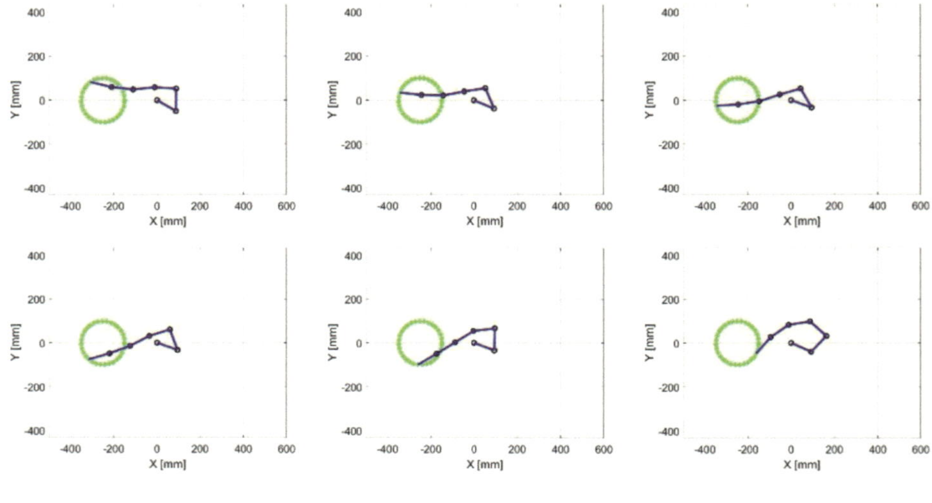

Figure 8.
Simulation of inverse kinematics for six-link manipulator.

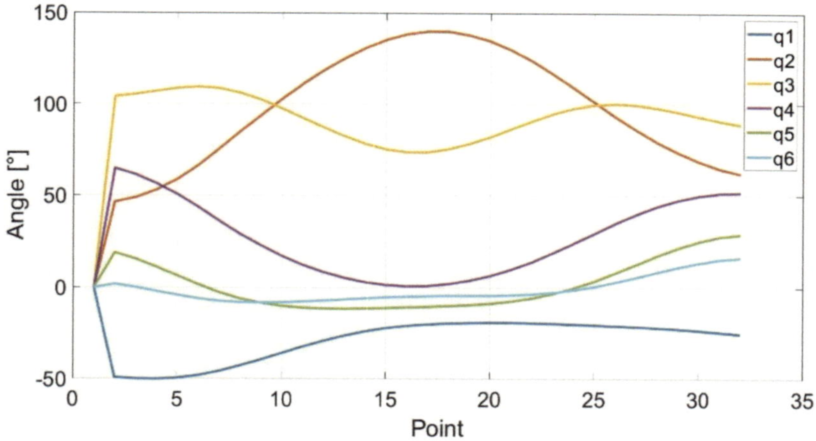

Figure 9.
Variations of joint angles.

where $\mathbf{W}$, $\mathbf{W}_c$, $\mathbf{W}_L$, and $\mathbf{W}_s$ are in our case 20×20 weight matrices of primary task, obstacle avoidance task, joint limit avoidance task, and kinematic singularities avoidance task, respectively. The setting of weight matrices is subjective. For example, if obstacle avoidance task should have higher priority above primary task by 100 times, the matrix $\mathbf{W}_c$ would consist of values 100 while $\mathbf{W}$ consist of values 1.

Let us consider individual secondary tasks. The joint limit avoidance task deals with the range of motion of individual manipulator links. How many joints will have their limit range is up to us. Of course, in real robots, usually all joints are limited in their motion. There are several ways on how to model the range of joint motion. In this case, an approach with changing of value of weight variable W_{li} based on joint position will be used. If the joint is in admissible range, the value of the weight variable is set to be zero. When the joint reaches the boundary of its range motion, the value of the weight variable increases. When the joint reaches a value outside its admissible range, the value of the weight variable increases to its maximum. This approach can be expressed by Eq. (49)

$$\mathrm{W_{li}} = \begin{cases} \mathrm{W_w} \leftarrow \mathrm{q_i} < \mathrm{q_{imin}} \\ \dfrac{\mathrm{W_w}}{2}\left\{1 + \cos\left[\pi\left(\dfrac{\mathrm{q_i} - \mathrm{q_{imin}}}{\rho_\mathrm{i}}\right)\right]\right\} \leftarrow \mathrm{q_{imin}} \le \mathrm{q_i} \le \mathrm{q_{imin}} + \rho_\mathrm{i} \\ 0 \leftarrow \mathrm{q_{imin}} + \rho_\mathrm{i} < \mathrm{q_i} < \mathrm{q_{imax}} - \rho_\mathrm{i} \\ \dfrac{\mathrm{W_w}}{2}\left\{1 + \cos\left[\pi\left(\dfrac{\mathrm{q_{imax}} - \mathrm{q_i}}{\rho_\mathrm{i}}\right)\right]\right\} \leftarrow \mathrm{q_{imax}} - \rho_\mathrm{i} \le \mathrm{q_i} \le \mathrm{q_{imax}} \\ \mathrm{W_w} \leftarrow \mathrm{q_i} > \mathrm{q_{imax}} \end{cases} \tag{49}$$

The value of the weight variable has to be set for every joint of the manipulator that needs to be limited in the range of motion. Individual weight variables W_{li} where $i \in \{0, \dots, N\}$ are parts of the final weight matrix of the joint limit avoidance task $W_l \in \mathbb{R}^{n \times n}$. The final weight matrix $\mathbf{W}_\mathrm{l}$ is the diagonal matrix [14]:

$$\mathbf{W}_{\mathrm{l}} = \begin{bmatrix} \mathrm{W}_{\mathrm{l}1} & & & & \\ & \mathrm{W}_{\mathrm{l}2} & & & \\ & & \mathrm{W}_{\mathrm{l}3} & & \\ & & & \cdots & \\ & & & & \mathrm{W}_{\mathrm{l}n} \end{bmatrix} \tag{50}$$

The weight matrix $\mathbf{W}_{\mathrm{l}}$ is used with the corresponding Jacobian matrix $\mathbf{J}_L \in \mathbb{R}^{n\times n}$. The Jacobian matrix for the joint limit avoidance task is $\mathbf{J}_L = \partial\mathbf{e}/\partial\mathbf{q}$. If a particular joint does not consider the joint limit avoidance task, the value of $\mathbf{J}_L$ is set to be zero; otherwise it is set to be one. The limit of all joints in motion of the manipulator investigated in this study is set to be $\pm 100°$. Different way of joint limit control is according to above mentioned Eq. (44).

During the obstacle avoidance task, the control system investigates the relation between manipulator links and obstacles in their environment. In general, this task can be solved from two views. At first, one group of obstacles can represent static obstacles or other robots in an investigated environment. The second group of obstacles can be represented by dynamic obstacles, which means that these obstacles change their position relating to global reference system in the time. Of course, the second group of obstacles is more difficult from the view of control in comparison with static obstacles. It is more difficult especially in the cases of requirements for real-time control [15].

The aim of obstacle avoidance is to prevent the collision between any part of the manipulator and potential obstacles, other robots, or collision with itself. Again, there are many methods on how to control robot motion at a safe distance from other objects, regardless of whether the obstacles have regular or irregular shape. For simplification, the irregular shapes are usually replaced by appropriate regular shape. This simplification can also significantly simplify the mathematical model and obtaining the numerical solution can be faster and the solution more stable. One of the methods on how to simplify irregular shapes is to replace all irregular shapes by a cylinder, with the obstacle being situated in the center of the cylinder, see **Figure 10**. The diameter of the cylinder determines the distance of influence of this obstacle.

At first, we set the coordinate of an obstacle in the task space as $\mathbf{s}_o$. The projection of the line from the i-th joint of the manipulator link to the center of a cylinder (obstacle) on the i-th link is:

$$\mathbf{p}_{\mathrm{i}} = \mathbf{e}_{\mathrm{i}}^{\mathrm{T}}(\mathbf{s}_0 - \mathbf{s}_{\mathrm{i}}) \tag{51}$$

The coordinate of the link point with potential to get into collision is:

$$\mathbf{s}_{\mathrm{ci}} = \mathbf{s}_{\mathrm{i}} + \mathbf{p}_{\mathrm{i}}\mathbf{e}_{\mathrm{i}} \tag{52}$$

The distance between the potential point of collision and the center of the cylinder is:

$$\mathbf{d}_{\mathrm{ci}} = \|\mathbf{s}_{\mathrm{ci}} - \mathbf{s}_0\| \tag{53}$$

Subsequently, the unit vector of the potential point of collision can be given by:

$$\mathbf{u}_{\mathrm{i}} = \frac{\mathbf{s}_{\mathrm{ai}} - \mathbf{s}_0}{\mathbf{d}_{\mathrm{ci}}} \tag{54}$$

Now, the Jacobian matrix for the obstacle avoidance task can be given. The i-th row of the Jacobian matrix can be expressed as

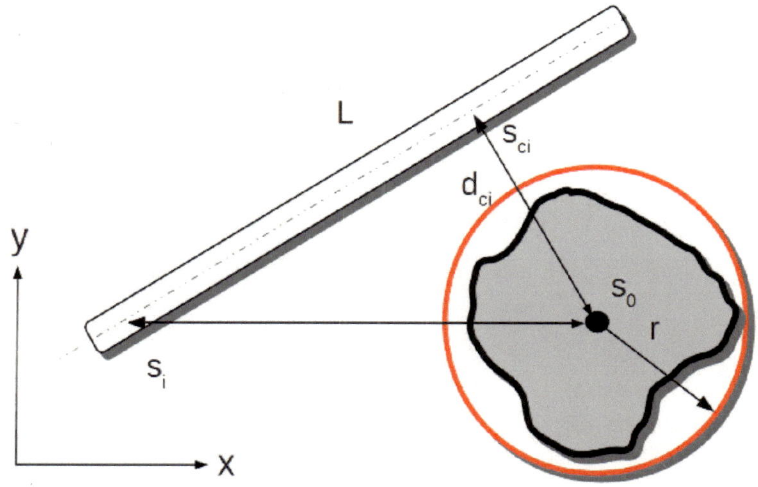

Figure 10.
Relation between manipulator link and obstacle.

$$\mathbf{J}_{ci} = -\mathbf{u}_i^T \mathbf{J}_{s_{ci}} \tag{55}$$

where matrix $J_{s_{ci}}$ is:

$$\mathbf{J}_{s_{ci}} = \frac{\partial \mathbf{s}_{ci}}{\partial \mathbf{q}} \tag{56}$$

The Jacobian matrix J_c consists of submatrices J_{ci}. The dimension of the Jacobian matrix is $J_c \in R^{c \times c}$, where c represents the number of manipulator links that could collide with the obstacles.

For numerical simulation, we will use following algorithm: Inverse kinematic model for 20-link manipulator [14].

Algorithm: Inverse kinematic model for 20-link manipulator

1: CYCLE WHILE 1
2: Determination of new required vector $x_d \in R^m$ from the matrix of planned path $P \in R^{r \times 2}$
3: CYCLE WHILE 2
4: Computation of Jacobian matrix J (damped least squares method)
5: Determination of actual end-effector position in the task space $x \in R^m$ with actual generalized variables $q \in R^n$
6: Computation of general equation
$\dot{q} = \left(J^{-1}WJ + J_c^{-1}W_cJ_c + J_l^{-1}W_lJ_l + W_s\right)^{-1}\left(J^TW\dot{x}\right)$
7: $q = q_{previous} + \dot{q}dt$
8: $q_{previous} = q$
9: IF $x_d = x$ THEN
END CYCLE WHILE 2
ELSE
CYCLE WHILE 2 continues
END IF
10: END CYCLE WHILE 2
11: END CYCLE WHILE 1

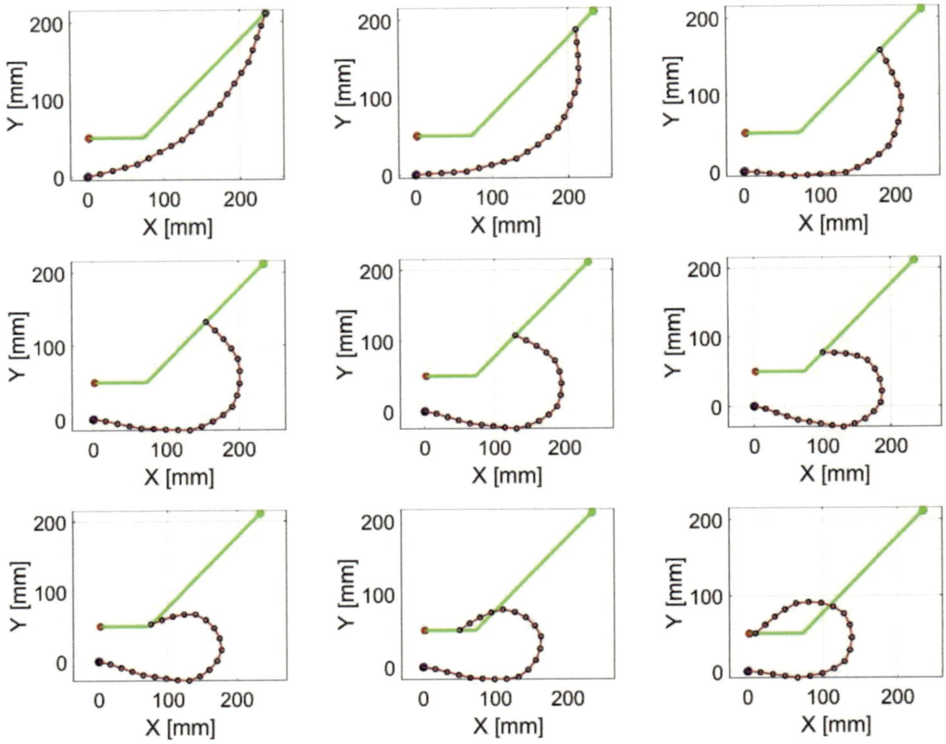

Figure 11.
Simulation of 20-link manipulator in free environment.

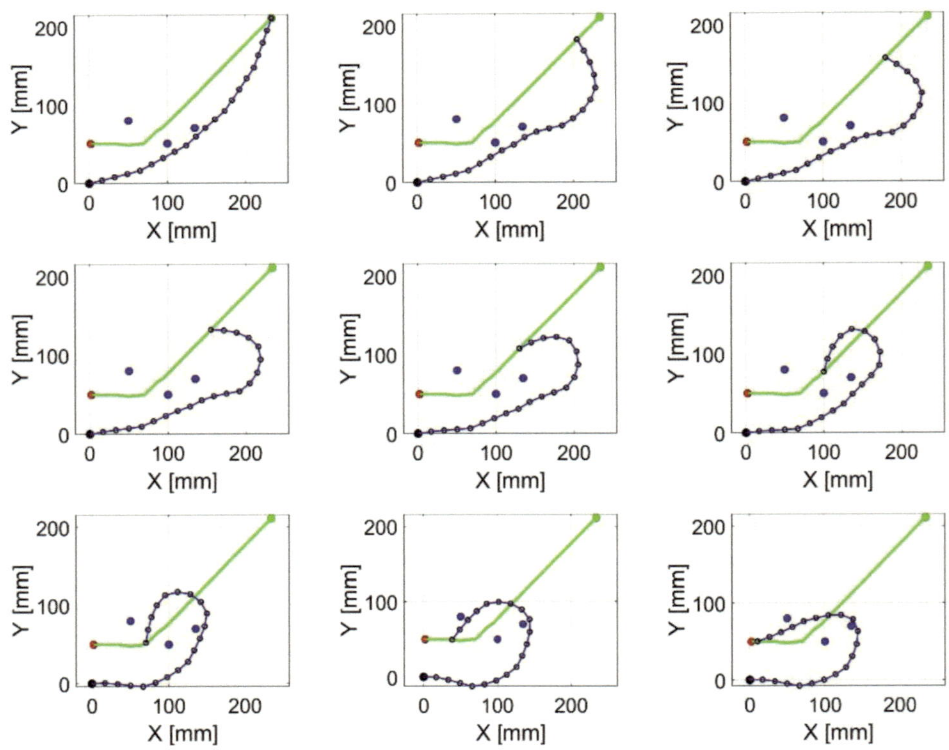

Figure 12.
Simulation of 20-link manipulator in constrained environment.

In **Figure 11** can be seen the simulation of motion of a 20-link manipulator, which moves in free environment without any obstacles. This case also does not consider any joint limit avoidance task. The aim is to move the end-effector of manipulator by predefined path (green color).

Figure 12 depicts a different situation. A 20-link manipulator moves according to predefined path between four obstacles. The solver also assumes joint limit avoidance task for all 20 joints. From this figure can be seen the difference in "self-motion" in joint space, while it does not affect the motion of end-effector in task space. The end-effector always tracks the same path (by neglecting end-effector orientation).

The second case is significantly difficult from the view of computational complexity in comparison with the case without any constraints in motion.

6. Conclusion

This chapter was focused on forward and inverse kinematics of open-chain mechanisms, namely manipulators with serial kinematic structures. We have introduced rigid motion and subsequently we have focused on forward kinematics. The chapter presents a kinematic model of SCARA by the well-known Denavit-Hartenberg convention. Within inverse kinematics was introduces several methods, including analytical, geometrical, and numerical. The last section dealt with modeling of kinematically redundant planar robots. At first, we introduced a six-link planar robot with focus on numerical solution using DLS. The last example presented a 20-link redundant manipulator moving between the obstacles. The solution includes, besides the primary solution, secondary tasks such as singularity avoidance, joint limit avoidance, and obstacle avoidance.

Acknowledgements

The authors would like to thank to Slovak Grant Agency VEGA 1/0389/18 "Research on kinematically redundant mechanisms" and to KEGA 018TUKE-4/2018 "Implementation of new technologies and education methods in area of controlling systems for improvement of educational level and practical experiences of graduate students of Mechatronics study program" and also to KEGA 030TUKE-4/2020 "Transfer of knowledge from the field of industrial automation and robotics to the education process in the teaching program mechatronics."

Author details

Ivan Virgala*, Michal Kelemen and Erik Prada
Technical University of Košice, Košice, Slovakia

*Address all correspondence to: ivan.virgala@tuke.sk

References

[1] Li XJ, Cao Y, Yang DY. A numerical-analytical method for the computation of robot workspace. In: The Proceedings of the Multiconference on "Computational Engineering in Systems Applications". Beijing; 2006. pp. 1082-1086. DOI: 10.1109/CESA.2006.4281805

[2] Dash AK, Chen IM, Yeo SH, Yang G. Workspace generation and planning singularity-free path for parallel manipulators. Mechanism and Machine Theory. 2005;**40**(7):776-805. DOI: 10.1016/j.mechmachtheory.2005.01.001

[3] Lynch KM, Park FC. Modern Robotics – Mechanics, Planning, and Control. Cambridge University Press; 2017. ISBN: 9781107156302

[4] Murray RM, Li Z, Sastry SS. A Mathematical Introduction to Robotic Manipulation. United States, Boca Raton, Florida: CRC Press; 1994. ISBN-13: 978-0849379819

[5] Trebuňa F, Virgala I, Pástor M, Lipták T, Miková Ľ. An inspection of pipe by snake robot. International Journal of Advanced Robotic Systems. 2016;**13**(5). DOI: 10.1177/1729881416663668

[6] Kucuk S, Bingul Z. Inverse kinematics solutions for industrial robot manipulators with offset wrists. Applied Mathematical Modelling. 2014;**38**(7–8): 1983-1999. DOI: 10.1016/j.apm.2013.10.014

[7] Siciliano B, Sciavicco B, Villani L, Oriolo G. Robotics: Modeling, Planning and Control. Springer; 2009. ISBN: 978-1-84628-641-4. DOI: 10.1007/978-1-84628-642-1

[8] Wampler CW. Manipulator inverse kinematic solutions based on vector formulations and damped least-squares methods. IEEE Transactions on Systems, Man, and Cybernetics: Systems. 1986;**16**(1):93-101

[9] Baillieul J. Kinematic programming alternatives for redundant manipulators. In: Proceedings of the IEEE International Conference on Robotics and Automation. St. Louis, MO, USA; 1985. pp. 722-728

[10] Powell MJD. A Hybrid Method for Nonlinear Equations. Gordon & Breach Science Publishers, Inc.; 1970

[11] Wang L-CT, Chen CC. A combined optimization method for solving the inverse kinematics problems of mechanical manipulators. IEEE Transactions on Robotics and Automation. 1991;**7**(4):489-499

[12] Aristidou A, Lasenby J. FABRIK: A Fast, Iterative Solver for the Inverse Kinematics Problem. Amsterdam, Netherlands: Elsevier; 2010. Submitted to Graphical Models

[13] Kelemen M, Virgala I, Lipták T, Miková Ľ, Filakovský F, Bulej V. A novel approach for a inverse kinematics solution of a redundant manipulator. Applied Sciences. 2018;**8**:2229

[14] Virgala I, Lipták T, Miková Ľ. Snake robot locomotion patterns for straight and curved pipe. Journal of Mechanical Engineering. 2018;**68**(2). DOI: 10.2478/scjme-2018-0020

[15] Fahini F. Autonomous Robots – Modelimg, Path Planning, and Control. Springer; 2009. ISBN: 978-0-387-09537-0. DOI: 10.1007/978-0-387-09538-7

Chapter 8

An In-Depth Analysis of Sliding Mode Control and Its Application to Robotics

İhsan Ömür Bucak

Abstract

In this study, a sliding mode control scheme with a bounded region and its convergence analysis are explained to the finest detail and are applied to robotic manipulators which represent the best examples for strongly coupled, highly nonlinear, time-varying dynamical systems. Simulation studies have been applied separately to two different control systems in order to demonstrate the feasibility, performance, and effectiveness of the proposed control methodology through the design of the sliding mode controller: firstly, the position control of an armature-controlled dc servo motor subject to a varying external disturbance, and secondly, a two-link robot manipulator that were also analyzed in terms of its robustness by adding extra mass to one of the joints to be able to maintain the trajectory in the sliding surface. Simulations show that a fast convergence rate, and therefore quick response, the ability to reject the varying external disturbances, and the robustness against the model uncertainty assumed to be unbounded and fast-varying have all achieved its purpose entirely. This study also examines the advantages of SMC and PID comparably. The results given here do not contradict the view that one can use it instead of the other without losing too much performance, and confirm the success of the presented approach.

Keywords: robotic manipulators, sliding mode control, convergence analysis, trajectory tracking, robustness, two-link planar robot manipulator, PID control

1. Introduction

Unknown plant parameters or, more generally, plant uncertainty and the preferences in which the system dynamics are purposely represented by simplifications, such as the use of linearized friction model, lead to model imprecision [1]. Control engineering classifies the model inaccuracies, which were used here as synonym of imprecision, into two main categories as structured and unstructured uncertainties. The first one implies inaccuracies within the model and the second one corresponds to inaccuracies on the system order (i.e., underestimated system order). Modeling inaccuracies can have adverse effect on nonlinear control systems [1].

Robotic manipulators represent the best examples for strongly coupled, highly nonlinear, time-varying dynamical systems [2]. These qualities alongside structured uncertainties caused by model imprecision of link parameters and payload variation, and unstructured uncertainties produced by unmodeled dynamics such as nonlinear friction, compliance in gearing, sensor noise, external disturbances, and the high-frequency part of the dynamics turn the motion control of rigid-link manipulators into a complex problem [2]. Particularly, robotic manipulators suffer so much from these structured and unstructured uncertainties. The consequence of having to deal with various uncertainties in their dynamics and the necessity to manage the various tools and, hence, the variation of dynamic parameters during operation make it difficult for robots to introduce a mathematical model suitable for employing model-based control strategies.

The theory of conventional sliding mode control (SMC) as a simple robust nonlinear control scheme has been applied to robotic manipulators successfully since the 1980s. In those studies, the advantages of the SMC properties such as its robustness against disturbances and variation of parameters, and its fast dynamic response have been utilized greatly. Two important approaches, such as robust control and adaptive control, can deal with modeling uncertainty [1]. Particularly, adaptive control is effective in solving the structured and unstructured uncertainties and is capable of maintaining a uniformly good performance over a limited range.

SMC as a special class of the variable structure systems (VSS) has been preferred in practical applications for over 50 years due to its simplicity and robustness against parameter variations and disturbances [3]. VSS concept was first evolved from the pioneering work of Emel'yanov and Barbashin in the early 1960s in Russia [4]. Especially, VSS and SMC have received a great attention by the control research community worldwide since the published 1977 article [4]. SMC methodology is used to design a control law that imposes all system trajectories to converge on a surface in the state space, the so-called sliding surface $S(t)$. The designer chooses the dynamics of this surface so that all trajectories will asymptotically converge to the set point. When the trajectory lies inside the sliding surface, the system operates in so-called sliding mode and is sensitive to parametric variations and external disturbances [5]. Control action in sliding mode is discontinuous by nature and can stimulate high-frequency dynamics [6]. Discontinuous nature of the control action serves to maintain a resulting superior system performance of VSS and SMC by switching between two incomparably different system structures such that sliding mode with this feature is also referred to as a new type of system motion in a manifold, or in another substantial terms, in the vicinity of a prescribed switching manifold, the velocity vector of the controlled state trajectories is always directed toward the switching manifold by such motion induced by imposing discontinuous control actions [3, 6]. This system performance is expected to exhibit insensitivity to parameter variations as well as demonstrate complete disturbance rejection [6].

Despite its advantages such as simplicity and robustness, SMC suffers from a rather widespread and well-known chattering problem, which is generally perceived as motion oscillating about the predefined switching manifold(s) [3, 6]. There are two reasons behind the chattering phenomenon: first, under the absence of switching nonidealities such as delays, that is, in a situation where the switching device ideally switches at an infinite frequency, the presence of parasitic dynamics in series with the plant causes a small amplitude high-frequency oscillation to occur around the sliding manifold. If the closed loop pole locations are well defined or the closed loop poles are well assigned with the aid of the pole placement design technique, these parasitic dynamics which represent the fast actuator and sensor dynamics are often neglected in the open loop model used for control design in control applications. In general, the motion of the real system is closer to an ideal system where the parasitic dynamics are neglected, and the difference between the ideal and the real motion, which is at negligible time constants, shows a rapid decline. However, the parasitic dynamics interacted with variable structure control

(VSC) in particular produce a nondecreasing oscillatory component with a finite amplitude and a frequency, referred previously to as chattering phenomenon already. Second, the switching nonidealities alone can cause such high-frequency oscillations around the sliding surface $S(t)$ [3, 6]. These may include small time delays due to sampling (e.g., zero-order hold), execution time required to calculate the control, and transmission delays in networked control systems [3]. As time delays cause the resulting chattering phenomenon, and the delay type of switching nonidealities is the most relevant to any electronic implementation of the switching device whether it includes analogue or digital circuits, delay type nonidealities are considered in general for the design approaches, and discrete-time control design techniques are the most commonly used design approaches by control engineers to mitigate the chattering caused by the switching delays [6]. Many design methodologies have been applied so far to mitigate or reduce the chattering. To eliminate the chattering, a boundary layer around the sliding surface $S(t)$ has been introduced [1]. Inside the boundary layer, the switching function is replaced by a high-gain proportional control [5]. In order to reduce the chattering level, several switching functions such as saturation functions, relay functions, hyperbolic functions, and hysteresis saturation functions have been used in literature. The use of these functions can minimize or, if desired, completely eliminate the chattering, but it turns perfect tracking into a tracking with a guaranteed precision problem, meaning that a steady-state error is always maintained.

SMC design procedure is split into two major steps corresponding to the two main phases [3]: reaching phase is defined to derive the system state from initial state to reach the switching manifolds in finite time; and sliding-mode phase is defined to induce the system into the sliding motion on the switching manifolds like an attractor.

No matter how active the research on SMC has been during the last 50 years, the key technical challenges such as chattering, the elimination of the effects caused by unmodeled dynamics, disturbances and uncertainties, adaptive learning, and improved robustness can still remain to be addressed to reach a perfect solution [3]. An ideal sliding mode can only be achieved when the dynamic equation governing the sliding mode is satisfied by the system state for all time. This implies an infinite switching to assure the sliding motion [3]. Although the switching rate of the switching control device of the SMC system (design) is infinite ideally, it is much lower than that in practice due to the physical limitations of switching [6]. Physical limitations of switching have been tried to be explained in the previous paragraphs.

Usually, intelligent control approaches can mitigate the effects of structured parametric uncertainty and unstructured disturbance with their effective learning ability without requiring a detailed knowledge of the controlled plant within the design process. SMC research has recently been integrated with intelligent control approaches such as neural networks, fuzzy logic, genetic algorithms, and probabilistic reasoning, just a few of them, to make it more intelligent [3, 7–11]. Another goal behind the combination of the intelligent control with the attractive features of this traditional control is to create more powerful control algorithms. Nevertheless, it appears that many intelligent control algorithms do not take into account actuator dynamics in robot control systems, which play a critical role in overall robot dynamics and their negligence can cause adverse effects, especially in the case of high-speed torque, respectable load variations, friction, and actuator saturations [2]. Electrical actuators are very much controllable than others and are more suitable for driving robot manipulators [2, 12].

Sliding mode control strategy is the simple approach to robust control. By intuition, controlling first-order systems is much easier than controlling general *nth*-order systems, even if they are nonlinear or uncertain. Therefore, an introduction of a notational simplification allows *nth*-order problems to be replaced by equivalent first-order problems. Thus, it is then easy to demonstrate in principle the achievement of perfect performance under the presence of arbitrary parameter inaccuracies for the transformed problems. However, such performance is only achieved at the expense of extremely high control activity. This is typically disputed by the other source of modeling uncertainty, such as the presence of neglected dynamics, which the high control activity can stimulate. This leads to a change in control laws aimed at achieving an effective trade-off between tracking performance and parametric uncertainty, given the acceptable control activity [1].

The concepts will first be presented for systems that have a single control input that allows us to develop an intuition about the fundamental aspects of nonlinear controller design.

2. Sliding surfaces

Consider the single-input dynamic system given below:

$$x^{(n)} = f(\boldsymbol{x}) + b(\boldsymbol{x})u, \tag{1}$$

where the scalar x is the output we are interested in, the scalar u is called the control input, and $\boldsymbol{x} = \left[x, \dot{x}, \dots, x^{(n-1)}\right]^T$ represents the state vector. While the function $f(\boldsymbol{x})$ in (1) is not exactly known, an upper bound on $f(\boldsymbol{x})$ is set by a known continuous function of $\boldsymbol{x}$ to limit the size of the imprecision. Similarly, the control gain $b(\boldsymbol{x})$ is not exactly known either, only its sign is known; therefore, it is also bounded by the known continuous functions of $\boldsymbol{x}$. The control problem is to ensure that the state $\boldsymbol{x}$ tracks an explicit time-varying state $\boldsymbol{x}_d = \left[x_d, \dot{x}_d, \dots, x_d{}^{(n-1)}\right]^T$ in the presence of model imprecision on $f(\boldsymbol{x})$ and $b(\boldsymbol{x})$. In order for the tracking task to be achieved using a finite control u, desired initial state $\boldsymbol{x}_d(0)$ must be such that it ultimately validates the following relationship:

$$\boldsymbol{x}_d(0) = \boldsymbol{x}(0) \tag{2}$$

For example, in a second-order system, the position or speed cannot bounce; as a result of this fact, any desired trajectory that can be workable from time $t = 0$ necessarily starts at the same position and speed as those of the plant.

2.1 A notational simplification

Let us define $\tilde{x} = x - x_d$ as the tracking error in the variable x, and $\tilde{\boldsymbol{x}} = \boldsymbol{x} - \boldsymbol{x}_d = \left[\tilde{x}, \dot{\tilde{x}}, \dots, \tilde{x}^{(n-1)}\right]^T$ as the tracking error vector. Let us also define a surface $S(t)$ that is time-varying in the state-space $\boldsymbol{R}^{(n)}$ with the scalar equation $s(\boldsymbol{x}; t) = 0$ expressed as:

$$s(\boldsymbol{x}; t) = \left(\frac{d}{dt} + \lambda\right)^{n-1} \tilde{x}, \tag{3}$$

where λ is a strictly positive constant. For example, if $n = 2$, Eq. (3) takes the following form:

$$s = \dot{\tilde{x}} + \lambda\tilde{x},$$

that is, it simply consists of a weighted sum of the position and the velocity errors; thus, we can express $\tilde{x}$ from Eq. (3) as follows:

$$\tilde{x} = \frac{s(\boldsymbol{x};t)}{\left(\frac{d}{dt}+\lambda\right)^{n-1}} \tag{4}$$

More specifically, a first-order stabilization problem in s is basically nothing other than a replacement for the problem of tracking the n-dimensional vector $\boldsymbol{x}_d$ (i.e., the original nth-order tracking problem in x). Indeed, s needs to be differentiated only once for the appearance of the input u due to the presence of the term $\tilde{x}^{(n-1)}$ in the expression s in (3).

Additionally, the bounds on s can be evaluated directly as the bounds on the tracking error vector $\tilde{\boldsymbol{x}}$, and therefore, the scalar s is considered a true measure of tracking performance. Specifically, under the assumption that $\tilde{\boldsymbol{x}}(0) = \boldsymbol{0}$ (in the meantime, we make a note that the effect of non-zero initial conditions in $\tilde{x}$ can be added separately), we write:

$$\forall t \geq 0, |s(t)| \leq \Phi \Rightarrow \forall t \geq 0, \left|\tilde{x}^{(i)}(t)\right| \leq (2\lambda)^i \varepsilon \quad i = 0, \dots, n-1 \tag{5}$$

where $\varepsilon = \Phi/\lambda^{n-1}$. Indeed, Eq. (3), or more precisely, Eq. (4) derived from (3) clearly indicates that the tracking error $\tilde{\boldsymbol{x}}$ can be obtained from s through a sequence of first-order low-pass filters (see **Figure 1**, where $p = (d/dt)$ is commonly known as the *Laplace operator*).

In general, a first-order low-pass filter's input-output relationship is given as follows:

$$\frac{Output_{lpf}}{Input_{lpf}} = K_f \frac{1}{1+p\tau},$$

where K_f is filter gain and τ is filter time constant. Let y_1 be the output of the first filter. We can express the output of the first filter in terms of a *convolution integral* in time domain by taking into account the fact that the input is already defined as s:

$$y_1(t) = \int_0^t e^{-\lambda(t-T)} s(T) dT.$$

Using $|s| \leq \Phi$, we get the following:

$$|y_1(t)| \leq \Phi \int_0^t e^{-\lambda(t-T)} dT = \frac{\Phi}{\lambda} e^{-\lambda(t-T)}\Big]_0^t = \frac{\Phi}{\lambda}\left(1 - e^{-\lambda t}\right) \leq \Phi/\lambda.$$

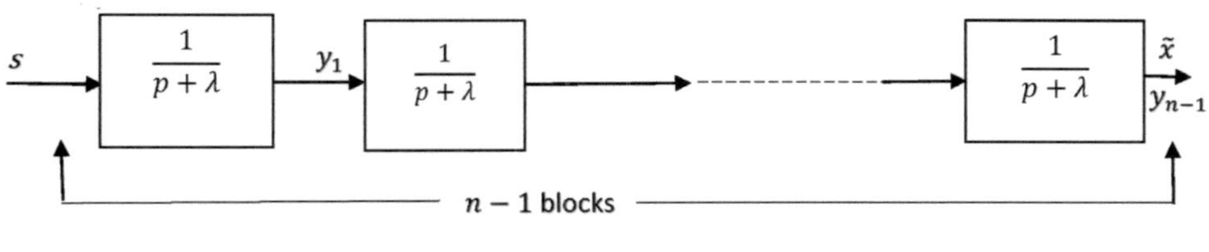

Figure 1.
Computing bounds on $\tilde{x}$.

We can apply similar reasoning to the second filter, et cetera, until we reach $y_{n-1} = \tilde{x}$. Ultimately, we get

$$|\tilde{x}| \leq \Phi/\lambda^{n-1} = \varepsilon.$$

Similarly, $\tilde{x}^{(i)}$ can be considered to be acquired through the sequence of **Figure 2**.

One can easily make another similar relationship, $|z_I| \leq \Phi/\lambda^{n-1-i}$ by referring to the previous result. Here, z_I is the output of the $(n-i-1)^{th}$ filter. It is worth noting, however, that

$$\frac{p}{p+\lambda} = \frac{p+\lambda-\lambda}{p+\lambda} = 1 - \frac{\lambda}{p+\lambda}.$$

One sees that the remaining i blocks right after $n-i-1$ blocks in the sequence of **Figure 2** include numerators of p as can typically be seen in practical filter applications, and that the sequence of **Figure 2** means that

$$|\tilde{x}^{(i)}| \leq \left(\frac{\Phi}{\lambda^{n-1-i}}\right) \underbrace{\left(1-\frac{\lambda}{p+\lambda}\right) \cdots \cdots \left(1-\frac{\lambda}{p+\lambda}\right)}_{i\ blocks},$$

where the first multiplier to the right of inequality sign includes the first $n-i-1$ blocks which do not have p as the numerator. When i blocks multipliers are arranged, the expression takes the form:

$$|\tilde{x}^{(i)}| \leq \left(\frac{\Phi}{\lambda^{n-1-i}}\right)\left(\underbrace{1\ldots\ \ldots\ \ldots\ \ldots.1}_{i\ times} + \frac{\lambda\ldots..\lambda}{\lambda^{i}}\right),$$

where term comes from the result derived for the sequential blocks each of which is represented by $1/(p+\lambda)$, and $1/(p+\lambda)^i \leq 1/\lambda^i$. Besides, the positive sign must be considered to obtain the prospective upper bound. In this case, the next step is:

$$|\tilde{x}^{(i)}| \leq \left(\frac{\Phi}{\lambda^{n-1-i}}\right)\left(1^i + \frac{\lambda^i}{\lambda^i}\right).$$

Finally, we can write,

$$|\tilde{x}^{(i)}| \leq \left(\frac{\Phi}{\lambda^{n-1-i}}\right)\left(1 + \frac{\lambda}{\lambda}\right)^i$$

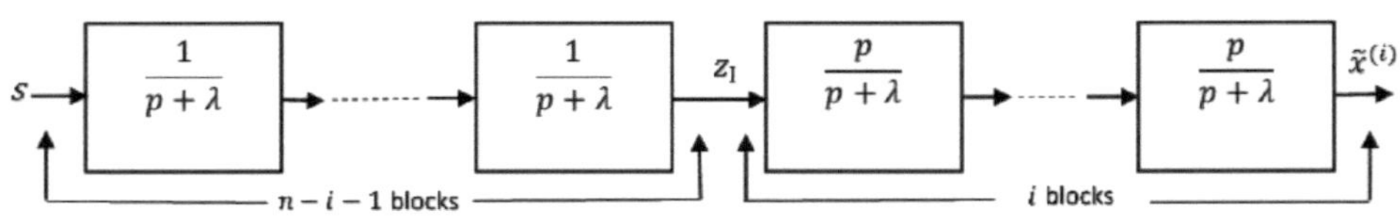

Figure 2.
Computing bounds on $\tilde{x}^{(i)}$.

since $\left(1^i + \frac{\lambda^i}{\lambda^i}\right) \le \left(1 + \frac{\lambda}{\lambda}\right)^i$. Thus, as a result, we reach the following statement:

$$|\tilde{x}^{(i)}| \le \left(\frac{\Phi}{\lambda^{n-1}}\right)\lambda^i(1+1)^i = \left(\frac{\Phi}{\lambda^{n-1}}\right)(2\lambda)^i = (2\lambda)^i\varepsilon,$$

i.e., the bounds of (5) are proven. Finally, in the case where $\tilde{\boldsymbol{x}}(0) \neq \boldsymbol{0}$, the bounds of (5) are obtained asymptotically, that is, within a short time constant $(n-1)/\lambda$. Please note that since the single filter block has a time constant equal to $1/\lambda$ as seen from the above analysis starting from (3), the sequence of $(n-1)$ filter blocks will have the time constants equal to $(n-1)/\lambda$.

Hence, we, indeed, have replaced an *nth*-order *tracking problem* by a first-order *stabilization problem*, and have quantified with inequality (5) the transformations in which the performance measures correspond to.

Keeping the scalar s at zero, which is a simplified first-order problem, can now be achieved by choosing the control law u of (1) such that apart from $S(t)$

$$\frac{1}{2}\frac{d}{dt}s^2 \le -\eta|s|, \tag{6}$$

where η is a strictly positive design constant and ensures inequality (6) which is called η- reachability condition. A fundamental requirement is that the sliding mode dynamics must be attractive to the system state and there are many reachability conditions defined in the literature [13–17]. Basically, inequality (6) indicates that the squared "distance" to the surface, as measured by s^2, decreases throughout entire system trajectories. Therefore, it restricts trajectories to head toward the surface $S(t)$, as depicted in **Figure 3**. Particularly, when they are on the surface, the system trajectories remain on the surface. In other words, the fact that the surface is an *invariant set* indicates that condition (6) (a.k.a., *sliding condition*) is satisfied. Moreover, as we shall notice, inequality (6) also suggests tolerating some disturbances or dynamic uncertainties while still holding the surface stationary (i.e., an invariant set). In **Figure 3**, this graphically means that the trajectories away from the surface can "move" while still denoting the surface. $S(t)$ verifying (6) is referred to as a *sliding surface*, and the system behavior that occurs on the surface is called *sliding regime* or *sliding mode*.

Another appealing feature of the invariant set $S(t)$ is that when the system trajectories are on the set, it is defined by the equation of the set itself, i.e.,

$$\left(\frac{d}{dt}+\lambda\right)^{n-1}\tilde{x} = 0.$$

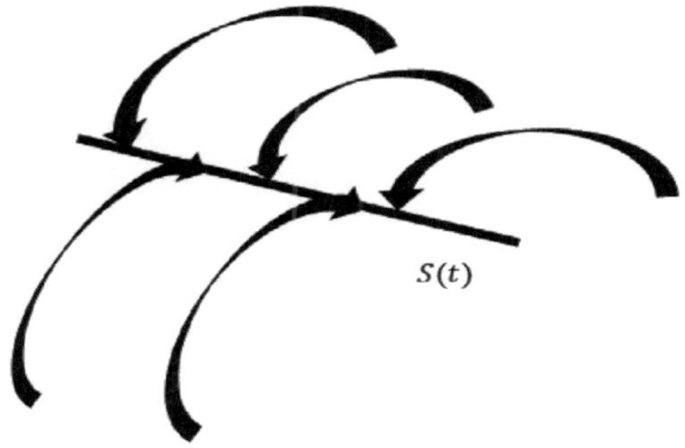

Figure 3.
The sliding condition.

That is to say, the surface $S(t)$ represents both a place and a dynamics. This fact is clearly the geometric interpretation of our previous statement that definition (3), in fact, allows us to substitute a first-order problem for an *nth*-order one.

Finally, if condition (2) is not fully validated, i.e., if $\boldsymbol{x}(t=0)$ is indeed away from $\boldsymbol{x}_d(t=0)$, then, nevertheless, satisfying (6) gives a guarantee for reaching the surface $S(t)$ at a *finite time* smaller than $|s(t=0)|/\eta$. Indeed, assume, for instance, that $s(t=0)>0$ and define t_{reach} as the time needed to hit the surface $s=0$. Integrating (6) between the points $t=0$ and $t=t_{reach}$ (i.e., in the interval $[0, t_{reach}]$) gives rise to:

$$0-s(t=0)=s(t=t_{reach})-s(t=0)\leq-\eta(t_{reach}-0),$$

which means that

$$t_{reach}\leq s(t=0)/\eta. \tag{7}$$

This result can simply be proven to be true by starting to integrate both sides of (6) between 0 and t_{reach} as follows:

$$\int_{t=0}^{t=t_{reach}}\frac{1}{2}\frac{d}{dt}s^2dt\leq-\int_{t=0}^{t=t_{reach}}\eta|s|dt.$$

Making the necessary simplifications within the integrals, we get the following:

$$\int_{t=0}^{t=t_{reach}}ds\leq-\int_{t=0}^{t=t_{reach}}\eta dt.$$

Now, the integrals are taken and evaluated for the lower and upper limits as shown below:

$$s(t=t_{reach})-s(t=0)\leq-\eta(t_{reach}-0).$$

Finally from here,

$$s(t=0)\geq\eta t_{reach}$$

is written, and the same result as (7) is hereby obtained. Even if $s(t=0)<0$, a similar result would be obtained, and thus, writing the above inequality as follows would be a correct representation:

$$t_{reach}\leq|s(t=0|/\eta.$$

Furthermore, definition (3) implies that once on the surface, the tracking error tends exponentially to zero, with a time constant $(n-1)/\lambda$.

The typical system behavior implied by satisfying sliding condition (6) is shown in **Figure 4** for $n=2$. A line with slope $-\lambda$ and containing the time-varying point $\boldsymbol{x}_d=[x_d\ \ddot{x}_d]^T$ represents the sliding surface in the phase plane. The state trajectory reaches the time-varying surface in a finite time smaller than $s(t=0)/\eta$ from any initial condition, and then slides across the surface towards $\boldsymbol{x}_d$ exponentially, with a time constant equal to $1/\lambda$.

In conclusion, the idea behind (3) and (6) is to obtain an appropriate function of the tracking error, s, in accordance with (3), and then choose the feedback control law u in (1) such that s^2 continues to be used as a Lyapunov-like function of the closed-loop system, in spite of the presence of model imprecision and of

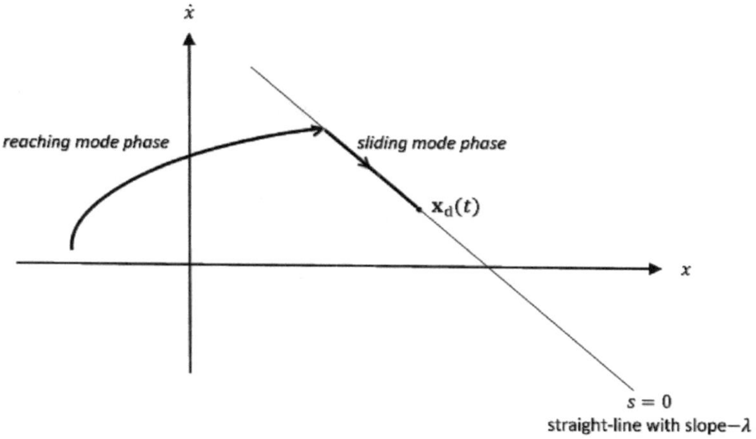

Figure 4.
Graphically represented Eqs. (3) and (6) for $n = 2$.

disturbances. Then, the controller design is a two-step procedure. First, the selection of a feedback control law u is performed to verify sliding condition (6). However, it is required that the control law be *discontinuous throughout* $S(t)$ to take into account the presence of modeling imprecision and of disturbances. Since the execution of the associated control switchings is not necessarily perfect (for instance, in practice, switching is not instantaneous, and the value of s is not known with infinite precision), this causes *chattering* (**Figure 5**). Now, with a few important exceptions, chattering is practically undesirable, because it contains high control activity and can trigger neglected high-frequency dynamics during modeling (such as unmodeled structural modes, neglected time-delays, etc.). Thus, in a second step, the *discontinuous control law u* is *smoothed accordingly to reach an optimal compromise between control bandwidth and tracking precision*: while the first step explains parametric uncertainty, the second step ensures robustness to high-frequency unmodeled dynamics.

As mentioned previously, the discontinuous control law causes chattering of the trajectories to take place around the surface $s = 0$. This problem can be eliminated

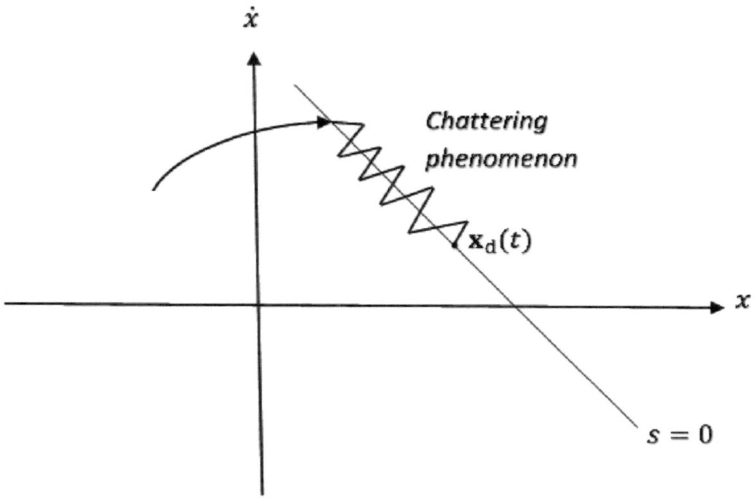

Figure 5.
Chattering caused by the switching delays.

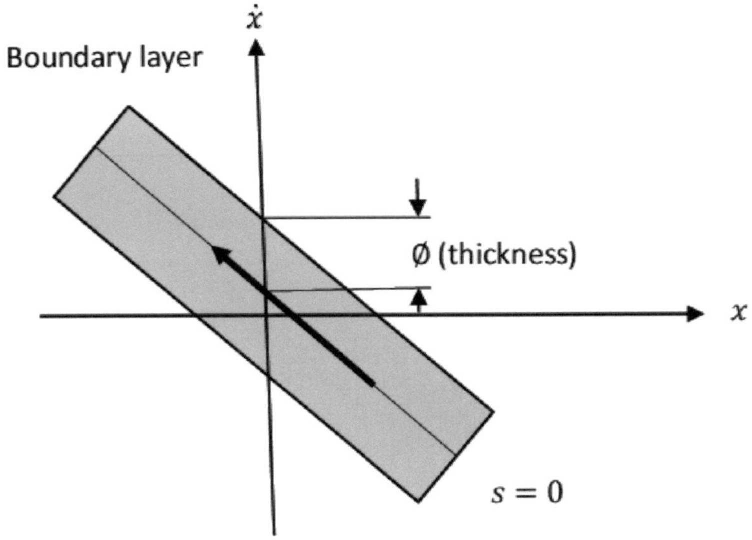

Figure 6.
Boundary layer with thickness ∅.

by smoothing out the discontinuities in the vicinity of the sliding surface through the introduction of a boundary layer thickness. An adaptation of saturation nonlinearity instead of signum nonlinearity in a position control system in which it is represented by Eqs. (83) and (84) in order to decrease the chattering phenomenon caused by sliding mode control law is the result of the same effort of smoothing out the discontinuities with the introduction of the boundary layer thickness as illustrated in **Figure 6**.

To maintain the system work in the sliding surface, a switching action term, u_{sw}, is added to the control law, and is defined by

$$u_{sw} = -Ksat(s),$$

and overall control law can be expressed as:

$$u = \hat{u} + u_{sw} = \underbrace{\left(-\hat{f} + \ddot{x}_d - \lambda\dot{\tilde{x}}\right)}_{Low\ frequency\ control} - \underbrace{Ksat(s)}_{High\ frequency\ control},$$

which will be explained in more detail in Section 5.1. Here, the nonlinear saturation function $sat(s)$, which is the replacement for nonlinear signum function $sgn\,(s)$, is defined by

$$sat(s) = \begin{cases} -1, & \frac{s}{\varnothing} < 0 \\ 0, & \left|\frac{s}{\varnothing}\right| < 1, \\ 1, & \frac{s}{\varnothing} < -1 \end{cases}$$

where ∅ is the boundary layer thickness.

3. Sliding mode control design

A continuous-time dynamical equation of an n-link robot manipulator is defined as follows:

$$\boldsymbol{M}(\boldsymbol{q})\ddot{\boldsymbol{q}} + \boldsymbol{C}(\boldsymbol{q}, \dot{\boldsymbol{q}})\dot{\boldsymbol{q}} + \boldsymbol{F}(\dot{\boldsymbol{q}}) + \boldsymbol{G}(\boldsymbol{q}) = \boldsymbol{\tau}, \tag{8}$$

where $\boldsymbol{q} \in \boldsymbol{R}^{nx1}$ denotes the joint configuation variables (translational or rotational) representing the generalized position coordinates (alias the joint positions) of the robot links. Similarly, $\dot{\boldsymbol{q}}, \ddot{\boldsymbol{q}} \in \boldsymbol{R}^{nx1}$ represent the joint velocity and acceleration of the robot links, respectively. $\boldsymbol{M}(\boldsymbol{q}) \in \boldsymbol{R}^{nxn}$ is the symmetric, bounded, positive definite inertia mass matrix. $\boldsymbol{C}(\boldsymbol{q}, \dot{\boldsymbol{q}}) \in \boldsymbol{R}^{nxn}$ is the matrix of Coriolis and centripetal forces and $\boldsymbol{F}(\dot{\boldsymbol{q}}) \in \boldsymbol{R}^{nx1}$ is the vector of viscous friction. Furthermore, the vector of $\boldsymbol{G}(\boldsymbol{q}) \in \boldsymbol{R}^{nx1}$ represents the gravity terms, and finally, $\boldsymbol{\tau} \in \boldsymbol{R}^{nx1}$ is called the control torque vector, or the vector of applied joint torques.

Sliding surface defined below is considered in the design of SMC controller:

$$\boldsymbol{s} = \dot{\boldsymbol{e}} + \boldsymbol{\lambda}\boldsymbol{e}, \tag{9}$$

where $\boldsymbol{e} = -\tilde{\boldsymbol{q}} = \boldsymbol{q} - \boldsymbol{q}_d$ represents the error vector and $\boldsymbol{\lambda}$ is assumed to be a symmetric positive definite matrix such that $\boldsymbol{s} = \boldsymbol{0}$ would evolve into a stable surface. The reference velocity vector $\dot{\boldsymbol{q}}_r$ is the same as the definition in [1]:

$$\dot{\boldsymbol{q}}_r = \dot{\boldsymbol{q}}_d - \boldsymbol{\lambda}\boldsymbol{e}. \tag{10}$$

Hence, one can define the sliding surface as follows:

$$\boldsymbol{s} = \dot{\boldsymbol{q}} - \dot{\boldsymbol{q}}_r. \tag{11}$$

Now, the following lemma refers to the sliding mode controller design.

Lemma 1. *Let us consider the system consisting of (8) through (10). If the following control rule is selected:*

$$\boldsymbol{\tau} = \hat{\boldsymbol{\tau}} - \boldsymbol{K}_{sg}\, sgn\,(\boldsymbol{s}) \tag{12}$$

such that

$$\hat{\boldsymbol{\tau}} = \boldsymbol{M}\ddot{\boldsymbol{q}}_r + \hat{\boldsymbol{C}}\ddot{\boldsymbol{q}}_r + \boldsymbol{G} \tag{13}$$

and

$$K_{sg_i} \geq \left\|\Delta\boldsymbol{C}\dot{\boldsymbol{q}}_r\right\| + \Gamma_i, \tag{14}$$

then the following sliding condition [1],

$$\frac{1}{2}\frac{d}{dt}\left[\boldsymbol{s}^T\boldsymbol{M}\boldsymbol{s}\right] < -\eta\left(\boldsymbol{s}^T\boldsymbol{s}\right)^{1/2}, \eta > 0 \tag{15}$$

is satisfied. In (14), K_{sg_i} stands for the element of sliding gain vector $\boldsymbol{K}_{sg}$ and $\boldsymbol{\Gamma}$ is a design parameter vector that must be chosen to ensure the inequality $\Gamma_i \geq F_{up} + \eta_i$.

Proof. Let Lyapunov function candidate be given as follows:

$$V = \frac{1}{2} s^T M s. \tag{16}$$

Since M is positive definite and s is different from zero ($s \neq 0$), V is always greater than zero ($V > 0$) and by taking time derivative of (16) and taking into account the symmetric property of M, it takes the following form:

$$\dot{V} = \frac{1}{2}\dot{s}^T M s + \frac{1}{2} s^T \left(\dot{M} s + M \dot{s}\right) = \frac{1}{2}\dot{s}^T M s + \frac{1}{2} s^T \dot{M} s + \frac{1}{2} s^T \dot{M} s = s^T M \dot{s} + \frac{1}{2} s^T \dot{M} s \tag{17}$$

Using (11), we get:

$$\dot{V} = s^T M(\ddot{q} - \ddot{q}_r) + \frac{1}{2} s^T \dot{M} s = s^T \left(M\ddot{q} - M\ddot{q}_r\right) + \frac{1}{2} s^T \dot{M} s \tag{18}$$

Taking $\ddot{q}$ from (8) and replacing it in (18), we have:

$$\dot{V} = s^T \left(M M^{-1}(\tau - C\dot{q} - G - F) - M\ddot{q}_r\right) + \frac{1}{2} s^T \dot{M} s.$$

Then, taking $\dot{q}$ from (11) and replacing it above yields:

$$\dot{V} = s^T \left(\tau - C\dot{q}_r - G - F - M\ddot{q}_r\right) + \frac{s^T \left(\dot{M} - 2C\right) s}{2}.$$

In the above equation, the second term is zero due to the *asymmetry property*; therefore, it disappears. In this new case,

$$\dot{V} = s^T \left(\tau - C\dot{q}_r - G - F - M\ddot{q}_r\right). \tag{19}$$

Next, applying (12) and (13) successively for τ and $\hat{\tau}$ in Eq. (19), and proceeding step by step, the following result is reached:

$$\dot{V} = s^T \left(\left(\hat{C} - C\right)\dot{q}_r - K_{sg} sgn(s) - F\right) = s^T \left(-\Delta C \dot{q}_r - K_{sg}\, sgn\,(s) - F\right). \tag{20}$$

In robot modeling, the terms $M(q)$ and $G(q)$ can be well and accurately determined, but in most cases it is not easy to have the parameters $C(q, \dot{q})$ and $F(\dot{q})$ precisely. Therefore, in this work, the matrix C is considered

$$C = \hat{C} + \Delta C, \tag{21}$$

where $C_{i,j}$ stands for the elements of the matrix C. Also, the vector F is assumed an external disturbance with the upper bound defined as,

$$\|F\| \leq F_{up}, \tag{22}$$

where the operator $\|.\|$ denotes Euclidian norm [2]. Now, by rearranging (20) as shown step by step below, we get the following:

$$\dot{V} = -s^T \left(\Delta C \dot{q}_r + F\right) - s^T K_{sg}\, sgn\,(s) = -s^T \left(\Delta C \dot{q}_r + F\right) - \sum_{i=1}^{n} K_{sg_i} |s_i|, \tag{23}$$

where $|s_i| = s_i^T\, sgn\,(s_i)$.

At this point, we can briefly verify that the terms on the right side of (14) are positive. First of all, it is easiest to say that the first term on the right, $\|\Delta C\dot{q}_r\|$, is positive in any case, because the Euclidian norm is used. The other term Γ_i is also positive. Because, as we have already stated in (14) that $\Gamma_i \geq F_{up} + \eta_i$, where F_{up} is an unknown upper bound defined as $\|F\| \leq F_{up}$ in (22), which also indicates that it is absolutely positive, and η is a strictly positive design constant and ensures inequality (6). Hence, $\Gamma_i \geq 0$. Now, if the inequality K_{sg_i} given by (14) is substituted in Eq. (23) under the acceptance of its equality sign for a moment, we can rewrite Eq. (23) by extending it as follows:

$$\dot{V} = -s^T(\Delta C\dot{q}_r + F) - \sum_{i=1}^{n}\|\Delta C\dot{q}_r\||s_i| - \sum_{i=1}^{n} F_{up}|s_i| - \sum_{i=1}^{n}\eta_i|s_i|.$$

The first, second and third terms on the right side of the equation above are negative in varying amounts and contribute to the final term, which is $\sum_{i=1}^{n}\eta_i|s_i|$, more negatively. Therefore, we can easily conclude that

$$\dot{V} \leq -\sum_{i=1}^{n}\eta_i|s_i|. \tag{24}$$

This shows that V is a Lyapunov function and the satisfaction of sliding condition in (15) is proven.

4. Achievement of the control law for robot manipulators for the adapted reaching mode

VSC systems include a group of different, generally fairly simple, feedback control laws and a decision rule. Depending on the system condition, a decision rule, usually called the *switching function*, determines which control law is "on-line" at any time. The transient dynamics of VSC systems consists of two modes: a "reaching mode" (or "non-sliding mode"), and a subsequent "sliding mode". Hence, VSC design involves two stages: the first one involves the design of the appropriate n-dimensional switching function $s(x)$ for a desired sliding mode dynamics. The second one involves a control design for the reaching mode where a reaching condition is met. The desired sliding mode dynamics usually includes a fast and stable error-free response without overshoot. In sliding mode, an asymptotic convergence to the final state will be accomplished. The desired response in the reaching mode, in general, is to reach the switching manifold defined as

$$s(x) = \psi^T x = 0, \tag{25}$$

in a finite time with a small amount of overshoot with regard to the switching manifold [18].

The reaching law is a differential equation that determines the dynamics of a switching function $s(x)$. If $s(x)$ is an asymptotically stable differential equation, then, it is solely a reaching condition. Further, the parameter selection in the differential equation controls the dynamic quality of the VSC system in the reaching mode. The reaching law can be expressed practically in general form as follows [18]:

$$\dot{s} = -Q sgn(s) - K h(s), \tag{26}$$

where $Q = diag[q_1, \ldots, q_n], q_i > 0$; $sgn\,(s) = [sgn\,(s_1), \ldots, sgn\,(s_n)]^T$; $K = diag[k_1, \ldots, k_n], k_i > 0$; $h(s) = [h_1(s_1), \ldots, h_n(s_n)]^T$; and $s_i h_i(s_i) > 0, h_i(0) = 0$.

The design principle of the SMC law for the plants of arbitrary order is to force a variable's error and its derivative to zero. Tracking of a desired motion $q^d(t)$ is the main task of the robot arm. Here, let us start first by defining a $2n$-dimensional error vector [18]:

$$e = \begin{bmatrix} e_1 \\ e_2 \end{bmatrix} = \begin{bmatrix} q^d - x_1 \\ \dot{q}^d - x_2 \end{bmatrix} = \begin{bmatrix} q^d - q \\ \dot{q}^d - \dot{q} \end{bmatrix}, \tag{27}$$

and then, an n-dimensional vector of switching function:

$$s(e) = \Psi e = [\Lambda \quad I] \begin{bmatrix} e_1 \\ e_2 \end{bmatrix} = \Lambda e_1 + \dot{e}_1, \tag{28}$$

where $\dot{e}$ represents the tracking speed error and:

$$\Lambda = diag[\lambda_1, \ldots, \lambda_n], \quad \lambda_i > 0,$$

that determines the system bandwidth. Next, the time derivative of (28) is taken as follows [18]:

$$\dot{s}(e) = \Lambda \dot{e}_1 + \dot{e}_2 = \Lambda \dot{e}_1 + \ddot{q}^d - \ddot{q}. \tag{29}$$

Now, constant plus proportional rate reaching law as represented by

$$\dot{s} = -Qsgn(s) - Ks \tag{30}$$

is adapted. Substituting (30) into (29) and setting $\ddot{q}$ apart yields:

$$\ddot{q} = Qsgn(s) + Ks + \Lambda \dot{e}_1 + \ddot{q}^d. \tag{31}$$

Finally, substituting (31) into the non-linear plant of continuous-time dynamic model of robot systems in (8) results in:

$$M(q)\left[Qsgn(s) + Ks + \Lambda \dot{e}_1 + \ddot{q}^d\right] + C(q, \dot{q})\dot{q} + F(\dot{q}) + G(q) = \tau. \tag{32}$$

This is also known as the *final control law*.

5. Proofs of the boundedness and convergence properties of smooth sliding mode controllers

In this section, the proofs of the boundedness and convergence properties of the smooth sliding mode controllers are introduced. In particular, the convergence analysis of smooth sliding mode controllers will be explained and discussed to the finest detail. Lyapunov's direct method is used to handle the finite-time convergence of the tracking error vector to the boundary layer. Also, once in the boundary layer, the tracking error vector is said to have exponentially converged to a bounded region, as proven analytically.

DOI: http://dx.doi.org/10.5772/intechopen.93027

5.1 Problem statement

Consider the following non-linear system class of n-th order:

$$x^{(n)} = f(\boldsymbol{x}) + b(\boldsymbol{x})u, \tag{33}$$

where u is the control input, $x^{(n)}$, is the n-th order derivative of the interested scalar output variable x with respect to time $t \in [0, \infty)$. Here, also, $\boldsymbol{x} = \left[x, \dot{x}, \dots, x^{(n-1)}\right]^T$ represents the system state vector, and both $f(x)$ and $b(x)$, such that $f, b : \boldsymbol{R}^n \to \boldsymbol{R}$, denote nonlinear functions.

The following assumptions will be made in terms of the dynamic system presented in (33).

Assumption 1. *f is an unknown function such that it is bounded by a known function $\boldsymbol{x}$, i.e., $\left|\hat{f}(\boldsymbol{x}) - f(\boldsymbol{x})\right| \leq F(\boldsymbol{x})$, where $\hat{f}$ is an estimate of f.*

Assumption 2. *Input gain $b(\boldsymbol{x})$ is an unknown function such that it is positive and bounded, i.e., $0 < b_{min} \leq b(\boldsymbol{x}) \leq b_{max}$.*

In the proposed state space control problem, the $\boldsymbol{x}$ state vector must be able to follow a desired trajectory $\boldsymbol{x}_d = \left[x_d, \ddot{x}_d, \dots, x_n^{(n-1)}\right]$, even under the presence of parametric uncertainties and unmodulated dynamics.

The following assumptions should also be made during the development of the control law.

Assumption 3. *The state vector $\boldsymbol{x}$ has availability.*

Assumption 4. *The desired trajectory $\boldsymbol{x}_d$ is differentiated once in time. Moreover, each element of the vector $\boldsymbol{x}_d$ as well as $\boldsymbol{x}_d^{(n)}$, is available and has known bounds.*

Now, let $\tilde{\boldsymbol{x}} = \boldsymbol{x} - \boldsymbol{x}_d$ be defined as the tracking error for the variable x, and $\tilde{\boldsymbol{x}} = \boldsymbol{x} - \boldsymbol{x}_d = \left[\tilde{x}, \dot{\tilde{x}}, \dots, \tilde{x}^{(n-1)}\right]$ as the tracking error vector.

Let us define a sliding surface S in the state space by the equation $s(\tilde{\boldsymbol{x}}) = 0$ in which s is the function mapping from n-dimensional real space $\boldsymbol{R}^n$ to one-dimensional real space $\boldsymbol{R}$, i.e., s: $\boldsymbol{R}^n \to \boldsymbol{R}$, and satisfying the following equation:

$$s(\tilde{\boldsymbol{x}}) = \left(\frac{d}{dt} + \lambda\right)^{n-1} \tilde{x},$$

which can be plainly rewritten as

$$s(\tilde{\boldsymbol{x}}) = \boldsymbol{c}^T \tilde{\boldsymbol{x}}, \tag{34}$$

where $\boldsymbol{c} = \left[c_{n-1}\lambda^{n-1} + \dots + c_1\lambda, c_0\right]$ with c_i representing binomial coefficients as follows:

$$c_i = \binom{n-1}{i} = \frac{(n-1)!}{(n-i-1)!i!}, i = 0, 1, \dots, n-1 \tag{35}$$

which makes $c_{n-1}\lambda^{n-1} + \dots + c_1\lambda, c_0$ a Hurwitz polynomial.

It can be easily verified from (35) that $c_0 = 1$, for $\forall n \geq 1$. Therefore, the time derivative of s will be expressed in the following form:

$$\dot{s} = \boldsymbol{c}^T\dot{\tilde{\boldsymbol{x}}} = \underbrace{\begin{bmatrix} c_{n-1}\lambda^{n-1} \\ \vdots \\ c_1\lambda \\ c_0 \end{bmatrix}}_{\boldsymbol{c}^T} \underbrace{\left[\dot{\tilde{x}}, \ddot{\tilde{x}}, \dots, \tilde{x}^{(n)}\right]}_{\dot{\tilde{\boldsymbol{x}}}} = \underbrace{\begin{bmatrix} 0 \\ c_{n-1}\lambda^{n-1} \\ \vdots \\ c_2\lambda^2 \\ c_1\lambda \end{bmatrix}}_{\triangleq \bar{\boldsymbol{c}}^T} \underbrace{\left[\tilde{x}, \dot{\tilde{x}}, \dots, \tilde{x}^{(n-1)}\right]}_{\dot{\tilde{\boldsymbol{x}}}} + \tilde{x}^{(n)}$$

i.e.,

$$\dot{s} = \boldsymbol{c}^T\dot{\tilde{\boldsymbol{x}}} = \tilde{x}^{(n)} + \bar{\boldsymbol{c}}^T\tilde{\boldsymbol{x}} \tag{36}$$

where, here, as used for the first time above, there is a definition in the form of $\bar{\boldsymbol{c}} = \left[0, c_{n-1}\lambda^{n-1}, \cdots, c_1\lambda\right]$. At this point, let us evaluate Eqs. (34) and (36) for $n = 3$, i.e.,

$$s(\tilde{\boldsymbol{x}}) = \left(\frac{d}{dt} + \lambda\right)^2 \tilde{x} = \ddot{\tilde{x}} + 2\lambda\dot{\tilde{x}} + \lambda^2\tilde{x},$$

from which $\boldsymbol{c}$ appears to be as $\boldsymbol{c} = \left[c_2\lambda^2, c_1\lambda, c_0\right]$. Then, $s(\tilde{\boldsymbol{x}}) = \boldsymbol{c}^T\tilde{\boldsymbol{x}}$ or to create a polynomial, $\boldsymbol{c}\tilde{\boldsymbol{x}}^T = \left[c_2\lambda^2, c_1\lambda, c_0\right]\begin{bmatrix} \tilde{x} \\ \dot{\tilde{x}} \\ \ddot{\tilde{x}} \end{bmatrix} = c_2\lambda^2\tilde{x} + c_1\lambda\dot{\tilde{x}} + c_0\ddot{\tilde{x}}$ with $c_0 = 1$ as always, and $c_2\lambda^2 + c_1\lambda + c_0$ is a Hurwitz polynomial. That is, it is defined as the polynomial with its coefficients (i.e., c_i) that are positive real numbers, and its zeros are located in the left half-plane –i.e., the real part of every zero is negative– of the complex plane.

Now, Let the problem of controlling the uncertain nonlinear system expressed by (33) be handled for review through the classical sliding mode approach that defines a control rule consisted of an equivalent control $\hat{u} = \hat{b}^{-1}\left(-\hat{f} + x_d^{(n)} - \bar{\boldsymbol{c}}^T\tilde{\boldsymbol{x}}\right)$ and a discontinuous term $-Ksgn(s)$ as follows:

$$u = \hat{b}^{-1}\left(-\hat{f} + x_d^{(n)} - \bar{\boldsymbol{c}}^T\tilde{\boldsymbol{x}}\right) - Ksgn(s). \tag{37}$$

where $\hat{b} = \sqrt{b_{max}b_{min}}$ represents the estimated value of b, and K represents a positive gain. Furthermore, the sign or signum function represented by $sgn\,(s)$ above

$$sgn\,(s) = \begin{cases} -1, & if\; s < 0 \\ 0, & if\; s = 0. \\ 1, & if\; s > 0 \end{cases}$$

Based on Assumptions 1 and 2 given above and taking into account the fact that $\beta^{-1} \leq \hat{b}/b \leq \beta$, where $\beta = \sqrt{b_{max}/b_{min}}$, the gain K must be determined in such a way as to ensure the following inequality:

$$K \geq \beta\hat{b}^{-1}(\eta + F) + (\beta - 1)\left|\hat{b}^{-1}\left(-\hat{f} + x_d^{(n)} - \bar{\boldsymbol{c}}^T\tilde{\boldsymbol{x}}\right)\right|, \tag{38}$$

where η is a strictly positive constant of the reaching time. Now, in this step, let us reaffirm the validity of the lower and upper bounds of b using the $\hat{b}$ and β

definitions given: first of all, Let the definitions of $\hat{b}$ and β be placed in the expression $\beta^{-1} \le \hat{b}/b \le \beta$ given above. In this case,

$$\frac{1}{\sqrt{b_{max}/b_{min}}} \le \frac{\sqrt{b_{max}b_{min}}}{b} \le \sqrt{b_{max}/b_{min}}.$$

If each side is multiplied by $1/\sqrt{b_{max}b_{min}}$, the following inequality is obtained:

$$\frac{1}{b_{max}} \le \frac{1}{b} \le \frac{1}{b_{min}}.$$

In this inequality, if inversion is applied to all terms, inequalities will be completely displaced, that is to say, it will become $b_{max} \ge b \ge b_{min}$. This is a necessary initial acceptance. Therefore, when we turn b's upper and lower bounds into inequality, we once again confirm the correctness of the definitions for $\hat{b}$ and β. Since the control rule will be designed to be robust against the inequality $\beta^{-1} \le \hat{b}/b \le \beta$, that is, a bounded multiplicative uncertainty, taking advantage of the similarity to the terminology used in linear control, we can call β the gain margin of the design.

In order to ensure $x \equiv x_d$ system tracking, we define a sliding surface $s = 0$ according to $s(\boldsymbol{x};t) = \left(\frac{d}{dt} + \lambda\right)^{n-1}\tilde{x}$, that is,

$$s = \left(\frac{d}{dt} + \lambda\right)\tilde{x} = \dot{\tilde{x}} + \lambda\tilde{x}.$$

When we derive the expression s, we obtain the following:

$$\dot{s} = \ddot{\tilde{x}} + \lambda\dot{\tilde{x}} = \ddot{x} - \ddot{x}_d + \lambda\dot{\tilde{x}} = f + u - \ddot{x}_d + \lambda\dot{\tilde{x}}.$$

For $\dot{s} = f + u - \ddot{x}_d + \lambda\dot{\tilde{x}} = 0$ to be realized, other terms outside of u must be determined equal to the opposite sign of $\hat{u}$, which is the best approximation of a continuous control rule u that can implement $\dot{s} = 0$, that is,

$$\hat{u} = -\hat{f} + \ddot{x}_d - \lambda\dot{\tilde{x}}.$$

In fact, to see this result, the first thing to do is to draw u from the equation $\dot{s} = f + u - \ddot{x}_d + \lambda\dot{\tilde{x}} = 0$. Hence, $u = -f + \ddot{x}_d - \lambda\dot{\tilde{x}}$ is obtained. Then, from here, in order to obtain the approximate value of u, searching for the approximation of the function on the right side of the equation, and representing this approximated function by $\hat{f}$ symbolically are sufficient to lead us to the correct result, as seen above.

The control rule $u = \hat{b}^{-1}[\hat{u} - ksgn(s)]$ with predefined s and $\hat{u}$, and k defined by the inequality $k \ge \beta(F + \eta) + (\beta - 1)|\hat{u}|$—as will be explained little below—meets the sliding condition. Indeed, when we substitute this control rule in the expression $\dot{s} = f + bu - \ddot{x}_d + \lambda\dot{\tilde{x}}$ by choosing the use of $\ddot{x} = f + bu$, which is more specific to this type of structure, instead of $\ddot{x} = f + u$ used only just two above, we obtain the following,

$$\dot{s} = f + b\hat{b}^{-1}[\hat{u} - ksgn(s)] - \ddot{x}_d + \lambda\dot{\tilde{x}}.$$

Once the previously determined $\hat{u}$ is replaced above, the following equation is reached:

$$\dot{s} = f + b\hat{b}^{-1}\left[-\hat{f} + \ddot{x}_d - \lambda\dot{\tilde{x}} - ksgn(s)\right] - \ddot{x}_d + \lambda\dot{\tilde{x}}.$$

The organized form of this statement will be as follows:

$$\dot{s} = \left(f - b\hat{b}^{-1}\hat{f}\right) + \left(1 - b\hat{b}^{-1}\right)\left(-\ddot{x}_d + \lambda\dot{\tilde{x}}\right) - b\hat{b}^{-1}ksgn(s).$$

Such that k should meet the following condition,

$$k \geq \left|\hat{b}b^{-1}f - \hat{f} + \left(\hat{b}b^{-1} - 1\right)\left(-\ddot{x}_d + \lambda\dot{\tilde{x}}\right)\right| + \eta\hat{b}b^{-1} \tag{39}$$

We can really achieve this condition by following the steps below:

$$\dot{s} = 0 = \left(f - b\hat{b}^{-1}\hat{f}\right) + \left(1 - b\hat{b}^{-1}\right)\left(-\ddot{x}_d + \lambda\dot{\tilde{x}}\right) - b\hat{b}^{-1}ksgn(s) \Longrightarrow$$

$$b\hat{b}^{-1}ksgn(s) = \left(f - b\hat{b}^{-1}\hat{f}\right) + \left(1 - b\hat{b}^{-1}\right)\left(-\ddot{x}_d + \lambda\dot{\tilde{x}}\right) \Longrightarrow$$

$$ksgn(s) = \hat{b}b^{-1}f - \hat{f} + \left(\hat{b}b^{-1} - 1\right)\left(-\ddot{x}_d + \lambda\dot{\tilde{x}}\right) \Longrightarrow$$

$$ksgn(s) = \hat{b}b^{-1}\left[\hat{f} + \left(f - \hat{f}\right)\right] - \hat{f} + \left(\hat{b}b^{-1} - 1\right)\left(-\ddot{x}_d + \lambda\dot{\tilde{x}}\right).$$

Here, it was previously described that $\left|f - \hat{f}\right| \leq F$. But, when determining k, it will be necessary to take into account the *reaching time* η. Therefore, we will allow $F + \eta$ to be written instead of F. Removing the term $sgn\ (s)$, k will be determined as follows as a result of our compensation through expressing, with an absolute value, the effect of its reciprocations being of which the negative s values relative to the positive s values and of which only the sign changed:

$$k \geq \hat{b}b^{-1}F + \eta\hat{b}b^{-1} + \left|\hat{b}b^{-1} - 1\right|\left|\hat{f} - \ddot{x}_d + \lambda\dot{\tilde{x}}\right|.$$

Note here that $F \geq 0$ and $\eta > 0$ (absolute positive). For this reason, there is no need to take absolute values of these terms. Here again, using the definition $\hat{b}b^{-1} \triangleq \beta$, for k, we get the expression,

$$k \geq \beta(F + \eta) + |\beta - 1||\hat{u}|. \tag{40}$$

Remark 1. To avoid any confusion, if we wanted to verify (40) by proceeding from (39), since η has already been taken into account in (39), we would not need to take $F + \eta$ instead of F. We would proceed directly with $\left|f - \hat{f}\right| \leq F$.

Remark 2. Considering the fact that the F value can be faced with moments where the estimation problem will be relatively large by nature and similarly with the moments when η reaching time will be relatively larger, it is possible to state precisely that it is the right choice or necessity to take the direction of inequality k greater than or equal to.

Thus, it can be easily verified that the control rule $u = \hat{b}^{-1}\left(-\hat{f} + x_d^{(n)} - \bar{c}^T\tilde{\boldsymbol{x}}\right) - Ksgn(s)$ is sufficient to impose the shift condition,

$$\frac{1}{2}\frac{d}{dt}s^2 \leq -\eta|s|,$$

which indeed guarantees the convergence of the tracking error vector to the sliding surface S and consequently its exponential stability in a finite-time. In response to the uncertainty of f on dynamics, we add a discontinuous term to $\hat{u}$ across the surface $s = 0$ to meet the slip condition given above, that is:

$$u = \hat{u} - ksgn(s).$$

Here we can now guarantee that the sliding condition will be verified by choosing $k = k(x, \dot{x})$ sufficiently large. Indeed,

$$\frac{1}{2}\frac{d}{dt}s^2 = \dot{s}s = \left[f + \hat{u} - ksgn(s) - \ddot{x}_d + \lambda\dot{\tilde{x}}\right]s = \left[f - \hat{f} - ksgn(s)\right]s.$$

This last operation is important; because we have reached this point by using the equations $u = -f + \ddot{x}_d - \lambda\dot{\tilde{x}}$, $\hat{u} = -\hat{f} + \ddot{x}_d - \lambda\dot{\tilde{x}}$, and $u = \hat{u} - ksgn(s)$ as follows:

$$\begin{aligned}\dot{s} &= \left[f + u - \ddot{x}_d + \lambda\dot{\tilde{x}}\right] = f + \hat{u} - ksgn(s) - \ddot{x}_d + \lambda\dot{\tilde{x}}\\ &= f + \left(-\hat{f} + \ddot{x}_d - \lambda\dot{\tilde{x}}\right) - ksgn(s) - \ddot{x}_d + \lambda\dot{\tilde{x}} = f - \hat{f} - ksgn(s).\end{aligned}$$

If we continue where we were,

$$\frac{1}{2}\frac{d}{dt}s^2 = \left[f - \hat{f} - ksgn(s)\right]s = \left(f - \hat{f}\right)s - ksgn(s)s.$$

Therefore, since $sgn\,(s)s = |s|$, the following expression is reached,

$$\frac{1}{2}\frac{d}{dt}s^2 = \dot{s}s = \left[f - \hat{f} - ksgn(s)\right]s = \left(f - \hat{f}\right)s - k|s|.$$

So that, when $k = F + \eta$ is selected, the above statement follows,

$$\frac{1}{2}\frac{d}{dt}s^2 = \dot{s}s = \left(f - \hat{f}\right)s - F|s| - \eta|s|.$$

However, although the definition of $\left|\hat{f} - f\right| \leq F$ is given at the beginning of the section, recalling that we prefer the form $\left|f - \hat{f}\right| \leq F$ to be used in the case study below, Let us give the statement its final form:

$$\frac{1}{2}\frac{d}{dt}s^2 = \dot{s}s = \left(f - \hat{f}\right)s - \left|f - \hat{f}\right||s| - \eta|s|. \tag{41}$$

In fact, note that here the expression $\left|f - \hat{f}\right|$, which we substitute for F represents the smallest value that F can take. We generally know that F is greater than this value.

We will now carry out the following case studies for the Eq. (41):

Case 1. *If* $\left(f - \hat{f}\right)$ *and* s *are both negative or positive, as such,* $\left(f - \hat{f}\right)s - \left|f - \hat{f}\right||s| = 0$. *However, it is known to be* $F \geq \left|f - \hat{f}\right|$, *hence,* $\left(f - \hat{f}\right)s - F|s| \leq 0$, *i.e., it will always be negative.*

Case 2. *However, if* $\left(f-\hat{f}\right)$ *and s opposite signs;* $\left(f-\hat{f}\right)s$ *will always be negative.* $-F|s|$ *will also be negative. Hence,* $\left(f-\hat{f}\right)s-F|s|$ *will always be more negative as compared to Case 1.*

As a result,

$$\frac{1}{2}\frac{d}{dt}s^2 \leq -\eta|s|,$$

is always true.

However, the presence of a discontinuous term (i.e., $-Ksgn(s)$) in the control rule leads to the well-known *chattering* effect. To prevent these unwanted high-frequency oscillations of the controlled variable, Slotine had proposed the idea of adopting a thin boundary layer $S_\varnothing$ around the switching surface [1]:

$$S_\varnothing = \{\tilde{\boldsymbol{x}} \in \boldsymbol{R}^n \,||s(\tilde{\boldsymbol{x}})| \leq \varnothing\}. \tag{42}$$

Here $\varnothing$ is an absolute positive constant, which represents the boundary layer thickness.

The boundary layer is accomplished by replacing the sign function with a continuous interpolation in $S_\varnothing$. It should be emphasized that this smooth approximation referring to the flatness or smoothness of the interpolating curve and its derivatives, which will be called $\varphi(s,\varnothing)$, here, will definitely act as a sign function outside the boundary layer.

Various options are available to smooth out the ideal switch. But the closest choices are the saturation function expressed by

$$sat\left(\frac{s}{\varnothing}\right) = \begin{cases} sgn\ (s),\ if\left|\frac{s}{\varnothing}\right| \geq 1 \\ \frac{s}{\varnothing},\ if\left|\frac{s}{\varnothing}\right| < 1, \end{cases} \tag{43}$$

and the hyperbolic tangent function expressed by $tanh\left(\frac{s}{\varnothing}\right)$. Thus, the smooth sliding mode control rule can be expressed as follows:

$$u = \hat{b}^{-1}\left(-\hat{f} + x_d^{(n)} - \bar{\boldsymbol{c}}^T\tilde{\boldsymbol{x}}\right) - K\varphi(s,\varnothing). \tag{44}$$

5.2 Convergence analysis

The attractiveness and invariance properties of the boundary layer are introduced in the following theorem:

Theorem 1. Consider four previously made assumptions with the uncertain nonlinear system given in (33). Therefore, the smooth sliding mode controller defined by (38) and (44) provides the finite-time convergence of the tracking error vector to the boundary layer $S_\varnothing$ defined by (42).

Proof. Let a positive-definite Lyapunov function candidate V be defined as,

$$V(t) = \frac{1}{2}s_\varnothing{}^2. \tag{45}$$

Here, as a measure of the distance of the current error to the boundary layer, $s_\varnothing$ can be computed as follows:

$$s_\varnothing = s - \varnothing sat\left(\frac{s}{\varnothing}\right). \tag{46}$$

Noting that $s_\varnothing = 0$ in the boundary layer, it is shown that $\dot{V}(t) = 0$ inside $S_\varnothing$. It is also possible to easily verify that $\dot{s}_\varnothing = \dot{s}$ outside the boundary layer through (43) and (46), and in this case, $\dot{V}$ can be written as follows:

$$\begin{aligned}\dot{V}(t) &= s_\varnothing \dot{s}_\varnothing = s_\varnothing \dot{s} = s_\varnothing\left(\boldsymbol{c}^T\dot{\tilde{\boldsymbol{x}}}\right) = s_\varnothing\left(\tilde{x}^{(n)} + \bar{\boldsymbol{c}}^T\tilde{\boldsymbol{x}}\right) = \left(x^{(n)} - x_d^{(n)} + \bar{\boldsymbol{c}}^T\tilde{\boldsymbol{x}}\right)s_\varnothing \\ &= \left(f + bu - x_d^{(n)} + \bar{\boldsymbol{c}}^T\tilde{\boldsymbol{x}}\right)s_\varnothing.\end{aligned}$$

Next, considering that the control rule given by (44) is written as

$$u = \hat{b}^{-1}\left(-\hat{f} + x_d^{(n)} - \bar{\boldsymbol{c}}^T\tilde{\boldsymbol{x}}\right) - Ksgn(s_\varnothing),$$

outside the boundary layer and noting that $f = \hat{f} - \left(\hat{f} - f\right)$, we get the following result:

$$\dot{V}(t) = -\left[\left(\hat{f} - f\right) - b\hat{b}^{-1}\left(-\hat{f} + x_d^{(n)} - \bar{\boldsymbol{c}}^T\tilde{\boldsymbol{x}}\right) - \hat{f} + x_d^{(n)} - \bar{\boldsymbol{c}}^T\tilde{\boldsymbol{x}} + bKsgn(s_\varnothing)\right]s_\varnothing.$$

Thus, by taking the Assumptions 1 and 2 into consideration, and defining K according to (38), $\dot{V}$ can be written as follows,

$$\dot{V}(t) \le -\eta|s_\varnothing|.$$

Because the Lyapunov function candidate, which we initially defined with (45) as positive definite, essentially inspired by the inequality in the form of $\frac{1}{2}\frac{d}{dt}s^2 \le -\eta|s|$ which we have always correctly demonstrated above, may well be represented by a similar structure to the form, $\frac{1}{2}\frac{d}{dt}{s_\varnothing}^2 \le -\eta|s_\varnothing|$. It will be seen from here that $\dot{V}(t) = s_\varnothing\dot{s}_\varnothing \le -\eta|s_\varnothing|$, as well. Hence, the inequality $\dot{V}(t) \le -\eta|s_\varnothing|$ will imply that $V(t) \le V(0)$ and therefore $s_\varnothing$ is bounded. Moreover, from the definitions of s and $s_\varnothing$ expressed in (35) and (46), respectively, it can be verified that $\tilde{\boldsymbol{x}}$ is bounded. Therefore, Assumption 4 and (36) imply that $\dot{s}$ is also bounded.

Finite-time convergence of the tracking error vector to the boundary layer can be shown remembering the expression,

$$\dot{V}(t) = \frac{1}{2}\frac{d}{dt}{s_\varnothing}^2 = s_\varnothing\dot{s}_\varnothing \le -\eta|s_\varnothing|.$$

Then, dividing both sides into $|s_\varnothing|$ above and integrating them between 0 and t will refer to the following result:

$$\int_0^t \frac{s_\varnothing}{|s_\varnothing|}\dot{s}_\varnothing d\tau \le -\int_0^t \eta d\tau.$$

Remark 3. Here, considering the ratio $s_\varnothing/|s_\varnothing|$ as the ratio of two numbers of the same size and therefore assuming it disappeared, that is, since it has no effect in

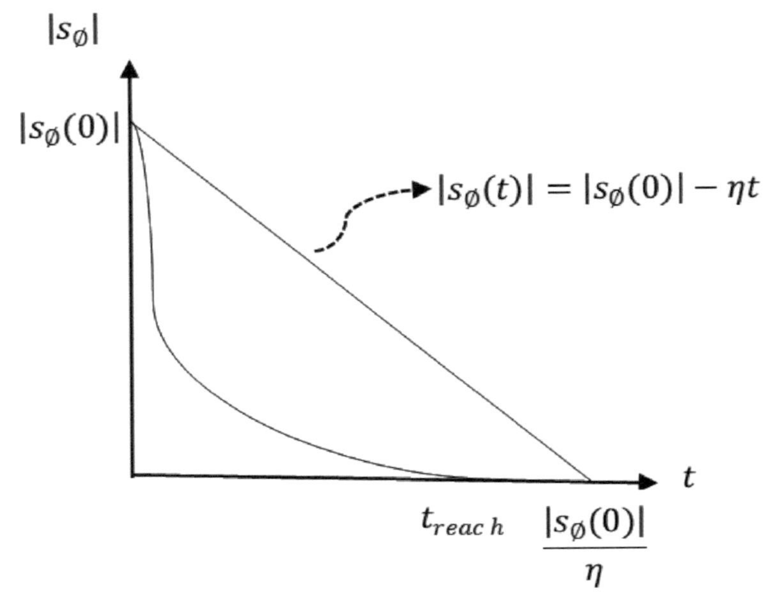

Figure 7.
Time evolution of the distance of the current tracking error to the boundary layer $|s_\varnothing|$.

size, substantially it is a correct approach to consider the integral as an equivalent to $\int_0^t \dot{s}_\varnothing d\tau$. This produces the result $|s_\varnothing(t)||_0^t$. Consequently, knowing the fact that in the situation before taking this approach, the product $s_\varnothing \dot{s}_\varnothing$ which appears in the numerator of the integral to the left of inequality is essentially equal to the derivative of the positive-definite V Lyapunov candidate function and is therefore positive again, it is essential to show the terms on the left side of the inequality with absolute value. That is to say, it is important to see that $\frac{s_\varnothing \dot{s}_\varnothing}{|s_\varnothing|} > 0$. Then, the next step to ensure this will turn into the form $|s_\varnothing(t)| - |s_\varnothing(0)| \leq -\eta t$. In this way, considering t_{reach} as the time required to reach $s_\varnothing$ and noting that $|s_\varnothing(t_{reach})| = 0$, we have the expression,

$$t_{reach} \leq \frac{|s_\varnothing(0)|}{\eta}$$

guaranteeing the convergence of the tracking error vector to the boundary layer in a time interval less than $|s_\varnothing(0)|/\eta$.

Remark 4. If both sides of $|s_\varnothing(t)| - |s_\varnothing(0)| \leq -\eta t$ are multiplied by -1, $|s_\varnothing(0)| - |s_\varnothing(t)| \geq \eta t$ is obtained, that is, briefly, the inequality is displaced. If t is left alone in the next step, $t \leq \frac{|s_\varnothing(0)| - |s_\varnothing(t)|}{\eta}$ is obtained. Hence, it is guaranteed to be $t_{reach} \leq |s_\varnothing(0)|/\eta$. That is, the right-hand side will act as the largest value achievable for t_{reach}. In other words, it will appear as a guaranteed upper value. Then, the value of $\frac{|s_\varnothing(0)| - |s_\varnothing(t)|}{\eta}$ is expected to be less than this guaranteed value of $|s_\varnothing(0)|/\eta$.

Therefore, to keep the reaching time, t_{reach}, as short as possible, the value of the positive constant η can be chosen appropriately. We clearly see from **Figure 7** that the time evolution of $|s_\varnothing|$ is bounded by the linear equation $|s_\varnothing(t)| = |s_\varnothing(0)| - \eta t$.

Lastly, the proof of the boundedness of the tracking error vector is based on Theorem 2.

Theorem 2. Let the boundary layer $S_\varnothing$ be defined according to (42). Then, once inside $S_\varnothing$, the tracking error vector will exponentially converge to an n-dimensional box defined according to $|\tilde{x}^{(i)}| \leq \zeta_i \lambda^{i-n+1} \phi, i = 0, 1, \cdots, n-1$, with ζ_i satisfying

$$\zeta_i = \begin{cases} 1, for\, i = 0 \\ 1 + \sum_{j=0}^{i-1} \binom{i}{j} \zeta_j, for\, i = 1, 2, \cdots, n-1. \end{cases} \tag{47}$$

Proof. Considering the fact that $|s(\tilde{\boldsymbol{x}})| \leq \varnothing$ can be rewritten as $-\varnothing \leq s(\tilde{\boldsymbol{x}}) \leq \varnothing$ with the definition of s given in (34), the expression below

$$s(\tilde{\boldsymbol{x}}) = \boldsymbol{c}^T \tilde{\boldsymbol{x}} = \left[c_{n-1}\lambda^{n-1} + \ldots + c_1\lambda, c_0\right] \begin{bmatrix} \tilde{x} \\ \dot{\tilde{x}} \\ \vdots \\ \tilde{x}^{(n-1)} \end{bmatrix}$$
$$= c_0 \tilde{x}^{(n-1)} + c_1 \lambda \tilde{x}^{(n-2)} + \cdots + c_{n-1}\lambda^{n-1}\tilde{x}.$$

Thus,

$$-\varnothing \leq s(\tilde{\boldsymbol{x}}) \leq \varnothing = -\varnothing \leq c_0 \tilde{x}^{(n-1)} + c_1 \lambda \tilde{x}^{(n-2)} + \cdots + c_{n-1}\lambda^{n-1}\tilde{x} \leq \varnothing, \tag{48}$$

or the following,

$$-\varnothing \leq \left(\frac{d}{dt} + \lambda\right)^{n-1} \tilde{x} \leq \varnothing \tag{49}$$

can be written. If (49) is multiplied by $e^{\lambda t}$, the following statement is reached:

$$-\varnothing e^{\lambda t} \leq \left(\frac{d}{dt} + \lambda\right)^{n-1} \tilde{x} e^{\lambda t} \leq \varnothing e^{\lambda t}.$$

In fact, this expression is equal to

$$-\varnothing e^{\lambda t} \leq \frac{d^{n-1}}{dt^{n-1}} \left(\tilde{x} e^{\lambda t}\right) \leq \varnothing e^{\lambda t}. \tag{50}$$

That is to say,

$$\left(\frac{d}{dt} + \lambda\right)^{n-1} \tilde{x} e^{\lambda t} = \frac{d^{n-1}}{dt^{n-1}} \left(\tilde{x} e^{\lambda t}\right). \tag{51}$$

We can confirm this form of (51) for small n values. Namely, if binomial expansion is applied for $\left(\frac{d}{dt} + \lambda\right)^{n-1}$,

$$\left(\frac{d}{dt} + \lambda\right)^{n-1} = \sum_{k=0}^{n-1} \binom{n-1}{k} \frac{d^k}{dt^k} \lambda^{n-1-k}$$

is written. Hence,

$$s(\tilde{\boldsymbol{x}}) = \left(\frac{d}{dt} + \lambda\right)^{n-1} \tilde{x} = \sum_{k=0}^{n-1} \binom{n-1}{k} \frac{d^k \tilde{x}}{dt^k} \lambda^{n-1-k}. \tag{52}$$

At this point, we can make a confirmation by taking $n = 3$:

For $n = 1$, it becomes $s(\tilde{\boldsymbol{x}}) = \binom{0}{0} \frac{d^0 \tilde{x}}{dt^0} \lambda^0 = \tilde{x}$. The binomial coefficient of this single term is 1, and this number is at the top of the Pascal triangle. For $n = 2$, it becomes $s(\tilde{\boldsymbol{x}}) = \binom{1}{0} \frac{d^0 \tilde{x}}{dt^0} \lambda^{1-0} + \binom{1}{1} \frac{d\tilde{x}}{dt} \lambda^{1-1} = \lambda\tilde{x} + \frac{d\tilde{x}}{dt}$. Here, the coefficients of both terms are 1. It gives the numbers of one-down row from the top of the Pascal triangle. For $n = 3$, it becomes

$$s(\tilde{\boldsymbol{x}}) = \binom{2}{0} \frac{d^0 \tilde{x}}{dt^0} \lambda^{2-0} + \binom{2}{1} \frac{d\tilde{x}}{dt} \lambda^{2-1} + \binom{2}{2} \frac{d^2 \tilde{x}}{dt^2} \lambda^{2-2} = \lambda^2 \tilde{x} + 2\lambda \frac{d\tilde{x}}{dt} + \frac{d^2 \tilde{x}}{dt^2}. \quad (53)$$

Here, the coefficients of the three terms from left to right are 1, 2, 1. This gives the elements of the two-down row from the top of the Pascal triangle. If the expression $-\varnothing \le s(\tilde{\boldsymbol{x}}) \le \varnothing$ is multiplied by $e^{\lambda t}$,

$$-\varnothing e^{\lambda t} \le s(\tilde{\boldsymbol{x}}) e^{\lambda t} \le \varnothing e^{\lambda t}$$

is obtained. If the result for $n = 3$ in Eq. (52), or its equivalent (53) is substituted above,

$$-\varnothing e^{\lambda t} \le \left(\lambda^2 e^{\lambda t} \tilde{x} + 2\lambda e^{\lambda t} \frac{d\tilde{x}}{dt} + e^{\lambda t} \frac{d^2 \tilde{x}}{dt^2} \right) \le \varnothing e^{\lambda t},$$

or the following expression is obtained:

$$-\varnothing e^{\lambda t} \le \frac{d^2}{dt^2} \left(e^{\lambda t} \tilde{x} \right) \le \varnothing e^{\lambda t}.$$

This verifies the multiplication of $s(\tilde{x})$ with $e^{\lambda t}$ for $n = 3$. In other words, the equation $\left(\frac{d}{dt} + \lambda\right)^2 \tilde{x} e^{\lambda t} = \frac{d^2}{dt^2} \left(\tilde{x} e^{\lambda t} \right)$ is satisfied. Once this statement is generalized for n, the validity of Eq. (51) is proven.

If the inequality (50) is integrated between 0 and t,

$$-\int_0^t \varnothing e^{\lambda \tau} d\tau \le \int_0^t \frac{d^{n-1}}{dt^{n-1}} \left(\tilde{x} e^{\lambda t} \right) d\tau \le \int_0^t \varnothing e^{\lambda t} d\tau,$$

and one step later,

$$-\frac{\varnothing}{\lambda} e^{\lambda \tau} \Big|_0^t \le \frac{d^{n-2}}{dt^{n-2}} \left(\tilde{x} e^{\lambda \tau} \right) \Big|_0^t \le \frac{\varnothing}{\lambda} e^{\lambda \tau} \Big|_0^t,$$

and finally the following expression is reached:

$$\frac{-\varnothing}{\lambda} e^{\lambda t} + \frac{\varnothing}{\lambda} \le \frac{d^{n-2}}{dt^{n-2}} \left(\tilde{x} e^{\lambda t} \right) - \frac{d^{n-2}}{dt^{n-2}} \left(\tilde{x} e^{\lambda t} \right) \Big|_{t=0} \le \frac{\varnothing}{\lambda} e^{\lambda t} - \frac{\varnothing}{\lambda}.$$

When the term $\frac{d^{n-2}}{dt^{n-2}} \left(\tilde{x} e^{\lambda t} \right) \Big|_{t=0}$ is added to each side of this expression, it takes the form below:

$$\frac{-\varnothing}{\lambda}e^{\lambda t}+\frac{\varnothing}{\lambda}+\frac{d^{n-2}}{dt^{n-2}}\left(\tilde{x}e^{\lambda t}\right)\bigg|_{t=0}\leq\frac{d^{n-2}}{dt^{n-2}}\left(\tilde{x}e^{\lambda t}\right)\leq\frac{\varnothing}{\lambda}e^{\lambda t}-\frac{\varnothing}{\lambda}+\frac{d^{n-2}}{dt^{n-2}}\left(\tilde{x}e^{\lambda t}\right)\bigg|_{t=0}. \tag{54}$$

Since we can always write,

$$\frac{d^{n-2}}{dt^{n-2}}\left(\tilde{x}e^{\lambda t}\right)\bigg|_{t=0}\geq-\left|\frac{d^{n-2}}{dt^{n-2}}\left(\tilde{x}e^{\lambda t}\right)\right|_{t=0}\text{ and }\frac{d^{n-2}}{dt^{n-2}}\left(\tilde{x}e^{\lambda t}\right)\bigg|_{t=0}\leq\left|\frac{d^{n-2}}{dt^{n-2}}\left(\tilde{x}e^{\lambda t}\right)\right|_{t=0}$$

as a result of replacing the derivative terms in the inequality (54) with their equivalents expressed with an absolute value one above, the inequality conditions will be preserved exactly as the term with the absolute value will be smaller than the term that satisfies the "less than or equal to" condition on the left and greater than the term that provides "greater than or equal to" condition on the right in the equality (54). Furthermore, aside from the absence of a violation, the conditions of inequality have been further reinforced. Therefore, it is possible to write the following under these conditions,

$$\frac{-\varnothing}{\lambda}e^{\lambda t}+\frac{\varnothing}{\lambda}-\left|\frac{d^{n-2}}{dt^{n-2}}\left(\tilde{x}e^{\lambda t}\right)\right|_{t=0}\leq\frac{d^{n-2}}{dt^{n-2}}\left(\tilde{x}e^{\lambda t}\right)\leq\frac{\varnothing}{\lambda}e^{\lambda t}-\frac{\varnothing}{\lambda}+\left|\frac{d^{n-2}}{dt^{n-2}}\left(\tilde{x}e^{\lambda t}\right)\right|_{t=0}.$$

Also, since both $\varnothing$ and λ are initially defined as positive definite constants, we take $-\varnothing/\lambda$ instead of the leftmost $\varnothing/\lambda$ and $\varnothing/\lambda$ instead of the rightmost $-\varnothing/\lambda$, which can be more safely adapted to existing inequality conditions without loss of generality will be preferred at this stage. Hence, the inequality (54) will turn into a rewritten appropriate form given below:

$$\frac{-\varnothing}{\lambda}e^{\lambda t}-\left(\left|\frac{d^{n-2}}{dt^{n-2}}\left(\tilde{x}e^{\lambda t}\right)\right|_{t=0}+\frac{\varnothing}{\lambda}\right)\leq\frac{d^{n-2}}{dt^{n-2}}\left(\tilde{x}e^{\lambda t}\right)\leq\frac{\varnothing}{\lambda}e^{\lambda t}+\left(\left|\frac{d^{n-2}}{dt^{n-2}}\left(\tilde{x}e^{\lambda t}\right)\right|_{t=0}+\frac{\varnothing}{\lambda}\right).$$

The same reasoning can be applied repeatedly until the $(n-1)th$ integral is reached on the inequality (50). Once (50) is integrated, recall that (54) is obtained. If we apply a second integral on (54) or, alternatively, a first integral to the form of (54) given immediately above, the following expression is obtained:

$$\underbrace{-\int_0^t\frac{\varnothing}{\lambda}e^{\lambda t}dt}_{Part(a)}-\underbrace{\int_0^t\left(\left|\frac{d^{n-2}}{dt^{n-2}}\left(\tilde{x}e^{\lambda t}\right)\right|_{t=0}+\frac{\varnothing}{\lambda}\right)dt}_{Part(b)}\leq\underbrace{\int_0^t\frac{d^{n-2}}{dt^{n-2}}\left(\tilde{x}e^{\lambda t}\right)dt}_{Part(c)}\leq\underbrace{\int_0^t\frac{\varnothing}{\lambda}e^{\lambda t}dt}_{Part(a)}$$
$$+\underbrace{\int_0^t\left(\left|\frac{d^{n-2}}{dt^{n-2}}\left(\tilde{x}e^{\lambda t}\right)\right|_{t=0}+\frac{\varnothing}{\lambda}\right)dt.}_{Part(b)}$$

In determining the generalized cases below, we would like to state in advance that we do not focus on other terms that will appear in the shape of increasing powers of t in the form of $\frac{t^n}{n!}$ especially in Parts (a) and (c), and we do not show them in the generalized statements. For that matter, as shown a little below, if Eq. (55) is divided by $e^{\lambda t}$ and the limit is taken as t goes to infinity (i.e., $t\to\infty$), those terms will eventually disappear completely, since the denominator will go to infinity faster than the numerator. After this essential explanation,

For *Part (a)*:

$$\int_0^t \frac{\varnothing}{\lambda} e^{\lambda t} dt = \frac{\varnothing}{\lambda^2} e^{\lambda t}\Big|_0^t = \frac{\varnothing}{\lambda^2} e^{\lambda t} - \frac{\varnothing}{\lambda^2}, n = 2(\textit{for the 2nd integral})$$

$$\frac{\varnothing}{\lambda^3} e^{\lambda t} - \frac{\varnothing}{\lambda^3}, n = 3(\textit{for the 3rd integral})$$

$$\vdots$$

$$\vdots$$

$$\frac{\varnothing}{\lambda^{n-2}} e^{\lambda t} - \frac{\varnothing}{\lambda^{n-2}}, n = n-2(\textit{for the}(n-2)\textit{th integral})$$

$$\frac{\varnothing}{\lambda^{n-1}} e^{\lambda t} - \frac{\varnothing}{\lambda^{n-1}}, n = n-1(\textit{for the } (n-1)\textit{th integral : Generalized form})$$

For *Part (b)*:

$$\int_0^t \left(\overbrace{\left| \frac{d^{n-2}}{dt^{n-2}} \left(\tilde{x} e^{\lambda t}\right) \right|_{t=0}}^{constant} + \overbrace{\frac{\varnothing}{\lambda}}^{constant} \right) dt = \left(\left| \frac{d^{n-2}}{dt^{n-2}} \left(\tilde{x} e^{\lambda t}\right) \right|_{t=0} + \frac{\varnothing}{\lambda} \right) * t, n$$

$$= 2(\textit{for the 2nd integral})$$

$$\frac{\left(\left| \frac{d}{dt^{n-2}} \left(\tilde{x} e^{\lambda t}\right) \right|_{t=0} + \frac{\varnothing}{\lambda} \right) * t^2}{2}, n = 3(\textit{for the 3rd integral})$$

$$\frac{\left(\left| \frac{d}{dt^{n-2}} \left(\tilde{x} e^{\lambda t}\right) \right|_{t=0} + \frac{\varnothing}{\lambda} \right) * t^3}{6}, n = 4(\textit{for the 4th integral})$$

$$\vdots$$

$$\vdots$$

$$\frac{\left(\left| \frac{d}{dt^{n-2}} \left(\tilde{x} e^{\lambda t}\right) \right|_{t=0} + \frac{\varnothing}{\lambda} \right) * t^{n-3}}{(n-3)!}, n = n-2(\textit{for the } (n-2)\textit{th integral})$$

$$\frac{\left(\left| \frac{d}{dt^{n-2}} \left(\tilde{x} e^{\lambda t}\right) \right|_{t=0} + \frac{\varnothing}{\lambda} \right) * t^{n-2}}{(n-2)!}, n = n-1(\textit{for the } (n-1)\textit{th integral: Generalized form})$$

For *Part (c)*:

Starting with (50), when the term in the middle of inequality, $\frac{d^{n-1}}{dt^{n-1}}\left(\tilde{x} e^{\lambda t}\right)$, is integrated $(n-1)$ times in a row, it is obvious that only the result, $\tilde{x} e^{\lambda t}$, will be found. Therefore,

$$\int_0^t \frac{d^{n-1}}{dt^{n-1}} \left(\tilde{x} e^{\lambda t}\right) dt = \frac{d^{n-2}}{dt^{n-2}} \left(\tilde{x} e^{\lambda t}\right)\Big|_0^t = \frac{d^{n-2}}{dt^{n-2}} \left(\tilde{x} e^{\lambda t}\right) - \frac{d^{n-2}}{dt^{n-2}} \left(\tilde{x} e^{\lambda t}\right)\Big|_{t=0}, n = 1(\textit{1st integral})$$

$$\int_0^t \frac{d^{n-2}}{dt^{n-2}} \left(\tilde{x} e^{\lambda t}\right) dt = \frac{d^{n-3}}{dt^{n-3}} \left(\tilde{x} e^{\lambda t}\right)\Big|_0^t = \frac{d^{n-3}}{dt^{n-3}} \left(\tilde{x} e^{\lambda t}\right) - \frac{d^{n-3}}{dt^{n-3}} \left(\tilde{x} e^{\lambda t}\right)\Big|_{t=0}, n = 2(\textit{2nd integral})$$

$$\vdots$$

$$\vdots$$

$$\int_0^t \frac{d^2}{dt^2}\left(\tilde{x}e^{\lambda t}\right)dt = \frac{d}{dt}\left(\tilde{x}e^{\lambda t}\right)\Big|_0^t = \frac{d}{dt}\left(\tilde{x}e^{\lambda t}\right) - \frac{d}{dt}\left(\tilde{x}e^{\lambda t}\right)\Big|_{t=0}, n = n-2((n-2)th\ integral)$$

$$\int_0^t \frac{d}{dt}\left(\tilde{x}e^{\lambda t}\right)dt = \tilde{x}e^{\lambda t}\Big|_0^t = \tilde{x}e^{\lambda t} - \tilde{x}(0), n-1((n-1)th\ integral)$$

is written. However, due to the reason we have explained above, we would like to remind that we do not take into account other terms that will appear in the shape of increasing powers of t in the form of $\frac{t^n}{n!}$ once again. Therefore, when the determinated generalized terms are put in place,

$$\frac{-\varnothing}{\lambda^{n-1}}e^{\lambda t} + \frac{\varnothing}{\lambda^{n-1}} - \frac{\left(\left|\frac{d}{dt^{n-2}}\left(\tilde{x}e^{\lambda t}\right)\right|_{t=0} + \frac{\varnothing}{\lambda}\right) * t^{n-2}}{(n-2)!} - \cdots \le \tilde{x}e^{\lambda t} - \tilde{x}(0) \le \frac{\varnothing}{\lambda^{n-1}}e^{\lambda t} - \frac{\varnothing}{\lambda^{n-1}} + \frac{\left(\left|\frac{d}{dt^{n-2}}\left(\tilde{x}e^{\lambda t}\right)\right|_{t=0} + \frac{\varnothing}{\lambda}\right) * t^{n-2}}{(n-2)!} + \cdots$$

is obtained. Based on the previous similar practice, the term $\tilde{x}(0)$ is added to each side, and once again reminding that $\varnothing$ and λ are positive definite constants and that $\tilde{x}(0) \ge -|\tilde{x}(0)|, \tilde{x}(0) \le |\tilde{x}(0)|$, if the inequalities, $\frac{\varnothing}{\lambda^{n-1}} \ge -\frac{\varnothing}{\lambda^{n-1}}, \frac{-\varnothing}{\lambda^{n-1}} \le \frac{\varnothing}{\lambda^{n-1}}$, are used,

$$\frac{-\varnothing}{\lambda^{n-1}}e^{\lambda t} - \frac{\left(\left|\frac{d}{dt^{n-2}}\left(\tilde{x}e^{\lambda t}\right)\right|_{t=0} + \frac{\varnothing}{\lambda}\right) * t^{n-2}}{(n-2)!} - \cdots - \left(|\tilde{x}(0)| + \frac{\varnothing}{\lambda^{n-1}}\right) \le \tilde{x}e^{\lambda t} \le \frac{\varnothing}{\lambda^{n-1}}e^{\lambda t} + \frac{\left(\left|\frac{d}{dt^{n-2}}\left(\tilde{x}e^{\lambda t}\right)\right|_{t=0} + \frac{\varnothing}{\lambda}\right) * t^{n-2}}{(n-2)!} + \cdots + \left(|\tilde{x}(0)| + \frac{\varnothing}{\lambda^{n-1}}\right) \tag{55}$$

is written. Also, if (55) is divided into $e^{\lambda t}$ and t is taken to infinity, the following result is obtained:

$$\frac{-\varnothing}{\lambda^{n-1}} - \frac{1}{e^{\lambda t}}\frac{\left(\left|\frac{d}{dt^{n-2}}\left(\tilde{x}e^{\lambda t}\right)\right|_{t=0} + \frac{\varnothing}{\lambda}\right) * t^{n-2}}{(n-2)!} - \cdots - \frac{1}{e^{\lambda t}}\left(|\tilde{x}(0)| + \frac{\varnothing}{\lambda^{n-1}}\right) \le \tilde{x}(t) \le \frac{\varnothing}{\lambda^{n-1}} + \frac{1}{e^{\lambda t}}\frac{\left(\left|\frac{d}{dt^{n-2}}\left(\tilde{x}e^{\lambda t}\right)\right|_{t=0} + \frac{\varnothing}{\lambda}\right) * t^{n-2}}{(n-2)!} + \cdots + \frac{1}{e^{\lambda t}}\left(|\tilde{x}(0)| + \frac{\varnothing}{\lambda^{n-1}}\right).$$

From here, it can be easily verified that

$$\frac{-\varnothing}{\lambda^{n-1}} \le \tilde{x}(t) \le \frac{\varnothing}{\lambda^{n-1}}. \tag{56}$$

Taking into account the $(n-2)th$ integral of (50),

$$\frac{-\varnothing}{\lambda^{n-2}}e^{\lambda t} - \frac{\left(\left|\frac{d}{dt^{n-2}}\left(\tilde{x}e^{\lambda t}\right)\right|_{t=0} + \frac{\varnothing}{\lambda}\right) * t^{n-3}}{(n-3)!} - \cdots - \left(|\dot{\tilde{x}}(0)| + \frac{\varnothing}{\lambda^{n-2}}\right) \le \frac{d}{dt}\left(\tilde{x}e^{\lambda t}\right) \le \frac{\varnothing}{\lambda^{n-2}}e^{\lambda t} + \frac{\left(\left|\frac{d}{dt^{n-2}}\left(\tilde{x}e^{\lambda t}\right)\right|_{t=0} + \frac{\varnothing}{\lambda}\right) * t^{n-3}}{(n-3)!} + \cdots + \left(|\dot{\tilde{x}}(0)| + \frac{\varnothing}{\lambda^{n-2}}\right), \tag{57}$$

and the derivative expression,

$$d\left(\tilde{x}e^{\lambda t}\right)/dt = \dot{\tilde{x}}e^{\lambda t} + \tilde{x}\lambda e^{\lambda t},$$

by having (56)'s bounds accepted to (57) and dividing it back into $e^{\lambda t}$ for $t \to \infty$,

$$\frac{-\varnothing}{\lambda^{n-2}} - (\bullet)\frac{t^{n-3}}{(n-3)!e^{\lambda t}} - \cdots - (\bullet)\frac{1}{e^{\lambda t}} \le \dot{\tilde{x}}(t) + \tilde{x}(t)\lambda \le \frac{\varnothing}{\lambda^{n-2}} + (\bullet)\frac{t^{n-3}}{(n-3)!e^{\lambda t}} + \ldots + (\bullet)\frac{1}{e^{\lambda t}},$$

and finally from here,

$$\frac{-\varnothing}{\lambda^{n-2}} \le \dot{\tilde{x}}(t) + \lambda\tilde{x}(t) \le \frac{\varnothing}{\lambda^{n-2}} \tag{58}$$

is obtained. However, in order to determine the bounds of (58) based on only $\dot{\tilde{x}}(t)$, the bounds corresponding to the term $\lambda x(t)$ in addition to $\dot{\tilde{x}}(t)$ must be determined exactly. For this, (56) is used and if each side in this inequality is multiplied by λ,

$$\frac{-\varnothing}{\lambda^{n-2}} \le \lambda\tilde{x}(t) \le \frac{\varnothing}{\lambda^{n-2}} \tag{59}$$

expression is obtained. Now then, if the effect of $\lambda\tilde{x}(t)$ in the inequality of (58) is substituted by the bound determination ascertained by (59) above, the inequality (59) turns into

$$-2\frac{\varnothing}{\lambda^{n-2}} \le \dot{\tilde{x}}(t) \le 2\frac{\varnothing}{\lambda^{n-2}}. \tag{60}$$

Similarly, taking into account the $(n-3)th$ integral of (50),

$$\frac{-\varnothing}{\lambda^{n-3}}e^{\lambda t} - \frac{\left(\left|\frac{d}{dt^{n-3}}\left(\tilde{x}e^{\lambda t}\right)\right|_{t=0} + \frac{\varnothing}{\lambda}\right) * t^{n-4}}{(n-4)!} - \cdots - \left(\left|\ddot{\tilde{x}}(0)\right| + \frac{\varnothing}{\lambda^{n-3}}\right) \le \frac{d^2}{dt^2}\left(\tilde{x}e^{\lambda t}\right) \le \frac{\varnothing}{\lambda^{n-3}}e^{\lambda t} + \frac{\left(\left|\frac{d^{n-3}}{dt^{n-3}}\left(\tilde{x}e^{\lambda t}\right)\right|_{t=0} + \frac{\varnothing}{\lambda}\right) * t^{n-4}}{(n-4)!} + \cdots + \left(\left|\ddot{\tilde{x}}(0)\right| + \frac{\varnothing}{\lambda^{n-3}}\right), \tag{61}$$

and the derivative expression,

$$d\left(\dot{\tilde{x}}e^{\lambda t} + \tilde{x}\lambda e^{\lambda t}\right)/dt = \ddot{\tilde{x}}e^{\lambda t} + \lambda e^{\lambda t}\dot{\tilde{x}} + \dot{\tilde{x}}\lambda e^{\lambda t} + \tilde{x}\lambda^2 e^{\lambda t} = \ddot{\tilde{x}}e^{\lambda t} + 2\dot{\tilde{x}}\lambda e^{\lambda t} + \tilde{x}\lambda^2 e^{\lambda t},$$

by imposing the bounds of (56) and (60) on (61) and dividing this expression once again to $e^{\lambda t}$ for $t \to \infty$, the following step is obtained first:

$$\frac{-\varnothing}{\lambda^{n-3}}e^{\lambda t} \le \ddot{\tilde{x}} + 2\dot{\tilde{x}}\lambda + \tilde{x}\lambda^2 \le \frac{\varnothing}{\lambda^{n-3}}e^{\lambda t}. \tag{62}$$

Now, the bounds for $\tilde{x}\lambda^2$ and $2\dot{\tilde{x}}\lambda$ are respectively determined with,

$$\frac{-\varnothing}{\lambda^{n-3}} \le \lambda^2\tilde{x}(t) \le \frac{\varnothing}{\lambda^{n-3}} \tag{63}$$

by multiplying each side of the inequality of (56) by the term λ^2, and with,

$$-4\frac{\varnothing}{\lambda^{n-3}}\le 2\dot{\tilde{x}}\lambda\le 4\frac{\varnothing}{\lambda^{n-3}} \tag{64}$$

by multiplying each side of inequality of (60) by the term 2λ. Once these bounds determined by the inequalities (63) and (64) are imposed on (62), the expression,

$$\frac{-\varnothing}{\lambda^{n-3}}-\frac{\varnothing}{\lambda^{n-3}}-4\frac{\varnothing}{\lambda^{n-3}}\le\ddot{\tilde{x}}\le\frac{\varnothing}{\lambda^{n-3}}+\frac{\varnothing}{\lambda^{n-3}}+4\frac{\varnothing}{\lambda^{n-3}},$$

and hence in brief, the result,

$$6\frac{\varnothing}{\lambda^{n-3}}\le\ddot{\tilde{x}}\le 6\frac{\varnothing}{\lambda^{n-3}} \tag{65}$$

is concluded. As in obtaining (56), (60) and (65), the following general conclusion is reached if the similar procedure is applied sequentially until the bounds of $\tilde{x}^{(n-1)}$ are achieved:

$$-\left(1+\sum_{i=0}^{n-2}\binom{n-1}{i}\zeta_i\right)\varnothing\le\tilde{x}^{(n-1)}\le\left(1+\sum_{i=0}^{n-2}\binom{n-2}{i}\zeta_i\right)\varnothing. \tag{66}$$

Here, the coefficients $\zeta_i(i=0,1,\cdots,n-2)$ are related to the pre-acquired bounds of each $\tilde{x}^{(i)}$ and are summarized in Theorem 2.

In this way, by examining Eqs. (56), (60), (65), and (66) and, as much as other skipped boundaries, the integrals of (50), the tracking error will be kept within the bounds of $\left|\tilde{x}^{(i)}\right|\le\zeta_i\lambda^{i-n+1}\varnothing$, $i=0,1,\cdots,n-1$, where ζ_i is defined by (47).

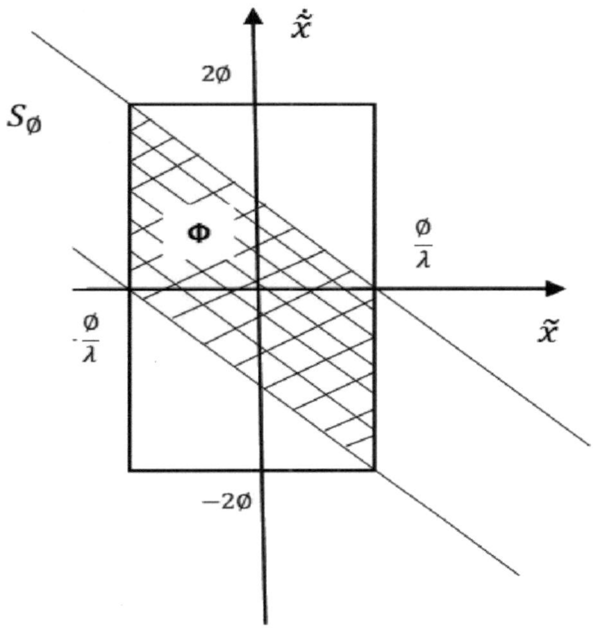

Figure 8.
Convergence region Φ in the case of a second-order system.

Remark 5. It should be noted that an n-dimensionally separated partition defined according to the boundaries mentioned earlier is not entirely within the boundary layer. Considering the attractiveness and invariant properties of $S_\varnothing$ proved in Theorem 1, the convergence region can be expressed as the intersection of an n-dimensional separated partition and boundary layer defined in Theorem 2. Thus, the tracking error vector will converge exponentially to a closed region $\Phi = \left\{x \in \mathbf{R}^n \| s(\tilde{x})| \le \varnothing \text{ and } |\tilde{x}^{(i)}| \le \zeta_i \lambda^{i-n+1}\varnothing, i = 0, 1, \cdots, n-1\right\}$, where ζ_i is defined by (47).

Figure 8 describes the Φ convergence region defined according to Remark 5 for a second-order ($n = 2$) system.

6. Numerical experimentation and simulation examples

6.1 Position control system by an armature-controlled dc servo motor

6.1.1 Positioning system description

The plant is an armature-controlled dc servo motor, the scheme of which is given in **Figure 9** [19].

In order to derive the state-space mathematical model from the physics of the motor, we first start by writing Kirchoff's voltage equation around the armature current:

$$e_a(t) = i_a(t)R_a + K_b \frac{d\theta_m}{dt}, \tag{67}$$

where $e_a(t)$ is the armature input voltage to the dc amplifier that drives the motor, $i_a(t)$ is the armature current, R_a is the armature resistance, K_b is a constant of proportionality called the back emf constant, or briefly, the armature constant, and θ_m is the angular displacement of the armature.

The torque, $T_m(t)$, delivered by the motor is expressed separately in relation to both the armature current and the load seen by the armature as follows:

$$T_m(t) = K_t i_a(t) = J_m + \frac{d^2\theta_m}{dt^2} + D_m \frac{d\theta_m}{dt}, \tag{68}$$

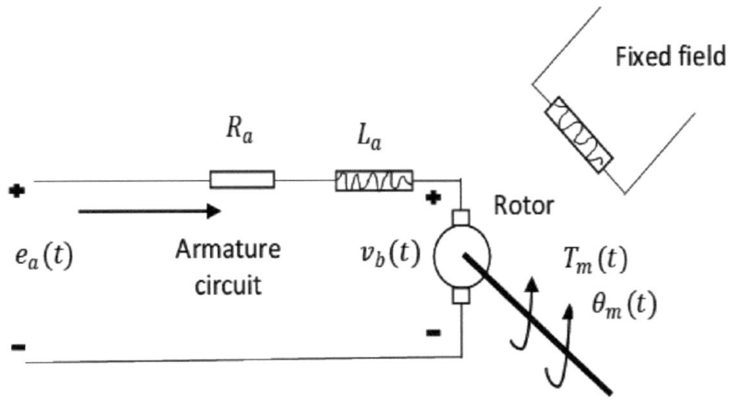

Figure 9.
DC motor schematic diagram.

where J_m and D_m both represent the equivalent inertia and viscous damping seen by the armature, respectively. Solving Eq. (68) for $i_a(t)$ and substituting the result into Eq. (67) yields

$$e_a(t) = \left(\frac{R_a J_m}{K_t}\right)\frac{d^2\theta_m}{dt^2} + \left(\frac{D_m R_a}{K_t} + K_b\right)\frac{d\theta_m}{dt}. \tag{69}$$

Defining the state variables x_1 and x_2 as

$$x_1 = \theta_m, \tag{70}$$

$$x_2 = \frac{d\theta_m}{dt}, \tag{71}$$

and substituting into Eq. (69), we get

$$e_a(t) = \left(\frac{R_a J_m}{K_t}\right)\frac{dx_2}{dt} + \left(\frac{D_m R_a}{K_t} + K_b\right)x_2. \tag{72}$$

Solving for dx_2/dt yields

$$\frac{dx_2}{dt} = \frac{-1}{J_m}\left(D_m + \frac{K_t K_b}{R_a}\right)x_2 + \left(\frac{K_t}{R_a J_m}\right)e_a(t). \tag{73}$$

Using Eqs. (70), (71) and (73), the state equations are written as

$$\frac{dx_1}{dt} = x_2 \tag{74}$$

$$\frac{dx_2}{dt} = \frac{-1}{J_m}\left(D_m + \frac{K_t K_b}{R_a}\right)x_2 + \left(\frac{K_t}{R_a J_m}\right)e_a(t) \tag{75}$$

Hence, in vector-matrix form,

$$\dot{\boldsymbol{x}}(t) = \begin{bmatrix} 0 & 1 \\ 0 & \frac{-1}{J_m}\left(D_m + \frac{K_t K_b}{R_a}\right) \end{bmatrix}\boldsymbol{x} + \begin{bmatrix} 0 \\ \frac{K_t}{R_a J_m} \end{bmatrix} e_a(t).$$

Now, let us consider a position control system and assume a case of varying external disturbance torque to the dc motor. In other words, we assume that a varying external disturbance can enter into the system in the form of varying torque $\tau(t)$ on the shaft of the motor. Therefore, in this derivation, taking account of the external disturbance term which will appear as an extra term in the state equation, we finally describe the state-space representation (a.k.a. the state differential equation) as follows:

$$\dot{\boldsymbol{x}}(t) = \begin{bmatrix} \dot{x}_1(t) \\ \dot{x}_2(t) \end{bmatrix} = \begin{bmatrix} 0 & 1 \\ 0 & -\alpha \end{bmatrix}\begin{bmatrix} x_1(t) \\ x_2(t) \end{bmatrix} + \begin{bmatrix} 0 \\ \kappa \end{bmatrix} e_a(t) - \begin{bmatrix} 0 \\ \gamma \end{bmatrix}\tau(t), \tag{76}$$

where $\boldsymbol{x}(t) = (x_1(t) \;\; x_2(t))^T$ has as components the angular position $x_1(t)$ and the angular velocity $x_2(t)$ and where $\alpha \triangleq \frac{1}{J_m}\left(D_m + \frac{K_t K_b}{R_a}\right)$ and $\kappa \triangleq \frac{K_t}{R_a J_m}$, and $\gamma = 1/J$, with J the moment of inertia of all the rotating parts. For the controller variable we choose the angular position:

$$x_1(t) = (1, 0)\boldsymbol{x}(t). \tag{77}$$

The numerical values of motor's parameters have been taken from the case study in [20]:

$$\alpha = 4.6\, s^{-1},$$
$$\kappa = 0.787\, rad/(Vs^2), \gamma = 0.1\, kg^{-1}m^{-2}.$$

Finally, we try several varying external disturbance torque (Nm) options such as

- $sin\,(at)$, $\gamma sin(at)$, or $msin(at)$, where $a = 3, 4, \cdots$ and m is any appropriate multiplier that is compatible and proportional to the size of the system variables and parameters.
- $sin\,(2\pi t)$, $\gamma sin(2\pi t)$, or $msin(2\pi t)$, where m is defined as one above.
- $sin\,(x_1)$, $\gamma sin(x_1)$, or $msin(x_1)$.

as examples, and we decide that the torque of $\tau(t) = sin(3t)\, Nm$ is the one we choose for the numerical experimentation and simulations as trade-off after we weigh the pros and cons upon several performance tests and considerations. In addition, the simulations given for the DC motor have been run in the form of script code using Matlab [21].

6.1.2 Design of the sliding mode controller for the position control system

Sliding mode surface is defined as:

$$s(t) = ce(t) + \dot{e}(t), \tag{78}$$

where $e(t) = r(t) - x_1(t)$, $\dot{e}(t) = \dot{r}(t) - x_2(t)$, and $r(t)$ is a desired input signal and defined as $r(t) = Asin(2\pi Ft)$in the simulations where A and F are suitably determined, and c is a positive constant. Therefore, once we get a derivative of the expression above, we get the following:

$$\dot{s}(t) = c\dot{e}(t) + \ddot{e}(t) = c\dot{e}(t) + \ddot{r}(t) - \dot{x}_2(t).$$

Taking the second row from Eq. (76) and replacing it with $\dot{x}_2(t)$ above yields:

$$\dot{s}(t) = c\dot{e}(t) + \ddot{r}(t) + \alpha x_2(t) - \kappa e_a(t) + \gamma\tau(t). \tag{79}$$

Next, we obtain the control law $e_a(t)$ by equalizing $\dot{s}(t)$ to zero and then using $\gamma = 1/J$ and $\tau(t) = Ksgn(s(t))$ in Eq. (79) as follows:

$$e_a(t) = \frac{1}{\kappa}[c\dot{e}(t) + \ddot{r}(t) + \alpha x_2(t) + \frac{Ksgn(s(t))}{J}]. \tag{80}$$

Once Eq. (80) is substituted into Eq. (79), we obtain the following:

$$\dot{s}(t) = \frac{\tau(t)}{J} - \frac{Ksgn(s(t))}{J}. \tag{81}$$

where K is a positive constant. Letting $K = max\,(|\tau(t)|) + \eta J$, where $\eta > 0$, and multiplying (81) by $s(t)$ yields:

$$s(t)\dot{s}(t) = \frac{\tau(t)}{J}s(t) - \frac{max\,(|\tau(t)|) + \eta J}{J}|s(t)|,$$

Here, the following can always be written:

$$s(t)\dot{s}(t) = \frac{\tau(t)}{J}s(t) - \frac{max\,(|\tau(t)|)}{J}|s(t)| - \eta|s(t)| \leq -\eta|s(t)|. \tag{82}$$

Remark 6. It is not difficult to briefly demonstrate the validity of the inequality (82) from a practical point of view: if $s(t)$ is greater than zero (i.e., $s(t) > 0$), then Z term, which we define as $\frac{\tau(t)}{J}s(t) - \frac{max\,(|\tau(t)|)}{J}|s(t)|$ from (82), becomes less than zero (i.e., $Z\ term < 0$); because $\frac{max\,(|\tau(t)|)}{J}|s(t)| > \frac{\tau(t)}{J}s(t)$. Since the larger term has a negative sign, the difference between the terms will also be negative. As a result, $(some\ negative\ term) - \eta|s(t)| \leq -\eta|s(t)|$ is always valid and correct. If $s(t) < 0$, then $\frac{\tau(t)}{J}s(t) < 0$; in addition, $\frac{-max\,(|\tau(t)|)}{J}\underbrace{|s(t)|}_{>0} < 0$, therefore the addition of two negative terms will make the overall addition more negative; hence, $(more\ negative\ term) - \eta|s(t)| \leq -\eta|s(t)|$ which means that the inequality is still valid and in the right direction. Therefore, $\frac{1}{2}\frac{d}{dt}(s^2(t)) \leq -\eta|s(t)|$.

In order to decrease the chattering phenomenon caused by sliding mode control law, saturation function is adapted in this work, and the controller becomes

$$e_a(t) = \frac{1}{\kappa}\left[c\dot{e}(t) + \ddot{r}(t) + \alpha x_2(t) + \frac{K}{J}\overbrace{sat(s(t),\varphi)}^{instead\ of\ sgn(s(t))}\right], \tag{83}$$

where saturation function $sat(s(t),\varphi)$ is defined as,

$$sat(s(t),\varphi) = \left[\begin{array}{l} \frac{s(t)}{\varphi}\ |s(t)| \leq \varphi \\ sgn\,(s(t))|s(t)| > \varphi \end{array}\right., \tag{84}$$

where φ is a very small positive constant (**Figures 10–18**).

Let us also see how the results will change if PID control is used as an alternative to the SMC. Although, in the comparisons given in the literature, the pros and cons of both strategies are mentioned, it is generally observed that SMC performs better than PID [22–25]. Nevertheless, PID control can still be used as an alternative to SMC. The results given here do not contradict the view that one can use it instead of the other without losing too much performance. In the case where only the PD control strategy is applied, let us state that we need to emphasize the following points for the tracking error performance indicated by **Figure 19**.

We can prefer PD control strategy mostly to advance faster between intermediate points of the entire trajectory, i.e. from waypoint to waypoint at which course is changed for following a reference trajectory in which we have to move end-effector along a predefined path. Speaking of which, the end-effector is crucial for the entire trajectory tracking problem in catching up a desired position within shortest time. In other words, accuracy is highly desirable for the end-effector to be positioned accurately under unknown disturbances and payload variations. Basically, the desired position is a function of time and continuously changes with respect to time. Therefore, conventional PD control strategy does not always exhibit good accuracy and robustness properties for trajectory tracking problem. However, we can still

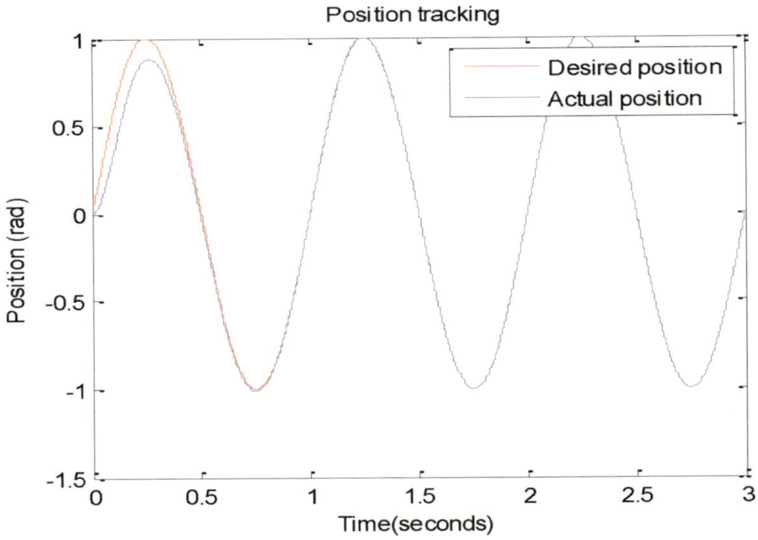

Figure 10.
Position tracking under SMC of the position control system.

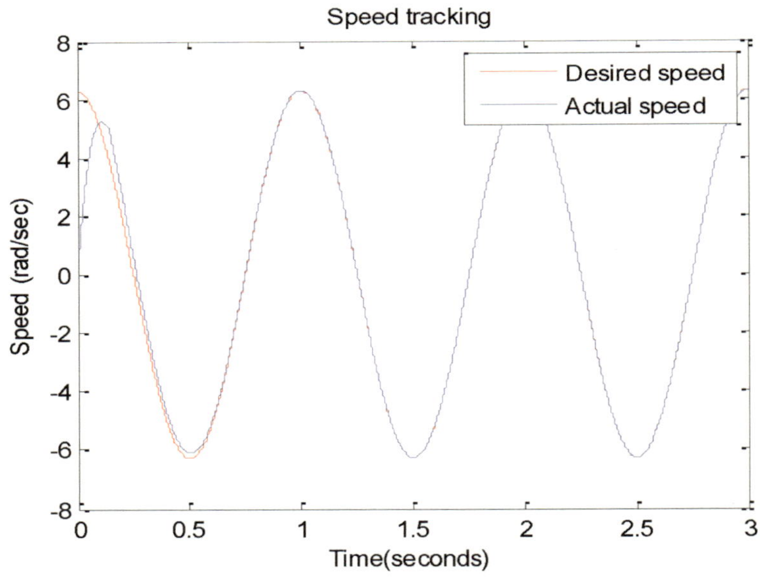

Figure 11.
Speed tracking under SMC of the position control system.

choose the PD control strategy because of the advantages it offers [26]. We should emphasize that the errors between the actual points and the waypoints, each of which can also be viewed as intermediate setpoints, do not necessarily have to be eliminated completely. As a result, we have decided to use the PD control because of the advantages it offers and to move faster between the waypoints by tolerating or neglecting the steady-state error computations that would bring extra computational burden (**Figures 20–22**).

6.2 Modeling and simulation of A two-link (2-DOF) planar robot manipulator

A manipulator consists of an open kinematic chain of rigid links. Power is supplied to each degree of freedom of the manipulator by independent torques. The

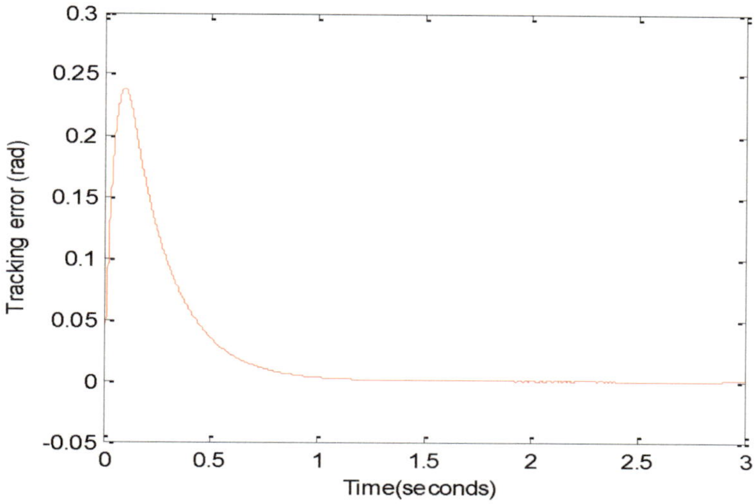

Figure 12.
Tracking error under SMC of the position control system.

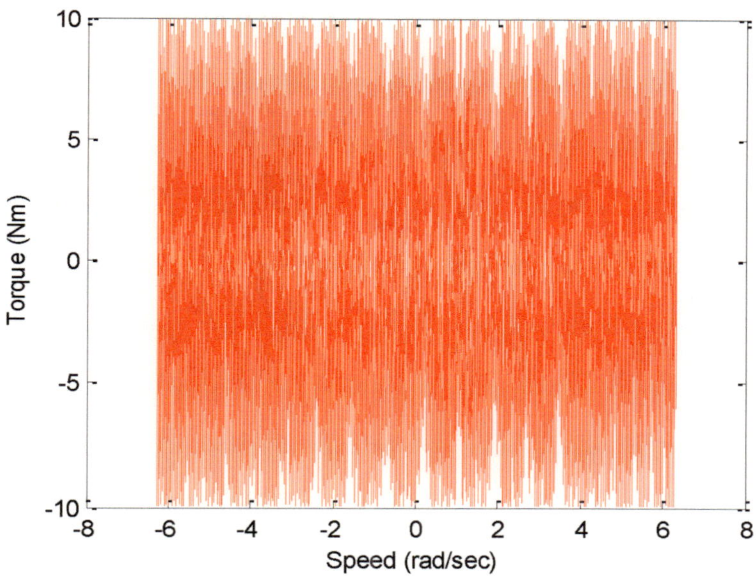

Figure 13.
Torque vs. speed curve under SMC of the position control system.

dynamical equations of motion of an n-link (i.e., n-degree-of-freedom) robot manipulator using the Lagrangian formulation has already been described in Section 3 by the Eq. (8). The robot model there is characterized by the following structural properties which are important for our sliding mode controller design of the tracking problem [27]:

Property 1. *A vector $\boldsymbol{\alpha} \in \boldsymbol{R}^m$ with components that depend on manipulator parameters (masses, moments of inertia, etc.) exists, such that*

$$\boldsymbol{M}(\boldsymbol{q})\dot{\boldsymbol{v}} + \boldsymbol{C}(\boldsymbol{q}, \dot{\boldsymbol{q}})\boldsymbol{v} + \boldsymbol{G}(\boldsymbol{q}) = \boldsymbol{\Phi}(\boldsymbol{q}, \dot{\boldsymbol{q}}, \boldsymbol{v}, \dot{\boldsymbol{v}})\boldsymbol{\alpha}, \tag{85}$$

where $\boldsymbol{\Phi} \in \boldsymbol{R}^{nxm}$ is called the regressor, $\boldsymbol{v} \in \boldsymbol{R}^n$ is a vector of smooth functions. This property implies that the dynamic equation can be linearized according to a specially selected manipulator parameter set, hence constituting the basis for the linear parameterization approach.

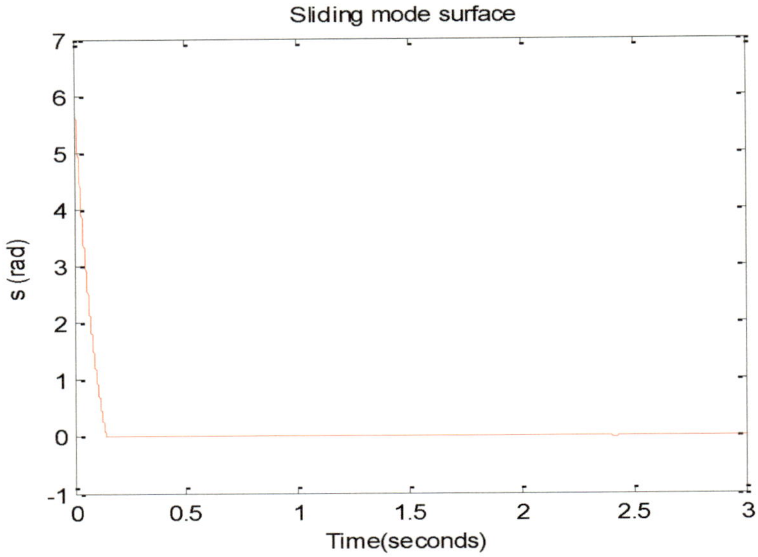

Figure 14.
Sliding mode surface under SMC of the position control system.

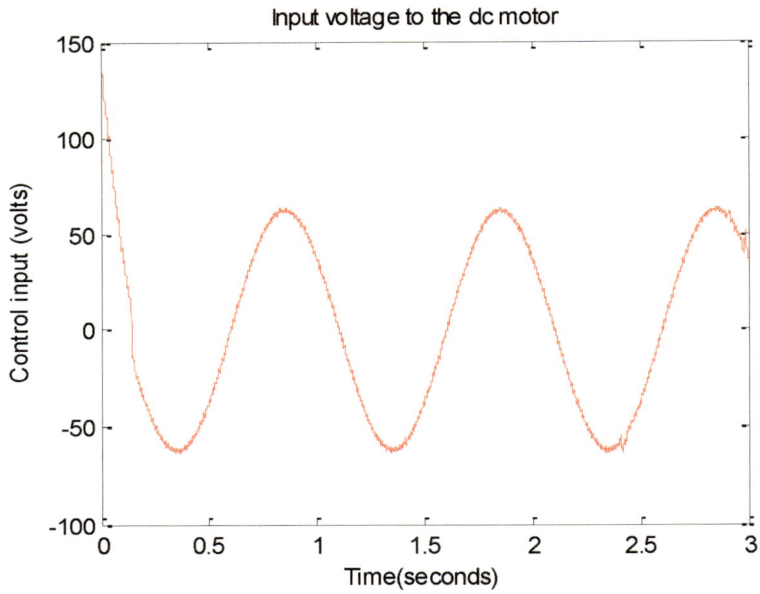

Figure 15.
Control input under SMC of the position control system.

Property 2. *Both* $\mathbf{M}(\boldsymbol{q})$ *and* $\mathbf{C}(\boldsymbol{q},\dot{\boldsymbol{q}})$ *in Eq. (8), using a properly defined matrix* $\mathbf{C}(\boldsymbol{q},\dot{\boldsymbol{q}})$*, satisfy*

$$\boldsymbol{x}^T\left(\dot{\boldsymbol{M}} - 2\mathbf{C}\right)\boldsymbol{x} = 0, \quad \forall \boldsymbol{x} \in \boldsymbol{R}^n \tag{86}$$

with $\boldsymbol{x}^T$ *the transposition of* $\boldsymbol{x}$*. That is,* $\left(\dot{\boldsymbol{M}} - 2\mathbf{C}\right)$ *is a skew-symmetric matrix. Property 2 simply states that the so-called fictitious forces, defined by* $\mathbf{C}(\boldsymbol{q},\dot{\boldsymbol{q}})\dot{\boldsymbol{q}}$*, do not work on the system.*

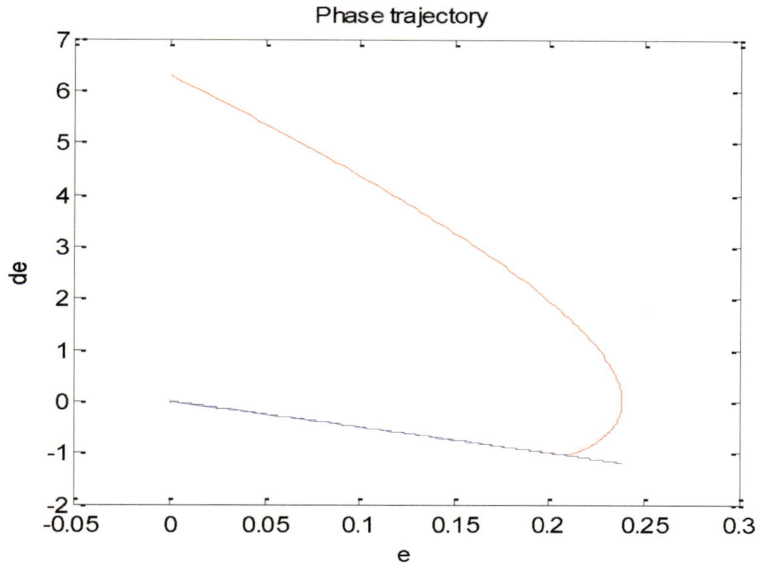

Figure 16.
Phase trajectory under SMC of the position control system.

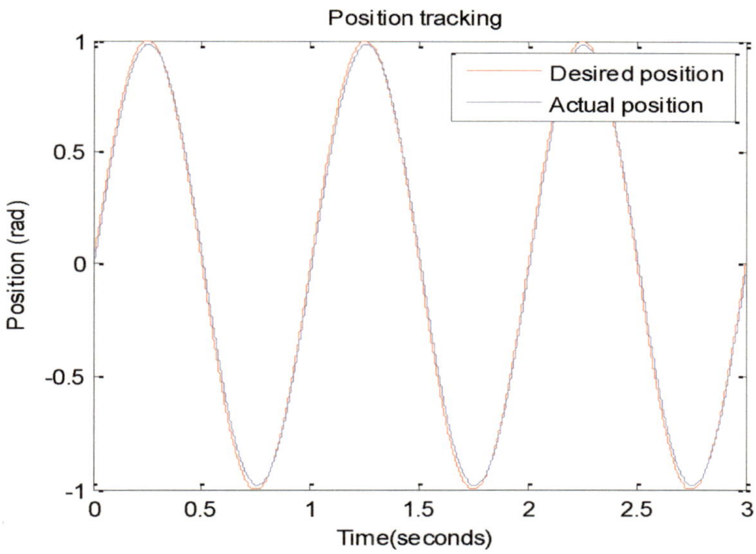

Figure 17.
Position tracking under PD control of the position control system.

The model given in **Figure 23** is known as a two-Link (2-DOF) planar robot, as it corresponds to the two-dimensional special case, where $n = 2$ is taken in the n-link robot manipulator [27].

The dynamic model chosen for the simulations is given by

$$\boldsymbol{M}(\boldsymbol{\theta})\ddot{\boldsymbol{\theta}}_r + \boldsymbol{F}(\boldsymbol{\theta}, \dot{\boldsymbol{\theta}}_r)\dot{\boldsymbol{\theta}}_r + \boldsymbol{G}(\boldsymbol{\theta}) = \boldsymbol{\tau},$$

and the dynamic equation is given by

$$\begin{bmatrix} M_{11} & M_{12} \\ M_{12} & M_{22} \end{bmatrix} \begin{bmatrix} \ddot{\theta}_1 \\ \ddot{\theta}_2 \end{bmatrix} + \begin{bmatrix} -F_{12}\dot{\theta}_2 & -F_{12}(\dot{\theta}_1 + \dot{\theta}_2) \\ F_{12}\dot{\theta}_1 & 0 \end{bmatrix} \begin{bmatrix} \dot{\theta}_1 \\ \dot{\theta}_2 \end{bmatrix} + \begin{bmatrix} G_1 g \\ G_2 g \end{bmatrix} = \begin{bmatrix} u_1 \\ u_2 \end{bmatrix},$$

where

$$M_{11} = (m_1 + m_2)r_1^2 + m_2 r_2^2 + 2m_2 r_1 r_2 \cos(\theta_2)$$
$$M_{12} = m_2 r_2^2 + m_2 r_1 r_2 \cos(\theta_2)$$
$$M_{22} = m_2 r_2^2$$

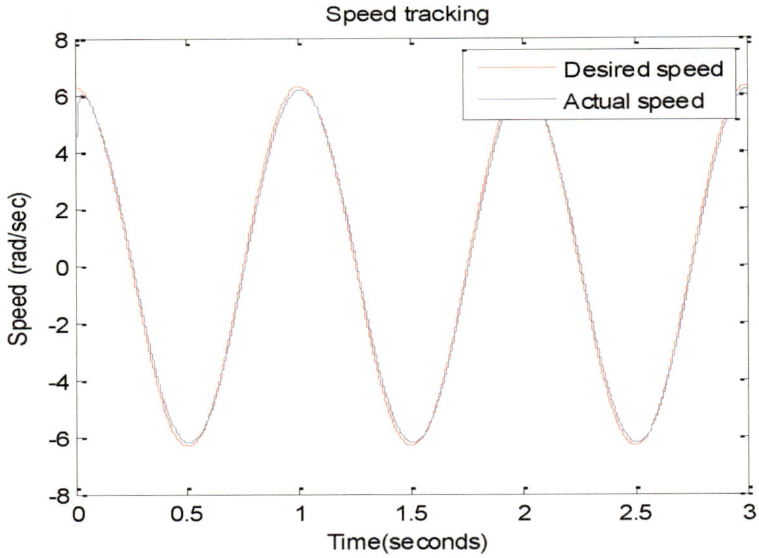

Figure 18.
Speed tracking under PD control of the position control system.

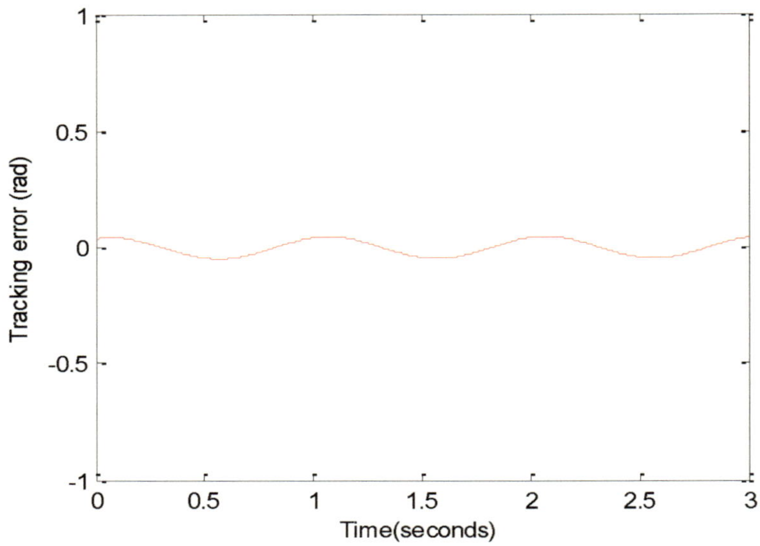

Figure 19.
Tracking error under PD control of the position control system.

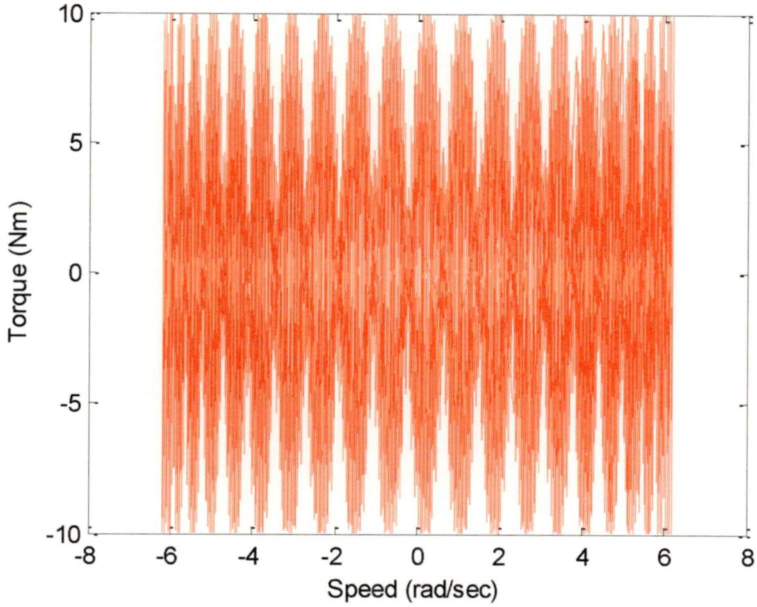

Figure 20.
Torque vs. speed curve under PD control of the position control system.

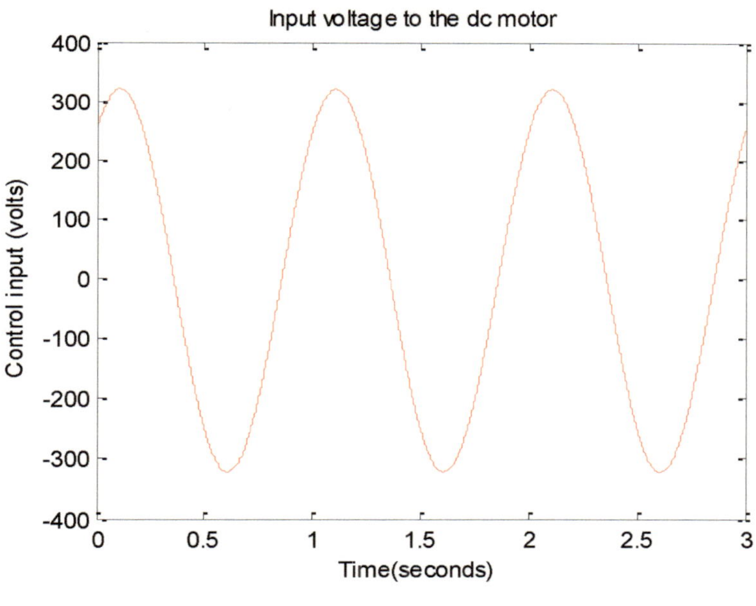

Figure 21.
Control input under PD control of the position control system.

$$F_{12} = m_2 r_1 r_2 \sin(\theta_2)$$
$$G_1 = (m_1 + m_2) r_1 \cos(\theta_2) + m_2 r_2 \cos(\theta_1 + \theta_2)$$
$$G_2 = m_2 r_2 \cos(\theta_1 + \theta_2)$$

6.2.1 No boundary layer

- The desired joint trajectory for each joint (i) is given by [27, 28] as:

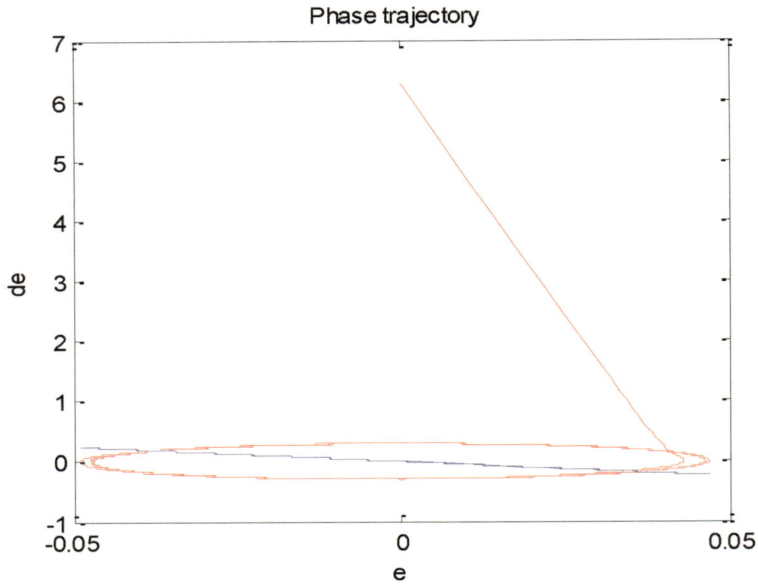

Figure 22.
Phase trajectory under PD control of the position control system.

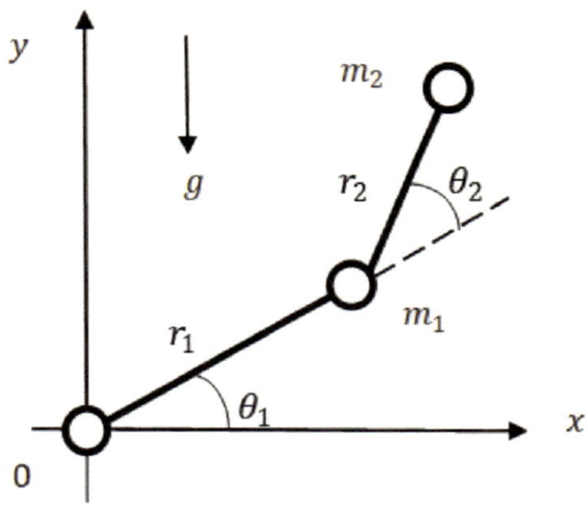

Figure 23.
A two-link robot manipulator model.

$$\theta_d(t) = -90^o + 52.5(1 - cos\,(1.26t))$$

- Initial conditions:

$$\theta_1(0) = -45° \text{ and } \theta_2(0) = -30°.$$

- The parameter values used are selected as in [27, 28]:

$$m_1 = 0.5\text{ kg}, m_2 = 0.5\text{ kg}$$
$$r_1 = 1.0\text{ m}, r_2 = 0.8\text{ m}$$

- Matlab-Simulink implementation options used in the simulations (**Figures 24–31**) as in [28]:

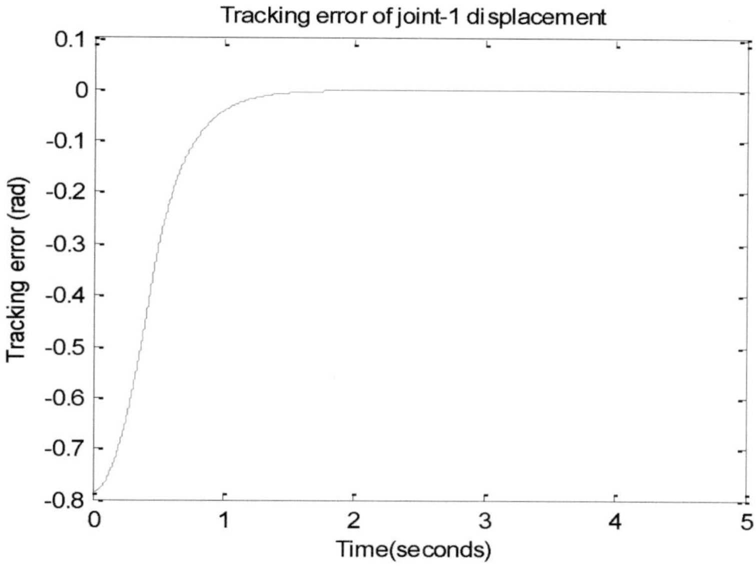

Figure 24.
Tracking error of Joint 1 displacement under SMC of the robot manipulator.

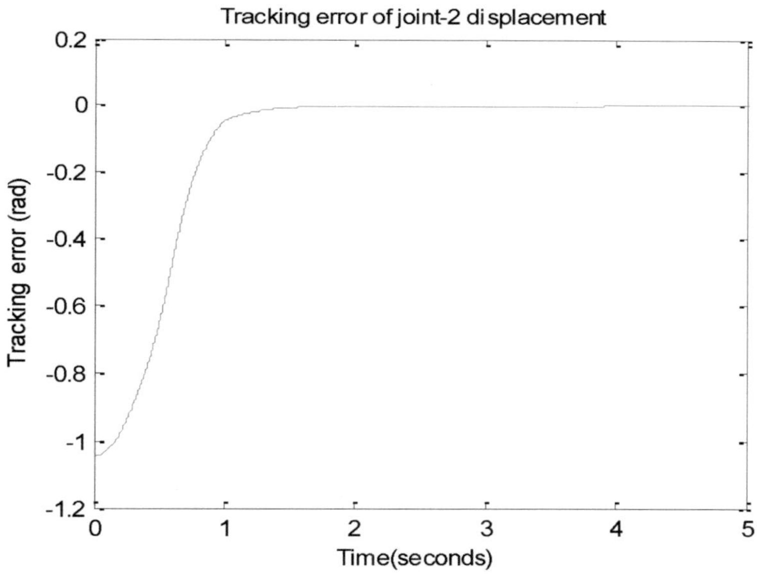

Figure 25.
Tracking error of Joint 2 displacement under SMC of the robot manipulator.

$$\text{Sampling time } T_s = 1\,\text{kHz}, \text{fixed} - \text{step}, \text{ode5}$$

- Joint 1: $K = 14$, $\lambda = 4$, $\phi = 0$.
- Joint 2: $K = 8$, $\lambda = 4$, $\phi = 0$.

6.2.2 Introducing boundary layer

The same parameters and initial conditions for the simulations have been chosen as in Section 6.2.1 except for the following ones which include the boundary layer thickness in particular:

- Joint 1: $K = 14$, $\lambda = 4$, $\phi = 0.02$.
- Joint 2: $K = 8$, $\lambda = 4$, $\phi = 0.02$.

Please note that due to the space constraint, we will be able to give only the figures whose effect is clearly observed, not eight figures as given in Section 6.2.1 (**Figures 32–34**).

For tracking a desired trajectory by two-link rigid planar robotic manipulator, PID control strategy will not work well under unknown disturbances and payload changes, and hence will not be represented here. In addition, the values of control input will get

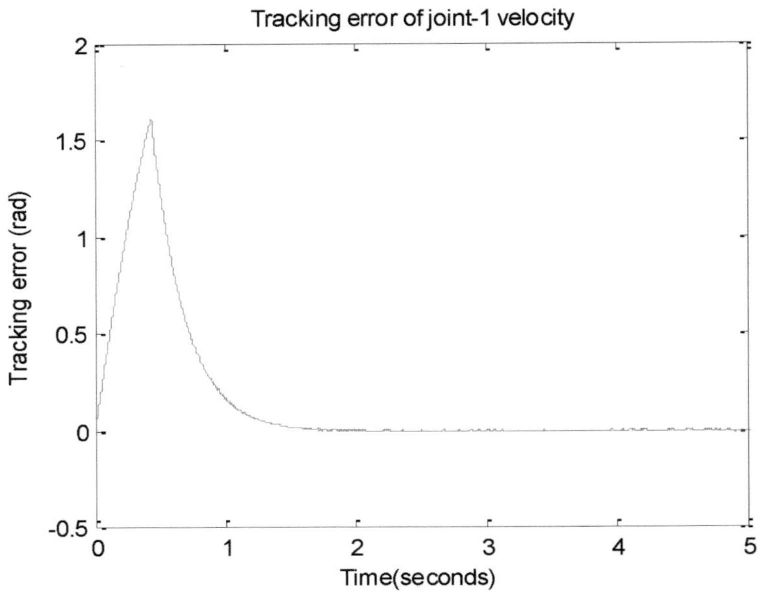

Figure 26.
Tracking error of Joint 1 velocity under SMC of the robot manipulator.

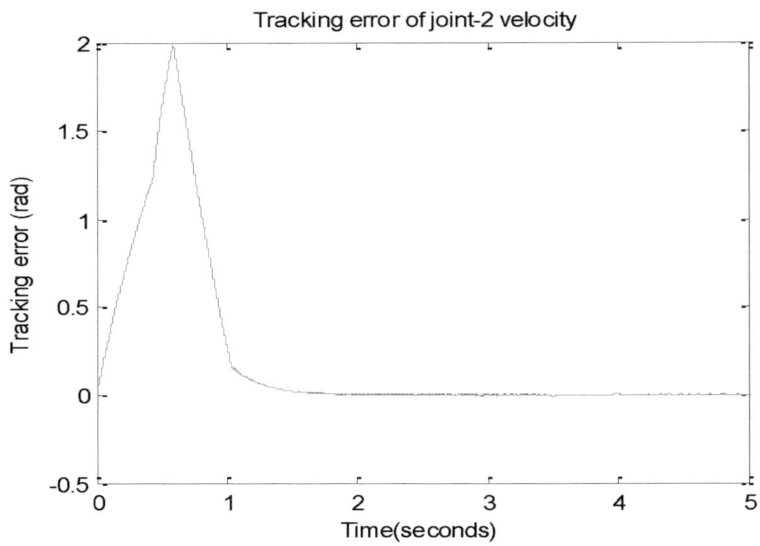

Figure 27.
Tracking error of Joint 2 velocity under SMC of the robot manipulator.

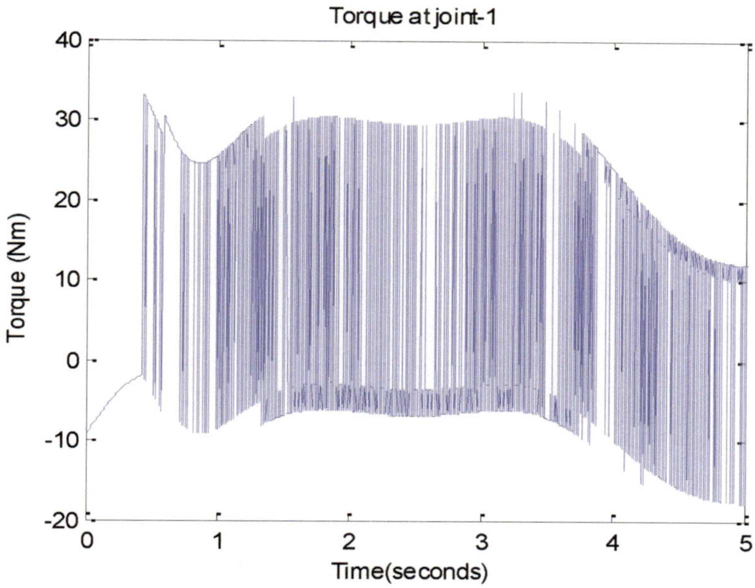

Figure 28.
Torque at Joint 1 under SMC of the robot manipulator.

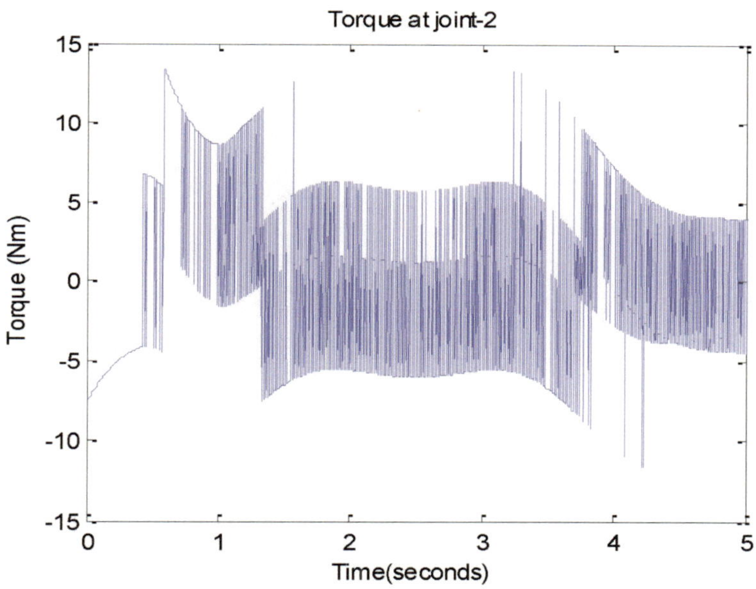

Figure 29.
Torque at Joint 2 under SMC of the robot manipulator.

higher as in the case of DC motor position control and that would complicate the realization of such high gains through the proper actuators. On the other hand, SMC provides robustness against parameter uncertainties and unmodeled disturbances so long as the observed undesirable chattering effect is overcome through some modifications by simply replacing nonlinear signum function with nonlinear saturation function and introducing boundary layer thickness in there as explained in earlier sections. In order to realize this, the boundary layer has been introduced for the first time in Section 6.2.2 simulations, and consequently, no switching or chattering effect has been observed as can be verified by the phase portrait in **Figure 34**.

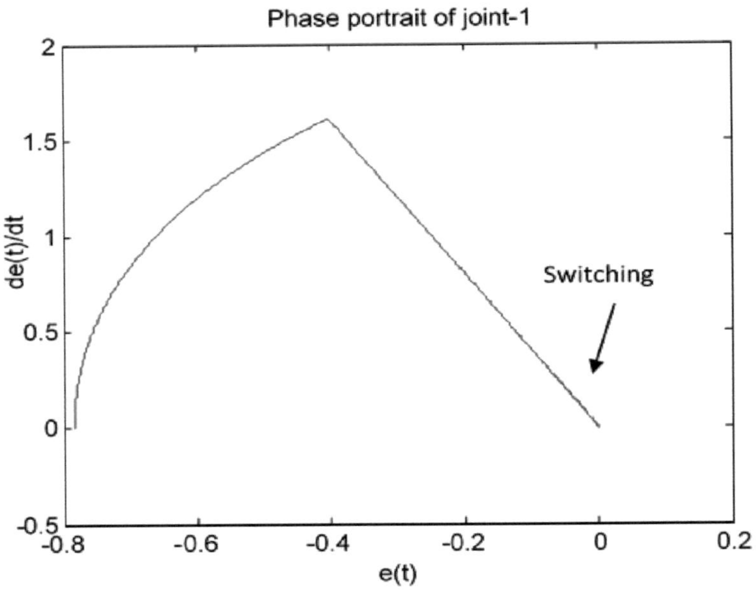

Figure 30.
Phase portrait of Joint 1 under SMC of the robot manipulator.

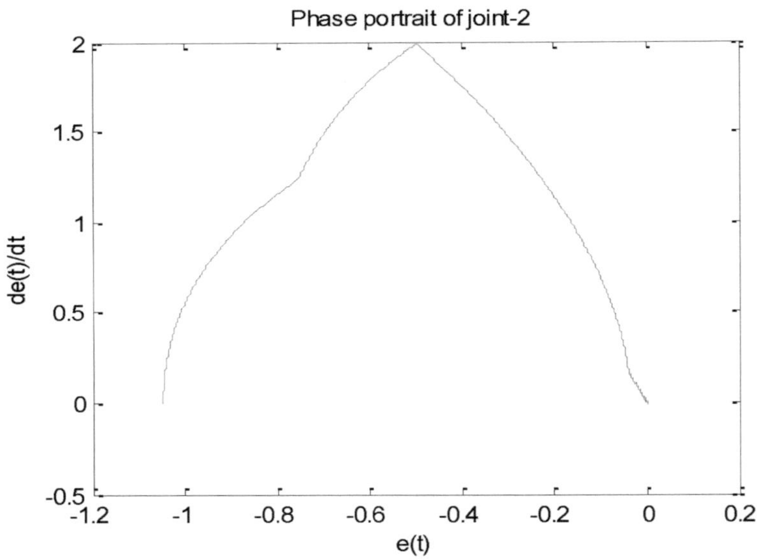

Figure 31.
Phase portrait of Joint 2 under SMC of the robot manipulator.

Later, the robustness of the SMC will be analyzed by adding an extra mass of 0.5 kg to Joint 2, and we have not observed any performance degradation of the trajectory to be maintained in the sliding surface. Therefore, the controller is said to be robust enough. However, it is expected that switching will reappear to maintain the trajectory in the sliding surface.

As a rule of thumb, It is possible to do tracking with more load by reducing the boundary layer to allow more switching to occur. Now, we reduce the boundary layer thickness from 0.02 to 0.005 and add the extra mass to Joint 2 by 0.75 kg to a final of 1.25 kg and we can still observe that SMC will be able to do the tracking by

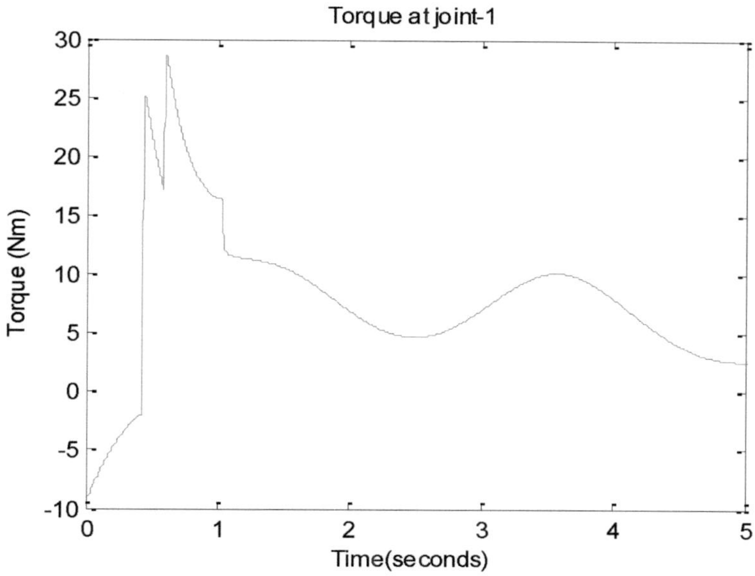

Figure 32.
Torque at of Joint 1 under SMC with a boundary layer.

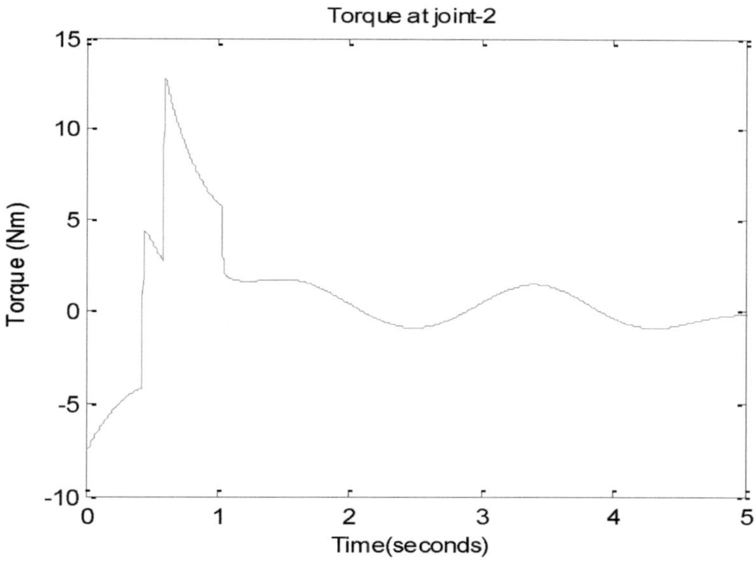

Figure 33.
Torque at of Joint 2 under SMC with a boundary layer.

observing the reemerged chattering effect as can be seen in the following simulations (**Figure 35**):

- Joint 1: $K = 14$, $\lambda = 4$, $\phi = 0.005$.
- Joint 2: $K = 8$, $\lambda = 4$, $\phi = 0.005$.
- $m_1 = 0.5$ kg, $m_2 = 1.25$ kg (an extra mass of 0.75 kg was added to Joint 2).
- The rest of the parameters and the IC's were kept the same as before.

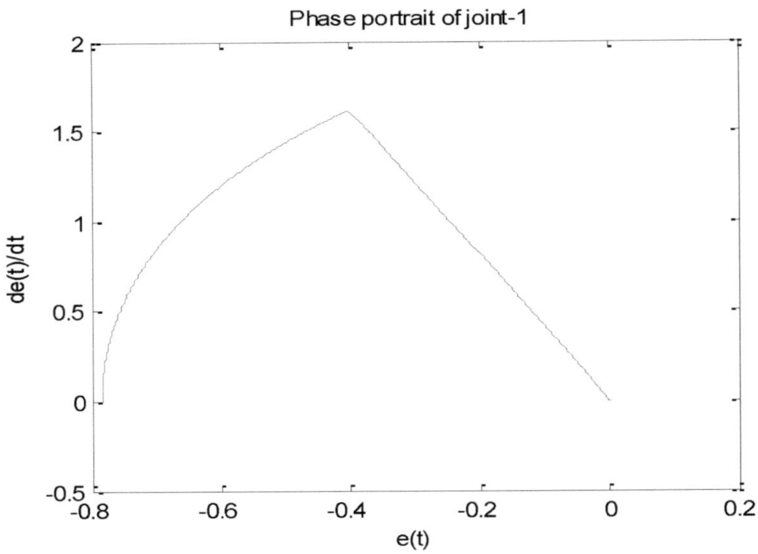

Figure 34.
Phase portrait of Joint 1 under SMC with a boundary layer.

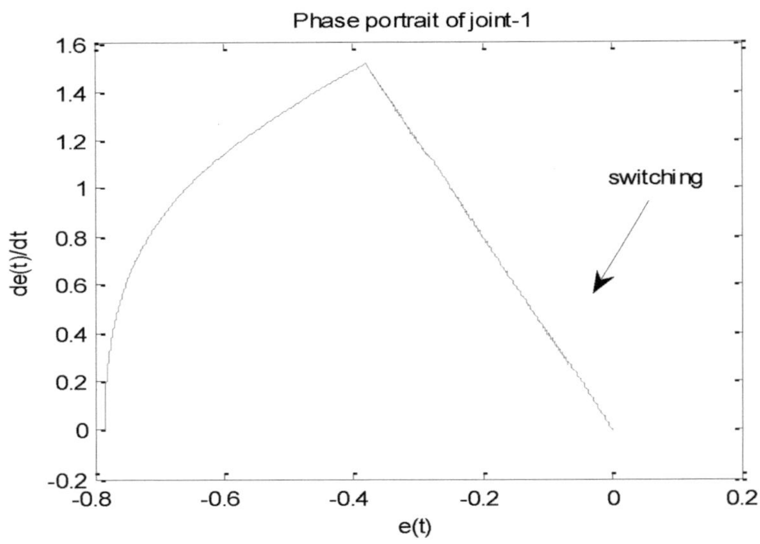

Figure 35.
Phase portrait of Joint 1 under SMC with a robustness test including more load.

7. Conclusions

In this study, a sliding mode control scheme with a bounded region and its convergence analysis are explained to the finest detail. In particular, it can easily be said that the work done here is a field study that specifically gives the relevant subject with such meticulous detail. It is our claim that this study has a guiding identity for the researchers who are interested in this control method or want to present it with the intelligent and modern control methodologies with its understandability and clarity targeted here. In this regard, the design of SMC including its finite-time convergence is handled by using Lyapunov's direct method. The tracking error vector converges exponentially to the bounded region once in the boundary layer as proven analytically. Two examples were used for simulation studies to demonstrate the feasibility and effectiveness of the proposed control problems, i.e., the position control of a dc motor subject to a varying external disturbance, and a two-link robot manipulator. Simulations show that a fast convergence rate, and hence quick response, the ability to reject the varying external disturbances, and the robustness against the model uncertainty assumed to be unbounded and fast-varying have all achieved its purpose entirely. Chattering is eliminated by using the boundary layer whose attractiveness and invariance properties of the boundary layer were also introduced. This study also examines the advantages of SMC and PID comparably. Although, in the comparisons given in the literature, the pros and cons of both strategies are mentioned, it is generally observed that SMC performs better than PID. Nevertheless, PID control can still be used as an alternative to SMC. When the PID control strategy does not work well under unknown disturbances and payload changes, SMC provides robustness against parameter uncertainties and unmodeled disturbances so long as the observed undesirable chattering effect is overcome through some modifications as described in the text. Robustness analysis has been performed and successfully applied to the two-link rigid planar robotic manipulator. We have not observed any performance degradation of the trajectory to be maintained in the sliding surface. The results given here do not contradict the view that one can use it instead of the other without losing too much performance. Finally, a two-step simulation has been carried out, testing all the features mentioned above, and the results have confirmed the success of the presented approach. However, it is meaningful and challenging to develop new SMC theories and methods for nonlinear systems due to its broad application potentials in today's world.

Conflict of interest

The authors declare no conflict of interest.

Author details

İhsan Ömür Bucak
Istanbul Rumeli University, Istanbul, Turkey

*Address all correspondence to: iomur.bucak@rumeli.edu.tr

References

[1] Slotine J, Li W. Applied Nonlinear Control. New Jersey: Prentice-Hall; 1991. p. 461. ISBN: 0-13-040890-5

[2] Shafiei SE. Sliding mode control of robot manipulators via intelligent approaches. In: Shafiei SE, editor. Advanced Strategies for Robot Manipulators. Rijeka: IntechOpen; 2010. pp. 135-172. DOI: 10.5772/10193

[3] Yu X, Kaynak O. Sliding-mode control with soft computing: A survey. IEEE Transactions on Industrial Electronics. 2009;**56**(9):3275-3285. DOI: 10.1109/TIE.2009.2027531

[4] Utkin VI. Variable structure systems with sliding modes. IEEE Transactions on Automatic Control. 1977;**22**(2): 212-222. DOI: 10.1109/TAC.1977. 1101446

[5] Tannuri EA, Agostinho AC, Morishita HM, Moratelli L. Dynamic partitioning systems: An experimental analysis of sliding mode control. Control Engineering Practice. 2010;**18**(10): 1121-1132. DOI: 10.1016/j. conengprac.2010.06.007

[6] Young KD, Utkin VI, Ozguner U. A control engineer's guide to sliding mode control. IEEE Transactions on Control Systems Technology. 1999;**7**(3):328-342. DOI: 10.1109/87.761053

[7] Köse E, Kızmaz H, Abacı K, Aksoy S. Control of SVC based on the sliding mode control method. Turkish Journal of Electrical Engineering and Computer Sciences. 2014;**22**:605-619. DOI: 10.3906/elk-1209-8

[8] Bessa WM, Paulo AS, Salvi MA. Sliding mode control with adaptive fuzzy dead-zone compensation for uncertain chaotic systems. Nonlinear Dynamics. 2012;**70**:1989-2001. DOI: 10.1007/s11071-012-0591-z

[9] Chang E. Study and application of intelligent sliding mode control for voltage source inverters. Energies. 2018; **11**(10):14. Article ID 2544. DOI: 10.3390/en11102544

[10] Sun XQ, Cai YF, Yuan CC, Wang SH, Chen L. Fuzzy sliding mode control for the vehicle height and leveling adjustment system of an electronic air suspension. Chinese Journal of Mechanical Engineering. 2018;**31**:13. Article ID 25. DOI: 10.1186/ s10033-018-0223-8

[11] Sharkawy AB, Salman SA. An adaptive fuzzy sliding mode control scheme for robotic systems. Intelligent Control and Automation. 2011;**2**(4): 299-309. DOI: 10.4236/ica.2011.24035

[12] Koubaa Y, Boukattaya M, Damak T. Intelligent control for nonholonomic mobile robot including actuator dynamics. In: Proceedings of the 15th International Multi-Conference on Systems, Signals & Devices (SSD); 19–22 March 2018; Hammamet. Tunisia: IEEE; 2018. pp. 1012-1016

[13] Jiang B, Gao C, Kao Y, Liu Z. Sliding mode control of Markovian jump systems with incomplete information on time-varying delays and transition rates. Applied Mathematics and Computation. 2016;**290**(C):66-79. DOI: 10.1016/j. amc.2016.05.038

[14] Leu VQ, Choi HH, Jung JW. LMI-based sliding mode speed tracking control design for surface-mounted permanent magnet synchronous motors. Journal of Electrical Engineering and Technology. 2012;**7**(4):513-523. DOI: 10.5370/JEET.2012.7.4.513

[15] Huynh VV, Tsai YW, Duc PV. Adaptive output feedback sliding mode control for complex interconnected time-delay systems. Mathematical

Problems in Engineering. 2015;**2015**:15. Article ID 239584. DOI: 10.1155/2015/239584

[16] Fallaha CJ, Saad M, Kanaan HY, A-Haddad K. Sliding-mode robot control with exponential reaching law. IEEE Transactions on Industrial Electronics. 2011;**58**(2):600-610. DOI: 10.1109/TIE.2010.2045995

[17] Zhang B, Yang X, Zhao D, Spurgeon SK, Yan X. Sliding mode control for nonlinear manipulator systems. IFAC-PapersOnLine. 2017;**50**(1):5127-5132. DOI: 10.1016/j.ifacol.2017.08.781

[18] Moldoveanu F. Sliding mode controller design for robot manipulators. Bulletin of the Transilvania University of Brasov, Series I: Engineering Sciences. 2014;7(2):97-104

[19] Nise NS. Control Systems Engineering. 7th ed. New Jersey: Wiley; 2015. p. 944. ISBN: 978-1-118-17051-9

[20] Kwakernaak H, Sivan R. Linear Optimal Control Systems. New York: Wiley; 1972. p. 595. ISBN: 0-471-51110-2

[21] The MathWorks Inc. MATLAB Version R2013a. Natick, Massachusetts: The MathWorks Inc; 2013

[22] Nasir ANK, Ismail RMTR, Ahmad MA. Performance comparison between sliding mode control (SMC) and PD-PID controllers for a nonlinear inverted pendulum system. International Journal of Computer, Electrical, Automation, Control and Information Engineering. 2010;**4**(10):1508-1513. Available from: scholar.waset.org/1999.4/1558

[23] Castillo-Zamora JJ, Camarillo-GóMez KA, PéRez-Soto GI, RodríGuez-ReséNdiz J. Comparison of PD, PID and sliding-mode position controllers for V–tail quadcopter stability. IEEE Access. 2018;**6**:38086-38096. DOI: 10.1109/ACCESS.2018.2851223

[24] Yousef AM. Experimental set up verification of servo DC motor position control based on integral sliding mode approach. WSEAS Transactions on Systems and Control. 2012;7(3):87-96. E-ISSN: 2224-2856

[25] Rafferty KJ, McGookin EW. A comparison of PID and sliding mode controllers for a remotely operated helicopter. In: Proceedings of the 12th International Conference on Control Automation Robotics & Vision (ICARCV); 5–7 Dec. 2012; Guangzhou, China: IEEE; 2013. pp. 984–989

[26] Xu J, Qiao L. Robust adaptive PID control of robot manipulator with bounded disturbances. Mathematical Problems in Engineering. 2013;**2013**:13. Article ID 535437. DOI: 10.1155/2013/535437

[27] Su CY, Leung TY. A sliding mode control with bound estimation for robot manipulators. IEEE Transactions on Robotics and Automation. 1993;**9**:208-213. DOI: 10.1109/70.238284

[28] A. Manurung, A Sliding Mode Control for Robot Manipulator [Internet]. 2010. Available from: http://sites.google.com/site/auraliusproject/a-sliding-mode-control-for-robot-manipulator [Accessed: 09 July 2019]

Section 4

Digital Image Processing

Chapter 9

Polygonal Approximation of Digital Planar Curve Using Novel Significant Measure

Mangayarkarasi Ramaiah and Dilip Kumar Prasad

Abstract

This chapter presents an iterative smoothing technique for polygonal approximation of digital image boundary. The technique starts with finest initial segmentation points of a curve. The contribution of initially segmented points toward preserving the original shape of the image boundary is determined by computing the significant measure of every initial segmentation point that is sensitive to sharp turns, which may be missed easily when conventional significant measures are used for detecting dominant points. The proposed method differentiates between the situations when a point on the curve between two points on a curve projects directly upon the line segment or beyond this line segment. It not only identifies these situations but also computes its significant contribution for these situations differently. This situation-specific treatment allows preservation of points with high curvature even as revised set of dominant points are derived. Moreover, the technique may find its application in parallel manipulators in detecting target boundary of an image with varying scale. The experimental results show that the proposed technique competes well with the state-of-the-art techniques.

Keywords: dominant point, projection position, iterative smoothing, minimal number of points, polygonal approximation

1. Introduction

Shape representation and shape classification are efficiently facilitated by polygonal approximation. This approach is popular due to its compact representation and insensitive to noise. These salient features are found useful in many applications [1–8]. The main objective of polygonal approximation is to approximate the shape of a curve using a polygon whose vertices are specified by a subset of points on the curve. These points are referred to as dominant points and are often the points with high curvature. An example is illustrated in **Figure 1**. A digital curve representing the shape of snowflake is displayed in **Figure 1(a)**, and its identified dominant points are shown in **Figure 1(b)**. The anticipated output of polygonal approximation using dominant point can be seen in **Figure 1(c)**. Broadly polygonal/closed curve approximation of a digital planar curve may be cast as min ε problem or min $\neq$ problem. In min ε problem, the techniques derive polygonal approximation with specified number of line segments or dominant points. These techniques

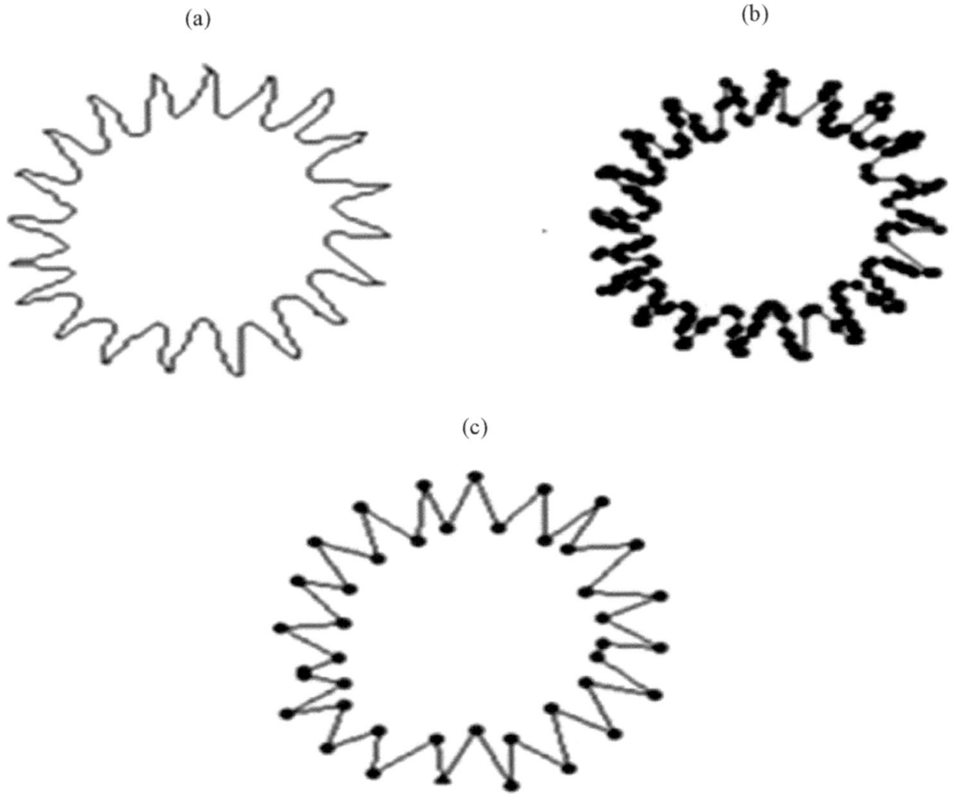

Figure 1.
(a) A digital curve representing the shape of a snowflake, (b) initial set of dominant points, and (c) suitable polygonal approximation are shown here.

ensure that the deviation between the curve and the approximate polygon is minimal, condition to the specified number of dominant points. Min # techniques derive polygonal approximation with a specified error. These techniques generate the approximate polygon with minimal number of dominant points while ensuring the measure of closeness is not larger than the specified error. In recent years, there are many dominant point-based polygonal approximation techniques that were presented in the literature [9–19].

And few older ones can be found in [20–22]. The techniques presented in [9, 10, 12, 20, 21] use reverse polygonization, where instead of detecting the real points the techniques make a search to detect redundant points and delete points iteratively. The methods in [11, 15] use breakpoint suppression, where the techniques apply criterion measure on the finest approximated set of points to suppress the redundant points and make the approximation. The methods in [3, 13, 16, 18] present a solution using dynamic programming, where the techniques make exhaustive search to detect points on curve, thereby making final approximation. The method in [14] makes polygonal approximation by detecting ADSS (Approximate Digital Straight Segment). The method in [17] uses MIP (mixed integer programming) model. The method in [19] uses vertex relocation procedure around neighbors. In this method, while approximating the output curve by detecting the dominant point, the technique allows neighborhood points to become a dominant point provided that new dominant point facilitates in reduction of approximation error. The method in [22] uses split and merge, where the method makes a search to find the points with maximum deviation in the splitting stage using the proposed criterion function and merge all the points identified in the splitting stage using the threshold value. Most of the dominant points [9–12] detecting methods use the magnitude of orthogonal projection of a point on the line segments, which connect adjacent high curvature points to influence the process of detecting dominant points. The methods in the literature [9–12, 14, 15, 20, 23] do not address the issue where the projection of point lies beyond its candidate line segment, where the situation may be often anticipated during approximation. The techniques that neglect to check this criterion may miss good curvature points, which are critical for shape representation. The technique proposed in this chapter measures the positions of projections of a point on the curve, thereby invoking different metrics for computing the significant measure of the dominant points. This practice makes the proposed technique to preserve the original shape of the curve even at very minimal number of dominant points. Such characteristic is very essential for compact representation. And it is very essential for object detection and shape classification applications. Especially, the proposed technique can facilitate the parallel manipulators in cutting and milling operations by preserving the actual shape of the target boundary points. The rest of the chapter is organized is as follows: Section 2 presents a brief review of some of the state-of-the-art methods along with an insight into their demerits wherever possible. Section 3 presents the proposed work. Section 4 summarizes the experimental results. Section 5 concludes the chapter.

2. Background

Several polygonal approximation techniques have been proposed in the recent decades. Some of them use various optimization approaches [3, 13, 16–19]. On the other hand, there are other techniques that use local/global geometric features of a curve to influence the process of determining the polygon with minimal number of line segments [9–12, 23–26], and these techniques prove its competence against many real-time datasets. Among these, this section briefly analyzes some of the bench mark techniques.

Prasad [23] proposed a non-parametric framework to detect points of high curvature. The framework uses the maximum deviation incurred between pixels from a digitized boundary as an upper bound to make approximation. The authors proved that the analytical bound can be incorporated by dominant point detection framework to get rid of specification in terms of the tolerable error (for min # approaches) or the number of points (for min ϵ approaches). The authors established the robustness of their framework against scaling invariance as well as noise tolerance. However, there are applications in which the curve needs to be approximated using a specified number of dominant points, which is not possible through this framework. Though the approximation bounded below to digitization value, points detected on the curve seem to be redundant for human visual perception. Prasad [24] used metrics such as precision and reliability as measures to fit the polygon edges. Depending upon the threshold values for these measures, the technique produces coarser or finer approximation. Thus, this technique can flexibly control the degree of smoothness required for an application. And also the paper suggests some performance metrics to quantify the techniques. Parvez [19] obtained the digital boundary using contour extraction techniques. The objective of the method was to produce approximate polygon with minimal error possible. To attain this goal, the method relaxes the criteria that dominant points need not be on the contour. The technique computes neighborhood points for every point p_i on the contour C_d and introduces a new point on the contour provided its presence should reduce the approximation error. The neighborhood points are not the ones computed using 4 connected graph or 8 connected graph; instead, the technique adaptively defines the width for every point on the curve, and thereby, it obtains the neighborhood points. Fernandez [25] produced symmetric approximation for symmetric contours. The technique obtains first initial point p_1 as the farthest in terms of distance from the centroid of the curve. The next point p_2 is the farthest to p_1. The method proceeds to find point p_4, which is farthest from p_2, and point p_3, which is farthest from p_1. Likewise, the technique obtains the all possible line segments such as $\{p_1, p_2\}$, $\{p_3, p_4\}$, until the maximum deviation from the curve does not exceed a threshold value that constitutes the boundary point set. The authors demonstrate that their method of choosing initial points ensures symmetricity. The technique then identifies all possible candidate points $(q_1, q_2, \ldots, q_m)$ from the boundary point set between every two initial points and computes a significant value by ensuring symmetry property. Additionally, the technique presents various thresholding methods to normalize the significant values of the boundary points. Though the technique produces symmetric approximation for symmetric curve, it did not establish geometric invariance. And in real-time data sets, in most of the cases, the points are always distributed asymmetrically on the planar curve. The main objectives of this chapter are to i) present a framework that considers the projection position of a point and thereby invokes the proper criterion measure to compute the contribution; ii) produce output polygon without missing significant points; iii) produce polygon with minimal possible number of points; and iv) present a technique that is reasonably strong enough against rotation invariance. These objectives are achieved and demonstrated through experimentations of the proposed technique using benchmarking data sets.

3. Proposed work

3.1 Problem formulation

The problem formulation is as follows: let $C_d = \{p_1, p_2, \ldots p_n\}$ where $p_i = (x_i, y_i)$ is a digital curve consisting of n points in clockwise direction in the discrete two-dimensional space. Such curves are the ones extracted from the boundaries of the digital images using contour detection or edge detection methods. The coordinates of these n points are integers since these points are extracted from the digital boundary. The objective of polygonal approximation of C_d is to derive a subset $D = \{p_1, p_2, \ldots, p_m\}$ from the super set of C_d, subject to the condition the polygon formed by the elements of D should represent the shape of the original curve. The technique starts with any three consecutive points p_i, p_j and p_k on the curve C_d, to detect the collinearity of these points (p_i, p_j, p_k), the distance measured from a point p_j to the line segment connecting p_i and p_k. The method shall conclude the three points are collinear, provided the measured distance is very minimal. On the other side, the method shall conclude non-collinearity, provided the measured distance is not very minimal and thus p_j becomes an element of D. Thereby, the polygonal approximation technique finds all the elements of D. With this problem formulation, our chapter focuses on the choice of the significant measure metric. Conventionally, the distance metric is the length of the line dropped from the point p_j on the line segment $p_i p_k$. This is being referred to as the perpendicular distance. This metric is generally good for smooth curves, but in some cases (explained later), it may miss significant points and reject sharp turn, which are essential in shape representation applications. Dunham [27] makes initial approximation using distance to a line segment. Ramaiah [28] uses distance to a line segment as a measure to make polygonal approximation, but the metric used in the technique to compute deviation is capable of preserving sharp turnings but fails to preserve the original shape of digital curve. Apart from the criterion measure proposed in any technique, the methodology is also an important factor to produce the output polygon without compromising its actual shape. This implies that the used metric in [28] is unsuitable for iterative smoothing. The framework proposed in this chapter automatically chooses the suitable significant measure metric based on the candidate point projection, as explained next.

3.2 Proposed technique

In this section, we present our proposed method to make polygonal approximation of C_d. The initial segmentation points are obtained using Freeman chain code [28], such as given in Algorithm 1. These initially segmented points are referred as initial set of dominant points. Example of initial segmentation for the snowflake curve is shown in **Figure 1(a)** and **(b)** where the dominant points are highlighted in bold markers and the final approximated curve is given in **Figure 1(c)**.

To compute the significant measure of every initial dominant point s_k, the proposed method uses the following steps: consider the scenario in **Figure 2(a)** where, namely s_{k-1}, s_k and s_{k+1} are three dominant points on the curve with the following traversing sequence: s_{k-1}- > s_k- > s_{k+1}. It may be interpreted as these three points are collinear by assuming the projections of a point s_k that lies on the line segment, which connects $(s_{k-1}s_{k+1})$. As a consequence, the approximation technique [9–12, 14, 15, 20, 23, 29, 30] may decide to drop s_k. In this scenario, the projection of a point (s_k) lies between its candidate line segment $(s_{k-1}s_{k+1})$. **Figure 3** shows the various anticipated position for possible projection of a dominant point (s_k) on the x-y plane. The proposed metric detects the position of projection. In order to predict the position of a projection, the proposed technique uses the following steps: translate the line segment connecting s_{k-1} and s_{k+1} so that the point S_i coincides with the origin of the x-y coordinate system and measures the amount of angle produced by the translated line segment with the x axis. In order to align the translated line segment with the x axis, rotate the line segment with a computed amount angle. The actual x-y coordinate system and new transformed coordinate systems are displayed in **Figure 2(a)** and **(b)**. In the next step, by checking transformed x coordinate of s_k', the method chooses metric to compute the significant measure. If the x coordinate of s_k' is less than 0, then the significant measure sig(s_k) is computed using Eq. (1) (see **Figure 3(a)**). If x_k' of s_k' lies between 0 and the x coordinate of s_i, then the significant measure is computed using Eq. (2) (see **Figure 3(b)**.

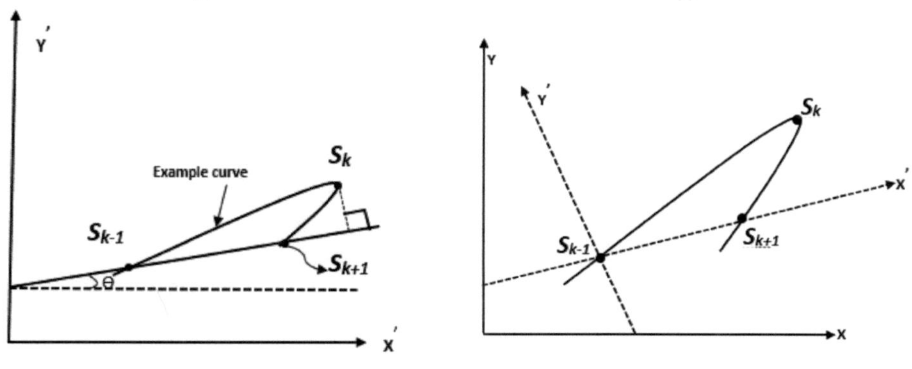

Figure 2.
Demonstration of the coordinate transform performed for the proposed self-adaptive significant measure computing metric for dominant point detection. (a) An example curve in the original x-y coordinate system is shown. (b) The transformed x'-y' coordinate system is shown in addition to the original x-y system.

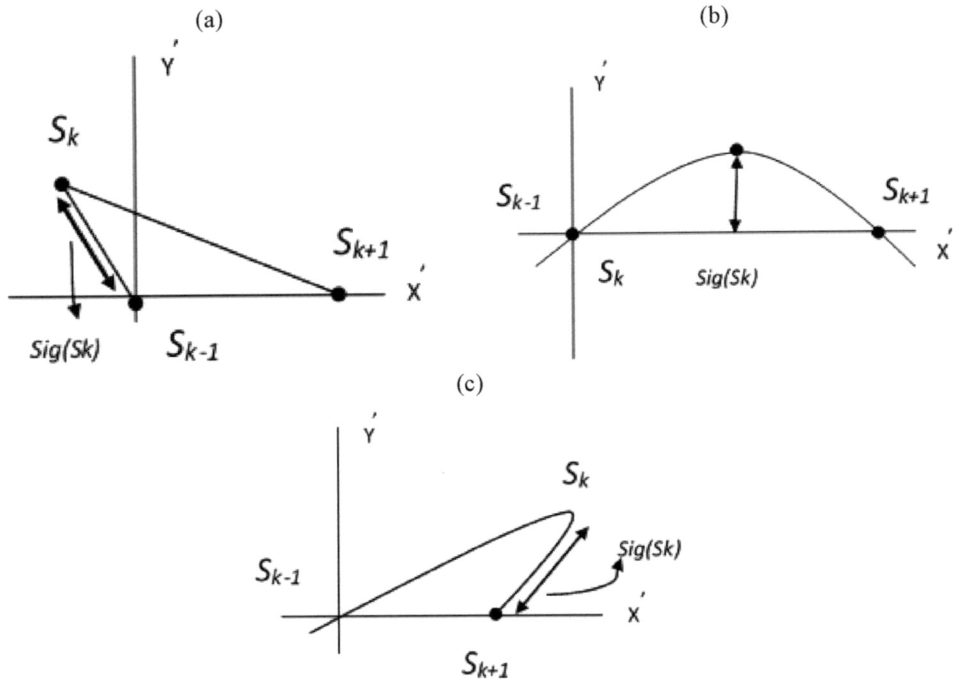

Figure 3.
Demonstration of computation of significant measure of the point sk from the line segment $s_{k-1}s_{k+1}$.

If the x_k' value is greater than x_j' of s_{k+1}, then the significant measure of s_k is computed using Eq. (3) (see **Figure 3(c)**).

$$sig(s_k) = \sum_{k=s_{k-1}}^{s_{k+1}} \sqrt{\left(s_{x_k} - s_{x_{k-1}}\right)^2 + \left(s_{y_k} - s_{y_{k-1}}\right)^2} \tag{1}$$

$$sig(s_k) = \sum_{k=s_{k-1}}^{s_{k+1}} \left|S_{y_{k'}}\right| \tag{2}$$

$$\boldsymbol{sig}(s_k) = \sum_{k=s_{k-1}}^{s_{k+1}} \sqrt{\left(s_{x_k} - s_{x_{k+1}}\right)^2 + \left(s_{y_k} - s_{y_{k+1}}\right)^2} \tag{3}$$

In all the three equations (Eqs. (1)–(3)), k range is k-1 < =k < =k + 1. (Note: the accent sign indicates the coordinates in the transformed coordinate system). While computing the significant measure associated with a dominant point, let us say s_k, the significant measure of every non-dominant point/boundary point lies between its candidate line segment and is accumulated to define the significance measure of s_k. These steps are repeated for each dominant point in the initial set, before making the decision to remove redundant dominant points in the next step. After measuring the significant measure of all initial dominant points, the proposed method removes the dominant point with minimal significant measure. If more than one dominant point has the same minimal significant measure, the dominant point appearing first in the order of sequence is removed. The steps to remove the dominant point and produce the final output polygon are given in Algorithm 2.

Algorithm 1. CIDP (Compute initial set of dominant points).

Input: The inputs are the coordinates of the boundary points.
C_d = p_i (x_i,y_i),i=1,2,3 … ..n; n boundary points.
Output: The outputs are the curve indices of initial dominant points.

```
Begin
Case 1: i=0
        If (x(0)-x(n-1) != x(1)-x(0)) or ((y(1)-y(0) != y(0)-y(n-1)) then
D[0]= 0;
Case 2: i=n-1
        If (x(n-1)-x(n-2) != x(0)-x(n-1)) or (y(n-1)-y(n-2) != y(0)-y(n-1))
        D[j]=i;
Default:
        While (i<n-1)
        If (x(i)-x(i-1) != x(i+1)-x(i)) or (y(i+1)-y(i) != y(i)-y(i-1))
          D[j] = i
End.
```

Algorithm 2. Polygonal approximation by computing the significant measure of IDP.

Input: Digital curve C_d, Number of dominant points (k) in the output polygon.
Output: Output polygon with the specified number of dominant points (k).
Begin.
Step 1: Invoke the function *CIDP*.
Step 2: Compute significant measure associated with all initial dominant points (s_k's).
Step 3: Repeat.

i. Identify the dominant point s_k with minimal significant measure in C_d.

ii. Remove the dominant point s_k and recalculate the significant measure of at s_{k-1} and s_{k+1}.

iii. Compute the performance measures with the available dominant points

Until (No. of DPs $= k$).
End

4. Experimental results

The proposed technique is tested on a variety of challenging curves to demonstrate its efficiency. The results are presented for two experiment sets. The experiment set 1 consists of synthetic curves usually used in the literature [9, 11, 16, 19, 24, 25, 31–38]. In experiment 2, the proposed method is tested extensively with images in MPEG data set [39]. We first present the quality assessment metrics for polygonal approximation of digital curves. Then, we present the results on the two experimental sets. Additionally, we include one experiment to demonstrate geometric invariance of the proposed technique.

4.1 Quality assessment

The best method to assess output of polygonal approximation is visual perception. Thus, we include extensive qualitative results. Moreover, we include quantitative performance measures as well for comparison of the performance of the tested methods, including the proposed technique. This chapter considers the following metrics to measure the goodness of the results: (i) compression ratio (CR), (ii) integral square error (ISE), (iii) figure of merit (FOM), (iv) weighted sum of square errors (WE), (v) modified version of WE (WE2). Details of these metrics are provided in **Table 1**. These metrics are taken from [9–11, 15, 19, 33, 36]. The readers interested in them are encouraged to read these articles and the references therein.

4.2 Experimental set 1

The quantitative performance measure for the synthetic curves chromosome, leaf, semicircle and infinity in experiment set 1 is given in **Table 2**. The visual shots are shown in **Figures 4–6**. The methods in [16–19, 34, 36, 37] present optimal solutions for the polygonal approximation. The proposed method output is close to optimal solution for all the curves and further supports reduction of the number of dominant points while retaining the shape information of the curve. **Table 2** summarizes the results from various articles [9, 11, 15–19, 23, 24, 26, 31–38] for the given input synthetic curves. For the chromosome curve display using 15 amount of dominant points, the proposed technique produces a low value for ISE than the method in [32–34]. The snapshot of chromosome curve at 6 number of points using the proposed method as well as by the methods [9, 23, 24] snapshots can be found in **Figure 4**. For the leaf curve, where the output curve at 21 number of dominant points, the proposed method produces the low value for ISE than [11, 24, 34] (in turn FOM value is high, which is appreciable) and high value than [19]. The snapshot for leaf output curve produced by the proposed method along with some of the state-of-the-art methods results is displayed in **Figure 5**. The final synthetic curve for this experiment set is a curve that intersects itself, that is, infinity-shaped curve. In the attempt of producing the output curve using 10 number of points, the proposed produce the minimal possible error than [11, 26]. And also the summarized results reveal that the proposed method output is better than [9, 11, 19, 24, 26, 33] in terms of ISE, WE and FOM. The graphic shots for the same can be found in **Figure 6**. According to human visual perception, four points are sufficient

Metric	Indicator of goodness	Mathematical representation
CR	Larger is better	$CR = \frac{n}{k}$, where n is the number of points in the initial segmentation, while k is the number of dominant points in the final polygonal approximation.
ISE	Smaller is better	$ISE = \sum_{k=1}^{n} e_k$, where e_k is the perpendicular distance of a point p_k on the original digital curve from the nearest line segment on the polygonal approximation.
FOM	Larger is better	$FOM = \frac{CR}{ISE}$
WE	Smaller is better	$WE = \frac{ISE}{CR}$
WE2	Smaller is better	$WE2 = \frac{ISE}{CR^2}$

Table 1.
Quality assessment metrics for comparing polygonal approximation methods.

Contour	Methods	*k*	*CR*	*ISE*	*WE*	*FOM*
Chromosome	Teh and Chin [32]	15	4.00	7.20	1.80	0.56
n = 60	Wu [33]	15	4.00	7.20	1.80	0.56
	Masood [9]	12	5.00	7.76	1.55	0.64
	Carmona et al. [11]	11	5.45	14.49	2.66	0.38
	Parvez [34]	10	6.00	14.34	2.39	0.42
	Madrid et al. [26]	12	5.00	5.82	1.16	0.86
	Nguyen and Debled-Rennesson [35]	25	3.33	4.06	1.22	0.82
	Nguyen and Debled-Rennesson [35]	15	4	5.69	1.42	0.70
	Parvez [19]	11	5.45	7.09	1.30	0.77
	Aguilera et al. [17]	10	6.00	8.07	1.35	0.74
	Lie et al. [18]	14	4.29	7.58	1.77	0.57
	Lie et al. [18]	12	5.00	7.96	1.59	0.63
	PRO0.6 [24]	11	5.45	11.00	2.02	0.50
	RDP2 [24]	8	7.50	59.99	8.00	0.13
	RDP3 [24]	6	10.00	91.18	9.12	0.11
	Proposed	15	4.00	4.87	1.22	0.82
	Proposed	**6**	**10.00**	**45.49**	**4.55**	**0.22**
Leaf	Teh and Chin [32]	29	4.14	14.96	3.61	0.28
n = 120	Wu [33]	24	5.00	15.93	3.19	0.31
	Marji and Siy [15]	17	7.06	28.67	4.06	0.25
	Carmona et al. [11]	21	5.71	17.97	3.15	0.32
	Parvez [34]	21	5.71	13.82	2.42	0.41
	Parvez [19]	21	5.71	11.98	2.10	0.48
	Nguyen and Debled-Rennesson [35]	33	3.64	5.56	1.53	0.65
	Backes and Bruno [36]	20	6.00	14.1	2.35	0.43
	Wang et al. [16]	20	6.00	13.9	2.32	0.43
	Madrid et al. [26]	22	5.45	11.16	2.05	0.49
	PRO0.6 [24]	21	5.71	21.70	3.80	0.26
	PRO1.0 [24]	18	6.67	36.70	5.50	0.18
	RDP1 [24]	22	5.45	19.17	3.51	0.28
	RDP2 [24]	16	7.50	65.46	8.73	0.11
	Proposed	21	5.71	13.25	2.32	0.43
	Proposed	**16**	**7.50**	**44.52**	**5.94**	**0.17**
Semicircle	Teh and Chin [32]	22	4.64	20.61	4.44	0.23
n = 102	Yin [37]	17	6.00	19.78	3.30	0.30
	Salotti [38]	14	7.29	17.39	2.39	0.42
	Wu [33]	27	3.78	9.01	2.38	0.42
	Marji and Siy [15]	15	6.80	22.70	3.34	0.30
	Masood [9]	21	4.86	9.82	2.02	0.49
	Carmona et al. [11]	26	3.92	4.91	1.25	0.80

Contour	Methods	*k*	*CR*	*ISE*	*WE*	*FOM*
	Parvez [34]	17	6.00	19.02	3.17	0.32
	Nguyen and Debled-Rennesson [35]	25	4.08	5.42	1.33	0.75
	Backes and Bruno [36]	14	7.29	19.80	2.72	0.37
	Wang et al. [16]	15	6.80	14.30	2.10	0.48
	Parvez [19]	15	6.80	18.22	2.68	0.37
	Aguilera et al. [17]	14	7.29	17.39	2.39	0.42
	Madrid et al. [26]	10	10.20	40.79	4.00	0.25
	Lie et al. [18]	14	7.29	29.30	4.02	0.25
	PRO 0.6 [24]	18	5.67	18.12	3.20	0.31
	Proposed	18	5.67	15.45	2.72	0.37
	Proposed	17	6.00	16.59	2.76	0.36
	Proposed	14	7.29	17.73	2.43	0.41
	Proposed	**12**	**8.50**	**40.62**	**4.78**	**0.21**
Infinity	Teh and Chin [32]	13	3.46	5.93	1.71	0.58
n = 45	Wu [33]	13	3.46	5.78	1.67	0.60
	Masood [9]	11	4.09	2.90	0.71	1.41
	Carmona et al. [11]	10	4.50	5.29	1.18	0.85
	Parvez [34]	9	5.00	7.35	1.47	0.68
	Parvez [19]	7	6.43	7.69	1.20	0.84
	Madrid et al. [26]	10	4.50	6.40	1.42	0.70
	PRO0.6 [24]	9	5.00	6.29	1.26	0.79
	PRO1.0 [24]	7	5.63	19.94	3.54	0.28
	RDP1 [24]	9	5.00	6.67	1.33	0.75
	RDP2 [24]	7	6.43	19.94	3.10	0.32
	RDP3 [24]	5	9.00	53.82	5.98	0.17
	Masood [9]	8	5.63	10.24	1.82	0.55
	Carmona et al. [11]	6	7.50	31.68	4.22	0.24
	Proposed	10	4.50	4.44	0.99	1.01
	Proposed	5	9.00	35.61	3.96	0.25

Table 2.
Comparative results of synthetic contour (chromosome, leaf, semicircle, infinity).

enough to represent the infinity curve; please see **Figure 6(g)**. On the outset, it is perceived that the proposed technique gives the best or second best ISE values for all the cases. This indicates competitiveness of the proposed technique.

4.3 Experiment set 2

In this section, the performance of the proposed methods has been demonstrated using image in MPEG database [39]. Fernandez [25] presents a technique to produce output polygon from a given digital boundary. Authors in [25] demonstrated the efficiency of their method by comparing their results with method [23],

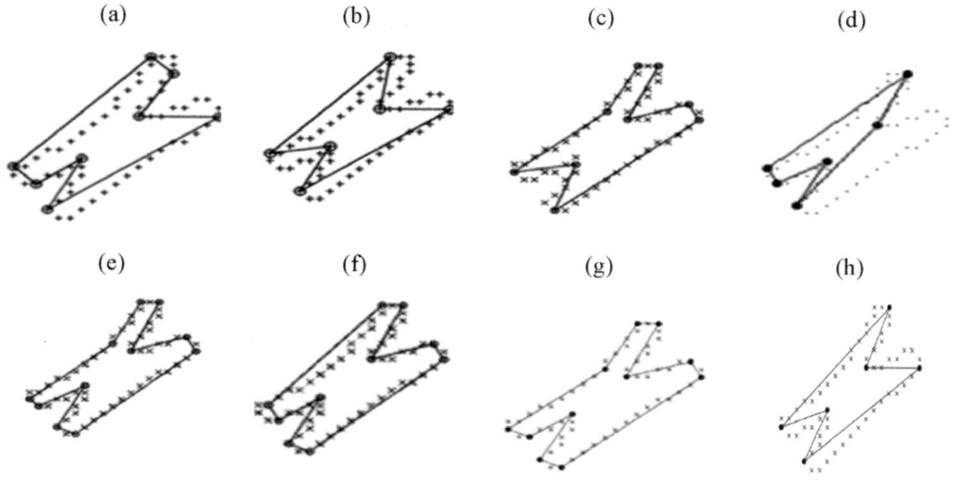

Figure 4.
Polygonal approximation of chromosome curve at varying amount of dominant points. (a) RDP2 [24] at 11 DPs, (b) RDP3 [24] at 6 DPs, (c) Masood [9] at 9 DPs, (d) Masood [9] at 6 DPs, (e) Prasad [23] Masood opt at 11 DPs, (f) Prasad [23] Carmona opt at 10 DPs, (g) Proposed method at 11 DPs, (h) Proposed method at 6 DPs.

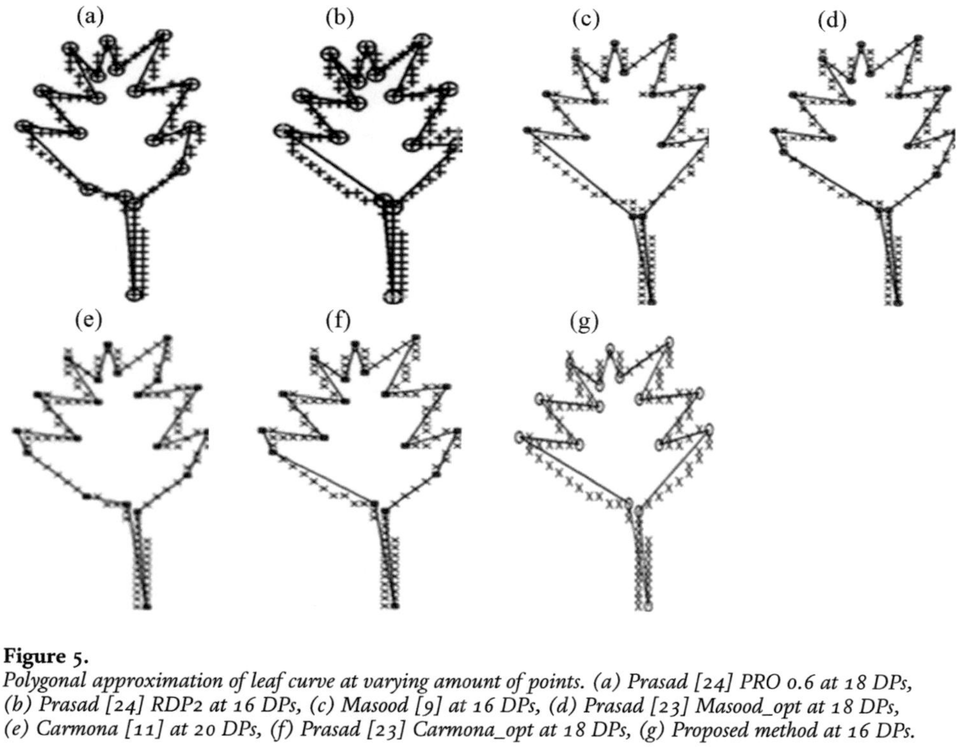

Figure 5.
Polygonal approximation of leaf curve at varying amount of points. (a) Prasad [24] PRO 0.6 at 18 DPs, (b) Prasad [24] RDP2 at 16 DPs, (c) Masood [9] at 16 DPs, (d) Prasad [23] Masood_opt at 18 DPs, (e) Carmona [11] at 20 DPs, (f) Prasad [23] Carmona_opt at 18 DPs, (g) Proposed method at 16 DPs.

which is capable of producing output polygon in non-parametric mode. So the better counterpart method to compare the proposed method is the one proposed in [25]. **Table 3** summarizes the results of the proposed method along with the results claimed as the best in [25] for the contours in MPEG database [39]. For the bell-7 contour, the snapshot at 23, 22, 20 and 7 number of dominant points, the proposed method produces a less approximation error in terms of ISE WE WE2 than others mentioned in [9, 11, 23, 25]. Especially the output approximation at 7 DPs, the

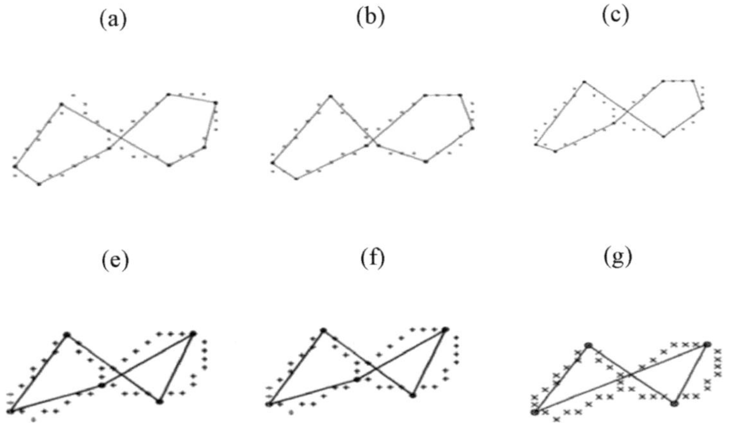

Figure 6.
Polygonal approximation of infinity curve at varying amount of DPs. (a) Masood [9] at 8 DPs, (b) Prasad [23] Masood _opt at 9 DPs, (c) Carmona [11] at 8 DPs, (d) Carmona [11] at 7 DPs, (e) Prasad [24] PRO 1.0 at 7 DPs, (f) Prasad [24]_RDP 3 at 5 DPs, (g) Proposed method at 6 and 4 DPs.

Contour	Methods	*k*	*CR*	*ISE*	*WE*	WE_2
Bell-7	Fernandez [25]	23	17.65	165.14	9.35	0.53
n = 407	Fernandez [25]	22	18.45	200.93	10.89	0.59
	Fernandez [25]	20	20.3	255.083	12.56	0.61
	Rosin [40]	7	58	2186.6	37.7	0.65
	Masood [9]	20	20.35	408.08	20.5	0.98
	Carmona [11]	23	17.69	332.563	8.84	0.23
	Prasad [23] RDP	28	14.53	97.60	6.71	0.46
	Proposed	22	18.5	176.54	9.54	0.51
	Proposed	20	20.35	210.16	10.32	0.50
	Proposed	7	**58.14**	**453.91**	**7.80**	**0.13**
Octopus-14	Fernandez [25]	79	15.33	236.62	15.44	1.00
n = 1211	Fernandez [25]	55	22.02	1270.17	57.69	2.62
	Fernandez [25]	50	24.22	1847.81	76.29	3.15
	Rosin [40]	43	28.16	2617.37	92.94	3.30
	Masood [9]	201	6.02	9268.43	1538.36	255.75
	Prasad [23] RDP	55	22.01	392.15	17.81	0.80
	Proposed	79	15.33	212.00	13.83	0.90
	Proposed	**43**	**28.16**	**1927.15**	**68.43**	**2.42**
Ray-17	Fernandez [25]	35	19.69	240.26	12.20	0.62
n = 689	Fernandez [25]	28	24.61	660.00	26.82	1.09
	Fernandez [25]	24	28.71	1152.83	40.16	1.40
	Rosin [40]	14	49.21	6999.71	142.23	2.89
	Masood [9]	24	28.71	749.01	26.09	0.91
	Masood [9]	14	49.21	8627.89	175.31	3.56

Contour	Methods	*k*	*CR*	*ISE*	*WE*	WE_2
	Prasad [23] RDP	54	12.75	342.36	26.93	2.10
	Proposed	35	19.69	208.48	10.59	0.53
	Proposed	**14**	**49.21**	**455.32**	**9.25**	**0.18**
Chicken-5	RDP [29, 30]	255	5.35	285.54	53.38	9.98
n = 1364	Masood [9]	401	3.40	147.86	43.47	12.79
	Carmona et al. [11]	134	10.18	906.52	89.06	8.74
	Fernandez [25]	54	25.26	2424.51	95.99	3.80
	Prasad [23] RDP	218	6.25	782.53	125.20	20.3
	Proposed	255	5.35	275.42	51.49	9.61
	Proposed	54	25.26	1994.15	78.95	3.12
Device 6–9	RDP [29, 30]	50	31.80	303.37	9.54	0.30
n = 1590	Masood [9]	84	18.93	189.89	10.03	0.53
	Carmona [11]	22	72.27	3395.17	46.98	0.65
	Fernandez [25]	33	48.18	348.22	7.23	0.15
	Prasad [23] RDP	38	41.84	741.416	17.02	0.42
	Proposed	84	18.93	216.24	11.42	0.60
	Proposed	**22**	**72.27**	**761.58**	**10.54**	**0.14**
Bell-10	RDP [29, 30]	110	10.92	181.25	16.59	1.52
n = 1202	Masood [9]	4	—	—		4.95
	Carmona [11]	104	11.78	549.52	46.64	3.96
	Fernandez [25]	42	28.61	687.56	24.03	0.84
	Prasad [23] RDP	81	14.83	326.47	22.01	1.48
	Proposed	110	10.92	241.45	22.06	2.02
	Proposed	**42**	**28.61**	**615.77**	**7.98**	**0.75**
Truck-07 n = 277	RDP [29, 30]	40	6.92	24.45	3.53	0.50
	Masood [9]	40	6.92	37.17	5.37	0.77
	Masood [9]	11	25.18	1133.29	45.00	1.78
	Carmona [11]	12	23.08	1132.45	49.06	2.11
	Fernandez [25]	40	6.92	24.15	3.48	0.50
	Prasad [23] RDP	33	8.39	59.17	7.05	0.84
	Proposed	12	23.08	319.24	13.83	0.59
	Proposed	**11**	**25.18**	**318.34**	**12.64**	**0.50**
Butterfly-13	RDP [29, 30]	344	5.19	383.30	73.85	14.23
n = 1786	Masood [9]	525	3.40	199.06	58.54	17.22
	Carmona-Poyato et al. [11]	171	10.44	1450.70	138.95	13.31
	Fernandez [25]	65	27.47	2195.88	79.93	2.91
	Proposed	525	3.40	197.58	58.11	17.09
	Proposed	**65**	**27.47**	**2063.91**	**75.13**	**2.73**

Table 3.
Comparative results for the MPEG database contours.

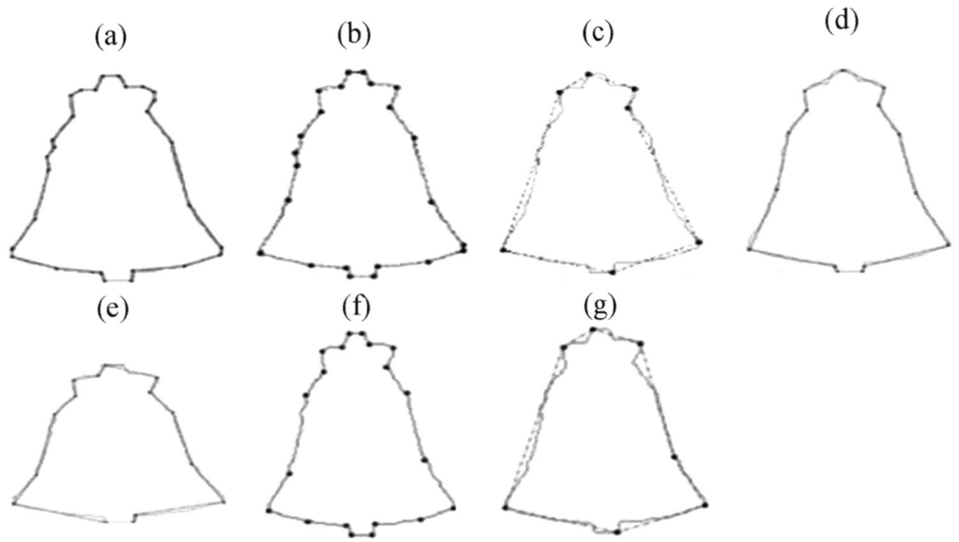

Figure 7.
The output approximation for the bell-7 contour by various methods: (a) Prasad [23] RDP at 28 DPs, (b) Fernandez [25] at 23 DPs, (c) Rosin [40] at 7 DPs, (d) Masood [9] at 15 DPs, (e) Carmona [11] at 15 DPs, (f) Proposed method 20 DPs, (g) Proposed method at 7 DPs.

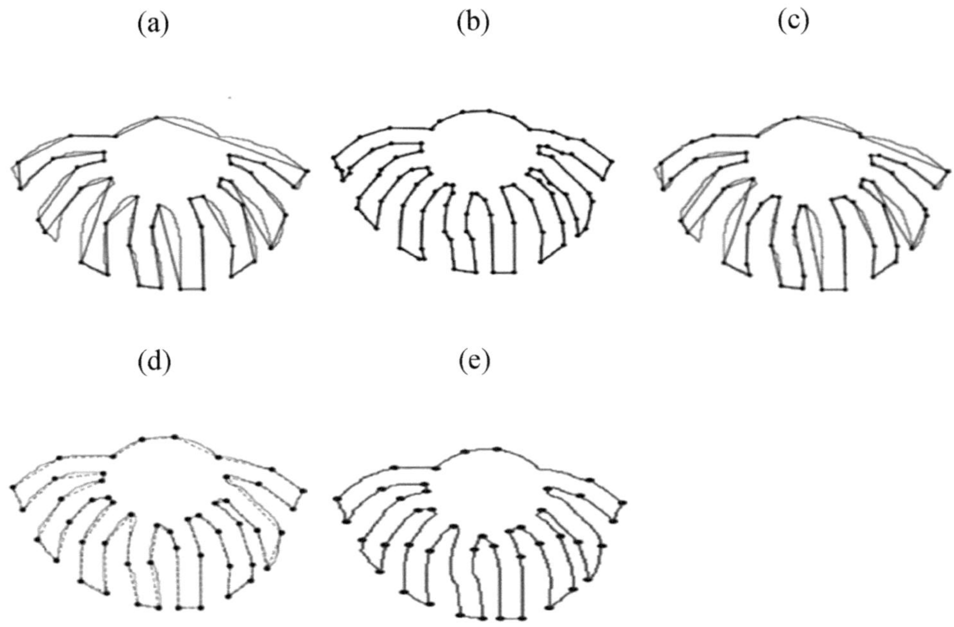

Figure 8.
The output polygon from octopus-17 by various methods: (a) Carmona [11] at 43 DPS, (b) Prasad [23] RDP at 55 DPs, (c) Prasad [23] Carmona_opt, (d) Fernandez [25] at 43 DPS, (e) Proposed method at 43 DPs.

proposed method and Rosin [40] method produce the curve with the mandatory points compared to others, but the proposed method produces minimal error measure than Rosin [40], and the output can be found in **Figure 7(c)** and **(h)**.

For the octopus-14 contour, the proposed efficiently produces the output curve with minimal deviation from the original curve compared to others. By observing **Figure 8(e)**, the proposed produces an outlying approximation that is visibly

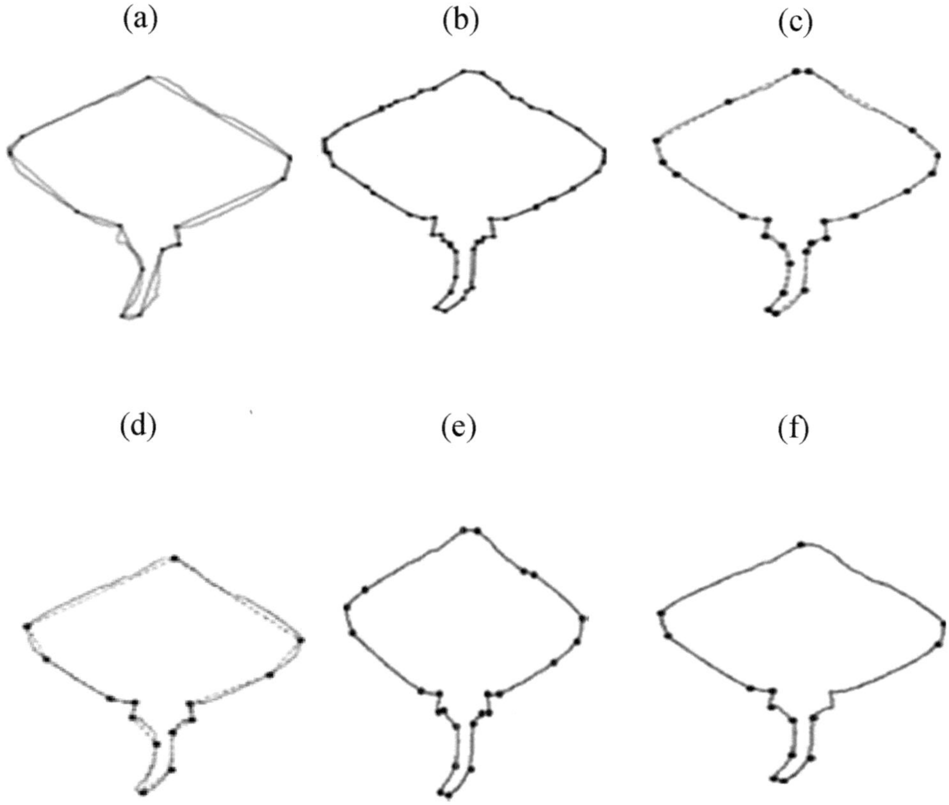

Figure 9.
Output approximated curve for ray-17 contour by various methods: (a) Carmona [11] at 14 DPs, (b) Prasad [23] RDP_opt at 54 DPs, (c) Fernandez [25] at 24 DPs, (d) Rosin [40] at 14 DPs, (e) Proposed method results at 24 DPs, (f) Proposed method at 14 DPs.

excellent than [25]. In order to support this claim, the output curve for octopus-14 can be found in **Figure 8** along with results of [11, 23, 25]. When the input is the ray-17 contour, at 14 number of DPs, the new proposal produces minimal error than the results of [9, 40], and then for the same curve at 35 DPs, the results are good than [25] in terms of ISE, WE and WE2. The graphic shots of the proposed method along with [11, 23, 25, 40] can be found in **Figure 9**. When the input for the proposed method is chicken-5 curve, the proposed method approximation error measures are compared with results produced by the techniques in [9, 11, 23, 25, 29, 30], and by using all the quantitative performance evaluators, the proposed work produces the output curve with minimal error possible, and the visual snapshots are shown in **Figure 10**. For the input curve device 6-9, the proposed method results are compared with the results in [9, 11, 23, 25, 29, 30], it is been conceived that the proposed one produces the minimal error (ISE ,WE) than the error produces by the methods in [9, 11, 23, 25]. The output curve for device 6–9 can be found in **Figure 11**. Then finally for the truck-07 curve, the results of the proposed method at 40, 12 and 11 dominant points are compared with the results of [9, 11, 23, 25]. In all iterations against the mentioned dominant points, the proposed method outperforms well than others. Especially output curve at 11 dominant points, the proposed method efficiently chooses the good curvature points in such a way that the output curve does not deviate much than the original input curve (please see the snapshot at **Figure 12(a)**, **(b)** with **(g)**).

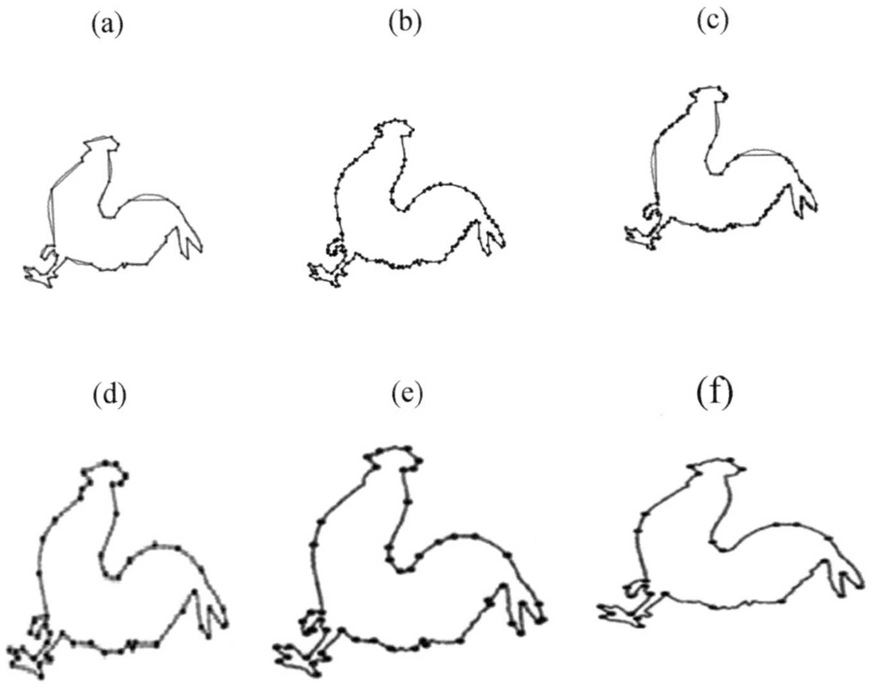

Figure 10.
Final approximation of chicken-5 contour by various methods: (a) Carmona [11] at 54 DPs, (b) Prasad [23] RDP_opt at 218 DPs, (c) Prasad [23] Carmona_opt 258 DPs, (d) Fernandez [25] at 54 DPs, (e) Proposed method at 54 DPs, (f) Proposed method at 29 DPs.

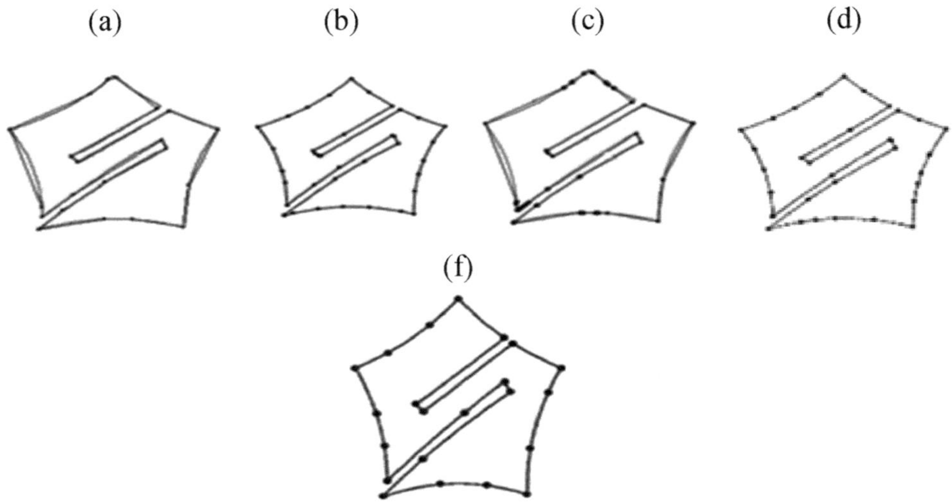

Figure 11.
Final approximation obtained from device 6–9 curve: (a) Carmona [11] at 22 DPs, (b) Prasad [23] RDP_opt at 38 DPs, (c) Prasad [23] Carmona_opt at 77 DPs, (d) Fernandez [25] at 33 DPs, (e) Proposed method at 22 DPs.

4.4 Rotation invariance

To test the efficiency of the proposed method against rotation invariance, bell-7 contour is rotated using varying amount angle. Then, the rotated contour is given as an input to the proposed method as well as to the technique in [9]. The results are

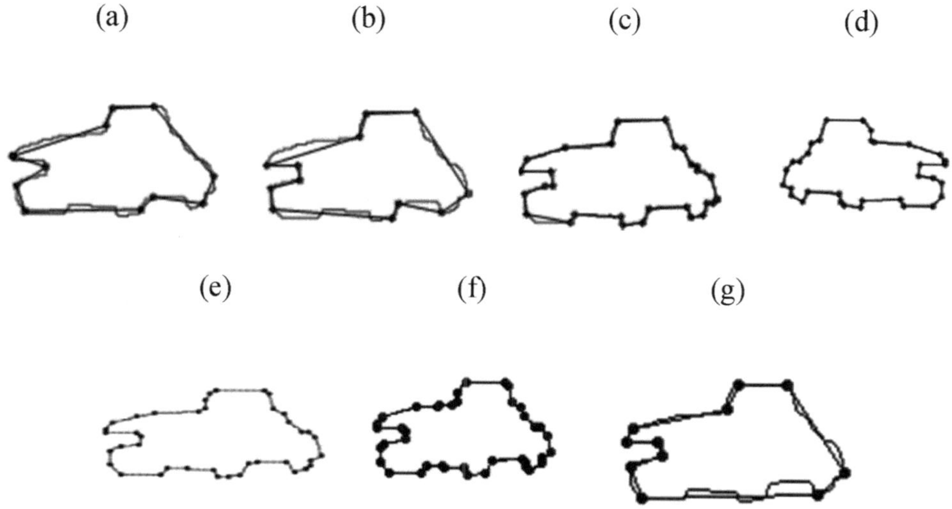

Figure 12.
Final approximation obtained from truck-07 curve: (a) Masood [9] at 11 DPs, (b) Carmona [11] at 12 DPs, (c) Prasad [23] Carmona_opt at 29 DPs, (d) Prasad [23] RDP_opt at 33 DPs, (e) Fernandez [25] at 40 DPs, (f) Proposed method at 44 DPs, (g) Proposed method at 11 DPs.

Contour	k	$\max(d_m)$	ISE	Area	Perimeter	Compactness
Bell-7	20	2.03	210.164	9231	299.13	0.10
Bell-7 at 20°		2.60	281.57	9.2475e+03	343.81	0.07
Bell-7 at 30°		2.70	348.868	9260	358.08	0.07
Bell-7 at 70°		2.91	325.29	9.254.5e+03	344.83	0.07
Bell-7 at 80°		2.59	319.50	9151	327.10	0.08
Bell-7 at 180°		2.03	210.164	9231	299.13	0.10

Table 4.
Robustness of the proposed method against rotation using quantitative measurement.

summarized for the reader's perusal. How do researcher determine a polygonal approximation is rotation invariant or not and what extent? The answer is the metrics such as area of polygon, perimeter and compactness may be suggested to use along with results from human perception. The authors in [41] use the above-mentioned metrics to prove whether the technique is able to produce the polygon with the same positioned points before as well as after the rotation. This can be measured using compactness metric. Moreover, the authors in [41] demonstrated that the techniques proposed in [9, 11, 12] are scaling as well as translation invariant using compactness metric.

The mathematical interpretation of compactness metric (*COMP*) has been mentioned in Eq. (2). **Table 4** summarizes the value obtained by using COMP for the bell-7 contour by the proposed method.

$$comp = Area/Perimeter^2 \tag{4}$$

To compare the robustness of the technique against rotation, the snapshots using bell-7 contour are displayed in **Figures 13** and **14**. The output polygon at 20

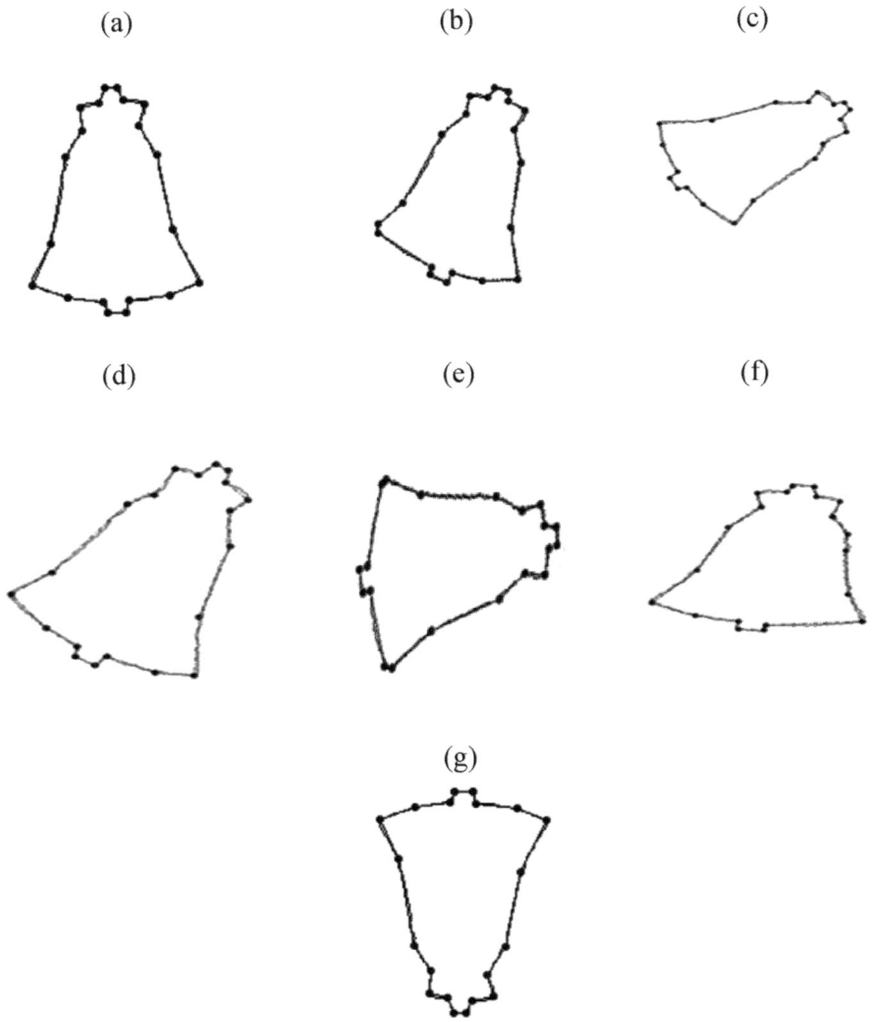

Figure 13.
The output polygon at 20 DPs by proposed methods in varying amount of angles: (a) Polygon at 20 DPs, (b) Polygon at 20°, (c) Polygon at 30°, (d) Polygon at 45°, (e) Polygon at 70°, (f) Polygon at 80°, (g) Polygon at 180°.

amounts of dominant points is used here to check if the technique is robust enough against rotation invariance. Most of the techniques considered in this chapter produce polygon in non-parametric mode. The best thing to compare the efficiency of rotation invariance is to compare the output at minimal possible amount of points since the input curve may contain more redundant points. So the result of the proposed method is compared with Masood [9]. By using [9], any researcher can produce a curve with specified number of dominant points. In **Table 4**, the value for geometric invariance assessment metrics (area of polygon, perimeter and compactness) reveals that the results by proposed method using rotated contours measure against compactness metric are more or less nearer to the value produced by the proposed method before rotation, and the visual snapshots in **Figure 13** also support the same. The results of Masood [9] in terms of quantitative measurements can be found in **Table 5**. Bell-7 at 30° value for compactness metric varies high while comparing the results obtained before rotation. In the remaining angles, the rotated contours compactness metric is more or less nearer to the value obtained by

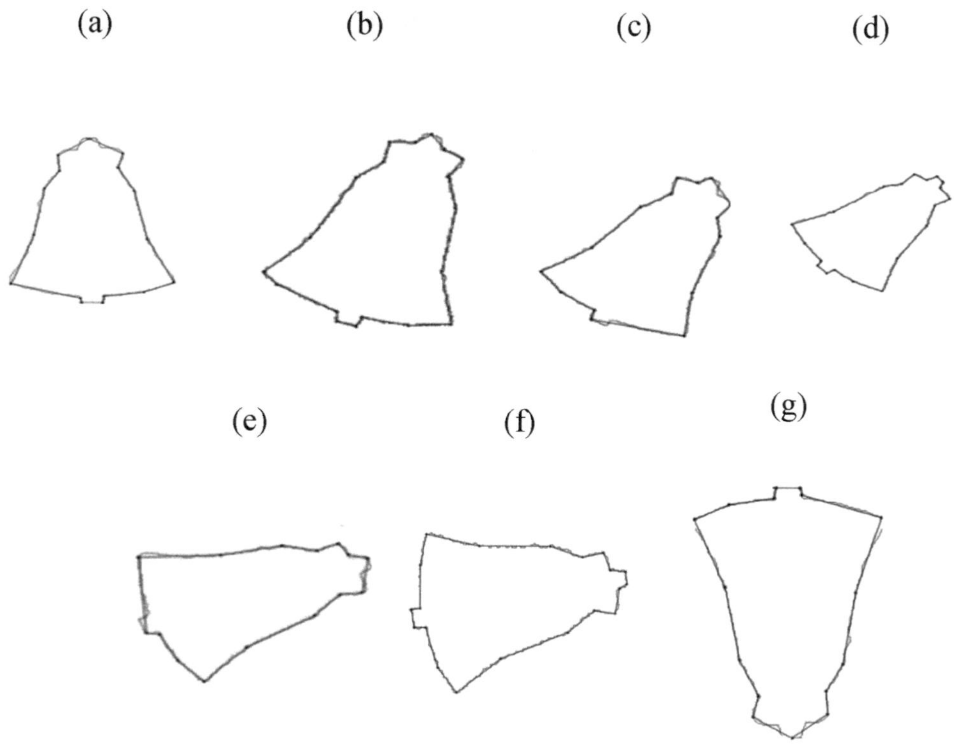

Figure 14.
The output polygon at 20 DPs by Masood [9] in varying amount of angles: (a) Polygon at 20 DPs, (b) Polygon at 20°, (c) Polygon at 30°, (d) Polygon at 45°, (e) Polygon at 70°, (f) Polygon at 80°, (g) Polygon at 180°.

Contour	k	max(d_m)	ISE	Area	Perimeter	Compactness
Bell-7	20	3.48	315.00	6835	321.78	0.06
Bell-7 at 20°		2.77	311.84	9130	333.32	0.08
Bell-7 at 30°		1.99	270.51	9.1255e+05	190.61	0.25
Bell-7 at 70°		3.79	381.35	9.1615e+03	343.59	0.07
Bell-7 at 80°		2	266.677	9.1585e+03	326.90	0.08
Bell-7 at 180°		3.487	315.00	6835	306.95	0.07

Table 5.
Robustness of Masood [9] against rotation using quantitative measurement.

the method before rotation. Masood [9] snapshots can be found in **Figure 14**. When the authors noticed that in the output curve produced by Masood [9], the position of the dominant point is heavily dislocated after rotation, whereas the proposed methods try to maintain the same positioned dominant points in the rotated contours too (see **Figure 13**).

5. Conclusion

The proposed significant measure computing metric predicts the position of a projection of every boundary point between its candidate line segment, thereby invoking suitable significant measure computing metric and accumulating its significant measure to define the significant value of every candidate of dominant points. The technique is demonstrated using wide variety of data sets, where the image contours are with different level details in terms of curvature as well as size. The proposed technique suits for parallel manipulators aspiring to produce the digital boundary with minimal number points without compromising its shape according to human perception as well as using benchmarking performance measuring metrics.

Acknowledgements

This research did not receive any specific grant from any funding agencies in the public, commercial or not-for-profit sectors. But the authors thank all anonymous reviewers for their comments on an earlier manuscript for improving the quality of the chapter.

Author details

Mangayarkarasi Ramaiah[1]* and Dilip Kumar Prasad[2]

1 School of Information Technology and Engineering, Vellore Institute of Technology, Vellore, India

2 Department of Computer Science, UiT, The Arctic University of Norway, Norway

*Address all correspondence to: rmangayarkarasi@vit.ac.in

References

[1] Lavallee S, Szeliski R. Recovering the position and orientation of free-form objects from image contours using 3D distance maps. IEEE Transactions on Pattern Analysis and Machine Intelligence. 1995;**17**:378-390. DOI: 10.1109/34.385980

[2] Elder JH, Goldberg RM. Image editing in the contour domain. IEEE Transactions on Pattern Analysis and Machine Intelligence. 2001;**23**:291-296. DOI: 10.1109/34.910881

[3] Kolesnikov A, Fränti P. Reduced-search dynamic programming for approximation of polygonal curves. Pattern Recognition Letters. 2003;**24**: 2243-2254. DOI: 10.1016/S0167-8655(03)00051-5

[4] Yang R, Zhang Eye Z. gaze correction with stereovision for video-teleconferencing. IEEE Transactions on Pattern Analysis and Machine Intelligence. 2004;**26**:956-960. DOI: 10.1109/TPAMI.2004.27

[5] Kolesnikov A, Fränti P. Data reduction of large vector graphics. Pattern Recognition. 2005;**38**: 381-394. DOI: 10.1016/j.patcog.2004.07.005

[6] Brunner D, Soille P. Iterative area filtering of multichannel images. Image and Vision Computing. 2007;**25**: 1352-1364. DOI: 10.1016/j.imavis.2006.09.002

[7] Mokhtarian F, Mackworth A. Scale-based description and recognition of planar curves and two-dimensional shapes. IEEE Transactions on Pattern Analysis and Machine Intelligence. 1986;**8**:34-43. DOI: 10.1109/TPAMI.1986.4767750

[8] Prasad DK, Leung MKH, Cho S-Y. Edge curvature and convexity based ellipse detection method. Pattern Recognition. 2012;**45**(9):3204-3221. DOI: 10.1016/j.patcog.2012.02.014

[9] Masood A. Dominant point detection by reverse polygonization of digital curves. Image and Vision Computing. 2008;**26**(5):702-715

[10] Masood A, Haq SA. A novel approach to polygonal approximation of digital curves. Journal of Visual Communication and Image Representation. 2007;**18**(3):264-274

[11] Carmona-Poyato A, Madrid-Cuevas FJ, Medina-Carnicer R, Muñoz-Salinas R. Polygonal approximation of digital planar curves through break point suppression. Pattern Recognition. 2010; **43**(1):14-25

[12] Ramaiah M, Ray BK. Polygonal approximation of digital planar curve using local integral deviation. International Journal of Computational Vision and Robotics. 2015;**5**(3):302-319

[13] Kolesnikov A, Fränti P. Polygonal approximation of closed discrete curves. Pattern Recognition. 2007;**40**(4): 1282-1293

[14] Bhowmick P, Bhattacharya BB. Fast polygonal approximation of digital curves using relaxed straightness properties. IEEE Transactions on Pattern Analysis and Machine Intelligence. 2007;**29**(9):1590-1602

[15] Marji M, Siy P. Polygonal representation of digital planar curves through dominant point detection—A nonparametric algorithm. Pattern Recognition. 2004;**37**(11):2113-2130

[16] Wang B, Brown D, Zhang X, Li H, Gao Y, Cao J. Polygonal approximation using integer particle swarm optimization. Information Sciences. 2014;**278**:311-326. DOI: 10.1016/j.ins.2014.03.055

[17] Aguilera-Aguilera EJ, Carmona-Poyato A, Madrid-Cuevas FJ, Marín-Jiménez MJ. Fast computation of optimal polygonal approximations of digital planar closed curves. Graphical Models. 2016;**84**:15-27. DOI: 10.1016/j.gmod.2016.01.004

[18] Liu Z, Watson J, Allen A. A polygonal approximation of shape boundaries of marine plankton based-on genetic algorithms. Journal of Visual Communication and Image Representation. 2016;**41**:305-313. DOI: 10.1016/j.jvcir.2016.10.010

[19] Parvez MT. Optimized polygonal approximations through vertex relocations in contour neighborhoods. Image and Vision Computing. 2015;**34**: 1-10. DOI: 10.1016/j.imavis.2014.10.012

[20] Pikaz A, Dinstein IH. An algorithm for polygonal approximation based on iterative point elimination. Pattern Recognition Letters. 1995;**16**(6): 557-563

[21] Zhu P, Chirlian PM. On critical point detection of digital shapes. IEEE Transactions on Pattern Analysis and Machine Intelligence. 1995;**17**(8): 737-748. DOI: 10.1109/34.400564

[22] Pavlidis T, Horowitz SL. Segmentation of plane curves. IEEE Transactions on Computers. 1974; **100**(8):860-870

[23] Prasad DK, Leung MK, Quek C, Cho SY. A novel framework for making dominant point detection methods non-parametric. Image and Vision Computing. 2012;**30**(11):843-859

[24] Prasad DK. PRO: A novel approach to precision and reliability optimization based dominant point detection. Journal of Optimization. 2013;1-15. DOI: 10.1155/2013/345287

[25] Fernández-García NL, Martínez LD, Carmona-Poyato A, Madrid-Cuevas FJ, Medina-Carnicer R. A new thresholding approach for automatic generation of polygonal approximations. Journal of Visual Communication and Image Representation. 2016;**35**:155-168. DOI: 10.1016/j.jvcir.2015.12.013

[26] Madrid-Cuevas FJ, Aguilera-Aguilera EJ, Carmona-Poyato A, Muñoz-Salinas R, Medina-Carnicer R, Fernández-García NL. An efficient unsupervised method for obtaining polygonal approximations of closed digital planar curves. Journal of Visual Communication and Image Representation. 2016;**39**:152-163. DOI: 10.1109/TPAMI.1986.4767753

[27] Dunham JG. Optimum uniform piecewise linear approximation of planar curves. IEEE Transactions on Pattern Analysis and Machine Intelligence. 1986;**Jan**(1):67-75. DOI: 10.1109/TPAMI.1986.4767753

[28] Ramaiah M, Ray BK. An iterative point elimination technique to retain significant vertices on digital planar curves. International Journal of Computational Vision and Robotics. 2016;**6**(4):354-368

[29] Ramer U. An iterative procedure for the polygonal approximation of plane curves. Computer Graphics and Image Processing. 01 November 1972;**1**(3): 244-256. DOI: 10.1016/S0146-664X(72)80017-0

[30] Douglas DH, Peucker TK. Algorithms for the reduction of the number of points required to represent a digitized line or its caricature. Cartographica: The International Journal for Geographic Information and Geovisualization. 01 October 1973; **10**(2):112-122. DOI: 10.3138/FM57-6770-U75U-7727

[31] Freeman H. On the encoding of arbitrary geometric configurations. IRE Transactions on Electronic Computers.

1961; **Jun**(2):260-8. DOI: 10.1109/TEC.1961.5219197

[32] Teh CH, Chin RT. On the detection of dominant points on digital curves. IEEE Transactions on pattern analysis and machine intelligence. August 1989; **11**(8):859-872. DOI: 10.1109/34.31447

[33] Wu WY. An adaptive method for detecting dominant points. Pattern Recognition. 01 october 2003;**36**(10): 2231-2237. DOI: 10.1016/S0031-3203(03)00087-6

[34] Parvez MT, Mahmoud SA. Polygonal approximation of digital planar curves through adaptive optimizations. Pattern Recognition Letters. 01 October 2010;**31**(13): 1997-2005. DOI: 10.1016/j.patrec.2010.06.007

[35] Nguyen TP, Debled-Rennesson I. A discrete geometry approach for dominant point detection. Pattern Recognition. 01 January 2011;**44**(1): 32-44. DOI: 10.1016/j.patcog.2010.06.022

[36] Backes AR, Bruno OM. Polygonal approximation of digital planar curves through vertex betweenness. Information Sciences. 10 February 2013; **222**:795-804. DOI: 10.1016/j.ins.2012.07.062

[37] Yin PY. Genetic algorithms for polygonal approximation of digital curves. International journal of pattern recognition and artificial intelligence. November 1999;**13**(07):1061-1082. DOI: 10.1142/S0218001499000598

[38] Salotti M. Optimal polygonal approximation of digitized curves using the sum of square deviations criterion. Pattern Recognition. 01 February 2002; **35**(2):435-443. DOI: 10.1016/S0031-3203(01)00051-6

[39] MPEG-7 Core Experiment CE-Shape-1Test set (Part B). Benchmarking Image Database for shape Recognition Techniques http://www.cis.temple.edu/~latecki/TestData/mpeg7shapeB.tar.gz

[40] Rosin PL. Unimodal thresholding. Pattern recognition. 2001 Nov 1;**34**(11): 2083-2096. DOI: 10.1016/S0031-3203(00)00136-9

[41] Ramaiah M, Ray BK. A brief review of closed curve approximation technique using iterative point elimination. In 2017 International conference on Microelectronic Devices, Circuits and Systems (ICMDCS) 2017 Aug 10 (pp. 1-6). IEEE. . DOI: 10.1109/ICMDCS.2017.8211729

Section 5

Electrical Grids

Chapter 10

Robust and Adaptive Control for Synchronous Generator's Operation Improvement

Jožef Ritonja

Abstract

Synchronous generators produce almost 95% of the world's electricity. Even a small improvement in their efficiency represents huge savings. Electromechanical oscillations of synchronous generators are harmful—they cause losses and can even lead to instability. An additional control system, called a power system stabilizer (PSS), is used to damp the oscillations of synchronous generators. The commercial realizations of the power system stabilizers are based on the use of the linear control theory. The effectiveness of these power system stabilizers is small, because of the nonlinear and time-varying characteristics of the synchronous generators. The application of robust and adaptive control represents an adequate theoretical basis for ensuring optimal damping of the electromechanical oscillations in a wide operating range. This work reviews the applicability of the advanced control theories to develop power system stabilizers. The work is focused on selecting the appropriate robust and adaptive control theories for the power system stabilizer implementation. The applicability and advantages are presented of the sliding mode control and the direct adaptive control, along with an evaluation of their impact on the operation improvement.

Keywords: sliding mode control, direct adaptive control, synchronous generator, power system stabilizer, operation of synchronous generator

1. Introduction

Synchronous generators are the most important electrical machines. They produce the majority of the world's electricity. In 2017, global electricity production was 25,721 TWh [1]. Assuming that the share of solar thermal sources is negligible compared to the share of solar photovoltaic sources, it can be estimated that about 98.2% of the total global energy is produced by electric generators. After analyzing the data, it can be estimated that synchronous generators produce 93.8% of the world's electricity and induction generators 4.4% of the total production of the world's electricity. The estimate is based on data for 2017. These, and also the following data, are obtained from statistics reports of the International Energy Agency [1].

An additional important point is that electricity trading and, thus, long-distance transmission of electricity are increasing significantly. In 2017, OECD countries produced 11,051 TWh of electricity, with a trading volume of 408 TWh, representing 3.7% of total production. Even more interesting is the growth rate of electricity trade. In the OECD, imports of electricity grew from 89 TWh in 1974 to 480 TWh in 2018, representing an average annual growth rate of 4.0%, compared to 2.1% growth in overall electricity supply.

The facts that the majority of the world's electricity is produced by synchronous generators and that a large amount of the world's electricity is transmitted over long distances result in significant oscillations of the produced and transmitted power. Despite the relatively small oscillations—the ratio of the amplitude of the oscillations of the transmitted power relative to the mean value of the transmitted power is mainly smaller than 10%—the total global losses due to the extremely large volume of production and transmission of electricity are not negligible. In terms of saving energy, it makes sense to reduce these losses.

The amount of the transmitted power oscillations can be affected by optimizing the topology of the new networks, by reconfiguring of the existing networks, by selection of the better damped new synchronous generators, and by replacement of the existing synchronous generators with the better damped ones. These solutions are expensive, and their realization also depends on other social and ecological factors. Therefore, it is a much more suitable solution to use a control system to damp the power system oscillations. In power systems, control systems called power system stabilizers (PSS) are used to suppress oscillations. PSS represent the best and the most economical solution for damping of the power systems' oscillations. PSS are simple to realize—they are mainly a part of the controller of the synchronous generator's static semiconductor excitation system. PSS, based on information of the oscillations of the transmitted power, rotor speed, rotor angle, or rotor acceleration, generate an additional reference signal for the rotor current control system. This additional reference signal represents the supplementary input to the static semiconductor excitation system, which is connected to rotor field winding.

Conventional PSS design is based on a linear control theory. Conventional PSS is simple to realize, but its application shows nonoptimal damping through the entire operating range; by varying the operating point, the synchronous generator's dynamic characteristics also vary; the fact is that PSS, which was determined for the nominal operating point, does not assure optimal damping in the entire operating range. Such a PSS reduces transmission losses optimally only at the operating point for which the PSS parameters are selected. Due to the large changes in the transmitted power and the large variations in power generation of the synchronous generators, conventional PSS are not satisfactory for use in modern power systems. To improve PSS performance, major modern control theories have been tested in the past decade for the purposes of PSS design. Of all the methods, robust and adaptive control has been implemented to be the most suitable for the design of PSS. Both control methods have been used in order to assure optimal damping through the entire operating range of the synchronous generators. The use of adaptive control is possible because the loading variations and, consequently, the variations of the dynamic characteristics of the synchronous generators are, in most cases, substantially slower than the dynamics of the adaptation mechanism [2].

Reduction of losses is not the sole and basic task of PSS. Even more important is that the PSS improves the stability of the power system and allows the transfer of power from the synchronous generator to the power system or between different points in the power system as near as possible to the stability limit of the transmission. In the presented work, however, we show the results of our study, which will show the applicability of the developed robust and adaptive PSS, mainly for the improvement of the damping of the power system oscillations.

For a detailed analysis of the benefits of the advanced PSS, a mathematical model of the synchronous generator is necessary. In this work, we first present the mathematical model of the synchronous generator connected to the power system, which is convenient for analysis of the physical characteristics of the power system and is appropriate at the same time for the controller design and synthesis. We focus our work on the analysis of a system where a single synchronous generator is connected to an infinite bus. In Section 3, we attempt to estimate the amount and dynamics of the oscillations in the power systems. A thorough analysis is made and presented for the first time. The conventional PSS control system is presented in Section 4. By means of the derived mathematical model of the synchronous generator, we estimate the improvement of the power system damping due to implementation of the conventional PSS. From the analysis, it is evident that conventional PSS does not assure optimal damping in the entire operating range. Therefore, advanced control theories for PSS design and synthesis are presented in Sections 5 and 6. In Section 5, the robust control system theory is used for PSS design. The suitable direct adaptive control theory is presented in Section 6. The PSS control system developed on the basis of the presented theories and the results of the implementation of the advanced control theories for PSS design and synthesis are shown in Section 7.

2. Mathematical model of the synchronous generator connected to the power system

The seventh-order nonlinear model of the synchronous machine connected to the infinite bus is the most detailed mathematical model of the synchronous generator connected to the large power system with constant frequency and constant voltage (=infinite bus) through the transmission line [3]. Park's matrix transformation is used to transform the origin windings' equations into a model with orthogonal axes. On this basis, the magnetic coupling of the stator, field, and damper windings is represented as a function of the position of the machine's rotor. The seventh-order model is represented in the form of a nonlinear state-space model [4]. The model's inputs are mechanical torque $T_m(t)$ and rotor excitation winding voltage $E_{fd}(t)$. The model's state-space variables are stator d-axis flux linkage $\lambda_d(t)$, stator q-axis flux linkage $\lambda_q(t)$, rotor excitation winding flux linkage $\lambda_F(t)$, rotor d-axis damper winding flux linkage $\lambda_D(t)$, rotor q-axis damper winding flux linkage $\lambda_D(t)$, mechanical rotor speed $\omega(t)$, and electric rotor angle $\delta(t)$. The seventh-order model is described by sets of algebraic equations (Eqs. (1)–(10)) [5]:

$$\lambda_{AD}(t) = L_{MD}\left(\frac{\lambda_d(t)}{l_d} + \frac{\lambda_F(t)}{l_F} + \frac{\lambda_D(t)}{l_D}\right) \tag{1}$$

$$\lambda_{AQ}(t) = L_{MQ}\left(\frac{\lambda_q(t)}{l_q} + \frac{\lambda_Q(t)}{l_Q}\right) \tag{2}$$

$$i_d(t) = \frac{1}{l_d}(\lambda_d(t) - \lambda_{AD}(t)) \tag{3}$$

$$i_q(t) = \frac{1}{l_q}(\lambda_q(t) - \lambda_{AQ}(t)) \tag{4}$$

$$i_F(t) = \frac{1}{l_F}(\lambda_F(t) - \lambda_{AD}(t)) \tag{5}$$

$$i_D(t) = \frac{1}{l_D}(\lambda_D(t) - \lambda_{AD}(t)) \tag{6}$$

$$i_Q(t) = \frac{1}{l_Q}(\lambda_Q(t) - \lambda_{AQ}(t)) \tag{7}$$

$$v_d(t) = -\sqrt{3}V_\infty \sin(\delta(t)) + R_e i_d(t) + \omega(t)L_e i_q(t) \tag{8}$$

$$v_q(t) = \sqrt{3}V_\infty \cos(\delta(t)) + R_e i_q(t) + \omega(t)L_e i_d(t) \tag{9}$$

$$T_e(t) = \frac{1}{3}\left(i_q(t)\lambda_d(t) - i_d(t)\lambda_q(t)\right) \tag{10}$$

and differential equations (Eqs. (11)–(17)):

$$\dot{\lambda}_d(t) = \omega_s\left(-R_s i_d(t) - \omega(t)\lambda_q(t) - v_d(t)\right) \tag{11}$$

$$\dot{\lambda}_q(t) = \omega_s\left(-R_s i_q(t) + \omega(t)\lambda_d(t) - v_q(t)\right) \tag{12}$$

$$\dot{\lambda}_F(t) = \omega_s(-R_F i_F(t) + E_{fd}(t)) \tag{13}$$

$$\dot{\lambda}_D(t) = \omega_s(-R_D i_D(t)) \tag{14}$$

$$\dot{\lambda}_Q(t) = \omega_s(-R_Q i_Q(t)) \tag{15}$$

$$\dot{\omega}(t) = \frac{1}{2H}(T_m(t) - T_e(t)) \tag{16}$$

$$\dot{\delta}(t) = \omega_s(\omega(t) - 1) \tag{17}$$

where $i_d(t)$ and $i_q(t)$ are stator d- and q-axis currents [pu]; $i_F(t)$ is field current [pu]; $i_D(t)$ and $i_Q(t)$ are damping d- and q-axis currents [pu]; $v_d(t)$ and $v_q(t)$ are stator terminal d- and q-axis voltages [pu]; $\lambda_{AD}(t)$ and $\lambda_{AQ}(t)$ are d- and q-axis mutual flux linkages [pu]; R_e and L_e are transmission line resistance and reactance [pu]; V_∞ is infinite bus voltage [pu]; $T_e(t)$ is electromagnetic torque [pu]; L_{MD}, L_{MQ}, L_{AD}, and L_{AQ} are mutual inductances [pu]; l_d, l_q, l_F, l_D, and l_Q are leakage inductances [pu]; R_S, R_F, R_D, and R_Q are stator, field, d-axis damping, and q-axis damping winding resistances [pu]; H is an inertia constant [s]; and ω_s is electric synchronous speed [rad s^{-1}]. All variables are normalized on the base quantities except the electric rotor angle $\delta(t)$ having unit [rad].

The seventh-order model is the superior one; although, on the other hand, it is too complicated to gain insight into the physical characteristics of the controlled plant [5]. It is also not suitable for the design and synthesis of control systems, since many control methods require linear mathematical models for the development of the control system. Many simplified models are derived from this seventh-order nonlinear model [6]. For a synchronous generator analysis and for the design of the PSS control system, a simplified linearized third-order model is still the most popular. It was presented for the first time in 1952 [7] and is, therefore, also called the Heffron-Phillips model.

The Heffron-Phillips model is obtained from the seventh-order nonlinear model by means of linearization for an every steady-state operating point (i.e., an equilibrium point). The Heffron-Phillips model describes the synchronous generator's dynamics in the proximity of the selected equilibrium point. The Heffron-Phillips model has two inputs and three state-space variables. The inputs are mechanical torque $T_{m\Delta}(t)$ and rotor excitation winding voltage $E_{fd\Delta}(t)$ deviations; the state-space variables are rotor angle $\delta_\Delta(t)$, rotor speed $\omega_\Delta(t)$, and voltage behind transient reactance $E'_{q\Delta}(t)$ deviations. Additional outputs are electric power $P_{e\Delta}(t)$ and terminal stator voltage $V_{t\Delta}(t)$ deviations. All the inputs and the state-space variables denote the deviations (subscript $_\Delta$) from the equilibrium state. The model is written as follows:

$$\begin{bmatrix} \dot{\delta}_\Delta(t) \\ \dot{\omega}_\Delta(t) \\ \dot{E}'_{q\Delta}(t) \end{bmatrix} = \begin{bmatrix} 0 & \omega_s & 0 \\ -\frac{K_1}{2H} & -\frac{D}{2H} & -\frac{K_2}{2H} \\ -\frac{K_4}{T'_{d0}} & 0 & -\frac{1}{K_3 T'_{d0}} \end{bmatrix} \begin{bmatrix} \delta_\Delta(t) \\ \omega_\Delta(t) \\ E'_{q\Delta}(t) \end{bmatrix} + \begin{bmatrix} 0 & 0 \\ \frac{1}{2H} & 0 \\ 0 & \frac{1}{T'_{d0}} \end{bmatrix} \begin{bmatrix} T_{m\Delta}(t) \\ E_{fd\Delta}(t) \end{bmatrix} \tag{18}$$

$$\begin{bmatrix} P_{e\Delta}(t) \\ V_{t\Delta}(t) \end{bmatrix} = \begin{bmatrix} K_1 & 0 & K_2 \\ K_5 & 0 & K_6 \end{bmatrix} \begin{bmatrix} \delta_\Delta(t) \\ \omega_\Delta(t) \\ E'_{q\Delta}(t) \end{bmatrix} + \begin{bmatrix} 0 & 0 \\ 0 & 0 \end{bmatrix} \begin{bmatrix} T_{m\Delta}(t) \\ E_{fd\Delta}(t) \end{bmatrix} \tag{19}$$

where $T_{m\Delta}(t)$ represents mechanical torque deviation [pu], $P_{e\Delta}(t)$ is electrical power deviation [pu], $\omega_\Delta(t)$ is rotor speed deviation [pu], $\delta_\Delta(t)$ is rotor angle deviation [rad], $E'_{q\Delta}(t)$ is the voltage behind the transient reactance [pu], $E_{fd\Delta}(t)$ is field excitation voltage deviation [pu], $V_{t\Delta}(t)$ is the terminal voltage [pu], H is an inertia constant [s], D is a damping coefficient representing total lumped damping effects from damper windings [pu/pu], ω_s is rated synchronous speed [rad s^{-1}], T_{do}' is a d-axis transient open circuit time constant [s], and K_1 through K_6 are linearization parameters. All parameters and variables in a Heffron-Phillips model are normalized, except for electric rotor angle $\delta_\Delta(t)$.

3. Analysis of the impact of the oscillations on the power system quality

The oscillations in the power system are due to the physical properties of the synchronous generator that operates parallel to the network. These properties are reflected in the dynamical mathematical model of the synchronous generator and appear as poorly damped dominant eigenvalues. Therefore, any changes in the synchronous generators' inputs (rotor field voltage and mechanical torque), in the network loads (changes in bus voltages) and disturbances, cause oscillations with relatively high amplitude and low damping. Oscillations in the power system are visible in several physical quantities of the system: in the synchronous generators' rotor speed, rotor angle, stator voltage, stator current, and produced power and in the power system's voltages, currents, frequency, and transmitted powers. These oscillations reduce the quality of the electricity and increase the stability risk of the power system.

It is very difficult to estimate the impact of oscillations on actual losses in a power system. In a real power system operation, it is problematic to evaluate how much of the losses is due to the rotor angle oscillations and how much of the losses are due to other factors. Therefore, in the first subsection, the influence of the amplitude and frequency of the oscillations on the amount of the losses in the transmission line and on the constancy of the transmitted power is discussed in more detail. The thoroughly steady-state analysis was made for this purpose. The dynamic analysis is presented in the second subsection. Dynamic analysis shows the vulnerability of the synchronous generator on the different input changes in different operation points.

3.1 Steady-state analysis

An analysis of the impact of the oscillations on losses and on the constancy of the transmitted power is made numerically. In the case of constant rotor speed, the induced voltage in the stator winding is sinusoidal, with constant amplitude and frequency. The swinging of the rotor speed results in induced voltage with variable frequency and amplitude. From the solution of the swing equation, it is evident that the swinging of the rotor is sinusoidal [8]. Therefore, in a steady-state analysis, we suppose that the rotor's speed oscillates about the constant synchronous speed with sinusoidal oscillations. This results in the stator's induced voltage, which varies sinusoidally in amplitude and frequency. The amplitude increases when the frequency increases, and vice versa. For such input signal, there is no transparent analytical solution for the evaluation of the losses and constancy of the transmitted power. Therefore, the numerical solution that is based on an equivalent circuit of the synchronous generator connected by transmission line to the infinite bus is used for the analysis. The equivalent circuit is shown in **Figure 1**.

where V_{1e} denotes the effective value of the generator's internal voltage [pu], V_{2e} is the effective value of the infinite bus voltage [pu], φ_1 is the generator's internal voltage angle [rad], φ_2 is the infinite bus voltage angle [rad], R_s is stator (armature) resistance [pu], X_s is synchronous reactance [pu], R_e is transmission line resistance [pu], and X_e is transmission line reactance [pu]. The equivalent circuit presented in **Figure 1** is a balanced symmetrical three-phase system. The impedance in any one phase is equal to that in either of the other two phases. Three voltages on the generator side are displaced 120° electrical degrees in time as a result of the phases being displaced 120° in space. Also, the three voltages on the infinite bus side are displaced 120° electrical degrees in time, so that the resulting phase currents are equal in amplitude and displaced in phase from each other by 120°. $v_{1a}(t)$, $v_{1b}(t)$, and $v_{1c}(t)$ denote instantaneous values of the synchronous generator's internal phase voltages; $v_{IBa}(t)$, $v_{IBb}(t)$, and $v_{IBc}(t)$ denote instantaneous values of the infinite bus phase voltages; and $i_a(t)$, $i_b(t)$, and $i_c(t)$ denote instantaneous values of the transmission line phase currents. Unless otherwise specified, $P_{IB}(t)$ represents the instantaneous three-phase power flow to the infinite bus [pu]; $P_{IBa}(t)$, $P_{IBb}(t)$, and $P_{IBc}(t)$ represent instantaneous power flow to the infinite bus for different phases [pu]; and P_{IB} represents the mean value of the instantaneous three-phase power flow to the infinite bus [pu]. $P_L(t)$ denotes instantaneous three-phase power losses

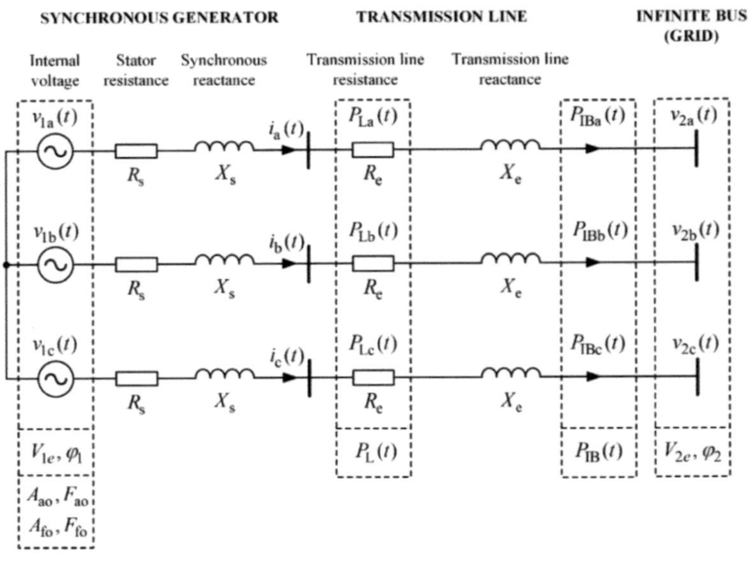

Figure 1.
Equivalent circuit of the synchronous generator connected by transmission line to the infinite bus used for the steady-state analysis.

in the transmission line [pu]; $P_{La}(t)$, $P_{Lb}(t)$, and $P_{Lc}(t)$ represent instantaneous power losses for different phases [pu]; and P_L represents the mean value of the instantaneous three-phase power losses in the transmission line [pu]. p_L indicates the relative value of the mean value of the three-phase power losses (P_L) compared to the mean value of the three-phase power flow to the infinite bus (P_{IB}) in [%]. Stator voltage amplitude and frequency oscillations resulting from rotor speed swing are described with the amplitude and frequency of both oscillations. A_{ao} and F_{ao} denote amplitude and frequency of amplitude oscillations. A_{fo} and F_{fo} denote the amplitude and frequency of frequency oscillations.

For the presented results, the impedances of the synchronous generator and transmission line are shown in **Table 1**.

Figures 2–5 show the time responses of the electrical quantities if there are no oscillations in rotor speed. The parameters of the generator's internal voltage and infinite bus voltage are shown in **Table 2**.

Figure 2 shows the instantaneous values of the synchronous generator's internal voltages for all three phases. Currents in the transmission lines are shown in **Figure 3**. Power flow to the infinite bus for all three phases separately and the sum of the phases' power flow are shown in **Figure 4**. As expected, the total three-phase power flow is constant. **Figure 5** shows power losses in the transmission line for all three phases separately and the three-phase power losses. Again, the total three-phase power losses are constant.

Figures 6–10 show the time responses of the electrical quantities if oscillations occur in the rotor speed. The parameters of the generator's internal voltage and infinite bus voltage are shown in **Table 3**.

R_a = 0.0011 [pu]	X_s = 0.9 [pu]
R_e = 0.02 [pu]	X_e = 0.4 [pu]

Table 1.
The synchronous generator's and transmission line's impedances.

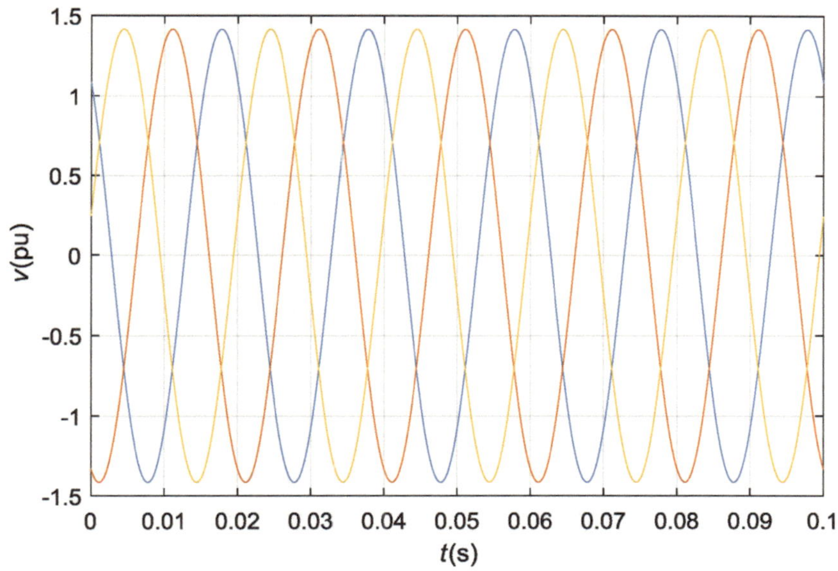

Figure 2.
Synchronous generator's internal voltages for all three phases: $v_{1a}(t)$, $v_{1b}(t)$, *and* $v_{1c}(t)$.

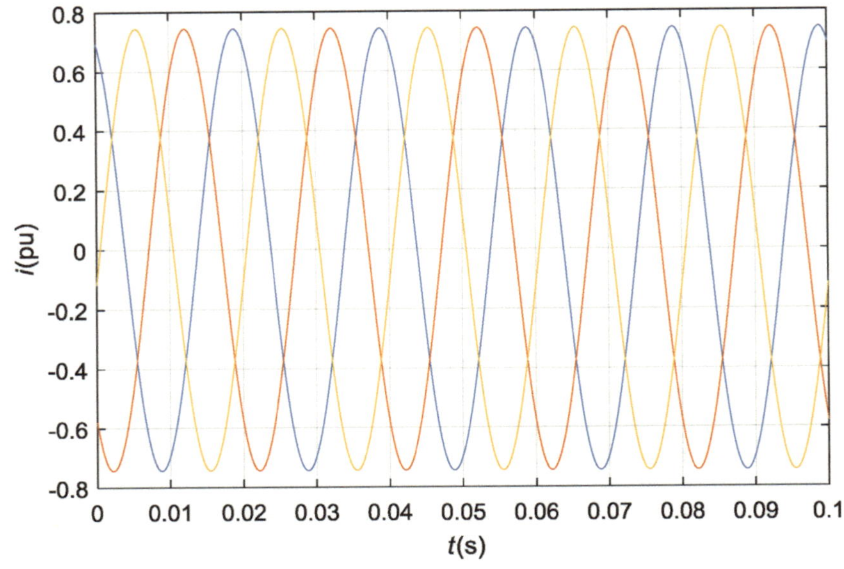

Figure 3.
Transmission line currents for all three phases: $i_a(t)$, $i_b(t)$, *and* $i_c(t)$.

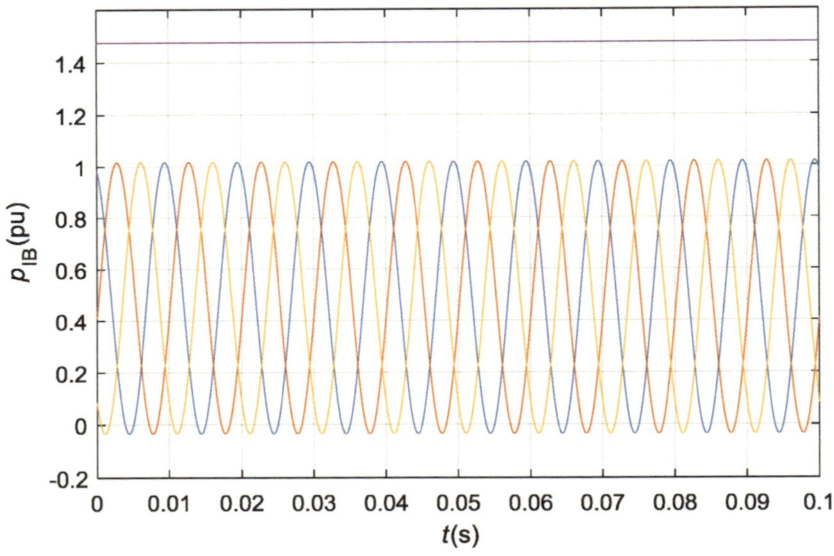

Figure 4.
Instantaneous three-phase power flow to the infinite bus $P_{IB}(t)$ *(constant), and instantaneous power flow to the infinite bus for different phases* $P_{IBa}(t)$, $P_{IBb}(t)$, *and* $P_{IBc}(t)$.

For better insight, **Figure 6** shows the instantaneous values of the synchronous generator's internal voltage for phase a. The oscillations are visible in amplitude and in frequency.

Figure 7 shows the instantaneous values of the synchronous generator's internal voltages for all three phases. Currents in the transmission lines are shown in **Figure 8**. Power flow to the infinite bus for all three phases separately and the instantaneous value of the three-phase power flow are shown in **Figure 9**. In this case, due to oscillations in rotor speed, and, consequently, oscillations in the internal voltages, the total three-phase power flow is not constant. **Figure 10** shows power losses in the transmission line for all three phases separately and the three-phase power losses. Again, total three-phase power losses are not constant.

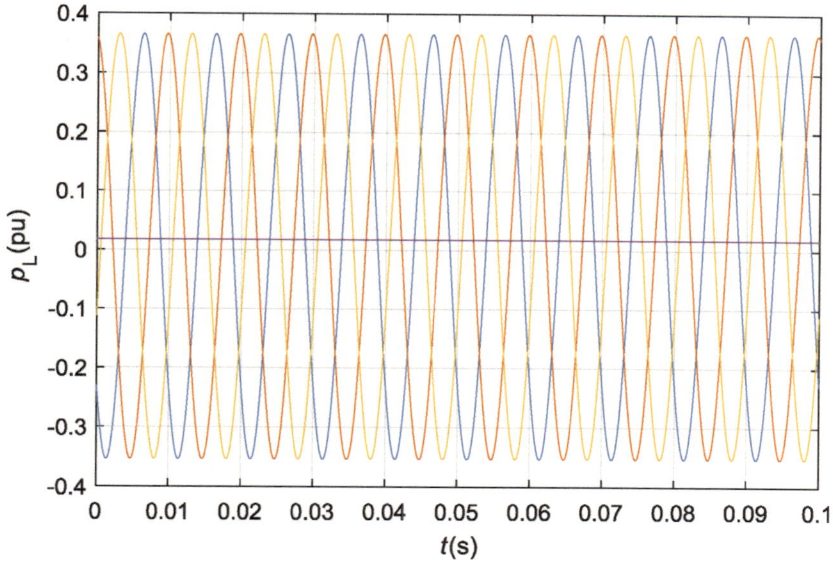

Figure 5.
Instantaneous three-phase power losses in the transmission line $P_L(t)$ (constant) and instantaneous power losses in the transmission line for different phases $P_{IBa}(t)$, $P_{IBb}(t)$, and $P_{IBc}(t)$.

V_{1e} = 1.0 [pu]	φ_1 = 40 [°]
V_{2e} = 1.0 [pu]	φ_1 = 0 [°]
A_{ao} = 0.0 [pu]	F_{ao} = 0 [Hz]
A_{fo} = 0.0 [pu]	F_{fo} = 0 [Hz]

Table 2.
The parameters of the internal voltage and infinite bus voltage, without oscillations.

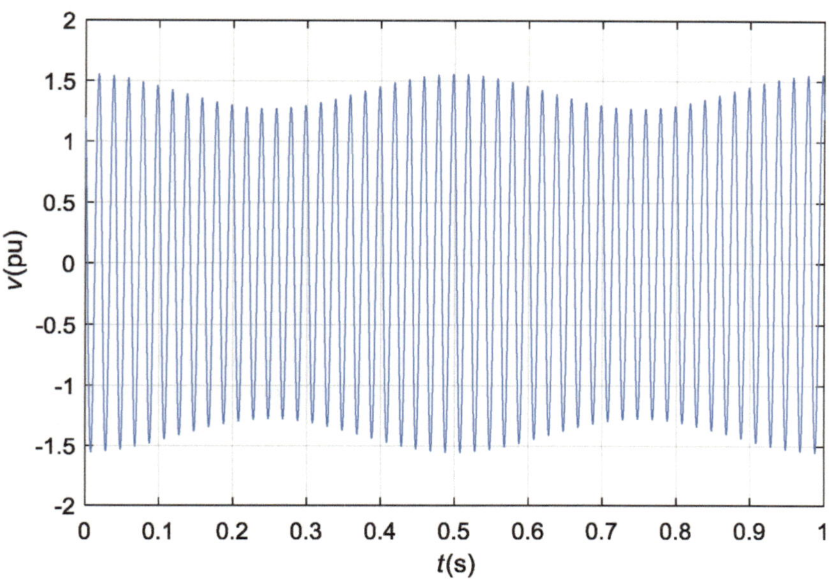

Figure 6.
Synchronous generator's internal voltages for phase a: $v_{1a}(t)$.

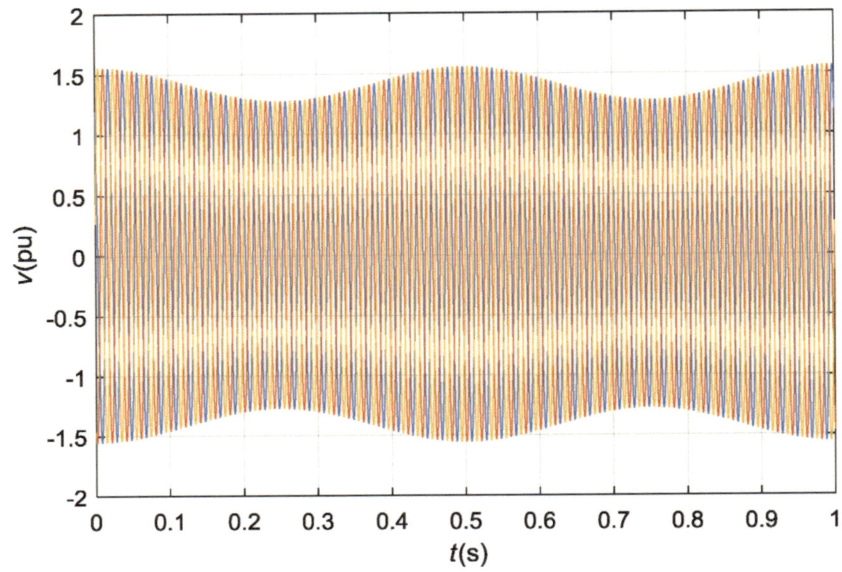

Figure 7.
Synchronous generator's internal voltages for all three phases: $v_{1a}(t)$, $v_{1b}(t)$, *and* $v_{1c}(t)$.

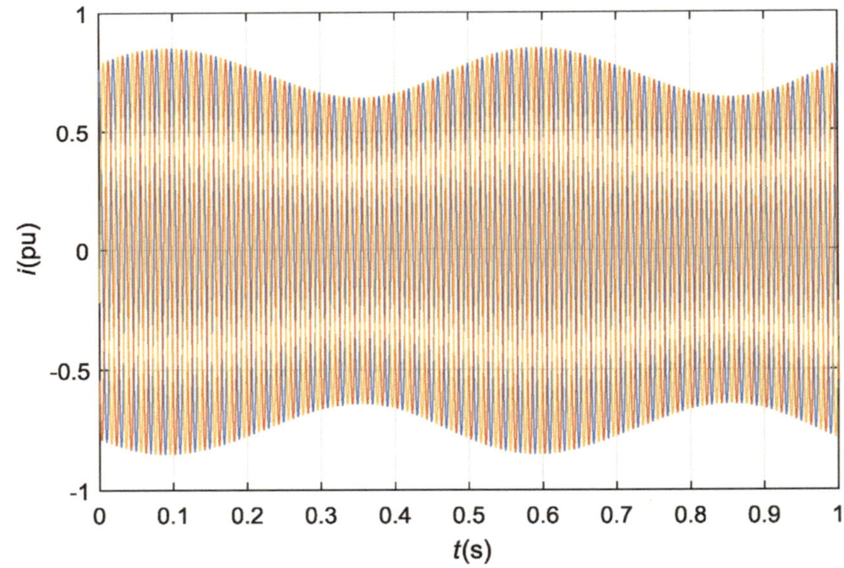

Figure 8.
Transmission line currents for all three phases: $i_a(t)$, $i_b(t)$, *and* $i_c(t)$.

From **Figures 2–10**, it is seen clearly that the rotor speed oscillations cause the oscillation in the transmitted power. The sum of the transmitted powers of the individual phases is no longer constant (**Figure 9**), as is the case for the balanced three-phase symmetric systems without oscillations (**Figure 4**). These oscillations reduce the transmission capability and quality. To ensure the power transmission with minimal power oscillations, it is necessary to reduce the rotor speed oscillations.

A thorough numerical analysis was performed to estimate the influence of the rotor speed oscillations on the power system losses. Some results are presented in **Tables 4** and **5**. **Table 4** shows the impact of the rotor speed oscillations on the

DOI: http://dx.doi.org/10.5772/intechopen.92558

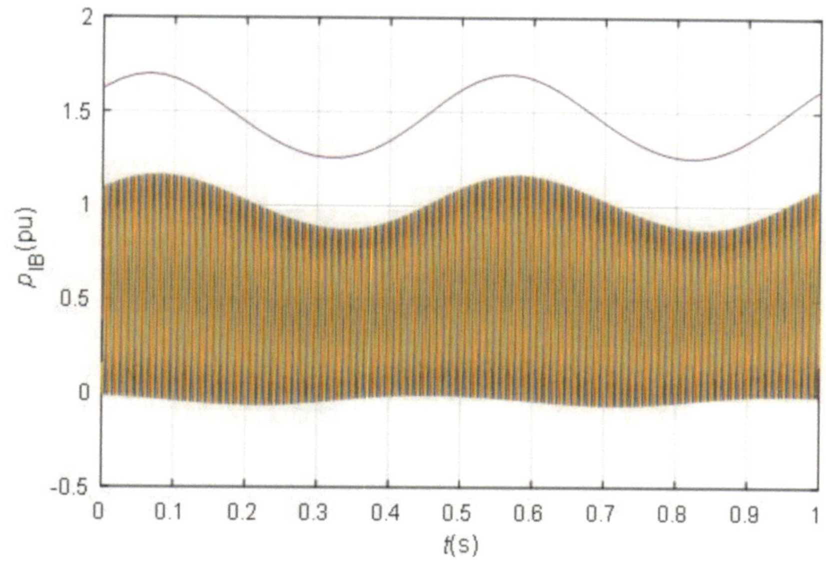

Figure 9.
Three-phase power flow to the infinite bus $P_{IB}(t)$ *(violet) and instantaneous power flow to the infinite bus for different phases* $P_{IBa}(t)$, $P_{IBb}(t)$, *and* $P_{IBc}(t)$.

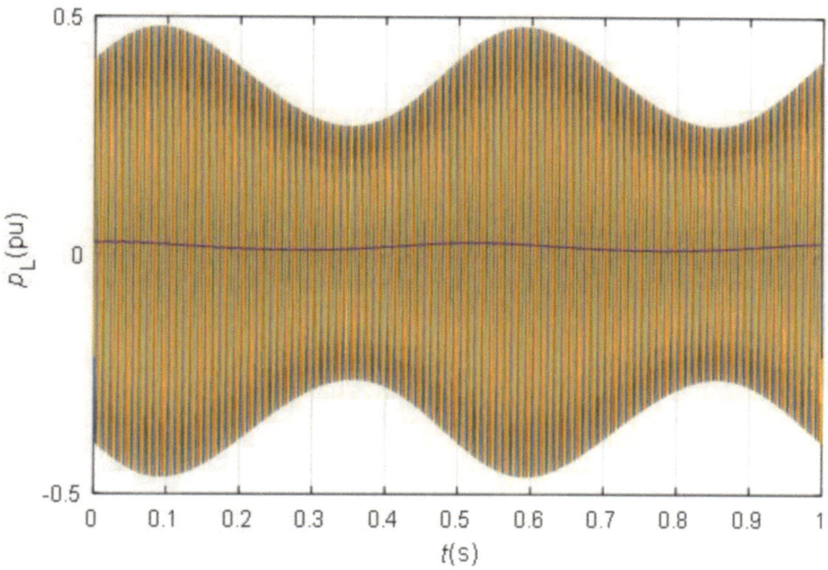

Figure 10.
Three-phase power losses in the transmission line $P_L(t)$ *(violet) and instantaneous power losses in the transmission line for different phases* $P_{IBa}(t)$, $P_{IBb}(t)$, *and* $P_{IBc}(t)$.

V_{1e} = 1.0 [pu]	φ_1 = 40 [°]
V_{2e} = 1.0 [pu]	φ_1 = 0 [°]
A_{ao} = 0.1 [pu]	F_{ao} = 2 [Hz]
A_{fo} = 0.1 [pu]	F_{fo} = 2 [Hz]

Table 3.
The parameters of the internal voltage and infinite bus voltage, with oscillations in internal voltage.

losses by active power generation. The results are presented in two different operating points. In the first one, the electric angle between generator internal voltage and infinite bus voltage was 10°, and, in the second one, the angle amounts to 40°. The results of the rotor speed oscillations' variations are presented in the amplitude range from 0 to 20% of the synchronous speed by frequencies 1 and 2 [Hz]. From the obtained results, it is evident that there is no influence of the oscillations on power losses in the transmission line.

Table 5 shows the impact of the rotor speed oscillations on the losses by reactive power generation. The results are presented at two different operating points. In the first one, the generator's internal voltage was 10% bigger than the infinite bus voltage, and, in the second one, the difference amounts to 50%. The results of rotor speed oscillations' variations are presented in the amplitude range from 0 to 20% of the synchronous speed by frequencies 1 and 2 [Hz]. From the results, it is obvious

V_{1e}	V_{2e}	φ_1	φ_2	A_{ao}	A_{fo}	F_{ao}	F_{fo}	P_{IB}	P_L	p_L
1.0	1.0	10	0	0.0	0.0	0	0	0.40	$1.1 \cdot 10^{-3}$	0.3
1.0	1.0	10	0	0.1	0.1	1	1	0.40	$1.5 \cdot 10^{-3}$	0.4
1.0	1.0	10	0	0.1	0.1	2	2	0.40	$1.5 \cdot 10^{-3}$	0.4
1.0	1.0	10	0	0.2	0.2	1	1	0.40	$2.6 \cdot 10^{-3}$	0.6
1.0	1.0	10	0	0.2	0.2	2	2	0.40	$2.5 \cdot 10^{-3}$	0.6
1.0	1.0	40	0	0.0	0.0	1	1	1.47	$1.8 \cdot 10^{-2}$	1.2
1.0	1.0	40	0	0.1	0.1	1	1	1.47	$1.8 \cdot 10^{-2}$	1.2
1.0	1.0	40	0	0.1	0.1	2	2	1.46	$1.9 \cdot 10^{-2}$	1.3
1.0	1.0	40	0	0.2	0.2	1	1	1.46	$1.9 \cdot 10^{-2}$	1.3
1.0	1.0	40	0	0.2	0.2	2	2	1.46	$1.9 \cdot 10^{-2}$	1.3

Table 4.
Mean values of three-phase transmitted power to the infinite bus (P_{IB}) and three-phase power losses in the transmission line P_L as a function of the angle between the generator's internal voltage and infinite bus voltage (φ_1-φ_2) by different amplitudes (A_{ao}, A_{fo}) and frequencies (F_{ao}, F_{fo}) of amplitude and frequency oscillations in the generator's internal voltage by active power transmission.

V_{1e}	V_{2e}	φ_1	φ_2	A_{ao}	A_{fo}	F_{ao}	F_{fo}	P_{IB}	P_L	p_L
1.1	1.0	0	0	0.0	0.0	0	0	$3.7 \cdot 10^{-3}$	$3.7 \cdot 10^{-4}$	10
1.1	1.0	0	0	0.1	0.1	1	1	$3.6 \cdot 10^{-3}$	$7.9 \cdot 10^{-4}$	22
1.1	1.0	0	0	0.1	0.1	2	2	$3.6 \cdot 10^{-3}$	$7.7 \cdot 10^{-4}$	21
1.1	1.0	0	0	0.2	0.2	1	1	$3.3 \cdot 10^{-3}$	$2.0 \cdot 10^{-3}$	61
1.1	1.0	0	0	0.2	0.2	2	2	$3.3 \cdot 10^{-3}$	$2.0 \cdot 10^{-3}$	61
1.5	1.0	0	0	0.0	0.0	0	0	$1.9 \cdot 10^{-2}$	$9.4 \cdot 10^{-3}$	50
1.5	1.0	0	0	0.1	0.1	1	1	$1.9 \cdot 10^{-2}$	$1.0 \cdot 10^{-2}$	54
1.5	1.0	0	0	0.1	0.1	2	2	$1.9 \cdot 10^{-2}$	$1.0 \cdot 10^{-2}$	54
1.5	1.0	0	0	0.2	0.2	1	1	$1.8 \cdot 10^{-2}$	$1.2 \cdot 10^{-2}$	66
1.5	1.0	0	0	0.2	0.2	2	2	$1.8 \cdot 10^{-2}$	$1.2 \cdot 10^{-2}$	66

Table 5.
Mean values of three-phase transmitted power to the infinite bus (P_{IB}) and three-phase power losses in the transmission line P_L as a function of the generator's internal voltage (V_{1e}) by different amplitudes (A_{ao}, A_{fo}) and frequencies (F_{ao}, F_{fo}) of amplitude and frequency oscillations in the generator's internal voltage by reactive power transmission.

that the influence of the oscillations on power losses in the transmission line is negligible.

3.2 Dynamic analysis

The Heffron-Phillips model described in Section 2 is used for the analysis of the dynamic characteristics of the synchronous generator. The impact of the inputs' variations on the oscillations is studied on the synchronous generator with the following data [4]:

New parameters in **Table 6** are as follows: S_N is nominal power [MVA], V_N is nominal voltage [kV], cos φ_N is a nominal power factor, and x'_d is unsaturated d-axis transient reactance [pu].

From the data in **Table 6**, the equilibrium state for the Heffron-Phillips model is calculated by means of a phasor diagram. Phasor equations permit the solution of the initial conditions that exist prior to the application of the inputs' variations. The linearization coefficients of the Heffron-Phillips model are calculated for the synchronous generator with data in **Table 6** and for the calculated equilibrium state's data. The linearization coefficients for nominal operating point (P_N = 1 [pu], cos φ_N = 0.85), and eigenvalues of the Heffron-Phillips model (λ_1, λ_2, and λ_3) are presented in **Table 7**.

The linearized Heffron-Phillips model of a synchronous generator has three eigenvalues. The damping ratio and frequency of rotor angle oscillation are seen directly from the dominant conjugate complex eigenvalues. Therefore, it is very transparent to investigate the dependence of the synchronous generator's oscillation dynamics from the operating point by means of eigenvalue analysis. By changing the operating point, the Heffron-Phillips model's eigenvalues also change.

Transient response of the synchronous generator with data in **Table 6** and nominal operating point with data in **Table 7** are shown in **Figures 11–13**. Step changes are simulated in both generator's inputs. **Figure 11** shows the simulated

S_N = 160 [MVA]	V_N = 15 [kV]	cos φ_N = 0.85
ω_s = 377 [rad s^{-1}]		
T'_{d0} = 5.9 [pu]	H = 2.37 [s]	D = 2.0 [pu]
R_e = 0.02 [pu]	L_e = 0.4 [pu]	V_{IB} = 1.0 [pu]
R_s = 0.0011 [pu]	R_F = 0.0007 [pu]	x'_d = 0.245 [pu]
L_d = 1.700 [pu]	L_q = 1.640 [pu]	L_F = 0.101 [pu]
L_D = 0.055 [pu]	L_Q = 0.036 [pu]	L_{AD} = 1.550 [pu]
l_d = 0.150 [pu]	l_q = 0.150 [pu]	L_{AQ} = 1.490 [pu]

Table 6.
Data of the synchronous generator used for dynamic analysis [4].

P_N = 1.0 [pu]	Q_N = 0.62 [pu]	cos φ_N = 0.85
K_1 = 1.4478	K_2 = 1.3174	K_3 = 0.3072
K_4 = 1.8052	K_5 = 0.0294	K_6 = 0.5257
$\lambda_1 = -0.3502 + 10.7270i$	$\lambda_2 = -0.3502 - 10.7270i$	$\lambda_3 = -0.2732$

Table 7.
Linearization parameters and eigenvalues of the Heffron-Phillips model in the nominal operating point.

trajectory of the turbine mechanical torque and rotor excitation voltage. Step changes are selected in order to maximize the excitation of oscillations.

The responses of the generated electrical power and the stator voltage are shown in **Figure 12**. The oscillations are seen clearly from the response of the generated electrical power.

Figure 13 shows the response of the rotor speed and rotor angle on the inputs' trajectories shown in **Figure 11**.

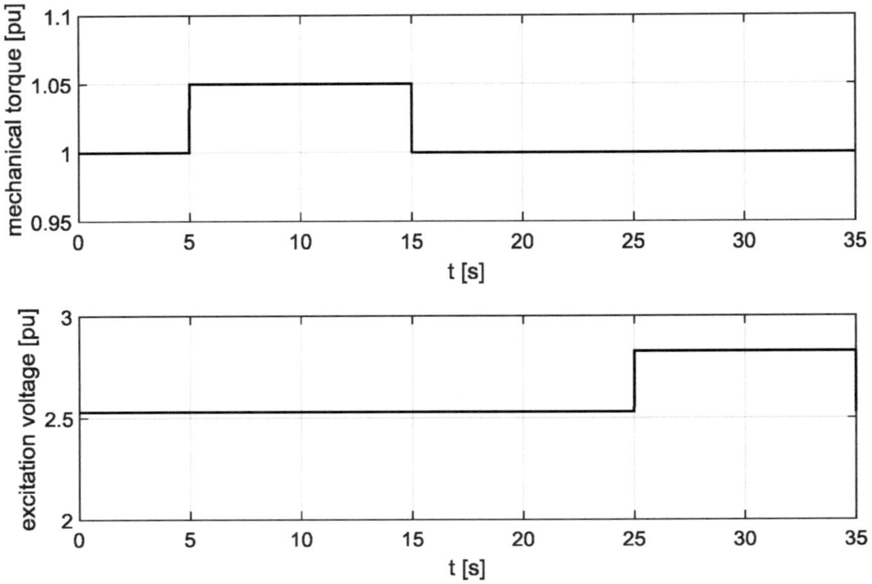

Figure 11.
Synchronous generator inputs' trajectories: Mechanical torque $T_m(t)$ *[pu] and rotor excitation voltage* $E_{fd}(t)$ *[pu], nominal operating point* $P = 1.0$ *[pu] and* $Q = 0.62$ *[pu].*

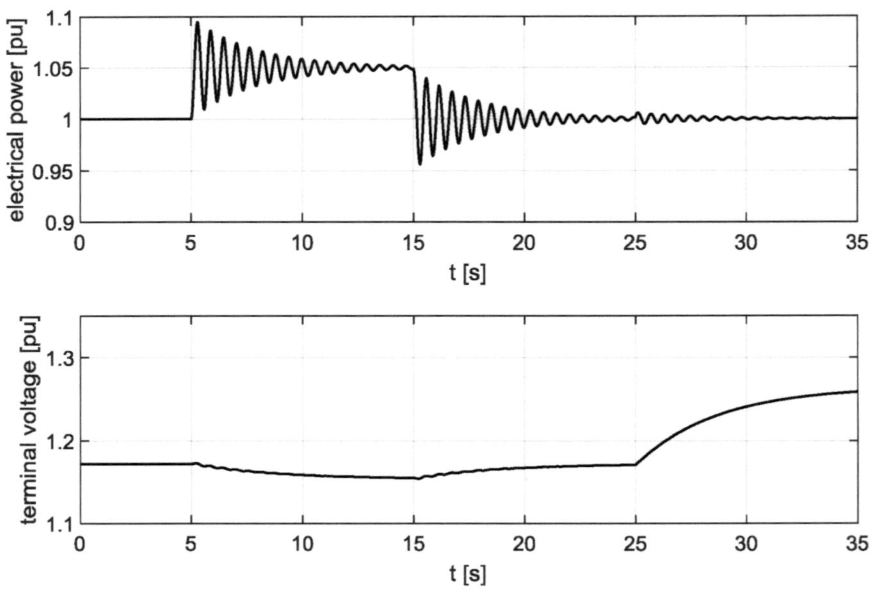

Figure 12.
Synchronous generator outputs' trajectories: Generated electrical power $P_e(t)$ *[pu] and stator terminal voltage* $V_t(t)$ *[pu], nominal operating point* $P = 1.0$ *[pu] and* $Q = 0.62$ *[pu].*

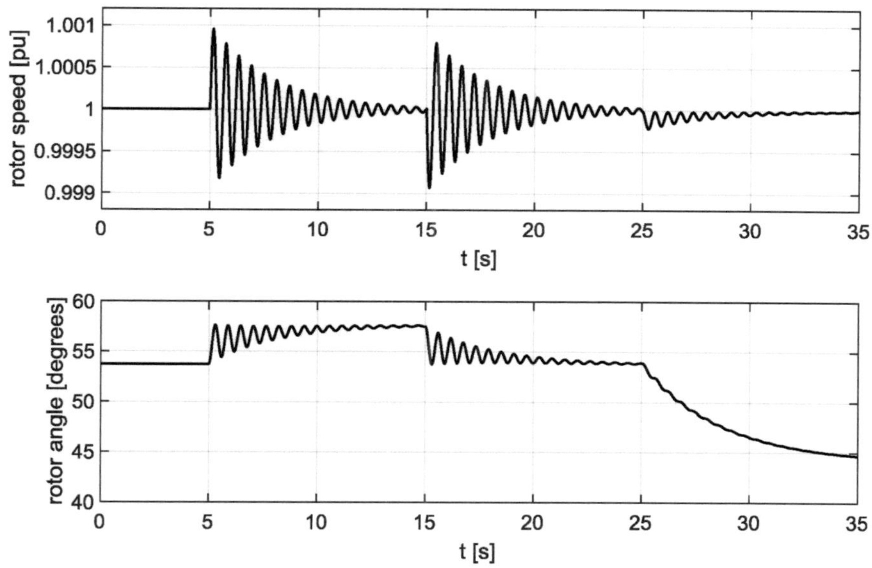

Figure 13.
Synchronous generator outputs' trajectories: Rotor speed ω(t) [pu] and rotor angle δ(t) [degrees], nominal operating point P = 1.0 *[pu] and* Q = 0.62 *[pu].*

$P = 1.0$ [pu]	$Q = 0.1$ [pu]	$\cos \varphi = 0.995$
$K_1 = 1.2506$	$K_2 = 1.5867$	$K_3 = 0.3072$
$K_4 = 2.2164$	$K_5 = -0.0730$	$K_6 = 0.3693$
$\lambda_1 = -0.4493 + 9.9715i$	$\lambda_2 = -0.4493 - 9.9715i$	$\lambda_3 = -0.0750$

Table 8.
Linearization parameters and eigenvalues of the Heffron-Phillips model at operating point P = 1.0 *[pu] and* Q = 0.1 *[pu].*

To evaluate the influence of the operating point on the synchronous generator's dynamics, the Heffron-Phillips model was calculated, and simulations were performed at different operating points for different types of generators with different nominal values. From the obtained results, it was obvious that the dynamics of the synchronous generator vary significantly.

In this text, results for the synchronous generator with data in **Table 6** are presented for the two most distinctive operating points:

- The operating point with high active power and low reactive power: $P = 1.0$ [pu] and $Q = 0.1$ [pu]. This point represents the strongly damped operating point and, therefore, the less problematic case for testing of the PSS.

- The operating point with low active power and high reactive power $P = 0.1$ [pu] and $Q = 1.0$ [pu]. This point represents the weakly damped real operating point and, therefore, the worst case for testing of the PSS.

Table 8 shows the linearization data and Heffron-Phillips model eigenvalues for the operating point where active power is generated with very high power factor $\cos \varphi$.

The results of the simulation of the synchronous generator at operating point $P = 1$ [pu] and $Q = 0.1$ [pu] are shown in **Figures 14–16**.

Table 9 shows the linearization data and Heffron-Phillips model eigenvalues for other interesting operating points. In this case, almost only reactive power is generated—Power factor cos φ is very small.

The results of the simulation of the synchronous generator in operating point $P = 0.1$ [pu] and $Q = 1.0$ [pu] are shown in **Figures 17–19**.

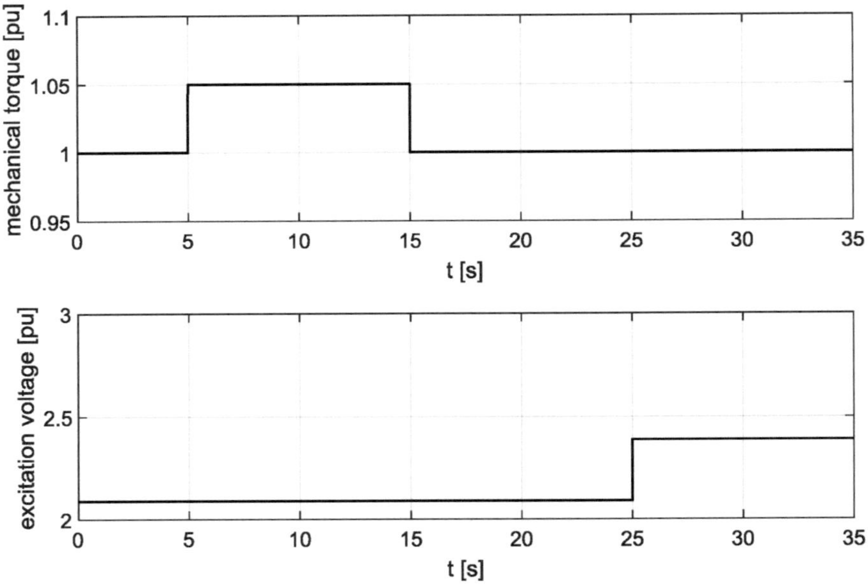

Figure 14.
Synchronous generator inputs' trajectories: Mechanical torque $T_m(t)$ *[pu] and rotor excitation voltage* $E_{fd}(t)$ *[pu], operating point* $P = 1.0$ *[pu] and* $Q = 0.1$ *[pu].*

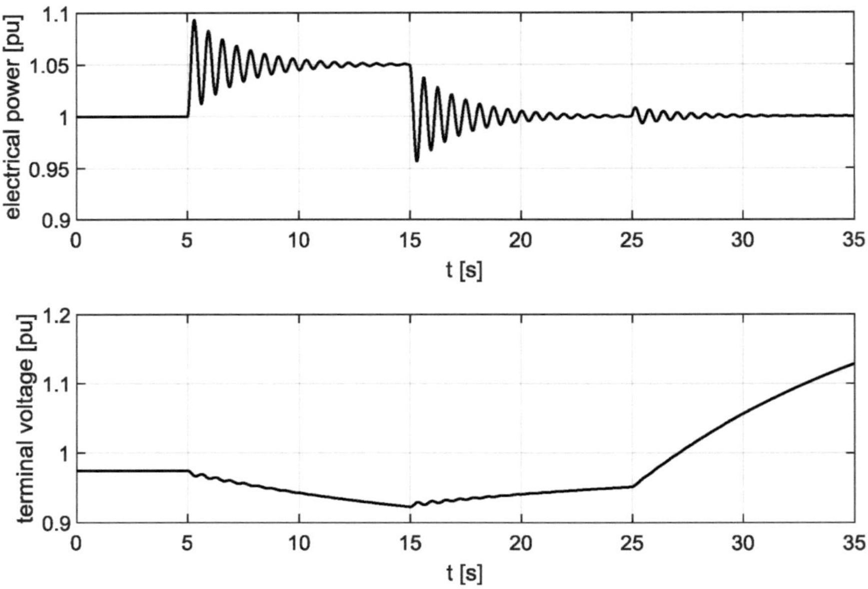

Figure 15.
Synchronous generator outputs' trajectories: Generated electrical power $P_e(t)$ *[pu] and stator terminal voltage* $V_t(t)$ *[pu], operating point* $P = 1.0$ *[pu] and* $Q = 0.1$ *[pu].*

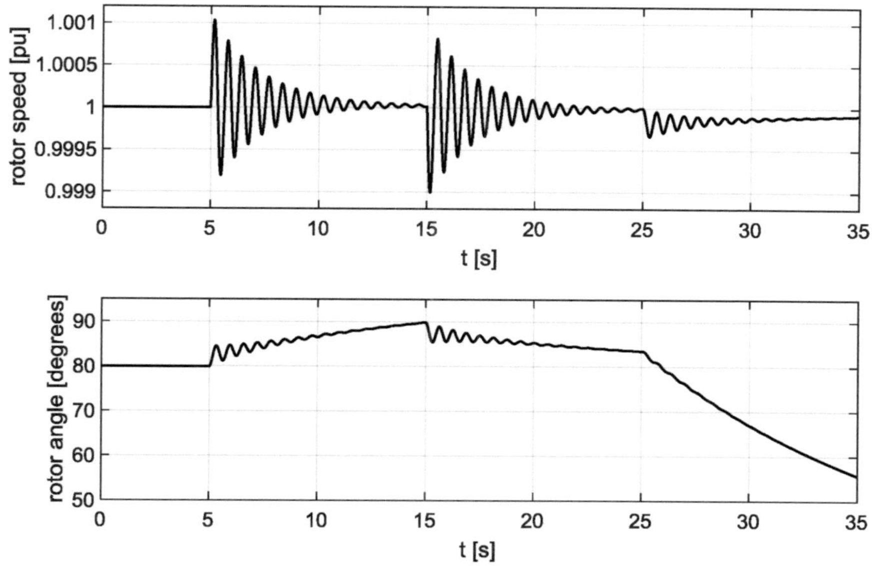

Figure 16.
Synchronous generator outputs' trajectories: Rotor speed ω(t) [pu] and rotor angle δ(t) [degrees], operating point P = 1.0 *[pu] and* Q = 0.1 *[pu].*

$P = 0.1$ [pu]	$Q = 1.0$ [pu]	$\cos\varphi = 0.099$
$K_1 = 1.2614$	$K_2 = 0.1631$	$K_3 = 0.3072$
$K_4 = 0.1219$	$K_5 = 0.0185$	$K_6 = 0.6207$
$\lambda_1 = -0.2123 + 10.0141i$	$\lambda_2 = -0.2123 - 10.0141i$	$\lambda_3 = -0.5490$

Table 9.
Linearization parameters and eigenvalues of the Heffron-Phillips model at operating point P = 0.1 *[pu] and* Q = 1.0 *[pu].*

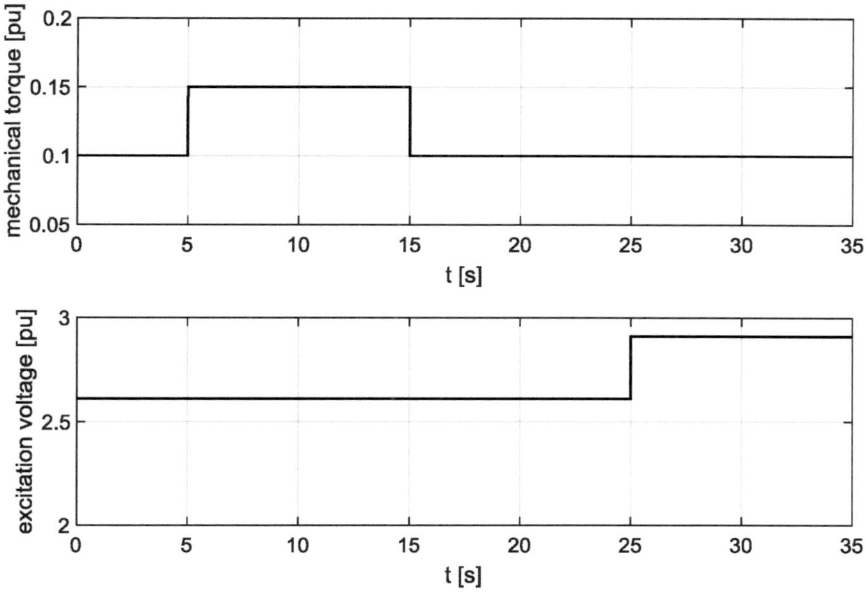

Figure 17.
Synchronous generator inputs' trajectories: Mechanical torque $T_m(t)$ *[pu] and rotor excitation voltage* $E_{fd}(t)$ *[pu], operating point* P = 0.1 *[pu] and* Q = 1.0 *[pu].*

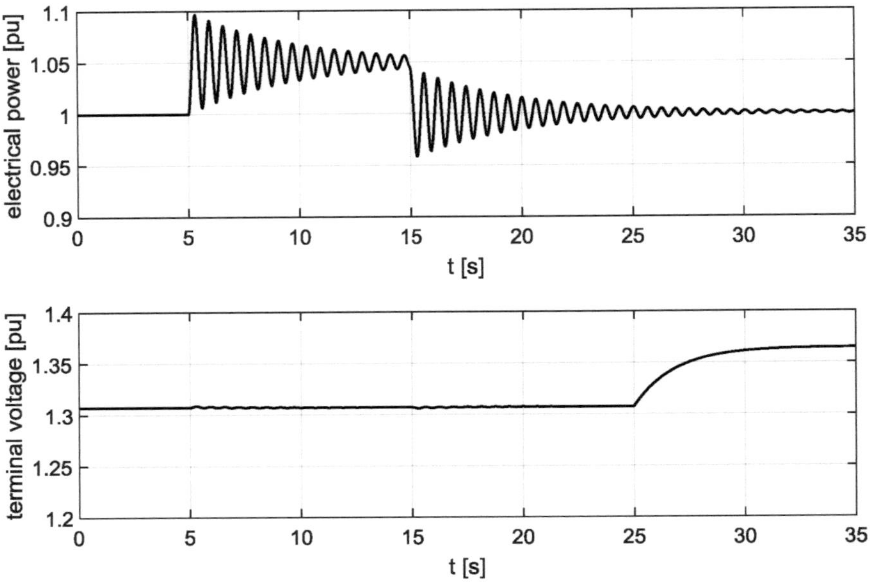

Figure 18.
Synchronous generator outputs' trajectories: Generated electrical power $P_e(t)$ *[pu] and stator terminal voltage* $V_t(t)$ *[pu], operating point* $P = 0.1$ *[pu] and* $Q = 1.0$ *[pu].*

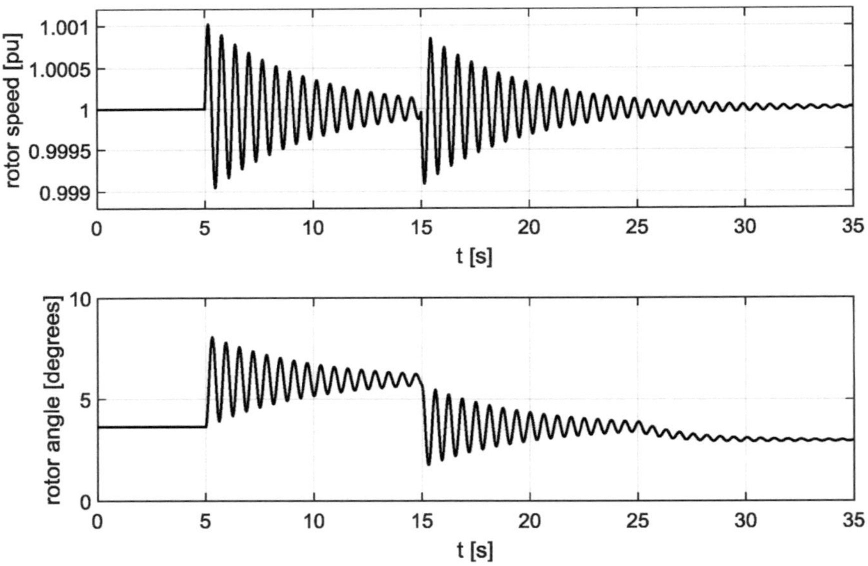

Figure 19.
Synchronous generator outputs' trajectories: Rotor speed $\omega(t)$ *[pu] and rotor angle* $\delta(t)$ *[degrees], operating point* $P = 0.1$ *[pu] and* $Q = 1.0$ *[pu].*

From the analysis of the effect of different loadings on the synchronous generator dynamic characteristics, it can be concluded that the variations in the machine dynamics are considerable in the entire operating range, and, therefore, a control system is necessary for damping of the oscillations. From the comparison of the responses across different operating points, it is obvious that the present system is nonlinear and that the conventional linear control theory does not provide adequate damping throughout the entire operating area. Therefore, the implementation of a robust or adaptive control theory is meaningful.

4. Conventional control system for synchronous generator's rotor excitation equipment

Two principal control systems affect a synchronous generator directly: a governor control system and an excitation control system. The governor control system controls the mechanical power from a steam or water turbine by opening or closing valves regulating the steam or water flow. The response of the governor control system is too slow to damp the synchronous generator's oscillations, which are mainly in the frequency range 0.5–2.5 Hz. Damping the oscillations is possible only with the excitation control system. The excitation control system (also called an automatic voltage regulator) changes the rotor field voltage (and current) in such a way that the generator's output voltage is the same (or close enough) to the reference voltage. In modern power plants, the thyristor or transistor rectifiers are used mainly to generate the required voltage for rotor winding. The electrical power flow from the excitation system is much smaller than the mechanical power flow. This, and the fact that semiconductor components are used in the excitation system instead of mechanical ones, is the reason that the excitation system is significantly faster than the governor system. Therefore, an exciter is used for the damping of the oscillations.

A conventional linear PSS approach is based on utilization of the static excitation system. Through this system, the PSS changes the field excitation voltage of a synchronous generator. An additional component of an electrical torque is generated as a consequence. This torque must be in phase with the rotor speed and thus increases damping of the synchronous generator [9]. **Figure 20** presents a block diagram of the Heffron-Phillips model of synchronous generator equipped with an excitation system, voltage controller, and power system stabilizer [10]. The generator's output voltage is compared with a reference voltage, and the calculated error is driven to the rectifier with an integrated voltage controller. The rectifier with the voltage controller is presented with the first-order model. The PSS input represents one or more signals in which oscillations are visible. The PSS generates an additional signal, which is added to the voltage error.

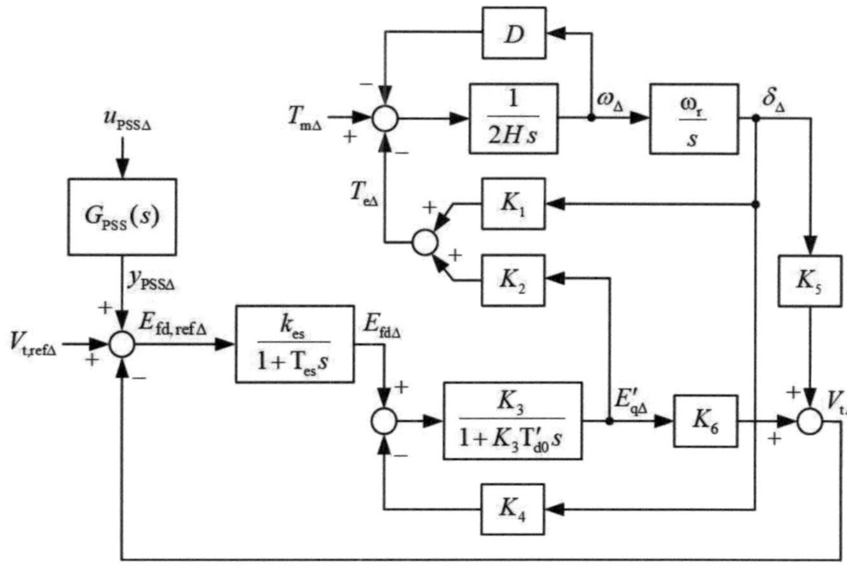

Figure 20.
Block diagram of the Heffron-Phillips model of synchronous generator equipped with excitation system, voltage controller, and power system stabilizer.

The symbols in **Figure 20** represent the following: k_{es} and T_{es} are the excitation system gain [pu.] and the time constant [s], respectively, $E_{fd,ref}$ is the reference for field excitation voltage E_{fd} (both in [pu]), while u_{PSS}, y_{PSS}, and $G_{PSS}(s)$ are the PSS input, the output (both in [pu]), and the transfer function, respectively. As for the variables of the Heffron-Phillips model, subscript $_\Delta$ denotes the deviation of the variables from the steady-state operating points, and s is the Laplace complex variable.

For PSS input $u_{PSS\Delta}$, the variables must be used which contain information about oscillations. These variables are electrical power, rotor angle, rotor speed, frequency, terminal voltage, and acceleration torque. The electrical power is selected commonly as the input to the PSS. The output of the PSS, $y_{PSS\Delta}$, is the control signal for the excitation system. A transfer function of a conventional linear PSS is represented as follows:

$$G_{PSS}(s) = \frac{y_{PSS}(s)}{u_{PSS}(s)} = k_{PSS}\left(\frac{sT_1+1}{sT_2+1}\right)\left(\frac{sT_3+1}{sT_4+1}\right)\left(\frac{sT_w}{sT_w+1}\right)G_{aaf}(s) \tag{20}$$

where k_{PSS} denotes the stabilizer gain [pu]; T_1, T_2, T_3, and T_4 are time constants of the stabilizers lead–lag compensators [s]; T_w is the time constant of the high-pass (washout) filter [s]; and $G_{aaf}(s)$ is the transfer function of the low-pass (antialiasing) filter.

Based on the block diagram in **Figure 20** and the transfer function in (Eq. (6)), the IEEE Association established the IEEE Standard for the PSS studies [11]. The Standard enables the unification of commercial applications of PSS. The Standard sets out four basic types of PSS, which differ mainly with regard to the available input and degree of the transfer function. Most of the commercial PSS are realized on the standardized proposals.

For synthesis of a PSS, knowledge is required of a mathematical model of a synchronous generator with an excitation system. The required model is calculated from the known data of a synchronous generator, or by means of identification. Usage of systematic methods for tuning parameters of conventional PSS assures effective damping for the nominal operating point, though with a significantly decreased damping for some non-nominal operating points. The other disadvantages of these methods are the requirement of the synchronous generator mathematical model's parameters and the time-consuming tuning. Therefore, in practice, the systematic methods are rarely implemented. Hence, neither optimal damping in the nominal operating point nor stable operation is secured in the entire operating range. The implementations of an incorrectly tuned PSS could be harmful. Such PSS are, in practice, often turned off [12].

Due to a mathematical model of a synchronous generator not being available, sophisticated and time-consuming synthesis of the conventional linear PSS, and its proven non-optimum damping in the entire operating range of synchronous generator, advanced control theories are recommended for the PSS implementation.

5. Robust PSS

Among many robust control approaches, the sliding mode control is one of the most interesting. The main advantages of this control are its insensitivity to parameter variations, rejection of disturbances, a decoupling design procedure, and simple implementations by means of power converters [13].

The fundamentals of the sliding mode control theory date back to the late 1950s. Since that time, new research directions emerged, due to the appearance of new classes of control problems, new mathematical methods, and new prospects of implementation [13–15].

A modification of the sliding mode control based on the decoupling principle will be used for the proposed PSS design. The mathematical model of the controlled plant must be transformed to a regular form:

$$\dot{\mathbf{x}}_{RF1}(t) = \mathbf{A}_{RF11}\mathbf{x}_{RF1}(t) + \mathbf{A}_{RF12}\mathbf{x}_{RF2}(t) \tag{21}$$

$$\dot{\mathbf{x}}_{RF2}(t) = \mathbf{A}_{RF21}\mathbf{x}_{RF1}(t) + \mathbf{A}_{RF22}\mathbf{x}_{RF2}(t) + \mathbf{B}_{RF2}\mathbf{u}_{p}(t) \tag{22}$$

where $\mathbf{A}_{RFij}$ (i,j = 1,2) and $\mathbf{B}_{RF2}$ are constant matrices of relevant dimensions, $\mathbf{x}_{RF1}(t) \in \Re^{n-m}$ and $\mathbf{x}_{RF2}(t) \in \Re^{m}$ are state-space vectors, and $\mathbf{u}_{p}(t)$ is a controlled plant input vector. Matrix $\mathbf{B}_{RF2}$ must be nonsingular.

For the PSS design being based on the simplified linearized model of synchronous generator, a state-space vector in regular form $\mathbf{x}_{RF}(t)$, where n = 3 and m = 1, could be selected as

$$\mathbf{x}_{RF}(t) = \begin{bmatrix} \mathbf{x}_{RF1}(t) \\ \mathbf{x}_{RF2}(t) \end{bmatrix} \quad \text{where}: \ \mathbf{x}_{RF1}(t) = \begin{bmatrix} \delta_{\Delta}(t) \\ \dot{\delta}_{\Delta}(t) \end{bmatrix} \quad \text{and} \quad \mathbf{x}_{RF2}(t) = \ddot{\delta}_{\Delta}(t) \tag{23}$$

Sliding mode control for implementation in PSS requires knowledge of all state-space variables of the synchronous generator's regular form model. Measurements of electrical power, rotor speed, and terminal voltage are feasible only at the synchronous generator. For the sliding mode control, the state-space variables for the regular form model need to be calculated from the measured variables. To calculate regular form state-space variables, firstly, variables $\delta_{\Delta}(t)$ and $E'_{q\Delta}(t)$ can be calculated by inverting (Eq. (19)), such as

$$\begin{bmatrix} \delta_{\Delta}(t) \\ E'_{q\Delta}(t) \end{bmatrix} = \frac{1}{K_1K_6 - K_2K_5} \begin{bmatrix} K_6 & -K_2 \\ -K_5 & K_1 \end{bmatrix} \begin{bmatrix} P_{e\Delta}(t) \\ V_{t\Delta}(t) \end{bmatrix} \tag{24}$$

Finally, state-space variables $\mathbf{x}_{RF1}(t)$ and $\mathbf{x}_{RF2}(t)$ can be calculated with transformation

$$\begin{bmatrix} \delta_{\Delta}(t) \\ \dot{\delta}_{\Delta}(t) \\ \ddot{\delta}_{\Delta}(t) \end{bmatrix} = \begin{bmatrix} 1 & 0 & 0 \\ 0 & \omega_r & 0 \\ -\dfrac{K_1\omega_s}{2H} & -\dfrac{D\omega_s}{2H} & -\dfrac{K_2\omega_s}{2H} \end{bmatrix} \begin{bmatrix} \delta_{\Delta}(t) \\ \omega_{\Delta}(t) \\ E'_{q\Delta}(t) \end{bmatrix} \tag{25}$$

In such a way, the state-space variables could be obtained without explicit differentiation.

A sliding surface was selected, such that the rotor's angle deviation and rotor's speed deviation converged exponentially to zero. For this aim, a linear equation of the sliding surface was selected:

$$\mathbf{s}(t) = \mathbf{D}\mathbf{x}_{RF1}(t) + \mathbf{x}_{RF2}(t), \quad \mathbf{s}(t) \in \Re^{m} \tag{26}$$

When the sliding mode appears on manifold $\mathbf{s}(t) = 0$ where $\mathbf{x}_{RF2}(t) = -\mathbf{D}\mathbf{x}_{RF1}(t)$, the system behavior is governed by $(n\text{-}m)$-th-order equation

$$\dot{\mathbf{x}}_{RF1} = (\mathbf{A}_{RF11} - \mathbf{A}_{RF12}\mathbf{D})\mathbf{x}_{RF1} \tag{27}$$

To obtain the required dynamic properties of the control system, we assigned eigenvalues of a closed-loop system with a linear feedback. For the controllable system described with (Eqs. (21), (22)), there exists matrix $\mathbf{D}$, which ensures the desired eigenvalues of the system in (Eq. (27)).

In the first stage of design of the sliding mode, we chose the desired eigenvalues of the system described with (Eq. (27)). From the desired eigenvalues, we determined matrix $\mathbf{D}$ as the solution to the (n-m)-th-order eigenvalue task. Matrix $\mathbf{D}$ determines the equation of discontinuous sliding surfaces (Eq. (27)).

The second stage of the design procedure is the selection of the discontinuous control law, such that the sliding mode always arises at manifold $\mathbf{s}(t) = 0$, which is equivalent to the stability of the origin in m-dimensional space $\mathbf{s}(t)$. The dynamics on the $\mathbf{s}(t)$ space are described by the equation

$$\begin{aligned}\dot{\mathbf{s}}(t) &= [\mathbf{D}\mathbf{A}_{RF11} + \mathbf{A}_{RF21}]\mathbf{x}_{RF1}(t) + [\mathbf{D}\mathbf{A}_{RF12} + \mathbf{A}_{RF22}]\mathbf{x}_{RF2}(t) + \mathbf{B}_{RF2}\mathbf{u}(t) \\ &= \mathbf{E}\mathbf{x}_{RF}(t) + \mathbf{B}_{RF2}\mathbf{u}_p(t)\end{aligned} \tag{28}$$

An appropriate choice of the control law represents the discontinuous control described with

$$\mathbf{u}_p(t) = -g|\mathbf{x}_{RF}(t)|\mathbf{B}_{RF2}^{-1}\,\mathrm{sgn}\,\mathbf{s}(t) \tag{29}$$

where $|\mathbf{x}_{RF}(t)|$ is the sum of vector $\mathbf{x}_{RF}(t)$ component moduli and g is the constant.

The selected discontinuous control leads to

$$\dot{\mathbf{s}}(t) = \mathbf{E}\mathbf{x}_{RF}(t) - g|\mathbf{x}_{RF}(t)|\,\mathrm{sgn}\,\mathbf{s}(t) \tag{30}$$

There exists such positive value of g that the functions $\mathbf{s}(t)$ and $\dot{\mathbf{s}}(t)$have different signs. It means that the sliding mode will occur on a discontinuity surface. The influence of discontinuity of the control signal is reduced by varying the amplitude of the control signal.

6. Adaptive PSS

Many examples with utilization of different adaptive techniques for realization of PSS can be found in publications. The majority of PSS realizations are based on usage of indirect adaptive control, where explicit identification of a mathematical model of a synchronous generator is needed to be carried out [16, 17]. A transparent structure of the adaptive control system with the separated identification algorithm and the control law represents an advantage of indirect adaptive PSS. There are significantly less publications available where usage of direct adaptive control for PSS is presented [18]. The methods of direct adaptive control are more difficult to be utilized for the conventional PSS structure than those for the indirect adaptive control. However, their advantage is in not requiring explicit identification of the SG, and they are, therefore, computationally less demanding. In this article, the developed robust PSS will be compared with direct adaptive PSS which was studied in detail in [2].

The theoretical foundation for the used direct adaptive PSS is represented by a theory of model reference adaptive control for almost strictly positive real plants.

The implemented direct adaptive control is considered for the controlled plant, which is described by

$$\dot{\mathbf{x}}_p(t) = \mathbf{A}_p\mathbf{x}_p(t) + \mathbf{B}_p\mathbf{u}_p(t) \tag{31}$$

$$\mathbf{y}_p(t) = \mathbf{C}_p\mathbf{x}_p(t) \tag{32}$$

where $\mathbf{x}_p(t) \in \Re^n$ is the controlled plant state-space vector, $\mathbf{u}_p(t) \in \Re^m$ is the controlled plant input vector, $\mathbf{y}_p(t) \in \Re^r$ is the controlled plant output vector, and $\mathbf{A}_p$, $\mathbf{B}_p$, and $\mathbf{C}_p$ are the matrices of the appropriate dimensions.

It is assumed that:

- The range of the plant matrices parameters is bounded.
- All possible pairs $\mathbf{A}_p$ and $\mathbf{B}_p$ are controllable and output stabilizable.
- All possible pairs $\mathbf{A}_p$ and $\mathbf{C}_p$ are observable.

The reference model is described by

$$\dot{\mathbf{x}}_m(t) = \mathbf{A}_m\mathbf{x}_m(t) + \mathbf{B}_m\mathbf{u}_m(t) \tag{33}$$

$$\mathbf{y}_m(t) = \mathbf{C}_m\mathbf{x}_m(t) \tag{34}$$

where $\mathbf{x}_m(t)$ is the model state vector, $\mathbf{u}_m(t)$ is the model command vector, $\mathbf{y}_m(t)$ is the model output vector, and $\mathbf{A}_m$, $\mathbf{B}_m$, and $\mathbf{C}_m$ are matrices of appropriate dimensions. The model is assumed to be stable. The dimension of the model state may be less than the dimension of the plant state.

The output tracking error is defined as

$$\mathbf{e}_y(t) = \mathbf{y}_m(t) - \mathbf{y}_p(t) \tag{35}$$

The control $\mathbf{u}_p(t)$ for the plant output vector $\mathbf{y}_p(t)$ to approximate "reasonably well" the output of the reference model $\mathbf{y}_m(t)$ without explicit knowledge of $\mathbf{A}_p$, $\mathbf{B}_p$, and $\mathbf{C}_p$ is generated by the adaptive algorithm:

$$\mathbf{u}_p(t) = \mathbf{K}_e(t)\mathbf{e}_y(t) + \mathbf{K}_x(t)\mathbf{x}_m(t) + \mathbf{K}_u(t)\mathbf{u}_m(t) \tag{36}$$

$$\mathbf{u}_p(t) = \mathbf{K}(t)\mathbf{r}(t) \tag{37}$$

where

$$\mathbf{K}(t) = [\mathbf{K}_e(t), \mathbf{K}_x(t), \mathbf{K}_u(t)] \tag{38}$$

$$\mathbf{r}^T(t) = \left[\mathbf{e}_y^T(t), \mathbf{x}_m^T(t), \mathbf{u}_m^T(t)\right]. \tag{39}$$

The adaptive gains $\mathbf{K}(t)$ are obtained as a combination of the "proportional" and "integral" terms

$$\mathbf{K}(t) = \mathbf{K}_P(t) + \mathbf{K}_I(t) \tag{40}$$

$$\mathbf{K}_P(t) = \mathbf{e}_y(t)\mathbf{r}^T(t)\mathbf{T} \tag{41}$$

$$\dot{\mathbf{K}}_I(t) = \mathbf{e}_y(t)\mathbf{r}^T(t)\overline{\mathbf{T}} - \sigma\mathbf{K}_I(t) \tag{42}$$

where σ term is introduced in order to avoid divergence of the integral gains in the presence of disturbance and $\mathbf{T}$ and $\overline{\mathbf{T}}$ are positive definite and positive semi-definite adaptation coefficient matrices, respectively.

The necessary condition for asymptotic tracking when $\mathbf{u}_m(t)$ is a step command is that the controlled plant is almost strictly positive real (ASPR) [19]. If the controlled plant is not ASPR, the augmenting of the plant with a feedforward compensator is suggested, such that the augmented plant is ASPR. In this case, the previously described adaptive controller may be utilized.

For the non-ASPR plant described by the transfer matrix

$$\mathbf{G}_{\mathrm{p}}(s) = \mathbf{C}_{\mathrm{p}}\left(s\mathbf{I} - \mathbf{A}_{\mathrm{p}}\right)^{-1}\mathbf{B}_{\mathrm{p}} \tag{43}$$

the feedforward compensator is defined by the strictly proper transfer function matrix $\mathbf{G}_{\mathrm{ff}}(s)$ with the realization

$$\dot{\mathbf{s}}_{\mathrm{p}}(t) = \mathbf{A}_{\mathrm{s}}\mathbf{s}_{\mathrm{p}}(t) + \mathbf{B}_{\mathrm{s}}\mathbf{u}_{\mathrm{p}}(t) \tag{44}$$

$$\mathbf{r}_{\mathrm{p}}(t) = \mathbf{D}_{\mathrm{s}}\mathbf{s}_{\mathrm{p}}(t) \tag{45}$$

Instead of the plant output $\mathbf{y}_{\mathrm{p}}(t)$, augmented output $\mathbf{z}_{\mathrm{p}}(t)$ is to be controlled:

$$\mathbf{z}_{\mathrm{p}}(t) = \mathbf{y}_{\mathrm{p}}(t) + \mathbf{r}_{\mathrm{p}}(t) \tag{46}$$

The augmented system is defined as

$$\mathbf{G}_{\mathrm{a}}(s) = \mathbf{G}_{\mathrm{p}}(s) + \mathbf{G}_{\mathrm{ff}}(s) \tag{47}$$

Feedforward compensator $\mathbf{G}_{\mathrm{ff}}(s)$ is an inverse of a (fictitious) stabilizing controller for the plant and must be selected such that the resulting relative degree of augmented plant $\mathbf{G}_{\mathrm{a}}(s)$ is indeed 1. For example, if SISO plant $\mathrm{G}_{\mathrm{p}}(s)$ is stabilizable by a PD controller, one can use its inverse in a manner that is just a simple first-order low-pass filter.

7. Results

The effectiveness of the proposed sliding mode PSS and direct adaptive PSS was tested with the simulations of the seventh-order nonlinear model of the synchronous generator in the entire operating range, numerically, as well as experimentally, in the laboratory.

7.1 Robust PSS

A block diagram of the sliding mode PSS is shown in **Figure 21**.

A sliding mode controller requires measurements of three synchronous generator's quantities: electrical power, rotor speed, and terminal voltage. Input filters are low-pass filters to eliminate the measured noise. From these measured variables, the state-space variables for the regular form model are calculated by means of state transformation. State transformation is carried out by Eqs. (24) and (25). The obtained regular form state-space variables are used in the control law described with Eqs. (26) and (29). The output of the discontinuous control law is conducted in the limiter. Hard type saturation of the PSS output was utilized, with a limited value of ±35% of the value of a nominal rotor excitation voltage. The set value represents a limitation in a real excitation system.

For a synchronous generator with the data listed in Section 3, we selected desired eigenvalues $\lambda_{1,2} = -2$ for the system in (Eq. (27)). The following control law parameters were calculated [2]:

$$\mathbf{D} = \begin{bmatrix} 4 & 4 \end{bmatrix} \quad \mathbf{B}_{\mathrm{RF2}}^{-1} = -0.06 \quad g = 350 \tag{48}$$

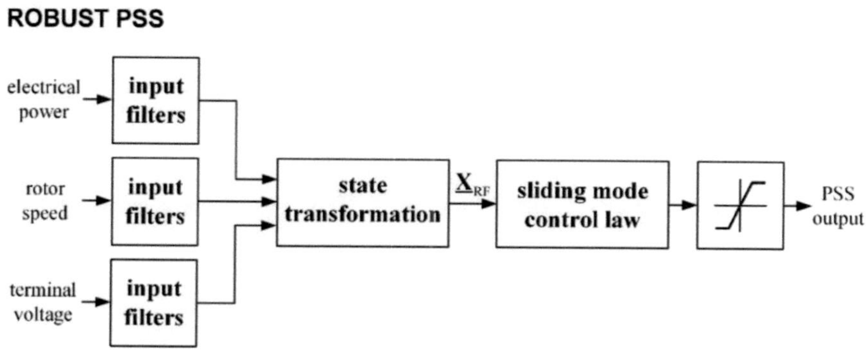

Figure 21.
Block diagram of the sliding mode PSS.

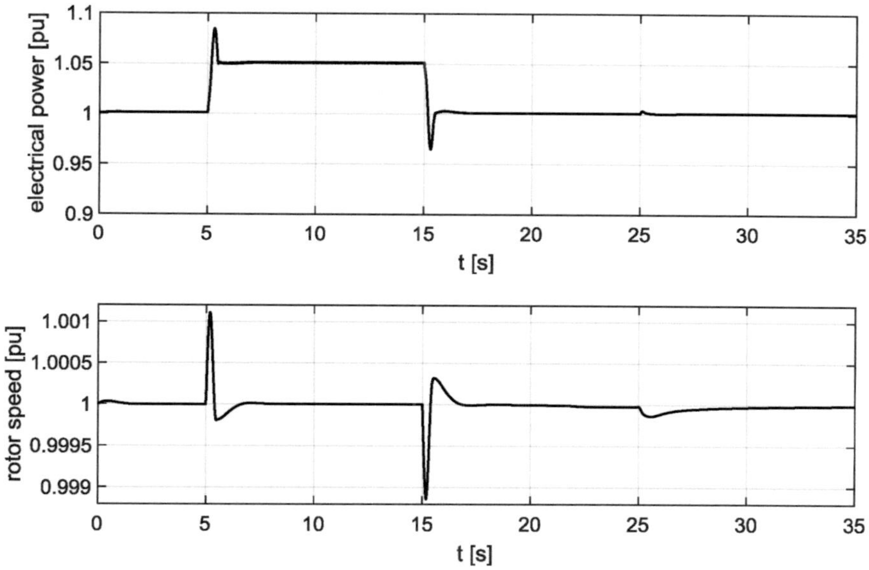

Figure 22.
Electrical power $P_e(t)$ [pu] and rotor speed $\omega(t)$ [pu] at nominal operating point $P = 1.0$ [pu] and $Q = 0.62$ [pu], with robust PSS.

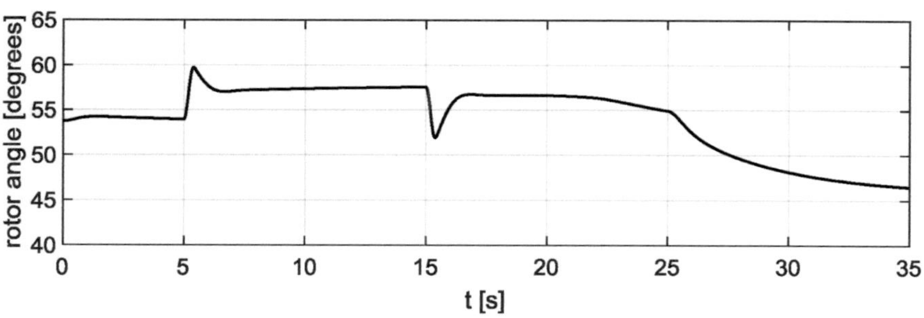

Figure 23.
Rotor angle $\delta(t)$ [pu] at nominal operating point $P = 1.0$ [pu] and $Q = 0.62$ [pu], with robust PSS.

7.1.1 Nominal operating point

Figures 22–24 show the responses of the seventh-order nonlinear model of the considered 160 MVA synchronous generator equipped with an excitation system and robust PSS to the step changes of the mechanical torque and the field excitation voltage, as shown in **Figure 11**. The synchronous generator operates in in the vicinity of the nominal operating point. **Figures 22** and **23** show the generated electrical power, rotor speed, and rotor angle at nominal operating point P = 1.0 [pu] and Q = 0.62 [pu].

Figure 24 shows the excitation voltage produced by a robust PSS at operating point P = 1.0 [pu] and Q = 0.62 [pu]. The limits of the limiters are seen clearly.

7.1.2 Influence of load disturbance

During the operation in the entire operating range, the dynamics of the synchronous generator vary significantly. The sliding mode controller with the calculated parameters was stable and robust and displayed the effective damping in all operating conditions. The theoretical analysis of the invariance of the proposed control system to the disturbances and the variation of the plant parameter are described in detail in [20].

In this work, the results of the two most extreme operating points are presented (the same operating points as described in Section 3.2):

- P = 1.0 [pu] and Q = 0.1 [pu]: This is a stable operation point with heavily damped oscillations.
- P = 0.1 [pu] and Q = 1.0 [pu]: This is the critical operating point with weakly damped oscillations.

Figure 25 shows the generated electrical power and rotor speed at operating point P = 1.0 [pu] and Q = 0.1 [pu], and **Figure 26** shows both quantities at operating point P = 0.1 [pu] and Q = 1.0 [pu].

7.1.3 Influence of parameter deviations

To analyze the impact of parameter variations on the damping efficiency of the proposed control systems, both control systems were tested at different operating points for synchronous generators of different types and nominal powers. In this work, the results are presented for a synchronous generator with nominal power 555 MVA. The data of the considered synchronous generator are shown in **Table 10** [4].

The linearization coefficients for nominal operating point (P_N = 1 [pu], cos φ_N = 0.9) and eigenvalues of the Heffron-Phillips model (λ_1, λ_2, λ_3) are presented in **Table 11**.

The transient response of the noncontrolled synchronous generator with data in **Table 10** and nominal operating point data in **Table 11** are shown in

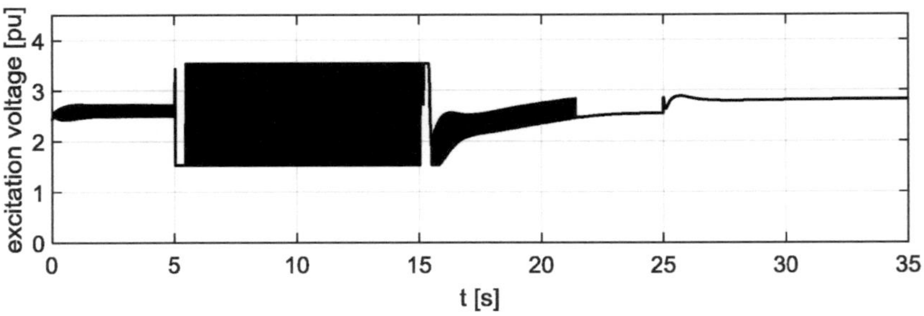

Figure 24.
Excitation voltage $E_{FD}(t)$ *[pu] at nominal operating point* P = 1.0 *[pu] and* Q = 0.62 *[pu], with robust PSS.*

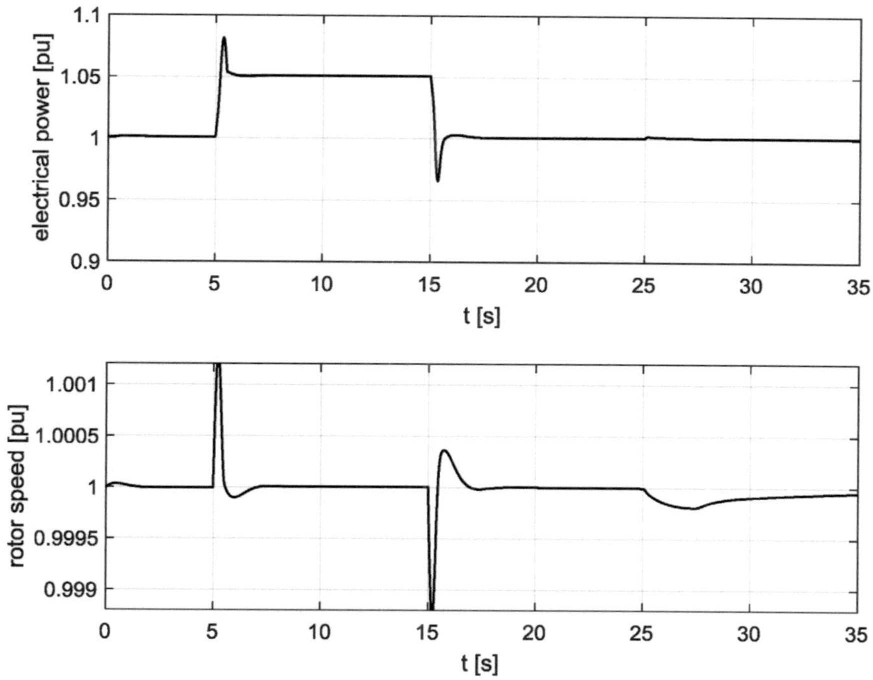

Figure 25.
Electrical power $P_e(t)$ *[pu] and rotor speed* $\omega(t)$ *[pu] at operating point* P = 1.0 *[pu] and* Q = 0.1 *[pu], with robust PSS.*

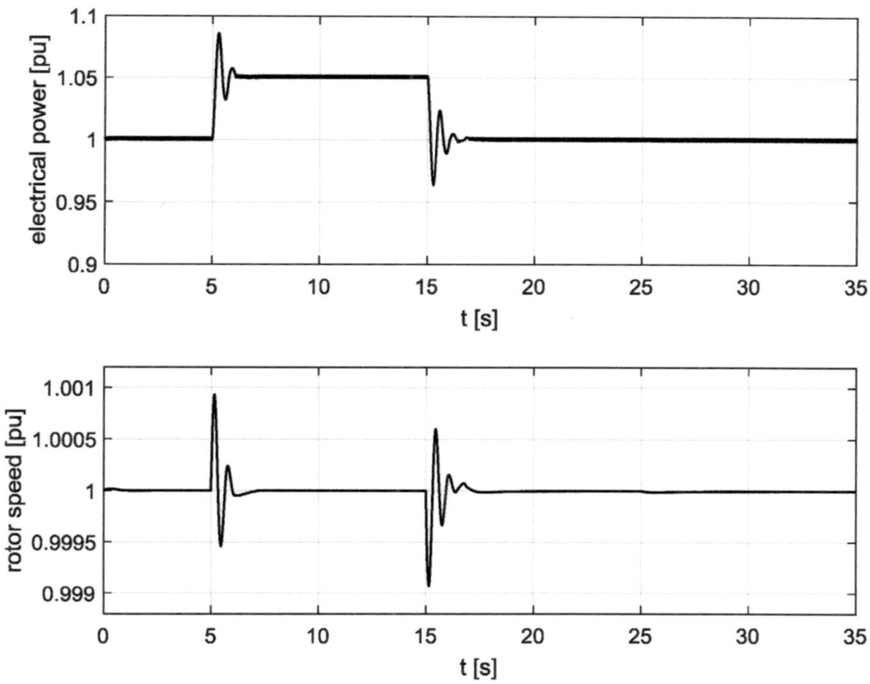

Figure 26.
Electrical power $P_e(t)$ *[pu] and rotor speed* $\omega(t)$ *[pu] at operating point* P = 0.1 *[pu] and* Q = 1.0 *[pu], with robust PSS.*

Figures 27 and **28**. Step changes are simulated in both generators' inputs. **Figure 27** shows the simulated trajectory of the turbine mechanical torque and rotor excitation voltage. Step changes are selected as the worst case in order to maximize the excitation of oscillations.

Figure 28 shows the response of the generated electrical power and rotor speed on the inputs' trajectories shown in **Figure 27**.

S_N = 555 [MVA]	V_N = 24 [kV]	$\cos \varphi_N$ = 0.90
ω_s = 377 [rad s^{-1}]		
T'_{d0} = 8.0 [pu]	H = 3.52 [s]	D = 2.0 [pu]
R_e = 0.02 [pu]	L_e = 0.4 [pu]	V_{IB} = 1.0 [pu]
R_s = 0.0030 [pu]	R_F = 0.0006 [pu]	x'_d = 0.300 [pu]
L_d = 1.810 [pu]	L_q = 1.760 [pu]	L_F = 0.165 [pu]
L_D = 0.171 [pu]	L_Q = 0.084 [pu]	L_{AD} = 1.660 [pu]
l_d = 0.150 [pu]	l_q = 0.150 [pu]	L_{AQ} = 1.610 [pu]

Table 10.
Data of the 555 MVA synchronous generator used for the analysis of the impact of parameter variations on the damping efficiency [4].

P_N = 1.0 [pu]	Q_N = 0.48 [pu]	$\cos \varphi_N$ = 0.90
K_1 = 1.3306	K_2 = 1.2988	K_3 = 0.3168
K_4 = 1.8578	K_5 = −0.0107	K_6 = 0.4545
λ_1 = −0.2554 + 8.4389i	λ_2 = −0.2554 - 8.4389i	λ_3 = −-0.1678

Table 11.
Linearization parameters and eigenvalues of the Heffron-Phillips model at the nominal operating point of the 555 MVA synchronous generator.

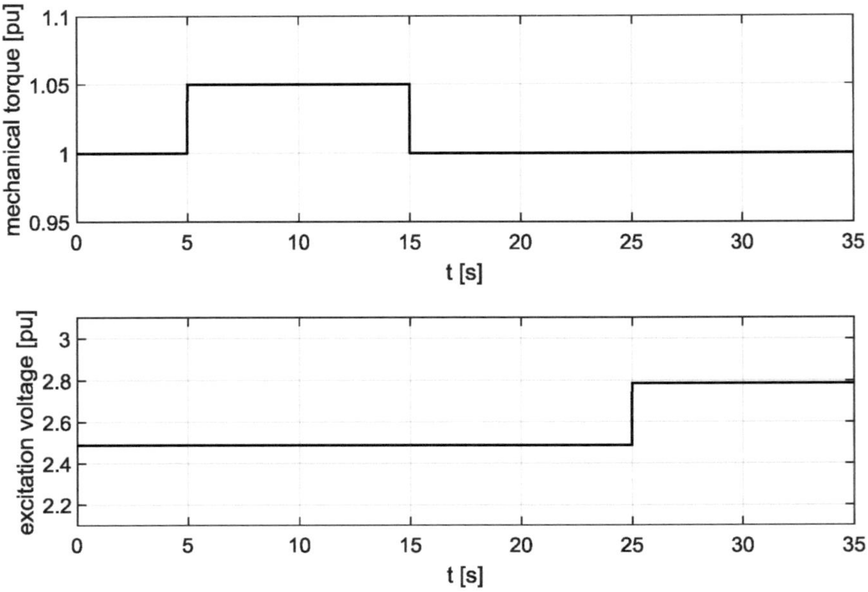

Figure 27.
The 555 MVA synchronous generator inputs' trajectories: Mechanical torque $T_m(t)$ *[pu] and rotor excitation voltage* $E_{fd}(t)$ *[pu], nominal operating point* P = 1.0 *[pu] and* Q = 0.48 *[pu].*

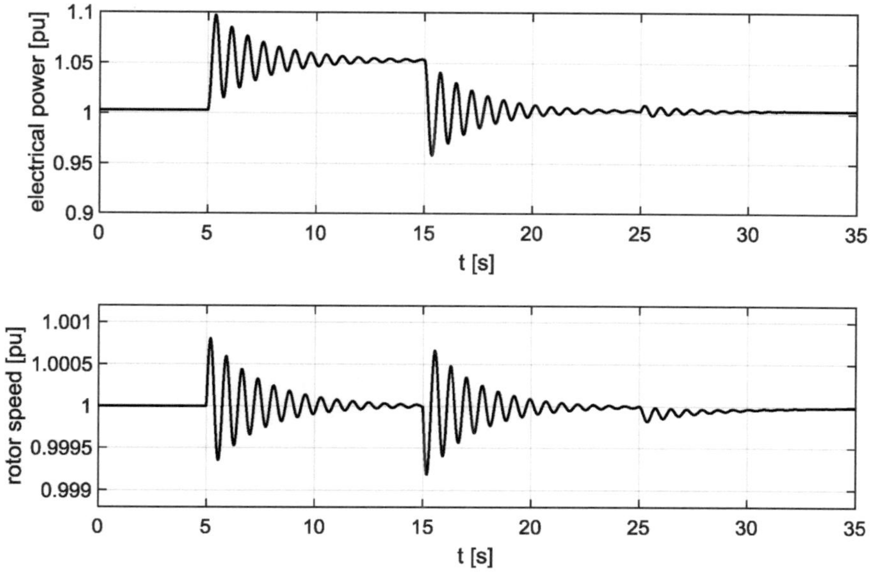

Figure 28.
The 555 MVA synchronous generator outputs' trajectories: Rotor speed ω(t) *[pu] and rotor angle* δ(t) *[degrees], nominal operating point* P = *1.0 [pu] and* Q = *0.48 [pu], without PSS.*

P_N = 1.0 [pu]	Q_N = 0.1 [pu]	cos φ_N = 0.995
K_1 = 1.1387	K_2 = 1.4710	K_3 = 0.3168
K_4 = 2.1296	K_5 = −0.1069	K_6 = 0.3281
λ_1 = −0.3140 + 7.8075i	λ_2 = −0.3140 - 7.8075i	λ_3 = −0.0506

Table 12.
Performance in linearization parameters and eigenvalues of the Heffron-Phillips model in operating point P = *1.0 [pu] and* Q = *0.1 [pu] of the 555 MVA synchronous generator.*

P_N = 0.1 [pu]	Q_N = 1.0 [pu]	cos φ_N = 0.099
K_1 = 1.2340	K_2 = 0.1533	K_3 = 0.3168
K_4 = 0.1204	K_5 = 0.0167	K_6 = 0.5720
λ_1 = −0.1430 - 8.1276i	λ_2 = −0.1430 + 8.1276i	λ_3 = −0.3927

Table 13.
Linearization parameters and eigenvalues of the Heffron-Phillips model at operating point P = *0.1 [pu] and* Q = *1.0 [pu] of the 555 MVA synchronous generator.*

The results are presented of the robust control at two operating points:

- P = 1.0 [pu] and Q = 0.1 [pu]: The linearization coefficients and eigenvalues of the Heffron-Phillips model (λ_1, λ_2, λ_3) are presented in **Table 12**, the step changes of the mechanical torque and the field excitation voltage are shown in **Figure 29**, and the generated electrical power and rotor speed are shown in **Figure 30**.

- P = 0.1 [pu] and Q = 1.0 [pu]: The linearization coefficients and eigenvalues of the Heffron-Phillips model (λ_1, λ_2, λ_3) are presented in **Table 13**, the step

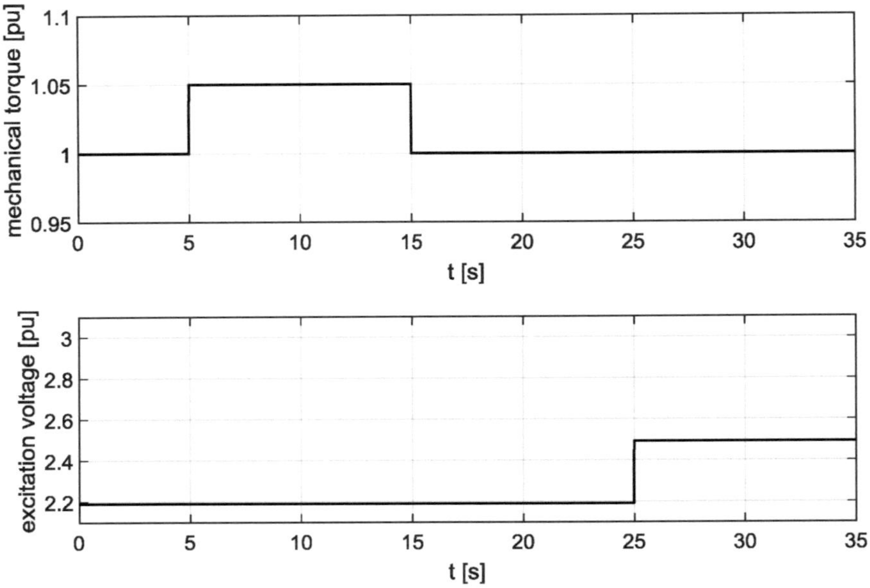

Figure 29.
The 555 MVA synchronous generator inputs' trajectories: Mechanical torque $T_m(t)$ *[pu] and rotor excitation voltage* $E_{fd}(t)$ *[pu], operating point* $P = 1.0$ *[pu] and* $Q = 0.1$ *[pu].*

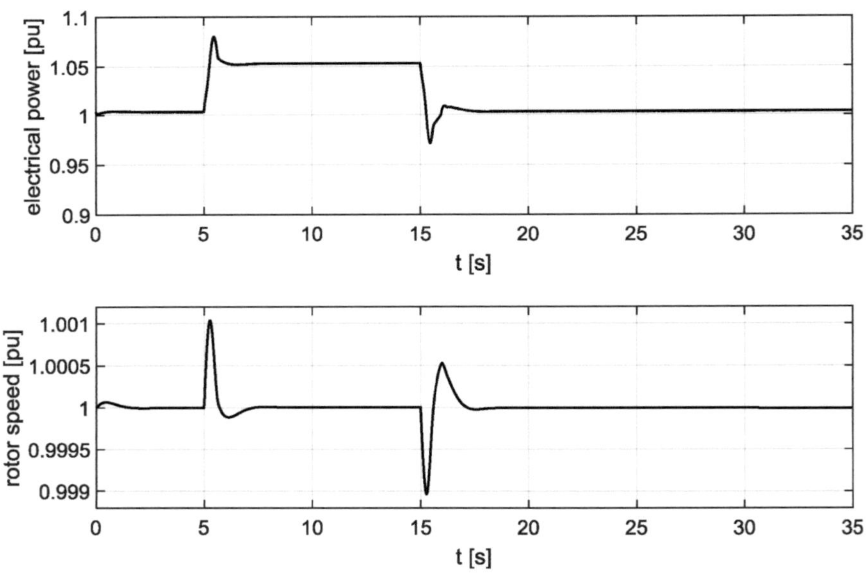

Figure 30.
The 555 MVA synchronous generator's electrical power $P_e(t)$ *[pu] and rotor speed* $\omega(t)$ *[pu] at nominal operating point* $P = 1.0$ *[pu] and* $Q = 0.1$ *[pu], with robust PSS.*

changes of the mechanical torque and the field excitation voltage are shown in **Figure 31**, and the generated electrical power and rotor speed are shown in **Figure 32**.

Figure 25 shows the generated electrical power and rotor speed at heavily damped operating point $P = 1.0$ [pu] and $Q = 0.1$ [pu], and **Figure 26** shows both quantities at weakly damped operating point $P = 0.1$ [pu] and $Q = 1.0$ [pu].

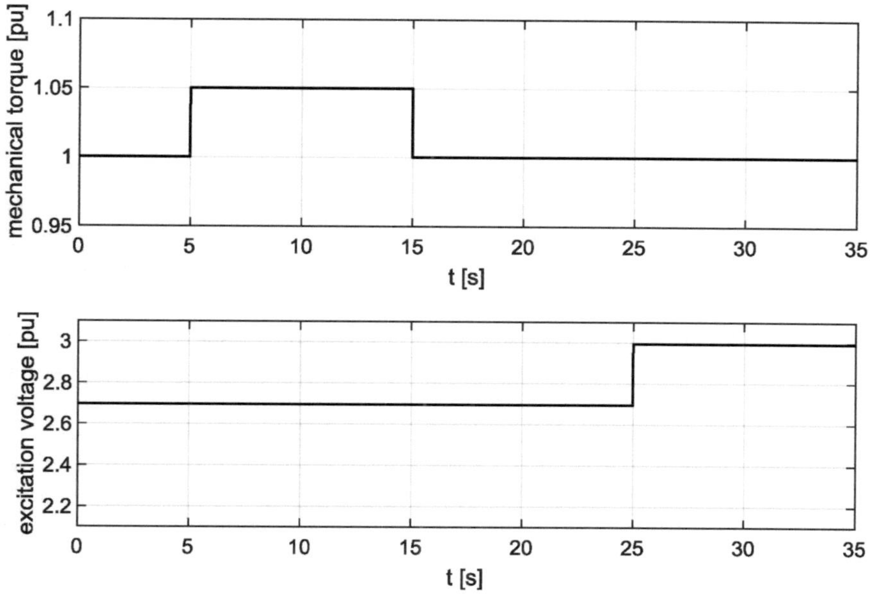

Figure 31.
The 555 MVA synchronous generator inputs' trajectories: Mechanical torque $T_m(t)$ *[pu] and rotor excitation voltage* $E_{fd}(t)$ *[pu], operating point* P = 0.1 *[pu] and* Q = 1.0 *[pu].*

7.2 Adaptive PSS

The proposed direct adaptive controller guarantees stability of any controlled plant that satisfies ASPR conditions. A synchronous generator with automatic voltage system does not satisfy the necessary ASPR conditions. Augmenting of the plant with a parallel feedforward compensator must be carried out to assure stable operation of the entire adaptive control system. The augmentation is performed such

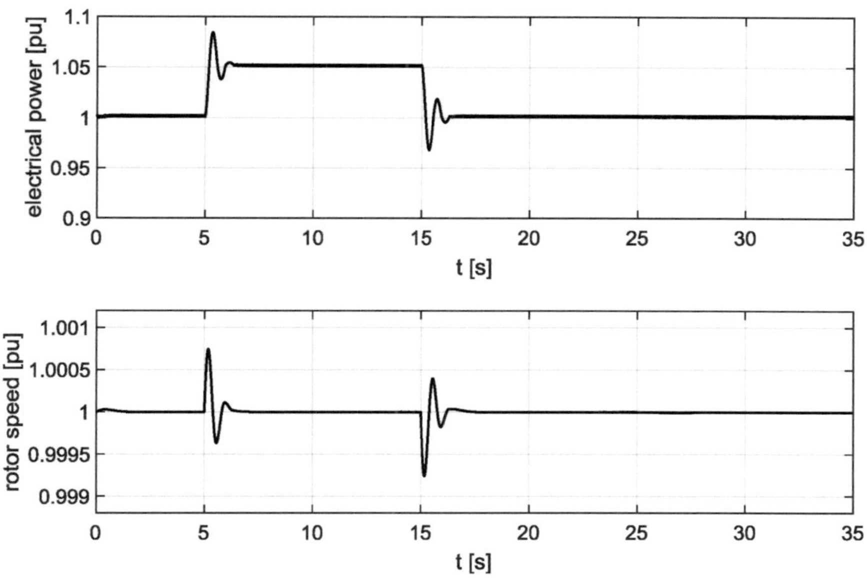

Figure 32.
The 555 MVA synchronous generator's electrical power $P_e(t)$ *[pu] and rotor speed* $\omega(t)$ *[pu] at nominal operating point* P = 0.1 *[pu] and* Q = 1.0 *[pu], with robust PSS.*

that the augmented plant fulfills ASPR conditions. The requirement is satisfied in the majority of cases with the introduction of a feedforward compensator $G_{ff}(s)$, which is connected in parallel to the basic controlled plant. The suitable feedforward stabilizer represents a first-order low-pass filter with feedforward compensator gain k_{ff} and feedforward compensator time constant T_{ff} [18].

A block diagram of the direct adaptive PSS is presented in **Figure 33**.

The benefit of the control diagram shown in **Figure 33**, if compared to other adaptive structures, is a very simple realization of the adaptation mechanism. The presented direct adaptive PSS is essentially simplified; namely, a reference model is not required because of the constant (zero) command signal.

The reference terminal voltage $V_{t,ref}$ and the mechanical torque T_m variables represent the main disturbances which affect the synchronous generator's dynamics. The variations of the synchronous generator loading can be treated as controlled plant parameters' perturbations. The washout block (input filters) serves as a high-pass filter, with the time constant T_w high enough to allow signals associated with oscillations in generator active power P_e to pass unchanged. Without it, steady changes in power would modify the terminal voltage. It allows the PSS to respond only to changes in generator active power. From the viewpoint of the washout function, the value of T_w is not critical and may be in the range of 1–20 s. The main consideration is that it would be long enough to pass stabilizing signals unchanged at the frequencies of interest. Direct adaptive control law is represented with (Eqs. (35)–(42)). The necessary feedforward compensator is described with Eqs. (44) and (45). The same model of the actuator saturation as in Section 7.1.1 was included in the simulations.

The parameters of the adaptation mechanism for the considered linearized controlled plant are determined with the rules described in [2], such as

$$\mathbf{T} = 0.1 \cdot 10^3 \quad \overline{\mathbf{T}} = 200 \cdot 10^3 \quad \sigma = 50 \cdot 10^{-3} \quad k_{ff} = 1 \cdot 10^{-3} \quad T_{ff} = 1 \cdot 10^{-3} \tag{49}$$

7.2.1 Nominal operating point

Figures 34–36 show the responses of the seventh-order nonlinear model of the considered 160 MVA synchronous generator equipped with an excitation system and adaptive PSS to the step changes of the mechanical torque and the field excitation voltage, as shown in **Figure 11**. **Figures 34** and **35** show the generated electrical power, rotor speed, and rotor angle at nominal operating point P = 1.0 [pu] and Q = 0.62 [pu].

Figure 36 shows the excitation voltage produced by an adaptive PSS at nominal operating point P = 1.0 [pu] and Q = 0.62 [pu].

Figures 34–36 are directly comparable with **Figures 22–25**.

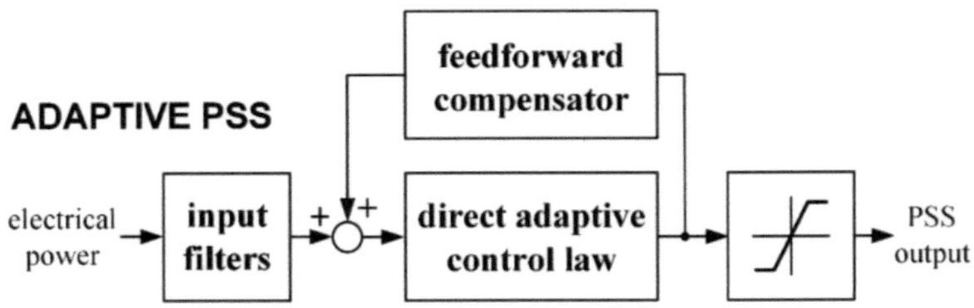

Figure 33.
Block diagram of the direct adaptive PSS.

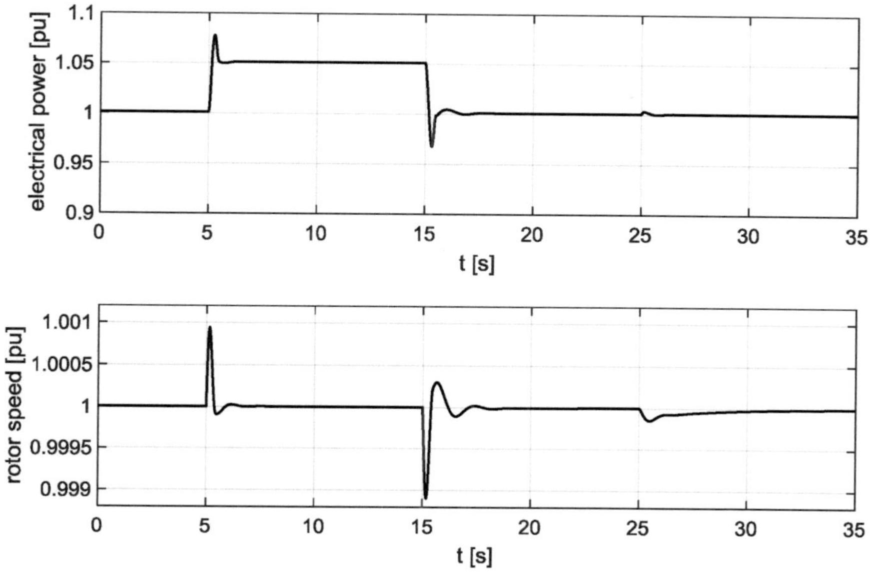

Figure 34.
Electrical power $P_e(t)$ *[pu] and rotor speed* $\omega(t)$ *[pu] at operating point* $P = 1.0$ *[pu] and* $Q = 0.62$ *[pu], with adaptive PSS.*

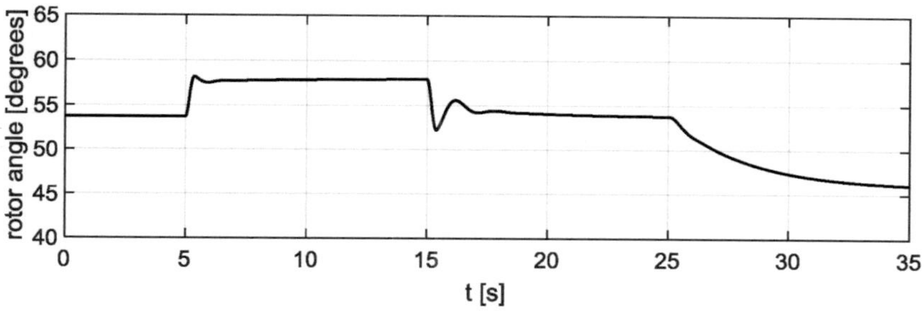

Figure 35.
Rotor angle $\delta(t)$ *[pu] at nominal operating point* $P = 1.0$ *[pu] and* $Q = 0.62$ *[pu], with adaptive PSS.*

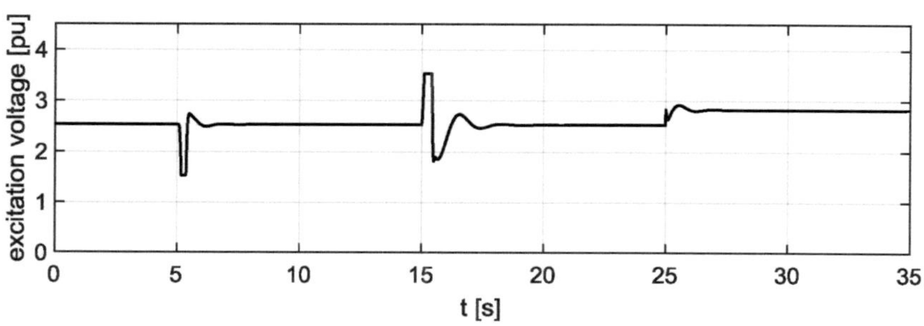

Figure 36.
Excitation voltage $E_{FD}(t)$ *[pu] at nominal operating point* $P = 1.0$ *[pu] and* $Q = 0.62$ *[pu], with adaptive PSS.*

7.2.2 Influence of load disturbance

The results of the two most extreme operating points are presented (the same as in Sections 3.2 and 7.1.2):

- $P = 1.0$ [pu] and $Q = 0.1$ [pu]: The generated electrical power and rotor speed are shown in **Figure 37**.

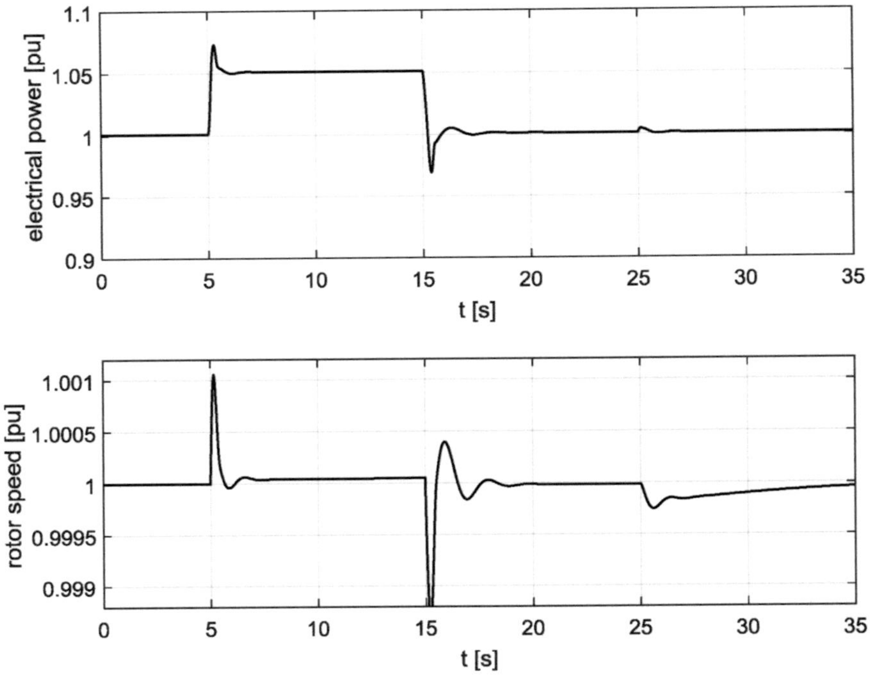

Figure 37.
Electrical power $P_e(t)$ *[pu] and rotor speed* $\omega(t)$ *[pu] at operating point* $P = 1.0$ *[pu] and* $Q = 0.1$ *[pu], with adaptive PSS.*

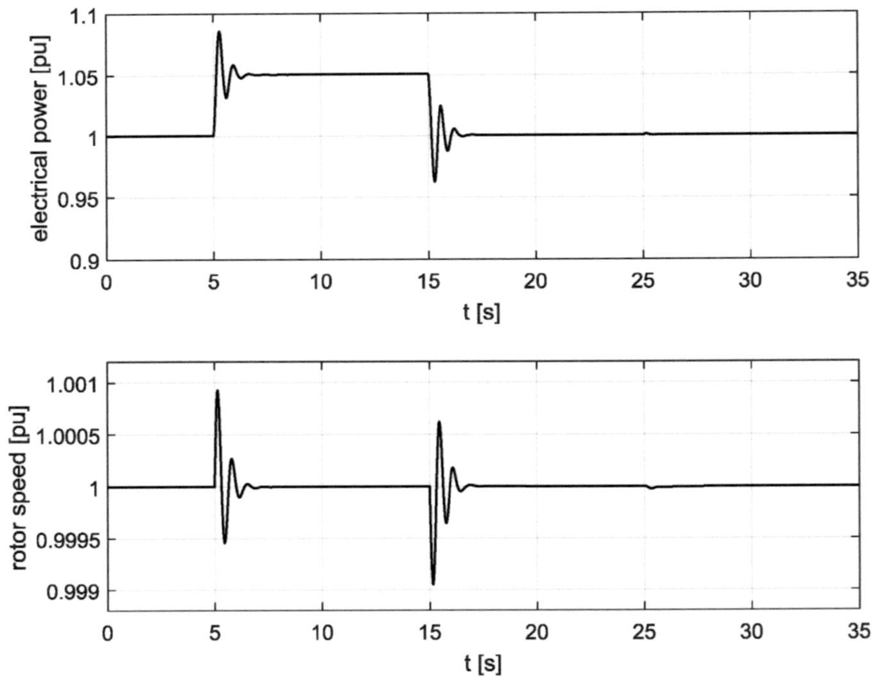

Figure 38.
Electrical power $P_e(t)$ *[pu] and rotor speed* $\omega(t)$ *[pu] at operating point* $P = 0.1$ *[pu] and* $Q = 1.0$ *[pu], with adaptive PSS.*

- P = 0.1 [pu] and Q = 1.0 [pu]: The generated electrical power and rotor speed are shown in **Figure 38**.

7.2.3 Influence of parameter deviations

The effectiveness of the adaptive PSS for oscillation damping in the presence of parameter deviations is shown with a test on the 555 MVA synchronous generator,

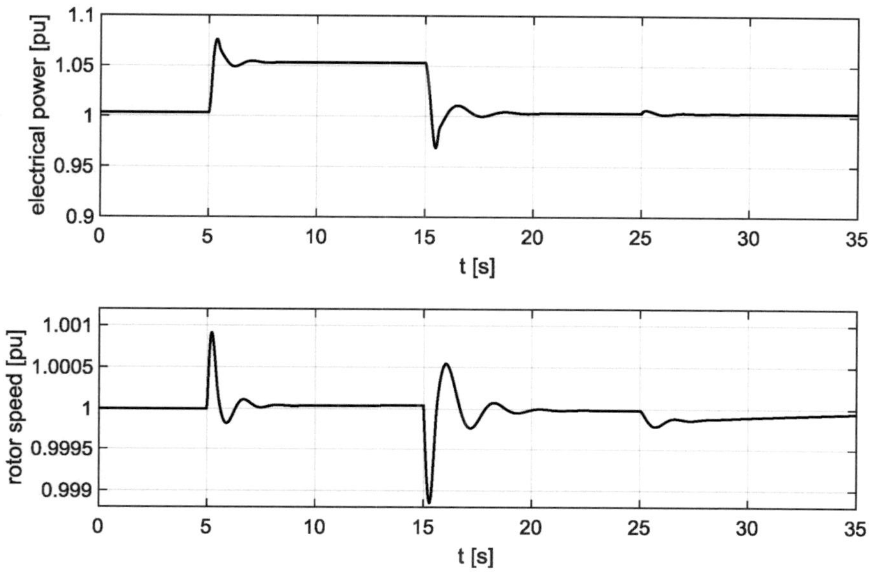

Figure 39.
The 555 MVA synchronous generator's electrical power $P_e(t)$ *[pu] and rotor speed* $\omega(t)$ *[pu] at operating point* P = *1.0 [pu] and* Q = *0.1 [pu], with adaptive PSS.*

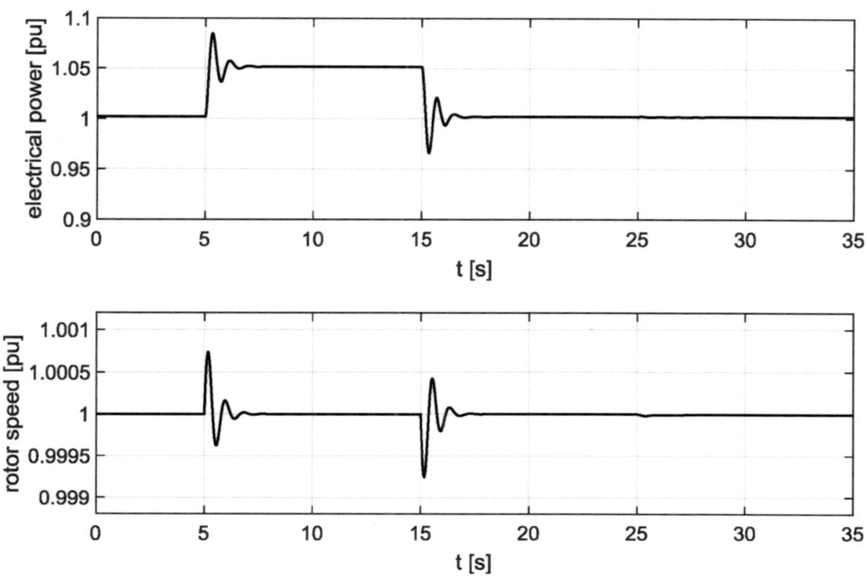

Figure 40.
The 555 MVA synchronous generator's electrical power $P_e(t)$ *[pu] and rotor speed* $\omega(t)$ *[pu] at operating point* P = *0.1 [pu] and* Q = *1.0 [pu], with adaptive PSS.*

as in Section 7.1.3. **Figure 39** shows the generated electrical power and rotor speed at operating point P = 1.0 [pu] and Q = 0.1 [pu], and **Figure 40** shows both quantities at operating point P = 0.1 [pu] and Q = 1.0 [pu].

7.3 Assessment of the presented power system stabilizers

From the results presented in Sections 7.1 and 7.2 presented in **Figures 22–32** and **34–40**, it is seen clearly that robust and adaptive PSS improve the damping of oscillations of the synchronous generators significantly (shorter time of oscillations, smaller overshoot, and better damping). The numerical assessment was done for better insight into the effectiveness of both control algorithms. The integral square root performance index of rotor speed oscillation ($\int \omega^2 dt$) was introduced for more objective numerical evaluation of the proposed control approaches. The time responses shown in **Figures 13, 16, 19, 22, 25, 26, 34, 37,** and **38** were considered

Synchronous generator with nominal power S_N = 160 [MVA]					
P [pu]	**Q [pu]**	**cos φ**	**Performance index**		
			Without PSS	**Robust PSS**	**Adaptive PSS**
1.0	0.1	0.995	$1.98\ 10^{-6}$	$0.96\ 10^{-6}$	$0.91\ 10^{-6}$
1.0	0.62	0.85	$0.99\ 10^{-6}$	$0.56\ 10^{-6}$	$0.44\ 10^{-6}$
0.1	1.0	0.099	$0.56\ 10^{-6}$	$0.41\ 10^{-6}$	$0.42\ 10^{-6}$

Table 14.
Integral square root performance index of rotor speed oscillation of the 160 MVA synchronous generator without PSS, with robust PSS, or with adaptive PSS at different operating points.

Synchronous generator with nominal power S_N = 160 [MVA]				
P [pu]	**Q [pu]**	**cos φ**	**Improvement of performance index regarding the case without PSS [%]**	
			Robust PSS	**Adaptive PSS**
1.0	0.1	0.995	51 [%]	54 [%]
1.0	0.62	0.85	43 [%]	55 [%]
0.1	1.0	0.099	26 [%]	25 [%]

Table 15.
The improvements of the integral square root performance index of rotor speed oscillation of the 160 MVA synchronous generator at different operating points following the use of robust or adaptive PSS.

Synchronous generator with nominal power S_N = 555 [MVA]					
P [pu]	**Q [pu]**	**cos φ**	**Performance index**		
			Without PSS	**Robust PSS**	**Adaptive PSS**
1.0	0.1	0.995	$1.73\ 10^{-6}$	$0.76\ 10^{-6}$	$0.92\ 10^{-6}$
1.0	0.62	0.85	$0.89\ 10^{-6}$	$0.51\ 10^{-6}$	$0.47\ 10^{-6}$
0.1	1.0	0.099	$0.33\ 10^{-6}$	$0.28\ 10^{-6}$	$0.28\ 10^{-6}$

Table 16.
Integral square root performance index of rotor speed oscillation of the 555 MVA synchronous generator without PSS, with robust PSS, or with adaptive PSS at different operating points.

Synchronous generator with nominal power S_N = 555 [MVA]				
P [pu]	*Q* [pu]	cos *φ*	Improvement of performance index regarding the case without PSS [%]	
			Robust PSS	Adaptive PSS
1.0	0.1	0.995	56 [%]	47 [%]
1.0	0.62	0.85	43 [%]	47 [%]
0.1	1.0	0.099	15 [%]	15 [%]

Table 17.
The improvements of the integral square root performance index of rotor speed oscillation of the 555 MVA synchronous generator at different operating points following the use of robust or adaptive PSS.

for a synchronous generator with nominal power 160 MVA. The calculated performance indexes are shown in **Table 14**. The improvements of the performance index by means of proposed control systems regarding the noncontrolled synchronous generator are presented in **Table 15**.

The time responses shown in **Figures 28, 30, 32, 39**, and **40** were considered for a synchronous generator with nominal power 555 MVA. The calculated performance indexes of the synchronous generator without PSS, with robust PSS, or with adaptive PSS are shown in **Table 16**. The improvements of the performance index by means of proposed control systems regarding the noncontrolled synchronous generator are presented in **Table 17**.

From the obtained numerical results, it is obvious that the proposed robust and adaptive PSS assure significant damping improvement in the entire operating range. The effectiveness of both proposed stabilizers are similar. They depend on the type of generator, largeness of the generator, operating point (loading), etc. In general, we can conclude that, according to the introduced performance index, their improvement is in the range of 10–60 [%].

8. Conclusion

Changes in construction of synchronous generators and the introduction of additional control systems into power systems have led to significant increase of oscillations in power systems and related stability problems. Conventional linear power system stabilizers are not able to solve these problems. Advanced control theories seem appropriate to design more powerful power system stabilizers.

Two power system stabilizers were developed based on robust control theory and adaptive control theory. The effectiveness of both stabilizers was evaluated as objectively as possible. The proposed control approaches were evaluated on a basis of a theoretical analysis and numerical simulations. The sliding mode stabilizer and direct adaptive stabilizer have the following advantages in comparison to conventional linear stabilizers:

- The proofs of the stability of the entire closed-loop system exist for both controllers presented.
- Both controllers require minimal preknowledge of the controlled plant structure and parameters.
- Both controllers have an uncomplicated tuning procedure.

- Both controllers do not require the measurement of the additional control plant variables.
- Both controllers are easy to implement (low computing demand and low sampling frequency).
- Both controllers damp oscillations in the entire operating range and assure stability.
- Both controllers do not calculate high output amplitudes and are insensitive to the actuator's saturations.

The comparison of the developed novel power system stabilizers shows the significant advantage of the modern concepts in all the considered ranges of the operation. Due to the actuality and importance of the issues tackled, the development of more effective power system stabilizers is inevitable. It is our estimation that the intensity of research in this field will increase in the future. In our evaluation, the robust and adaptive controls emerged as the most prospective concepts for implementation in power system stabilizers.

Author details

Jožef Ritonja
Faculty of Electrical Engineering and Computer Science, University of Maribor, Maribor, Slovenia

*Address all correspondence to: jozef.ritonja@um.si

References

[1] International Energy [Internet]. 2020. Available from: https://www.iea.org/data-and-statistics?country=WORLD&fuel=Energy%20supply&indicator =Electricity%20generation%20by%20source [Accessed: January 19, 2020]

[2] Ritonja J. Adaptive stabilization for generator excitation system. COMPEL. 2011;**30**(3):1092-1108. DOI: 10.1108/03321641111111022

[3] Boldea I. Synchronous Generators. Boca Raton: Taylor & Francis Group; 2016

[4] Anderson PM, Fouad AA. Power System Control and Stability. Ames, Iowa: The Iowa State University Press; 1977

[5] Ritonja J, Petrun M, Černelič J, Brezovnik R, Polajžer B. Analysis and applicability of Heffron-Phillips model. Elektronika ir Elektrotechnika [Print ed.]. 2016;**22**(4):3-10. DOI: 10.5755/j01.eie.22.4.15905. Ilustr. ISSN: 1392-1215

[6] Kundur P. Power System Stability and Control. New York: McGraw-Hill, Inc.; 1994

[7] Heffron WG, Phillips RA. Effect of a modern amplidyne voltage regulator on underexcited operation of large turbine generators. AIEE Transactions. 1952;**71**: 692-697

[8] Bergen AR, Vittal V. Power Systems Analysis. Upper Saddle River, New Jersey: Prentice-Hall Inc.; 2000

[9] Demello FP, Concordia C. Concepts of synchronous machine stability as affected by excitation control. IEEE Tran. Power Appar. Syst. 1969;**88**(4):316-329

[10] Machowsky J, Bialek JW, Bumby JR. Power System Dynamics, Stability and Control. West Sussex, United Kingdom: John Wiley and Sons, Ltd.; 2008

[11] IEEE Std. IEEE recommended practice for excitation system models for power system stability studies. In: IEEE Std 421.5–2005, IEEE Power Engineering Society by Energy Development and Power Generation Committee; 21 April 2006

[12] Ritonja J, Dušak M. Analysis of the Slovenian Power System Stabilizers. Journal of Electrical Engineering and Computer Science. Ljubljana: Electrotechnical Society of Slovenia; 2014

[13] Utkin VI. Sliding mode control design principles and application to electric drives. IEEE Transactions on Industrial Electronics. 1993;**40**(1):23-36

[14] Bartolini G, Fridman L, Pisano A, Usai E. Modern Sliding Mode Control Theory. New York: Springer Verlag; 2008

[15] Slotine JJE, Li W. Applied Nonlinear Control. Englewood Cliffs, New Jersey: Prentice Hall Inc.; 1991

[16] Landau YD. Adaptive Control. New York: Marcel Dekker Inc.; 1979

[17] Eichmann A, Kohler A, Malik O P, Taborda J. A prototype self-tuning adaptive power system stabilizer for damping of active power swings. In: Power Engineering Society Proceedings of the Summer Meeting in Seattle, WA, Vol. 1. IEEE; 2000. pp. 122-126

[18] Ritonja J, Dolinar D, Grčar B. Simple adaptive control for a power-system stabiliser. In: IEEE Proceedings Control Theory and Applications [Print ed.]; 2000. pp. 373-380

[19] Kaufman H, Bar-Khana I, Sobel K. Direct Adaptive Control Algorithms. New York: Springer Verlag; 1993

[20] Šabanovic A. Variable structure systems with sliding modes in motion control—A survey. IEEE Transactions on Industrial Informatics. 2011;**7**(2): 212-223

Chapter 11

Automatic Control of the Structure of Dynamic Objects in High-Voltage Power Smart-Grid

Andrey Petrovich Nikiforov

Abstract

The control and protection algorithms for the considered class of dynamic facilities with a tunable structure are considered. Their relevance follows from the concepts of the development of the electric power industry—*smart grid*, *digital substation* and *outsourcing services*. The properties of this class of dynamic objects are spatial distribution, many options for changing the structure and motion in vibrational circuits at natural frequencies from ultra-high to ultra-low. The input coordinate of the dynamic objects is the vector of the change in the structure of the object. The output coordinate is the power of a special semantic signal about the state of the facilities. Their management is carried out as part of the stabilisation system of the normal operation of the facility. Stabilisation is achieved by the criterion of the minimum deviation of the power of the semantic signal from the 'normal mode' setting of operation. Executive bodies rebuild the structure of the dynamic facility in a pulsed, programmatic way, using the possibilities of self-healing and resource reservation. If stabilisation is not possible, the damaged area is excluded from the facility. The problem of the stability of the system turned out to be the lack of sufficient information about the state of the object and the similarity of the structure and significance of unrecognisable semantic situations to the main situations. Control algorithms are synthesised by the developed structural-informational (SI) method of dynamic pattern recognition.

Keywords: relay protection, automation, Petersen's coil, smart grid, protection against single-phase earth faults, structural-informational method, modelling

1. Introduction

Generalised and developed solutions to the problems of synthesis and ensuring the stability of the operation of automatic control and protection algorithms for a class of objects with a tunable internal structure. At the present stage of development, it is necessary to operate with information flows at a higher level of abstraction than traditional methods of building networks. The concept of information flow refers to a set of demodulated signal parameters from transient signals (transients) in an object. Connected network elements are considered as dynamic facilities (objects) of the control and protection (OCP). Within the framework of the smart grid concept, all the necessary algorithms for relay protection and automation (RPA) devices are in the foreground, and the equipment that controls the operation of the OCP (**Figure 1**) is at the second level. It implies the possibility of describing the properties of OCP electrical circuits through control algorithms [1–9]. Such a structural-informational (SI) method is developed by the author further [8–15]. The methods of applying the available results of the SI method to describe the internal structure of high-voltage equipment of the OCR network are shown.

In the traditional construction of networks, a great deal of experience has been gained in describing and calculating OCP [6–7]. Usually it is enough to indicate the established terminology and mathematical description of the OCP in the technical literature [15–18]. A method is necessary for describing the transient process in the OCP from the place of formation to the resulting text messages at a more abstract level, without loss of information and with the possibility of engaging the achievements of the traditional description. This allows a compact description of the problem and a focus on control algorithms [9–11].

The task was posed in a series of published works [7–11], based on the results achieved by improving recognition algorithms in devices for protection against single-phase earth faults in networks with a Petersen's coil. Studies show [8–13] that, for example, to maintain the stability of the operation of RPA device algorithms in distribution networks with a voltage of 6–35 kV, the amount of input information is insufficient [6, 8, 13, 16]. Faults are often and continuously present in externally operable high-voltage distribution networks, and information about the occurrence of such a situation cannot be distinguished during visual observation by the means currently available at substations [11, 13, 15]. Identification of the causes of situations is difficult and is associated with the lack of reliable information about the actual state of the network sections among consumers, configuration, state of high-voltage equipment, deviations in the technological processes of network load operation, etc. [15–18].

Replenishment of the lack of information is drawn from the capabilities of intelligent search algorithms for pattern recognition, which are able to extract semantic states from transient signals occurring in networks [5, 8–16].

Further, when developing the SI method, the patterns of structural diagrams, which are stages in solving the problems of synthesis of automatic control and protection algorithms, are summarised in a general sequence [5, 8–16], namely, from information structures describing transients in a network, through a formal presentation of information in high-voltage network equipment, then dynamically changing network operating modes, to the synthesis of the necessary algorithms for structural processing of information in automatic devices.

The tasks of numerical assessment of the sufficiency of the signal information volume for automatic decision-making on shutting down a damaged part

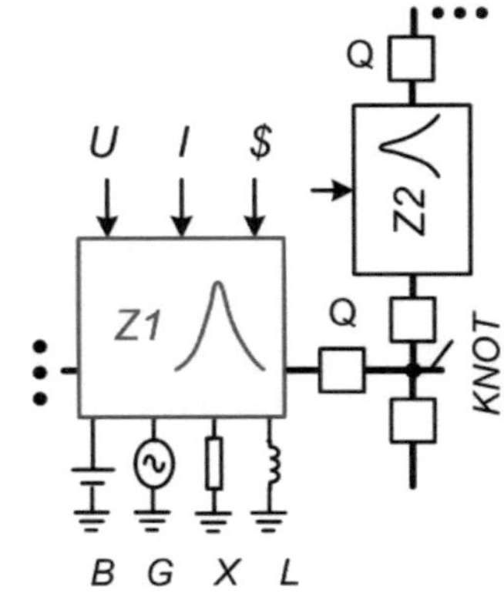

Figure 1.
Unified circuit of smart grid Driver–EU–OCP.

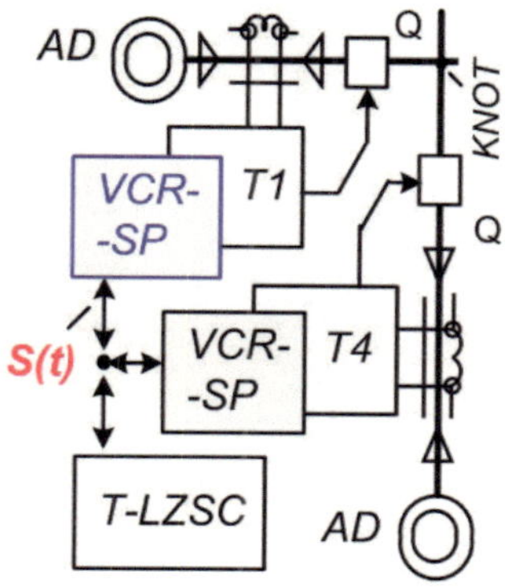

Figure 2.
Diagram of a smart grid that is unique to a local measure and a signal S(t). *T—RPA terminal.*

of an object are analysed (**Figure 2**). Algorithms for controlling the amount of information are implemented in the VCR, with built-in general purpose software information sensors and an arbitration function of RPA algorithms. The device is located in each OCP cell [7–9, 15–16]. A method is proposed for obtaining the necessary additional information based on the sequence of steps of the SI method, namely, structured separation of information, control of the passage of elementary information components according to the structural diagrams of the control and OCP protection object, then through relay protection and automation devices RPA, and to the output of the SCADA control system [5, 11].

2. Separation of dynamic objects with variable structure into a separate class

Dynamic OCP management and control objects are those in which their internal information state develops sequentially in time, a new state replacing the previous one. We will call the elementary state the semantic situation *SN*, for example, *S1* 'normal operation' or 'NM'. One *SN* gives rise to another or many possible known developments. Such transitions to a fixed *SN* are similar to the operation of automaton models. A sequential change in time of the parameters, structure, dynamic properties and individual elements of the OCP is possible. The removal of sections from the OCP structure for various reasons or combining them into a common structure during the operation of automatic control devices is typical.

The OCP is characterised primarily by a change in time of the internal structure. Parametric changes are the result of such a change. For example, the task of adjusting the coordinate parameter *L* to changing the coordinate parameter *C* is successfully solved (**Figure 3**). Changes in parameter *C* result in changes of different types—(A) a change in the structure of the OCP by disconnecting the *QN* switches of the separated sections of the OCP; (B) there is a test change of the parameter *L* during operational work in the 'NM' mode, while the structure of the OCP does not change; and (C) the damaged section is disconnected by the *QN* switch by selective relay protection. In some modes of structural change, when damage occurs, blocking changes in the parametric coordinate *L*.

An analysis is made of the methods for describing well-known typical schemes of RPA systems to justify the possibility of operating with such a concept. So, on well-known typical schemes, a single-line OCP scheme, EU executive bodies,

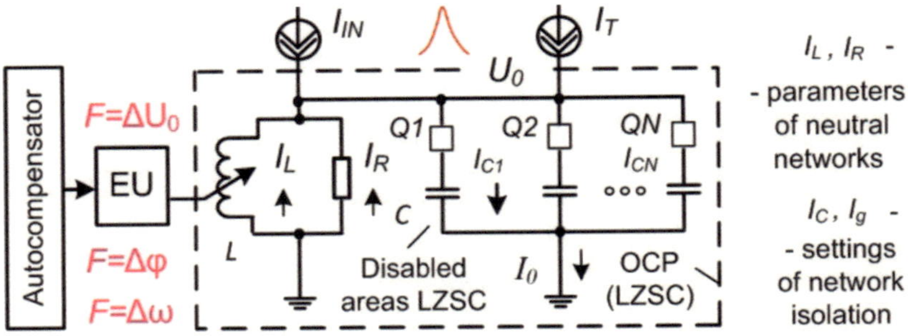

Figure 3.
Structural scheme OCP (LZSC) with a Petersen's coil L.

measuring transformers and the measuring part of the relay are indicated (**Figure 3**). The relay output was initially replaced with a multiple control result—from immediate shutdown to alarm.

In RPA tasks, the perturbation control principle is common. It is known that this principle is characterised by the presence of fatal errors and is relevant in the event of catastrophic failures in the OCP equipment. But most of the insulation damage to the network (75–98%), according to well-known statistics, can self-destruct. To do this, RPA algorithms must steadily determine such a transition process. Case studies have found that, in such *SN* situations, robustness of the RPA systems was not achieved.

3. Description of the SI method for working with dynamic OCP

The SI method is universal enough to describe various tasks and implement modern concepts of technology development [6–13]. It consists of a description of the sequence of patterns as steps to solve the problem. The steps are due to the dynamic development of processes, control algorithms and moments of issuing control commands. Patterns are generalised equivalent structure (GES) schemes. They allow us to uniformly describe known and new devices within the framework of network development concepts (**Figure 4**). The GES scheme consists of automata A—morphological MorphA, syntactic (SyntA) and semantic (SemA). The SyntA machine controls the correct sequence of primary terminal symbols (TS) in the TS chains in the input information. SyntA states are nonterminal symbols (NTS) that appear when TS chains are known to SyntA recognition engines. The basis of GES circuits is the sensing S-detector introduced. It detects the semantic signal *S(t)*, which is formed sequentially in the GES schemes of the object, recognition and control algorithms and is used at all hierarchical levels of information processing.

The quality of processing dynamic information flows in RPA systems in energy networks is controlled by the semantic signal *S(t)* (**Figure 4**). Such a signal is formed at each hierarchical level of processing the total amount of information in the system—morphological, syntactic and semantic. The amount of information about the state of the OCP is allocated and controlled by RPA devices based on demodulation of the information components in transient signals (**Figure 5**).

The relationship of TS and NTS in the GES scheme is described by the rules *P*. Rules *P* are divided into groups of selectivity *PS* and blocking *PB*. Rules *P* are assigned weights *KSN* and *KBM*, according to the contribution to the overall semantic output

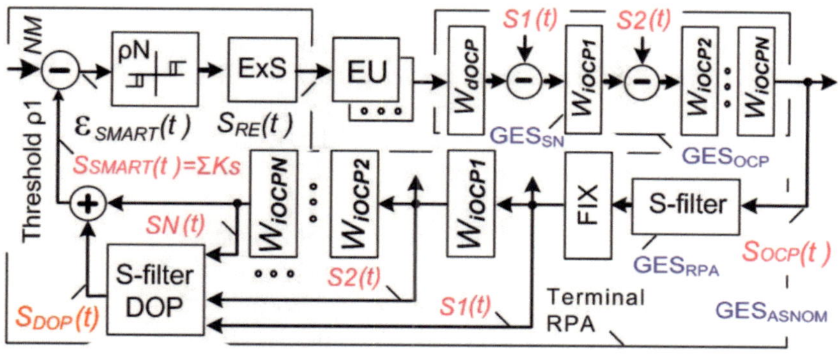

Figure 4.
Diagram of the GES_{ASNOM} of system stabilisation ASNOM based on $S_{SMART}(t)$.

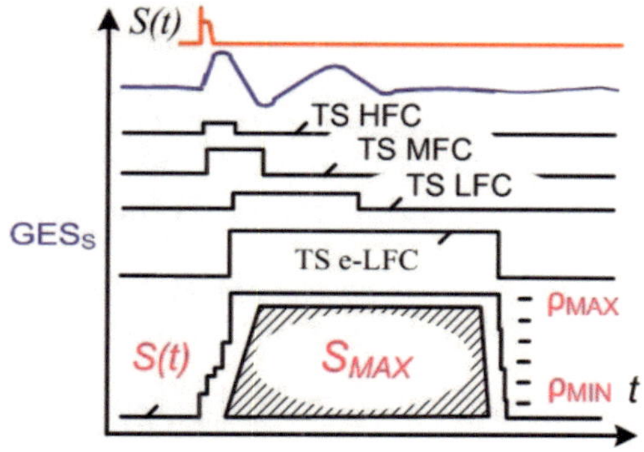

Figure 5.
Formation of a signal with an S(t) *signal—maximum amount of information.*

$S_{SMART}(t)$. So, the information part of the S-filter is highlighted. The SemA output generates a semantic signal $S_{SMART}(t) = For(t)–Against(t)$ (1). The $S_{SMART}(t)$ signal reflects changes in semantic information in the OCP and in the system of automatic stabilisation of the normal operating mode (ASNOM) of the network (**Figure 4**).

$$S_{SMART}(t) = \mathrm{FIX}[\beta \times K_S SB \times (S_S S(t) - S_S B(t))] = \mathrm{FIX}[\beta \times K_S SB \times f(\Sigma KSN(t) - \Sigma KBM(t))], \quad (1)$$

where $N = 1, 2, \ldots$; $M = 1, 2, \ldots$; Σ is the sum of all weight coefficients *KSN* or *KBM*, which establish the significance of the rules *PS* and *PB* of the SyntA automaton and the weight coefficient $K_S SB$ of the resulting root rule $P_S SB$ of the SemA automaton; and β is the general scaling factor. The FIX function describes the operation of the fixation unit, in which, during the development of transients, the activated rules *PS* and *PB* are stored for a while to accumulate the $S_{SMART}(t)$ value. This signal can vary between '0%' and '100%' (**Figure 6**). The FIX function is similar to the operation of the emergency recorder and acts when the 'activating' TS appears, which is set in the settings of the RPA device. Such a TS is commonly known for OCP.

The GES scheme is mathematically described by a list—grammar *G* as shown in Eq. (2).

$$G_O \rightarrow (TSSN, TSBM, NTSSN, NTSBM, PSN, PBM, P_S S, P_S B, P_S SB), \quad (2)$$

where *O* are objects, for example, OCP, GES, TS, RPA, ASNOM, SCADA, *PS* and *PB*—the rules for connecting TS and secondary NTS—and $P_S SB$ is the resulting root rule of the GES scheme. The executive bodies EU of the ASNOM system (**Figure 4**) are the outputs of the RPA algorithms, combined by the block of the ExS expert system. The final semantic conclusions are formed in ExS, and decisions are made on the automatic removal of deviations from ρ1 'NM' of the normal mode (NM) of OCP operation.

The OCP is presented on a single information field of elementary information components of TS. From these TS, an *SN* formation tree, a TS terminal symbol tree describing the *SN* structure and signal control $S_{OCP}(t)$ are built.

The SI method allows you to describe *SN* semantic situations by passing them through the GES_{OCP} information block diagram. This is controlled by the sense signal $S(t)$. Its area (otherwise, power in the interval of control duration) can be considered the meaning of events in the OCP circuit (**Figure 5**). By the amount of information that is formed in the structural diagram TS, we mean the area $S(t)$. The authors of [6–16] studied the construction of a tree that recognises systems in RPA devices based on the generation and control of the $S_{RPA}(t)$ signal. The $S_{RPA}(t)$ signal in recognition algorithms controls the course of processes in the OCP in meaning. General control of all OCP loops is performed similarly within the framework of the ASNOM and SCADA systems with the corresponding G_{ASNOM}, G_{SCADA} grammars and S_{ASNOM} and S_{SCADA} signals (**Figure 4**).

Thus, the structure of the OCP can be described by a set of GES_{RPA} schemes, i.e. the sum of Σ components and their G_O grammars, namely,

$$G_{OCP} \approx \Sigma G_{GES} = \Sigma G_{TS} = \Sigma G_{RPA} = \Sigma G_{ASNOM} = G_{SCADA}, \Sigma\text{—the sum of the components.} \quad (3)$$

The OCP objects are managed within the ASNOM system (**Figure 4**). Stabilisation is achieved by the criterion of the minimum deviation of the power of the semantic signal $S(t)$ from the setpoint of the 'normal mode' system. The class of objects is controlled by issuing commands to the RPA device. The formation of commands is controlled on the time axis of the emergency file as a reaction of the recognition algorithm to the input signals from the OCP.

Figure 6 shows the GES_{COMAND} diagram as a command generation template in the ASNOM system. Teams manage not only the settings but primarily the structure of the OCP. Therefore, the ASNOM system differs from adaptive systems, which are characterised by a change in the parameters or structure of the controller depending on changes in the parameters of the object or external disturbances.

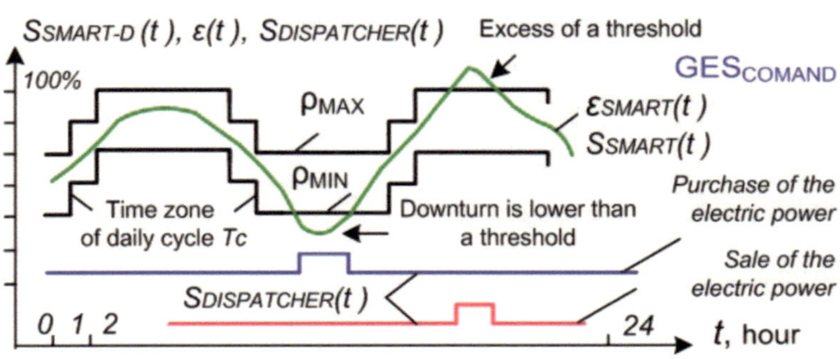

Figure 6.
GES_{COMAND} scheme. Illustration of change of a signal S(t) *and thresholds ρN.*

The operation of technological and other processes in the network is considered further along the OCP–SCADA chain, where $G_{SCADA} = \Sigma G_{ASNOM}$ (**Figure 4**).

4. Research problem description

Identifying the site or location of damage in the OCP from the indicated types of damage is a difficult task to put into practice (**Figures 2** and **3**).

Selective relays from the main phase to ground faults are separately available in large numbers. Relays are widespread but are characterised by instability. By device stability, we mean the quality of recognition of *SN* semantic situations in transient signals in the OCP, which is necessary for the correct operation of devices in the ASNOM system. To ensure the stability of the relay protection and automation devices, it turns out that in some cases the traditional choice of the settings for the operation of these devices is not enough. The same type of device with the same algorithm is installed in different networks. In each network, high-voltage equipment has its own specifics and operating history, exhaustion factors, etc. Therefore, it is possible to have complex interfering extraneous signals at the installation sites of devices that are present for long periods of time and for which the use of devices is not designed.

The flow of information develops sequentially in the object and can be interrupted and resumed for various reasons. This property of the object during operation can lead the ASNOM stabilisation system to instability.

Studies show that recognising RPA algorithms is often built on the analysis of only one or several *SNs* [16–18]. In the remaining *SN*, the algorithms are blocked during the development of damage to the OCP and do not recognise the change of *SN* and *SN* scenarios. In this case, the available information is not used. However, the change of *SN* and *SN* scenarios allows one to highlight this additional information and use the additional rules of the $_{DOP}$ in recognition algorithms (**Figure 4**).

The reasons for the instability of the work are insufficient information to make a decision on the management of OCP, the similarity in structure and content of the main situations of *SN* with interference and unrecognisable *SNs*. It was revealed that volume restrictions may be a consequence of: (1) minimisation and shortcomings of synthesised recognition algorithms; (2) the influence of various conditions of specific sections of the network when introducing devices with the same methods; and (3) the lack of redundancy in the reception and processing of signals. It is known that redundancy of information allows you to work with distortions, extraneous signals, interference and more. This requirement may not be implemented, despite the fact that the amount of information should correspond to the complexity of the task. We will consider ways to address these causes of violations of the recognition of dynamically changing information.

If, when modelling in CAD on a joint model of information sensors TS and an object, it turns out that the signal $S(t)$ does not track the gradation of changes in the transient in the OCP and changes roughly, without distinguishing close *SN* situations, this is a sign of insufficient information for recognition. For the success of solving the recognition problem, it is necessary to provide a sufficient amount of information. If it turns out that it is necessary to use the internal coordinate of the object, then it is necessary to ensure the formation of such information. To do this, an additional TS sensor is needed for direct or indirect information acquisition.

Studies show that a residual amount of information is present in the situation recognition tree [16–18]. Information is captured by the FIX unit and evaluated by other algorithms. In the case of continued development of the processes in the OCP, algorithms for additional recognition of information in the DOP block are involved (**Figure 4**). Additional algorithms are also used to clarify the recognition and completion of the amount of information as well as for self-monitoring and partial diagnostics of the object, evaluating its performance, equipment and algorithms that make up the ExS expert system.

The purpose of further study is to obtain an additional amount of information from the OCP for its use in the GES_{RPA} recognition tree to increase the stability of RPA algorithms. To do this, the passage of the elementary component of interest is tracked along the template structures, which are the GES_{OCP}, GES_{SN}, GES_{ASNOM} schemes, a separate *SN* situation, a scenario of *SN* changing in time and a tree of possible *SN* developments. It can also be considered from the place of origin to the exit to the OCP through the coordinates of the OCP to the input of the recognition system and then according to the patterns of the GES schemes to the control system output of interest (**Figure 4**).

5. The development of well-known methods of working with dynamic OCP

There are several types of description of object models OCP:

A. Only by electrical parameters. The description of the models is based on the coordinates of currents, voltages, conductivities, etc. It is typical for him to study the parameters of movements in time using mathematical dependencies and identification by *RLC* elements [7], performed on the basis of functional modelling (integration of differential equations). Each test impact leads to a transient in time. A general solution is sought as attractors and hodographs of unit calculations. But these methods do not eliminate the fragmentation of the obtained results, the loss of compatibility for high-frequency components (HFC) and super-slow information (s-LFC) and informal empirical data preparation. Therefore, most RPA devices have been developed by heuristic, expert methods.

B. A structural-operator (SO) method. It establishes a connection between electrical and information parameters [7]. It is used to describe network circuits as a control object and control algorithms for the network zero sequence circuit. However, when solving related problems, namely, RPA, it turned out that the means of this method are not enough.

C. A structural-informational (SI) method that works only with informational semantic components. The establishment of interconnections between elementary information components (terminal symbols, TS) is considered, which can be detected from electrical parameters. The movements in dynamics are considered, similarly to the automaton model, as transitions between states. Modelling leads to visual methods for displaying the results in the form of attractors and hodographs. Equilibrium points of solutions can be found by the game theory method. This allows you to work with electric power OCP on a more abstract level. The use of the SI method of pattern recognition removes such restrictions on the processing of only information without the participation of parametric components.

Studies show that the first two methods for describing OCP models are limited to obtaining a parametric OCP control system.

Fuzzy logic and neural network methods are well known. Common to these methods are the following actions: the decomposition of information into elementary components; multiplication by weights; summation of the result; and comparison of the result with many threshold values. However, the extent to which these methods are used is limited by the specifics of the tasks they are to solve, for example, the dynamic development of the transition process in an OCP.

According to the SI method, the OCP structure is represented by transient signals [11–13]. The description is based on the fact that the total amount of information is supplied by all of the oscillatory circuits. These circuits constitute the OCP (**Figure 7**). Such an OCP model allows one to operate only on semantic components while improving RPA device algorithms. The OCP also appears to be a list of *SN* semantic situations.

An *SN* situation is understood as a dynamic change in the output of the OCP block diagram as a reaction to the appearance of changes at any point of the OCP (input or internal) [8–9]. That is, the *SN* modulates the industrial frequency signals in the OCP. Demodulation of the signals makes it possible to recognise the *SN* situation among the well-known and rather limited number of *SN*. Recognition is performed by automatic machines (A). Cases of a complex, simple *SN* situation and dynamic *SN* change [9–11] are taken into account. The signals at the output of the OCP are characterised by the consistent development in time of information components. These components are distinguished by information sensors [8]. The sensor outputs are terminal symbols and are grouped in TS chains. The totality of a TS is a morphological automaton (MorphA).

6. Description of the OCP by frequency components

According to the SI-method, all available information about the OCP is monitored. First of all, the OCP is replaced by oscillatory circuits with the corresponding frequency components (FC), that is, from the HFC to the super-LFC (**Figure 7**). Low-quality contours, in which only one half-wave of oscillation develops, are also vibrational. It is known that in RPA and SCADA devices, input information is sampled by time in the ADC block. The sampling frequency of the ADC is selected based on the presence of the highest-frequency component in the OCP. So, the ADC block will fix in the input signal all the components with a lower natural frequency (i.e. all the LFCs). So with the well-known operator description of the OCP, $y(t) = W(D) \cdot x(t)$. Constants in *W* with indices n and m describe the highest frequencies with which movements in the OCP are possible. We assume further that for the ADC, these constants a_N, b_M will be the initial a_0, b_0. Then all other frequencies in the new *W* will be slower; therefore a_N, b_M will be the slowest s-LFCs. This

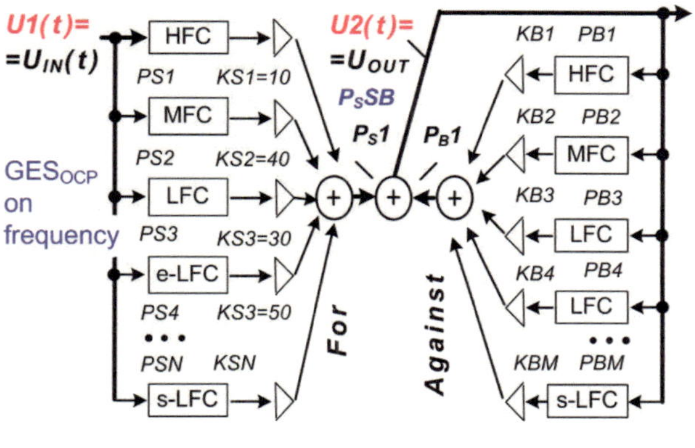

Figure 7.
Equivalent structural scheme of GES_{OCP} with division of movements by frequencies.

allows us to provide a description of the OCP transfer function *W* for the highest-frequency component, which can be recorded by the ADC. You can control all the slow oscillatory circuits of the OCP. Very low-frequency motions may be present in the OCP. The lowest-frequency components and movements in the OCP can be considered informational (**Figure 7**).

This record allows you to highlight the presence of super-LFC in the OCP using the signals of emergency files of RPA devices. Such frequency components are generated in the OCP (**Figure 7**) by input shock effects of overvoltage, short circuit, OPG, operational switching, etc. Motions in super-LFC circuits of the OCP can be in different time ranges—second, monthly, annual and so on, for example, envelope industrial frequency response (e-LFC). e-LFC mathematically summarises signal circuits in OCP. s-LFC is only an information loop.

The OCP structure is represented by the GES_{OCP} scheme in the common field of TS information sensors that recognise *SN* situations. Each *SN* is divided into elementary components—nonterminal components of NTS. Information components are controlled by a logical change in TS outputs. For example, the amplitude parameters of the signals of the OCP loops are monitored and represented by the corresponding TS. Each OCP oscillation circuit is controlled by RPA devices. Devices at a morphological hierarchical level will be further represented by a GES_{TS} scheme with a G_{TS} grammar (**Figure 9**).

Extending the discussion about the presence of information loops to the features of higher-frequency loops, we can separate the signal description of processes in the OCP from the information description of the essence of the processes. This implies the task of searching for vibrational components in the ranges of the LFCs of super-LFCs by means of analysis in CAD of the frequency components of the alarm file signals. Solving the problem of separating parametric and information loops will help fill the lack of information when recognising *SN* situations in OCP.

7. Description of OCP equipment by information components

It is proposed to implement the stable working of recognition devices based on an algorithm for the selective search (SP) for a sufficient amount of information to perform RPA functions. The amount of information is accumulated based on the control of a number of information components, for example, the type of damage, the steadiness of the development of damage, the location of the damage, the presence of selective and blocking signs for damaged and undamaged areas, the absence of extraneous semantic situations, etc.

Further, attention is paid to the formation of OCP grammar for the tasks of improving the algorithms of RPA devices. It is necessary to develop consistent structural trees of OCP and RPA with respect to the semantic signal $S(t)$ as well as a method for the end-to-end mathematical description of transients in the 'network equipment-RPA devices' hierarchical chain in the automatic control system (**Figure 4**). This is part of the developed SI method.

The SI method is based on the application of sequential graphic transformations from a circuit diagram based on electrical parameters and the transition to a circuit based on information components. The structural-operator method is involved for the formal description of the OCP by parametric components [7]. Then the transition to the description of the SI method is carried out only by information components. The transition is performed by introducing terminal symbol sensors at the control points of the OCP scheme (**Figure 9**). In the internal OCP scheme, these can be imaginary points, since most of the structure of the OCP is not divided into control parts.

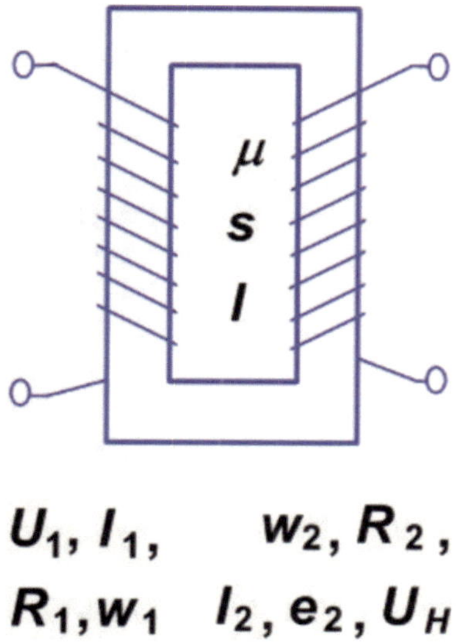

Figure 8.
Hardware example diagram.

For example, **Figure 8** shows the SO description in the form of electrical parameters of one of the elements of the OCP—transformer. Information flows are generated by inertial fields and are described by the operators $D = d/dt$, $1/D = S$. The input to the circuit is the coordinate of the perturbation signal $SN(t)$ for OCP. According to the SI method, at each control point of the circuit, the output of the elementary transfer function must be monitored by the TS sensor with the corresponding weight coefficient. TS consists of a GES_{TS} scheme. It includes a filter and a threshold element ρN. The TS acts as an observer for the operation of the structural-operator diagram of the OCP. Naturally, this is an imaginary and often unrealisable scheme in connection practice. But the implementation of such a solution is performed in CAD. This makes it possible to implement a theoretical description of the OCP scheme. In OCP, all TSs are combined into a more general S-filter, which generates the signal $S_{OCP}(t)$ and is controlled by the threshold ρN. Thus, a rigorous mathematical description of the OCP is feasible.

This allows you to build a mathematical model of OCP in CAD for the selective search algorithm SP to reach the maximum amount of information. Functional modelling in CAD will provide an additional amount. The passage of the informational components of the *SN* semantic situation through the structural-informational models of electrically connected equipment (**Figure 8**) of the OCP using the signals of real emergency files supplied by modern devices is simulated. If theoretically and practically received amount of information from real OCP is sufficient for recognition, then the problem is solved by practically recognisable algorithms. OCP is observed and controlled by recognition algorithms. If the theoretical amount of information is not sufficient for distinguishing *SN* and making decisions in particular cases or in the general case, then indirect methods of detecting errors in management are necessary.

8. Representation of OCP by semantic components

GES_{OCP} scheme description. The structural tree of $SN \rightarrow$TS formation for the OCP uses the internal coordinates and OCP contours not observed by the GES_{RPA} device circuits (**Figure 9**). The construction of the GES_{OCP} scheme proceeds from three foundations:

a. *Structural-operator description.* We will compile descriptions of the OCP and high-voltage equipment based on **Figure 9**. We will divide the description elements into alternative streams 'For' and 'Against' in relation to the formation of the OCP output.

b. *Separation of movements by frequencies.* We transform the structural-operator description of the OCP to the canonical form of a digital filter with an infinite impulse response (IIR) according to the rules of structural transformations (**Figure 10**). This possibility follows from the theory of filtration, according to which any inertial model can be described by an IIR or FIR filter. We will focus on the 'central' elements of the GES scheme. The 'Cons' group includes flows initiated by the U_{OUT} output and arriving with a minus sign at the central internal adder. The elements of the 'Cons' group belong to the *Filter* block on the basis that the *Filter* passes only a selective signal.

c. *Mutual grammar mapping of* $G_{OCP} \approx \Sigma G_{RPA}$. In the OCP structural models obtained, when the situation is *S1* 'NM', the output information stream 'balances' the input stream. If the 'balance' is violated, a transient occurs in the OCP. The transition to a new state of the SyntA automaton occurs as the transition process develops and is determined by the *SN* situation. Since the physically non-separable design of the OCP and equipment has its own movements, the IIR filter scheme is more consistent with the OCP scheme.

The OCP model is built on the basis of a bundle relative to the internal adder. In the IIR filter scheme, each rule *P* is assigned a weight coefficient *KN* or *KM* (**Figure 10**). In the structure of digital filters, the inertia is set discretely by the SyntA structural automaton, and the dynamic behaviour of the filters (their own transient process)

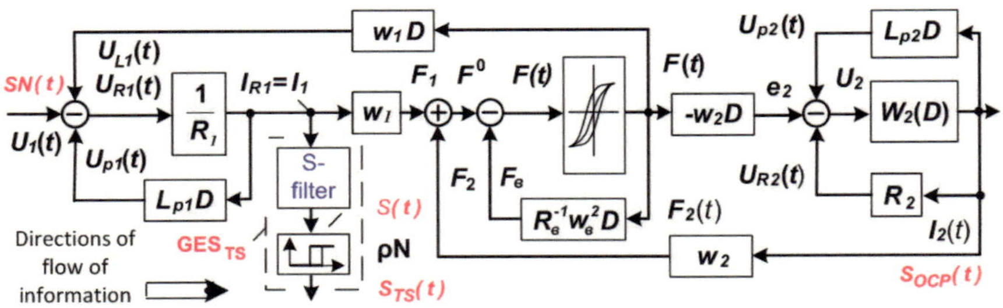

Figure 9.
Structural schemes of ownership, prompted by the operator method.

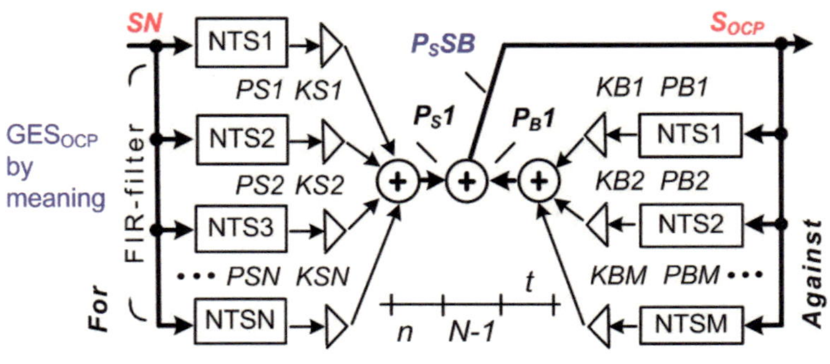

Figure 10.
GES_{OCP} scheme for dividing SN *situations into elementary components of NTS.*

depends on the information received from the input and from the output of the OCP circuit. The G_{OCP} grammar differs from G_{RPA} in (2, 3) in that its elements have an oscillatory, harmonic output, while G_{RPA} elements have a threshold output.

Separation of movements by frequencies: The OCP representation is introduced as a description of the G_{OCP} grammar, including a list of *SN* situations (root *S* characters). *SNs* are interconnected by the semantic rules of *PN* (**Figure 11**) and characterise the work of OCP. In turn, the *SNs* themselves are represented by semantic information components (terminal and nonterminal symbols) and the *PN* rules of their relationship. Thus, the lists of the grammar *G* and the GES schemes form the semantic model of OCP, composed of semantic components. The start and end vertex is the *SN* 'NM' situation.

By dividing the movements according to frequencies, we mean to clarify the number and parameters of the oscillatory circuits in the OCP (**Figure 7**). Based on the descriptions obtained by the structural-operator method [7], it is possible to represent the structure of the OCP definition tree in the form of the relationship of a number of oscillatory or inertial circuits (**Figures 7** and **9**). The method of dividing movements by frequencies from the point of view of the SI method is as follows.

According to the SI method, when frequency motions are divided, all the frequency circuits of the OCP are controlled [9–11]. If the selective GES_{RPA} part (synchronous detector) controls one circuit (**Figure 7**), then the blocking part of the GES_{RPA} monitors the other circuits. For information to appear in the slowest OCP loop, a series of events in the transition process must occur, and a sufficiently large amount of information should have accumulated. But the goal of the RPA is to minimise the development of events in the OCP. As a result of this, movement in super-slow circuits rarely occurs; the appearance of transient information in the OCP can be interrupted. This leads to the indicated conflicts 'a' to 'c'.

For the OCP scheme: (A) The number and parameters of the oscillatory circuits in the OCP are determined from the mathematical descriptions and signals of the emergency files (**Figures 7** and **8**). (B) An OCP is drawn up in the form of an IIR filter (**Figure 9**). (C) The GES_{OCP} scheme is compiled in the form of an FIR filter as the inverse of the GES_{RPA} scheme according to $G_{OCP} \approx \Sigma G_{RPA}$ relative to ΔSN (**Figure 10**). (D) The elements MorphA, SyntA and SemA stand

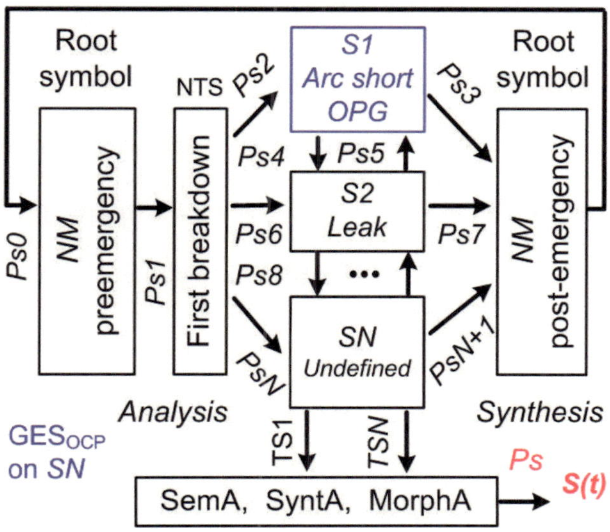

Figure 11.
Scheme of the 'GES_{OCP} by the SN *situation'. Shaping* SN *and signal situations* S(t).

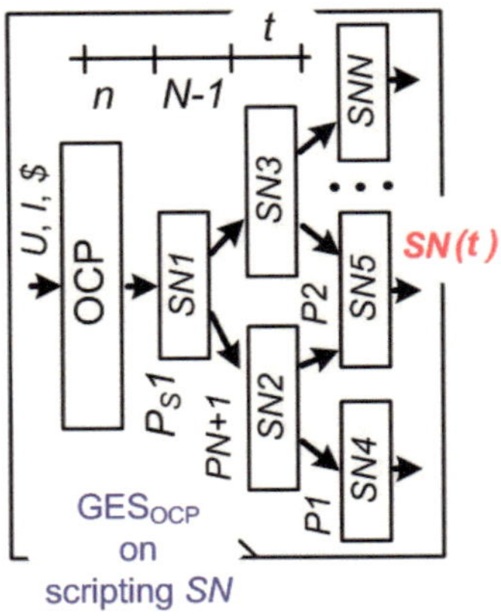

Figure 12.
Tree scripting SN *in dynamics by signal* S(t).

out in G_{OCP} and are implemented in separate computing parts. (E) W_{TABL} is filled out to represent the OCP. Training and supervising samples of *SN* situations are formed.

Such a volume of simultaneously solved problems leads to a limited capability to divide movements by frequencies for modelling in the OCP and solving smart grid problems.

Separation of movements by meaning: Further solution of the problems leads to using the method of separation of movements by meaning [9–11]. We will understand the separation of the dynamic flow of information into hierarchically subordinate structural elementary parts and the operation of these parts (**Figures 10–12**). Additionally, the division into alternative information flows 'For-Against' is performed.

The relationship of the methods of separation of movements: The separation of motions in frequencies determines the structure of oscillatory motions in the OCP (**Figure 7**), but the meaning of these motions is determined by the division in meaning (**Figure 10**). To clarify the relationship of the methods, an analogy can be introduced. The method for separating movements by frequencies describes the relationship of all the available circuit elements. This description is located within a single plane. At the same time, the LFC processes take up the resources of the computing system while the HFC processes are being calculated. The method of separation of movements by meaning represents the relationship of the elements in the circuit, not in the plane but in space [11–13].

Within the framework of the SI method, we can talk about the method of separating movements by frequencies as an additional preliminary hierarchical level of the method of separating movements by meaning. The application of the above methods can be as follows—structural-operator, identification, separation of movements by frequencies to describe the OCP and then the SI method with structural-morphological, structural-syntactic and structural-semantic steps.

9. Structuring OCP semantic *SN* situations

The parameters of the electrical signals are only the *SN* carriers. The characteristic features of the input coordinates of the RPA are the natural spatio-temporal sequence of consideration of the information components. The SI method is applied to the description of the OCP. All *SN* components are located in a certain way relative to each other and relative to the general synchronising time axis [10, 11]. The semantic information about the state of the OCP can be represented in the form of separate elementary *SN* situations. Each *SN* characterises the corresponding known (classical) state of the OCP. The problem arises of separation of the information set of states and recognition of the *SN*. In the limit, $S_{OCP}(t)$ can be specified by a single *SN*. This becomes the minimum information for building an OCP model, which reduces the time required for a single calculation in CAD.

The OCP scheme can be in dynamic consideration in one of the semantic situations of *SN*. Suppose that *S1* 'NM' corresponds to the steady-state values of the OCP internal parameters (operator outputs, NTS symbols, *P* rules). Thus, the concept of semantic situation *SN* means the appearance of the reaction ΔU_{OUT} of the OCP structural scheme in response to a change in Δ in any OCP coordinate. We also understand the structural, logical relationship of the individual TS control points in the structural-operator model of the OCP or equipment that form the transient signals (**Figure 9**). And also *SN* means a part of the OCP formation tree with activated root symbols *PS*, *PB*, then NTS and TS. The steady state of the OCP parameters can be distinguished and named *SN*. By the *SN* situation template is meant the sequence of characters *PS*, *PB*, TS, NTS of the OCP tree established for this *SN*, which is formally described by the G_{OCP} grammar. Thus, it is necessary to systematise all *SN* situations. These *SN* definitions are based on the OCP. The definition of *SN* relative to RPA devices is introduced as an analogy—each *SN* corresponds to a stencil with the corresponding sequence of 'selectivity windows'. Compilation of the OCP generation tree can be planned using the table transfer function W_{TABL} = TSN/*SN* [9, 11]. W_{TABL} is populated based on emergency files as well as by calculating OCP models and GES_{RPA} schemes in CAD.

The *SN* tree consists of a series of *SN*, replacing one another in time logically sequentially as the transition process develops (**Figure 12**). A scenario of *SN* development is formed.

A special role in the operation of the OCP is the situation of *SN* 'NM'. Although this *SN* does not apply to single-phase insulation faults to ground (OPG), it is from this that it begins, and it ends with its analysis of the transient into semantic and structural-informational components (**Figure 13**). The normal *SN* mode includes: (a) *SN* 'NM' itself; (b) *SN* 'processes not related to OPG'; and (c) *SN* 'neutral displacement', caused by the exceeding the normalised levels of emergency situations and resulting from the operation of the technological equipment of the distribution network.

Characteristic *SN* situations can be highlighted (**Figure 12**). The initial OPG breakdown (the first damage to the network insulation until a pronounced neutral reaction of the network), then the network reaction to the initial breakdown, subsequent OPG breakdowns and the restoration of *SN* 'NM' after OPG elimination are different stages of the transition process, of which OPG is a particular case. Of these named elementary *SNs*, the 'GES_{OCP} by *SN* situation' formation tree is composed. Recognition algorithm in RPA devices restores exactly such *SNs*, *SN* scripts and *SN*

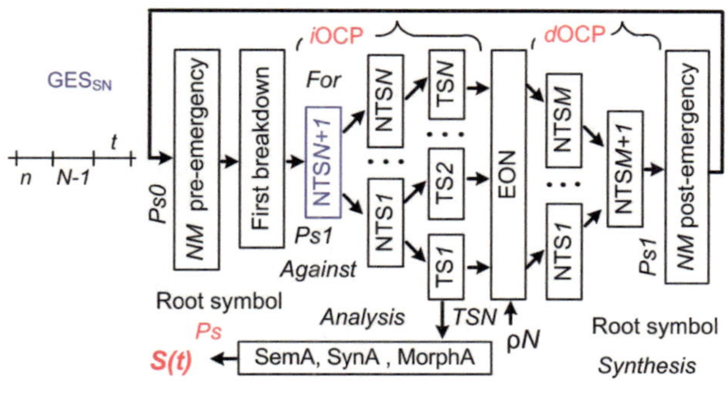

Figure 13.
Structural schemes of GES_{SN} semantic situation SN *in OCP by signal* S(t).

script tree. Therefore, the description of the OCP obtained by the SI method should return all possible relationships in *SN* in the *SN* tree.

In the dynamics, a sequential change of the elementary *SN* occurs. As a result, a scenario for the development of the *SN* appears. The OCP can be in one *SN* state for a long time (e.g. *SN* 'NM', *SN* 'metallic OPG', *SN* 'skew'), for a short time (*SN* 'metallic') or very briefly (*SN* 'first half-wave OPG'). Scenarios of *SN* development can also be sequentially described by the logical change of certain elementary *SNs*. Such scenarios are typical for OCP and should have their own names (e.g. *SN* 'intermittent arc', *SN* 'self-eliminating multiple breakdowns'). In RPA algorithms, such *SN* scenarios are restored at a semantic level of recognition. It is possible to create an *SN* transition tree (**Figure 12**). Both mutually inverse formation and recognition trees make up the 'GES_{OCP} scheme for the *SN* situation'. We can distinguish 'GES_{OCP} along oscillatory circuits' schemes for considering the separation of movement and a 'GES_{OCP} scheme for *SN* semantic situations', which shows movement in the meaning of *SN* situations.

Thus, the recognition algorithm determines not only the *SN* situation but also the essence of the transition process in the OCP. At this recognition level, a significant part of the non-selective operation of RPA devices occurs due to the limited or lack of appropriate *PN* rules in RPA algorithms.

The task associated with RPA and OCP management has the peculiarity of intersecting *SN* semantic situations. Generally speaking, in order to increase the stability of the operation of RPA algorithms for each area of overlapping situations *SN*, it is desirable to define its own information sensor TS. The more it is necessary to determine the current *SN*, the greater the number of TS that should be laid in the structure of the algorithms. Reducing this intersection is the ultimate goal of a theoretical description of the OCP. For this, according to the theory of information, it is necessary to supply an excessive amount of information for rechecking, determining errors and recovering information from errors for a recognition system. In practice, the implementation of this important property does not occur. This leads to recognition errors.

In a real OCP, there are third-party processes unrelated to typical *SN* situations and *SN* scenarios. These may be present for a long time and be registered by some TS in the RPA. They need to be discovered and their named list compiled. Formation can be in CAD based on mathematical functional modelling, with real emergency files. In the presence of a generated mathematical model of the OCP or individual equipment, it is possible to introduce situational changes from the signal source at a certain moment during the transition process, which interfere with *SN* 'NM', for example, by short-circuiting or shunting a single element, or a series of elements in the structural-operator model of OCP, or equipment (**Figure 8**). This leads to observable transients. Similarly, it is possible to repair the cause of damage through transient signals in real emergency files.

10. Drawing up a combined GES scheme

The combined scheme in **Figure 14** allows one to obtain a high-quality *SN* recognition model in view of the mutual reflection of $G_{OCP} \approx \Sigma G_{RPA}$ and the possibility of comparing $S_{OCP}(t)$ and $S_{RPA}(t)$. From a comparison of the parts of **Figure 14**, it follows that the synchronous detector is a 'central' element in the GES_{OCP} scheme. It is known that a synchronous detector consists of a comparing element based on the multiplication or summation, a zero component filter and a threshold device. The filter turns out to be feedbacks from the U_{OUT} output to the common adder. When replacing the elements covered by feedbacks in **Figures** 7 and **10** with the transfer function, a more general OCP scheme is obtained in the form of a unidirectional *SN*→TS formation tree (**Figure 14**). The scheme makes it possible to solve modelling problems and smart grid information tasks [7–16].

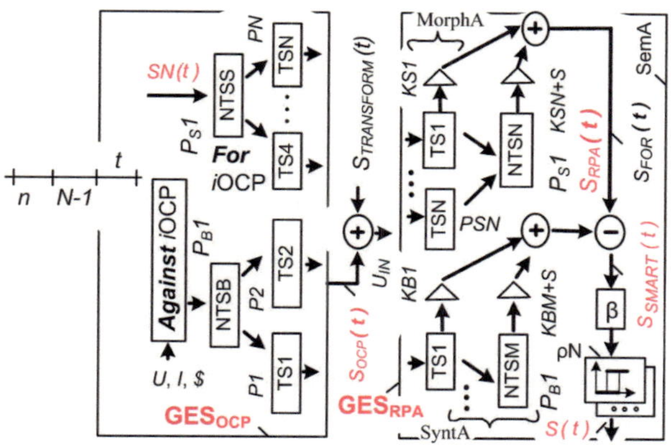

Figure 14.
Structural schemes of GES_{OCP} *and* GES_{RPA} *by signal* S(t) *and* $G_{OCP} \approx \Sigma G_{RPA}$.

The minimum information for constructing an OCP model are two elements of the *i*OCP information part—a selective TS and a blocking TS. Also, the bundle—the signal source $S_{OCP}(t)$ and the controlled frequency generator—will be the minimum OCP model [11–13]. The generator can have amplitude, frequency and phase modulation of the output signal from the super-LFC to HFC range, depending on the problem being solved. Such OCP models minimise the time required for a single calculation. Thus, models of the OCP, RPA devices and S-detector were built. The $S_{OCP}(t)$ recovery algorithm from the OCP model can be used to decrypt the accumulated emergency files (**Figures 11–13**). For this, models of the ExS expert system and the diagnostic message generator are used (**Figure 4**).

11. Balancing structures and algorithms using game theory

A 'balancing' task arises when considering GES schemes (**Figure 14**). According to $G_{OCP} \approx \Sigma G_{RPA}$, it is possible to set the 'balance' task for the GES_{OCP} scheme as well. The solution of the 'balancing' problem is a development of the theory of the SI method [8–11]. A numerical determination of the equilibrium point of GES schemes is proposed when considering the 'For-Against' problems for OCP and 'selectivity-blocking' using the game theory method [14]. The result is compared with the published solution through the method of dynamic functional modelling [11, 17]. This section briefly discusses the algorithm for solving the problem, designed in the form of a computing module. The problem when using computational modules is the possibility of a multivariate interpretation in the formulation of the problem and in the analysis of the results. The following aspects of the problem have emerged—(a) the need to be an expert in the specific problem being solved; (b) being sure of the possibility of using the module to solve the particular problem; and (c) it requires the presence of training and supervising samples of real semantic situations for testing and verification of results. Experience of using the module can be gained on the basis of modelling in CAD.

In the 'balancing' game, the competitors are the two root weights, K_SS *selectivity* and K_SB *blocking*, where the corresponding root characters are S_SS and S_SB. The contest is won (victory) V, when there is a clear victory for K_SS or K_SB—the correct operation of the threshold element ρN. It is lost L if there is 'error excessive response ρN'. The competitors have the following features, aimed at implementing the requirement to increase the stability of RPA devices in complex *SN* situations. The K_SB *blocking* group always fights 'compete K_SS and K_SB' until they win: 'The $S_SB(t)$ signal is greater than $S_SS(t)$' of the *blocking* group and retreats with 'error excessive

←	↓	*Selectivity, 1–z, For, (Doves, D), K_SS*	*Blocking, z, K_SB, Against (Hawk, H)*
$S_D(z)$	**Selectivity,** *For (Doves), K_SS*	$S(S,S) = (V–L)/2–E$ $S(S,S) = 30$	$S(S,B) = –L$ $S(S,B) = –5$
$S_H(z)$	**Blocking,** *Against (Hawk), K_SB*	$S(B,S) = V$ $S(B,S) = 75$	$S(B,B) = (V–W)/2$ $S(B,B) = –37.5$

Table 1.
Payment matrix of the participants in the 'balancing' game with example.

operation ρN' in the case of serious injuries, with 'error excessive failure ρN'. The K_SS *selectivity* group is limited by the threats of winning: 'The $S_SS(t)$ signal is greater than the threshold ρN'; however, if it gets to a 'complex *SN*' bout, the *selectivity* group recedes, with 'error excessive ρN failure'.

Thus, there are four game options. These appear more pronounced in a difficult *SN* situation, when there is little information for the correct operation of the threshold element ρN. The results of the game are evaluated in the form of arbitrary units (points) received or lost by the participants.

In the 'grammar *G* of one of the devices' population, we denote the share of the *blocking* group with z, and then the fraction of the *selectivity* group will be 1–z. If two competitors K_SS and K_SB are randomly participating in a 'one of the *SN* situations' clash, then with a probability of z × 2, they are two K_SB; with a probability of (1–z) × 2, they are two K_SS; and with a probability of 2 × z × (1–z), they are K_SS versus K_SB.

Let us designate the parameters of the 'balancing' game as damage from injury *W*, 'error excessive failure of ρN' and energy consumption *E* for the opposition 'operation-failure of ρN'. We will assume that in the 'balancing' game, the winning 'correct work ρN' is estimated as follows (**Table 1**)—victory $V = 75$ points, loss $L = 5$ points, damage from injury $W = 150$ points and energy consumption for opposing $E = 5$ points. Let us find the average number of points that competitors K_SS and K_SB get as a result of the 'one of the trainings *SN*' fight. The results of the 'balancing' game can be visualised in the form of a payment matrix (**Table 1**). Based on **Table 1**, on average for the *blocking* group, we get

$S_B(z) = S_H(z) = 0.5 \times (V–W) \times z + V\times(1–z) = V–0.5 \times (V + W) \times z = 75–112.5 \times z.$

Similarly, for the *selectivity* group, we get.

$S_S(z) = S_D(z) = –L \times z + [0.5 \times (V–L)–E] \times (1–z) = 0.5 \times (V–L)–E–[0.5 \times (V + L)–E] \times z = 30–35 \times z.$

The graphs of these equations in the *S*–z coordinate axes are shown in **Figure 15**. As can be seen, the winning lines 'correct work of ρN' for the groups *selectivity* K_SS and *blocking* K_SB intersect at the equilibrium point Ω defined by the relation:

$\Omega = 75–112.5 \times z = 30–35 \times z, z = 45/77.5 = 0.58.$

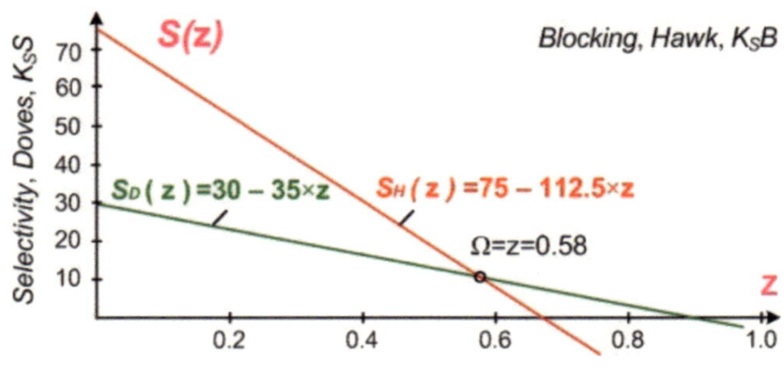

Figure 15.
Graph of the winning line of the participants in the 'balancing' game.

In **Figure 15**, we choose the reliability point $z_{RELIABILITY} = 0.8 \times z = 0.8 \times 0.58 = 0.464$. At this point, we put $S_S(z)/S_B(z) = K_SS/K_SB$. The ratio $k_R = K_SS/K_SB = 13.76/22.8$. Then.

$K_SS = k_R/(k_R + 1) = 0.356$, $K_SB = 1/(k_R + 1) = 0.624$.

12. The specifics of managing dynamic objects

An object is considered to be managed when it can be transferred from any initial state using an input action to a given state within a finite time. In this case, the ASNOM system stability requirement must not be violated. Stability of operation is understood as the ability of the system to return to the initial *SN* 'NM' situation when OCP parameters change within specified limits in the absence of malfunctions in the system (**Figure 4**).

A latent malfunction should lead to exceeding, by one or several components of the vector of parameters Ξ, the limits acceptable for the normal operation of the OCP. The fact of exceeding by Ξ the boundaries of the parameters will be considered a sign of a latent or developing *SN* 'unrecognised' situation. A change in the steady-state value of one parameter Ξ leads to a change in the steady-state value of the parameters Ξ of all the components of the boundary vector.

In dynamics, some *SN*, given the random, complex, spontaneous development of damage, or the irregular, unstable behaviour of OCP elements, can replace one another very quickly, according to the GES_{SN} or GES_{OCP} tree *SN* scenarios. Such situations are 'difficult'. A more complete definition and separation of the *SN* 'violation NM' is possible based on automatic control of the ASNOM system's controllability parameters [13]. These parameters are controlled if a disturbance appears. Natural functional disturbances of the OCP are not a sufficiently complete search signal of self-control from the point of view of determining and separating the *SN* 'violation NM'. So, for example, in the case of malfunctions of the elements of the formation of the information signal *3uo*, the error ε mismatch in the ASNOM system will still remain in the zero zone (**Figure 3**). Although natural resonance detuning is not a sufficient search signal, its use can increase the frequency of self-checking. For a more complete definition and separation of the *SN* 'violation NM', an artificial search action is introduced into a closed ASNOM system (**Figure 4**).

It can be proposed to use 'additional' information components in the selective search algorithm SP of sufficient information in cases where there are no 'main' information components, despite the fact that this proposal tends to solve the problem in the direction of 'complicating' the devices. For this, additional rules for the recognition of P_{DOP} and an excess of information and additional recognition time are required. The device solves the following problems within the framework of the automatic stabilisation system: (1) eliminating the ambiguity of determining the essence of transients; (2) control of the amount of information necessary for the operation of relay protection and automation algorithms; (3) monitoring the current possibility of self-destruction of the damage site; (4) monitoring the effectiveness of measures over long time intervals; and (5) implementation of the requirements of self-monitoring and diagnostics when 'energised'.

13. Simulation in CAD of OCP control algorithms

All the variety and 'complexity' of real-world network signals is controlled by the high-frequency emergency files of RPA device registrars. These registrars supply emergency files for outsourcing analysis. [8, 11, 15]. To solve the control problems and eliminate the instability of the algorithms, we will continue to use the dynamic synthesis method in CAD. A presentable sample of signals from the OCP emergency files is used. In the synthesis and development of real-time algorithms in CAD systems, real circuit diagrams are used. The physical implementation of tasks is monitored in a continuous improvement mode, from 'simple' to 'essential'.

A model is being developed in *Matlab* to study the properties of a class of dynamic objects for the synthesis of algorithms for managing them over time [11–16]. Monitoring the operation of the system and algorithms is performed at a given time interval in the form of alarm file signals (**Figures 4, 7, 9, 13**, and **14**). Device models are also represented by GES_{TS} and GES_{RPA} schemes. The more structured and detailed the description of such schemes is obtained, the more qualitatively it is possible to build stable recognition algorithms. Sources of interference signals and damage are used for 'hacking', connected to control points of the device. Hacking signals are generated by arcing and interference burning models for various types of development of insulation damage, in emergency operation modes and in normal operation of an object. The training and control samples of signals of real emergency files are generated.

The purpose of the simulation is to develop ways to search for additional information. The modelling problems were solved by dividing the project into hierarchically subordinate computational parts with the result reduced in the generalising part [11–13]. Each part is calculated separately. Groups of preparatory, main and resulting elements are allocated. Each subsequent group is less active and does not consume computing resources. The more elements in the circuit, the more computing parts it can be divided into. As a result, the time required for a single calculation is reduced while maintaining the stability of the calculation. For example, the group of preparatory elements (MorphA) may include the signal sources, both generating and controlling, of measuring transformers, etc. In the group of basic elements (SyntA) are the inertial circuits of equipment, etc. (**Figure 4**). The resulting group (SemA) includes elements of the executive bodies, of operational switching, etc. Thus, in addition to modelling, first a certain third-party algorithm for preparing the project for calculation follows. Such an algorithm can be automated. The main general synchronising parameters for the separation of movements are the well-known preliminary settings of the CAD project—the calculation time and control points for displaying information.

14. Examples of dynamic OCP control algorithms

The practical significance of the work comes from solving the problems of protection against earth faults in medium-voltage networks of 6–35 kV [7–9, 16–18]. The results are used to improve the ASNOM system devices (**Figures 1–4**), namely, a selective search for the SP of a damaged part of the 'P-VCR-SP' network with the functions of a high-frequency recorder VCR, an RPA terminal 'T-LZSC-ARC' with the function of auto-compensation of ARC capacitive currents and a widget 'W-LZSC' for a window graphically representing of the terminal information on the computer display of the automatic control system.

An analysis of the waveforms of the real high-frequency transient emergency files in the OCP shows that as the transient develops, the ASNOM system will in most cases enter a state of information sufficiency [11–13]. Therefore, if the ASNOM system is not blocked in the case of insufficient information, as is usually done in known devices, and continues to work according to the appropriate algorithms, then the solution of the problems will be achieved.

Automatic resolution of practical conflicts: The important practical significance of the whole work is automatic conflict resolution in the work of RPA algorithms, the lack of solutions of which has led to a decrease in the reliability of OCP operation, but could not be resolved by other methods. In resolving conflicts previously, there was a reliance on technical and economic optimisation methods, with which it is difficult to practically eliminate doubts about the correctness of the method chosen for dealing with the development of damage in the network. Conflicts were found during practical research in real networks.

Known management conflicts for OCP are: (A) The requirement to quickly disconnect the damaged section of the OCP and realise the network's ability to isolate itself for self-repair; (B) The possibility of damage to the weakened isolation area of high-voltage equipment caused by a voltage increase in healthy phases in cases of the malfunctioning of protective equipment, and the recorded efficiency of the resonant network's neutral tuning; (C) Automation of the control algorithms of technological processes and the forced opening (transfer to alarm) of the output of an unstable selective relay protection; (D) The tradition of full control over the operation of the network, but a lack of control over the long-term effectiveness and correct functioning of the protective equipment of the network and of the devices that implement the selected type of neutral grounding in the tasks under consideration. Here are some examples of resolving some conflicts:

Conflict resolution A: 'temporary compromise'. This is based on the formulation of the general problem of stabilising the normal operating mode of the OCP (**Figure 16**). It is based on a change in the response time t_{OFF} of the selective search relay SP depending on the value of the semantic signal $S(t)$. Change t_{OFF} is not prohibited for the OCP and may be in the range 0.1 s to 4 h. **Figure 17** shows a diagram illustrating the change in the t_{OFF} value depending on the appearance of various (structural) information (TS, NTS) components of the transition process in the OCP. If, during the transition process, counting from the appearance of 'LFC3u_O', the indicated TS and NTS have appeared, and then the damaged section will be disconnected after the corresponding

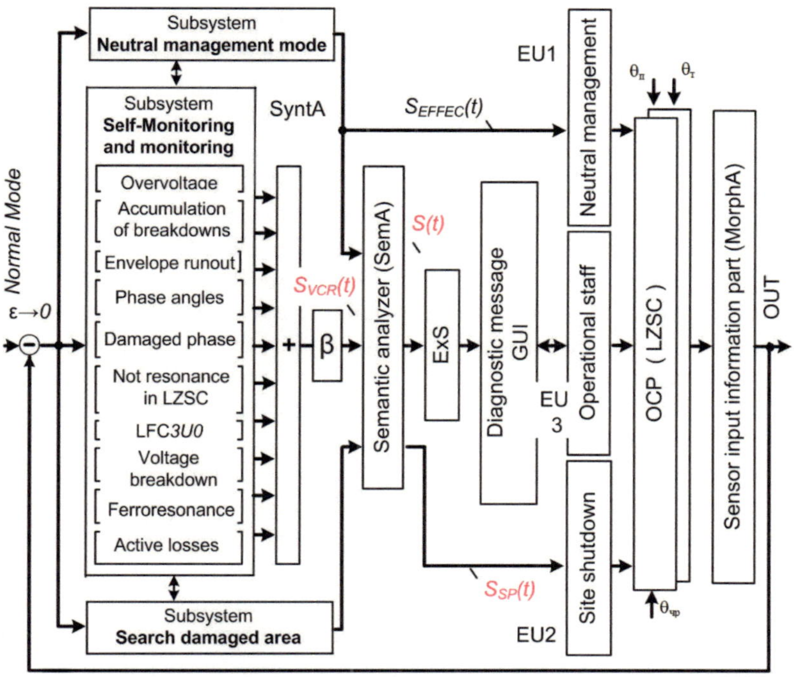

Figure 16.
Method for automatically resolving a temporary conflict between SP devices and resonant tuning of a network zero sequence loop (LZSC).

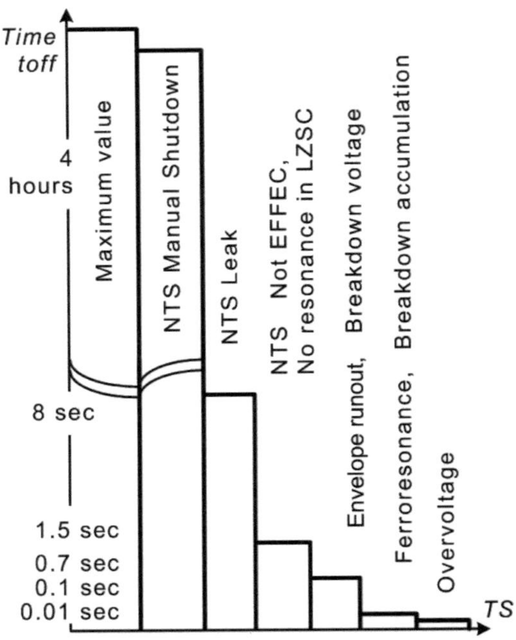

Figure 17.
Changes in time tOFF *in the ASNOM system depending on* S(t).

time interval t_{OFF}. If a longer t_{OFF} time has elapsed from the start of the transient than indicated, and then TS and NTN appear, then disconnection occurs immediately.

Conflict resolution C: 'automation of algorithms'. Analysis of the solution of the problem at the modelling stage shows that for correct and effective working of the operational staff in the event of the appearance of *SN* 'undetectable', it is necessary to display the operation of the ASNOM system in the 'scanner-analyser' mode (**Figures 4** and **16**). The screen displays the line of the entire transient process in the network, starting from the background and ending with the current moment. This will help the operating staff in the absence of automatic repair of the damaged network section to eliminate it in manual mode.

Conflict resolution D: 'on the coincidence of exits'. A method is proposed for the selective search of the required amount of information to detect equipment failures, failures in RPA devices and then through the RPA devices OCP faults (**Figures 4** and **15**). The algorithm generalises the functions of self-monitoring, monitoring and diagnostics 'under voltage'. The criteria of active and passive selective SP search are used with the control of natural and artificial transient processes in the OCP [11, 13]. For example, the active methods may include phase-voltage imbalance in a network with a resonantly tuned neutral or phase shunting to ground in a network with an isolated neutral (**Figure 3**).

An outsourcing method for investigating *SN* situations using alarm file signals is being implemented. For this, specialised organisations or individual specialists may be involved. Firstly, it is assigned that the output of the TS is formed as efficiently as possible. If the recognition ability is preserved and the amount of information is sufficient, then the second stage of the investigation begins—the establishment and refinement of the parameters of the TS, their weight coefficients of significance and the quality of the TS structural elements. If the problem is not solved, we must investigate further along the information formation chain. The structural tree of the formation of information inside the control object (high-voltage network equipment) should be studied.

15. Expanding the scope of the SI method

The SI method is universal enough for tasks that can be reduced to informational. Information flows can describe the control and protection algorithms of various control objects not only in the energy sector but also in other areas of control. The problems of automatic control in the classical setting are being solved. For example, coordinate control, search, extreme, adaptive, optimal, automatic, automated, expert, dispatch control systems, etc. The problems of pattern recognition in linguistics are solved: decision-making tasks based on morphological features, for example, biological, mechanical, etc., and recognition of information carried by sound vibrations, for example, cochlear and sound pattern recognisers in medicine.

The SI method allows to improve control systems with a lack of initial information in situations that are difficult in meaning when operating the control object. When the input information is changed, impulse rarely appears, consisting of separate diverse components.

Consistent application of SI method templates allows you to not ignore the necessary stages of obtaining results when solving management problems. The statement of management is carried out in the dynamics of the development of processes with the issuance of management commands, for example, up-down, buy-sell, etc. So, first, when solving problems, a workplace is built that allows you to realise the final desired result in CAD, for example, a device issuing commands and receiving input signals. The final result is immediately controlled. In case of its inconsistency, part of the control algorithm changes. Further, as the problem is solved, control algorithms are improved. The development of algorithms proceeds along the path of improvement: 'from the simple to the necessary'.

16. Conclusions

1. A class of dynamic control and protection objects has been allocated for solving problems of improving control algorithms. The development of the SI method for dynamic objects with minimal reference to the specifics of objects is shown. A generalisation of the results is carried out for application to a wider range of tasks in relation to the initial electrical tasks.

2. Problem-solving is obtained using the instrument of formation and control of a semantic signal. It allows you to operate with a minimum of information but to control the necessary amount of information and solve problems at a more abstract level as well as simulate the operation of specific recognition algorithms, taking into account the specific operation of objects.

3. A sequence of patterns is presented that allows one to obtain solutions of problems by the SI method. Templates are the reference points for the solution. By templates, you can explore the passage of information components as on navigation charts through interconnected structures. Examples are considered for OCP, individual equipment, transient recognition algorithms and automatic control systems, which can be reduced to dynamic ones.

4. A method of restructuring the general task of describing the OCP into elementary information components is shown. The principles of the separation of movements by frequencies and meaning for the analysis of the flow of information through the elements of the OCP and recognition algorithms by

discrete steps are shown. Operating the structure of information flows will help to improve recognition algorithms and ensure the stability of their work.

5. Solving the problems of analysing the operation of algorithms and recognising situations is performed at the workplace in CAD in two ways—by functional modelling and by using game theory. The functional modelling takes into account the inertia of movements in the contours of the object as well as the movement of information in the recognition algorithms. Real-time algorithms are presented on a temporary scan of alarm file signals received by device registrars from real objects. Situations make up a library of test signals. An album of the results of modelling emergency files with the results of 'hacking' algorithms and countering 'hacking' is accumulating. Game theory makes it possible to get a static solution on the points of equilibrium and recognition, which may be the recognition algorithm. This also applies to the object of control and protection. A computing module for use in relay protection and automation tasks is described. The methodology for generating information for the module should prevent misuse of the module and facilitate the interpretation of the results.

6. According to the SI method, control algorithms are synthesised based on the equality of the GES forming and recognising grammars. The chapter focuses on the formative grammar of the OCP, high-voltage equipment, the structure of the *SN* semantic situation, the passage of the *SN* script through the GES_{OCP} and GES equipment circuits as the input of control and monitoring algorithms. Methods for the analysis and synthesis of the recognising grammar are described in detail in the articles from the list of published sources. The recognition algorithms are written off to the finished devices. They are improved for each network, taking into account the reaction of the algorithms to the signals from real emergency files, and the network model in CAD.

7. The general algorithm for the automatic control of dynamic objects is developed in the form of the selective search for sufficient information for recognition. The SP algorithm makes it possible to synthesise new and analyse well-known algorithms for relay protection and automatic control of 'complex' objects with a tunable structure, distributed parameters and the spontaneous development of damage.

8. It is possible to compensate for the lack of information in difficult semantic situations through new ways of obtaining information, additional information sensors and increasing their intellectual capabilities. Separate sensors are required for all the parameters of the input signals—frequency, phase, transit time and the ratio of special moments in the transient signals in the network.

9. Several practical examples of the application of the theoretical part of the description are given. This is, first of all, the solution of known conflict situations of automatic control, manifested in the dynamics of the development of semantic situations, namely: (A) The operating time of the automatic system for compensating capacitive currents and the operation of selective relays can and should be variable according to the volume of the semantic signal. (B) Distinguishing between a working and a faulty system in the absence of output commands can be based on analysis of the semantic situations under natural and test influences. (C) Recognition systems need to be built as a dynamic sequence of the opening 'windows of selectivity' as one of the moves through the semantic situation tree and the situation scenario tree.

10. The approach considered is proposed for the joint analysis and synthesis not only of algorithms and software products of semantic processing of information at different levels of the hierarchy of control systems but also for constructive technical solutions. In this analysis, the method is applied in the 'bottom-up' direction, that is, starting from the processing of information by a multitude of information sensors; next, the processing of information is concentrated in the 'branches' of the processing of the structural-semantic relationships and then in the 'root' of automatic decision-making on the outcome. It is clear that, for synthesis problems, this method is best applied 'bottom-up'.

Author details

Andrey Petrovich Nikiforov
Education and Research Institute of Energetics, Automation and Energy Efficiency, NNI EAE (ERIEA), National University of Life and Environmental Sciences of Ukraine, NUBiP, Kyiv, Ukraine

*Address all correspondence to: a.p.nikiforov@i.ua

References

[1] Kozyrskyi V, Petrenko A, Trehub M, Charyev Y. Renewable Energy and Power Supply Challenges for Rural Regions, ch009. Pennsylvania, United States: IGI Global; 2019. pp. 197-228. DOI: 10.4018/978-1-5225-9179-5.ch009

[2] Bodunov V, Kulko T, Prystupa A, Gai A. Topological task of distributed generation placement using a Pareto optimization. In: Proceedings of IEEE 3rd International Conference on Intelligent Energy and Power Systems (IEPS), art. no. 8559502. Kharkiv, Ukraine: IEEE; 2018. pp. 183-188. DOI: 10.1109/IEPS.2018.8559502

[3] Tygai Y, Besarab A. The mathematical model of voltage transformers for the study of ferroresonant processes. In: IEEE International Conference on Intelligent Energy Power Systems (IEPS). Kyiv, Ukraine: IEEE; 2014. pp. 77-80

[4] Beliaev V, Panenko H. Determination of insulation parameters of current transformers at multiple measurements in monitoring systems under working voltage. Journal of Electrical Engineering & Electromechanics. 2016;**5**:40-46. DOI: 10.20998/2074-272X.2016.5.06

[5] Kaplun V, Osypenko V, Makarevych S, Lurie I. Forecasting electricity pricing of energy islands with renewable sources. International Journal of Intelligent Systems and Applications. 2018;**10**(8):1-6. DOI: 10.5815/ijisa.2018.08.01 (published online in MECS, mecs-press.org)

[6] Sivokobylenko V, Nikiforov A, Burlaka V, Podnebennaya S. Analysis of the 0,4 kV smart grid islanding prevention method. Journal of Energetika. Proceedings of CIS Higher Education Institutions and Power Engineering Associations. 2015;**6**(2):26-34. ISSN: 1029-7448, E-ISSN: 2414-0341 (In Russ.) https://www.scopus.com/authid/detail.uri?authorId=57191825673

[7] Obabkov V, Nikiforov A. Accuracy of automatic trimming frequency of free oscillations in symmetrical networks with resonantly earthed neutral. Journal of Elektrichestvo. 1996;**12**:8-16. DOI: 10.13140/RG.2.1.4204.7127. (In Russ.) Available from: https://www.scopus.com/authid/detail.uri?authorId=57191825673

[8] Sivokobylenko V, Nikiforov A. Development of a generalized equivalent scheme of morphological level of transient recognition in smart grid distribution networks. In: Scientific Papers of Donetsk National Technical University. Series "Electrical and Power Engineering" №1(19)-2(20)'2018; Pokrovsk, Ukraine: Donetsk National Technical University; 2019. pp. 47-62, 2019. ISSN 2074-2630, DOI: 10.13140/RG.2.2.14513.94568. (In Russ.)

[9] Nikiforov A. Unified smart-detector for electrical power smart-grid networks. In: 13th International Conference on Industrial Informatics (INDIN), Anglia Ruskin University. Cambridge, UK: IEEE; 2015. pp. 1032-1039. IEEE Catalog Number: CFP15INI-15, ISBN: 978-1-4799-6648-6/15, DOI: 10.13140/RG.2.1.1812.6487

[10] Nikiforov A. Theorem on the existence of semantic signal in relay protection systems. Journal of "Tekhnichna Elektrodynamika", PPE-2010: Thematic issue. 2010:73-76. DOI: 10.13140/RG.2.1.3713.1921, 10.13140/RG.2.2.24718.66886. Available from: http://lib.nmu.org.ua/catalog/site/view.html?doc_id=407519. https://www.scopus.com/authid/detail.uri?authorId=57191825673

[11] Nikiforov A. Application of the theorem of 'about the unity of the structural description of RPA devices' for the simulation of a power network smart-grid. In: 3rd Renewable Energy and Green Technology International

Conference (REEGETECH), Jakarta, Indonesia. Journal of Telecommunication, Electronic and Computer Engineering. 2017;**9**(1-5):39-48. ISSN: 2180-1843, e-ISSN: 2289-8131. Available from: http://journal.utem.edu.my/index.php/jtec/article/view/1831

[12] Nikiforov A. The through simulation of devices on the basis of the structural linguistic method. In: 7th International Conference-Workshop Compatibility and Power Electronics (CPE 2011); IEEE. 2011. IEEE Catalog Number: CFP11851-CDR, ISBN: 978-1-4244-8804-9, pp. 50-55, 978-1-4244-8807-0/11/S26.00. Identifiers book ISBN: 978-1-4244-8806-3, e-ISBN: 978-1-4244-8807-0. DOI: 10.1109/CPE.2011.5942206. Available from: http://ea.donntu.org/handle/123456789/19294

[13] Nikiforov A. Modelling infraslow circuits in real-time systems smart-grid on the basis of separation of motions in frequency and sensing. In: 2016 2nd International Conference on Intelligent Energy and Power Systems (IEPS-2016); Kyiv, Ukraine: IEEE; 2016. pp. 67-72. IEEE Catalog Number: CFP1605X-PRT, ISBN: 978-1-5090-1767-6, 978-1-5090-1769-0/16/S31.00. DOI: 10.1109/IEPS.2016.7521853. Available from: https://www.scopus.com/authid/detail.uri?authorId=57191825673, https://publons.com/researcher/2303561/andrey-petrovich-nikiforov

[14] Nikiforov A. Solving the problem of balancing algorithms for relay protection and automation using game theory. In: IEEE 6th International Conference on Energy Smart Systems (2019 IEEE ESS). Kyiv, Ukraine: IEEE; 2019. pp. 177-182. 978-1-7281-2159-8/19/$31.00©2019 IEEE, CFP19U02-USB, ISBN: 978-1-7281-2159-8

[15] Nikiforov A. Optimization of control systems in a smart grid the power grid on the basis of generalization of information flows. Journal of Tekhnichna Elektrodynamika, PPE-2016. 2016;**5**:64-66. Kiyv, Ukraine, ISSN 1607-7970. (In Russ.) Available from: https://www.scopus.com/authid/detail.uri?authorId=57191825673; https://elibrary.ru/item.asp?id=26638920, http://techned.org.ua/index.php?option=com_content&view=article&id=1105&Itemid=77

[16] Sivokobylenko V, Nikiforov A, Kollarov O, Kardash D. Improving current protection against short circuits in the smart-grid 6-35 kV network using the structural-informational method. Part 1. Analysis. In: Series "Electrotechnics and Energy". Vol. 1(18). Ukraine: Science and Technology of the Donetsk National Technical University; 2017. pp. 5-15. ISSN: 2074-2630. DOI: 10.13140/RG.2.2.19547.11044. (In Russ.) Available from: https://drive.google.com/file/d/1f5VhLSuXLbUIeIEI4H26Eic5_u_42u16/view

[17] Grebchenko N, Osipov D, Teslya A, Bozok J, Koval I. Adaptive current short-circuit protection in electric systems with distributed generation. In: International Symposium on Power Electronics, Electrical Drives, Automation and Motion. Capri, Italy: SPEEDAM; 2016. pp. 1279-1283

[18] Sivokobylenko V, Lysenko V. Numerical Simulation of Transient Ground Faults in Medium Voltage Networks, In: IEEE International Conference on Modern Electrical and Energy Systems (MEES), Kremenchuk, Ukraine: Kremenchuk Mykhailo Ostrohradskyi National University; 2017. pp. 184-187. DOI: 10.1109/MEES.2017.8248884

Section 6

Artificial Intelligence

Chapter 12

Evaluation of Algorithmic Management of Digital Work Platform in Developing Countries

Olalekan Samuel Ogunleye and Billy Mathias Kalema

Abstract

The economy of the Modern Work Platform is becoming increasingly relevant due to the spread of information and communication technology. As a result, digital work has gained popularity as a source of employment, especially in an economy where finding decent work is becoming increasingly difficult. Computer algorithms are now being used to alter and change the way people operate in increasing job specialization, handling large-scale human labour in a distributed manner. In these structures, human works are delegated, supplemented, and analyzed using tracked data and algorithms. Building on emerging algorithmic literature and qualitative examination, this article assesses the mechanisms by which the digital network manages staff in the sense of Uber, Bolt (formerly Taxify). It describes the difference in the degree to which such platforms limit freedoms over schedules and activities relevant to gig work. Based on in-depth interviews with 41 respondents working on different digital media and a survey of 105 staff on the same platform, the study finds that while all digital work platforms use algorithm management to delegate and assess work, substantial cross-platform variation. Uber, the largest network for ride-sharing, exercises a type of control called "algorithmic despotism" that controls the time and activities of staff more strictly than other network distribution firms. We end with a debate on the implications for the future of work of the spectrum of algorithmic power. It also addresses how algorithmic management and data-driven systems can be developed to build an improved workplace with intelligent machines, with implications for future work.

Keywords: algorithmic management, digital platform, intelligent systems, on-demand work, performance evaluation

1. Introduction

The capacity to work enables human beings to engage in conscious constructive and imaginative practices that alter and define nature so that human beings can create their means of existence to fulfill their needs, which in themselves constitute the creation of material life [1]. An online job is a term that has been a vital ground for debates in the political economy of internet technologies [2]. The Automated Job Network is a framework in which businesses such as Uber, Bolt (formerly Taxify), Takealot, JumiaEat, Otlob, and others use cloud-based technologies to "match" staff on the customer platform [3]. However, the spread of interactive work networks has been one of the significant revolutions in work over the last decade. Moreover, the Network Capitalism [4] is part of a broader transition from usual job security to variable employment conditions, including contract, contractual and part-time work.

The amount of data generated from several organizations' processes and activities made it possible for software algorithms to accumulate and interpret data, making it possible to contribute to management and decision-making processes [1]. As a result, data-driven algorithms allow digital work systems to handle transactions between thousands of network staff automatically. These algorithms assign, refine, and analyze the jobs of different platform workers [2].

We apply to automated algorithms that perform managerial decision-making and institutional instruments that assist algorithms in the practice of algorithmic management. Algorithmic management helps companies like Uber, Bolt, and many more oversee the platform's multitudes of employees in an improved way, on a wide scale. Algorithmic management is one of the key technologies that facilitate virtual organization management. A variety of ride-sharing systems, such as Uber and Bolt, have helped connect independent, dispersed drivers with customers who need systems within minutes or seconds. Simultaneously, service rates change rapidly, based on how the demand increases from applications installed on their mobile phones.

Algorithmic management allows individual human administrators to oversee drivers at any place where the ride-sharing systems run, including on a global scale. As a result, drivers had little to no prior contact with the company's members. However, they should communicate via online platforms (such forums) to enhance their social awareness of ride-share programs. This scenario helps one research what happens when algorithms delegate jobs, refine work activity through information analysis, and measure or measure job performance.

This leads to some key testing questions: are human workers (i.e., drivers) engaging and agreeing with work that is algorithmically delegated to them? Are human workers inspired or distracted by algorithmic optimization, and if so, by how much? How successful and accurate is the data-driven assessment of this algorithmic administration, and how do human employees feel about it?

The first move in answering these questions was to interview 41 drivers with Bolt and Ubers. We then triangulated their answers, interviewed 19 passengers, and studied the online driver forums' archived discussion.

Our findings highlight difficulties and opportunities in the architecture of human-centered algorithmic job management, evaluation, and knowledge. The results further underscore the value of fostering sensemaking around social algorithmic structures. We use the results to explore how data-driven algorithmic management can create a safer working atmosphere for intelligent machines while providing potential work recommendations.

2. Research background: Uber and Bolt ride-sharing services

In Africa and most developed nations, Uber and Bolt are probably the prominent peer-to-peer ride-sharing firms. Uber has been operating in over 100 cities across 37 countries between 2009 and 2013, and Bolt operates in more than 80 cities across 33 countries. Bolt is trying to create a social atmosphere among its clients, inviting riders to occupy their front seat and embracing the driver in a generous act of celebration. Uber creates an environment for more specialized drivers where social experience with the driver is not emphasized. Everyone older than 21 with a driving history of at least one year and a valid driver's authorization will become a driver on the network. Apart from this, inexperienced drivers are forced to attend short online video orientations via background checks. As soon as a new driver is admitted, the candidate becomes a self-employed agent, not workers. You are wholly controlled by where, when, and how far you travel.

2.1 Algorithmic management within the digital work platform

Under Uber's and Bolt's platforms, the management of ratings and ratings with information, decision-making, and appraisal functions of human managers under organizations is focused on three algorithmic processes: passenger-driver matched, interactive display of price-priced areas, and data-driven assessments [5].

2.2 Allocation of work on the platform: driver-passenger pairing

Drivers must activate their platform app to perform their function. This allows drivers to get their work done and execute it. According to Uber, "the nearest driver to this rider immediately gets the tour order in a 15-second timeframe to receive it," until the user orders a trip across the portal (i.e., the smartphone app) [2]. But both Uber and Bolt deny specifics of the pairing algorithm based on proximity. Our analysis, however, showed that the algorithm could also have other variables. The passengers' request contains information on the location, image, name, and rating of passengers.

When the driver agrees with the request, it will notify the passenger, and the driver will be able to drive to the position where the passenger is waiting for the driver to begin the journey. Both Uber and Bolt do not have the option or drivers' desires, whether they choose to obtain on the network for those passengers or trips. Both platforms facilitate passive dismissal of allocated passengers only, i.e., if the driver does not want to approve the request, the accepted 15-second window must wait before a passenger's new request is sent for the app.

2.3 Dynamic platform display of surge-priced areas

The regular fare defines the price on the network, which fluctuates according to a complex pricing algorithm, which dictates how the pricing is processed. The businesses are working together to make this feature clear to drivers and the public. This chapter would adopt the word Uber, i.e., "surge price" as demand exceeds supply, competitive pricing algorithms tend to rise, and price increases to support market equilibrium [2]. "Dynamic pricing" is used by Bolt Surge pricing will play a vital role in form driver sales, as the 80% commission stays stable during these high price cycles. As of July 2019, the two organizations, with a chart with areas shaded in various colors depending on the real-time price measured, high-priced display areas inside the network. To allow the demand and optimize the overall number of purchases, drivers can switch to locations with higher driving demands (and prices) than available drivers.

2.4 Rating and acceptance rates system used to monitors drivers' performance

Once the ride is completed, all riders and drivers can evaluate each other through a ranking system built into the platform. The ranking system is a five-star rating system. Bolt points out that when a driver is scored, it is critical that passengers consider whether their drivers are generally polite and polite, that they navigate well, that they are safe, keep the car clean, and want to use Bolt again [2]. The driver also has the approval score, which is determined by the number of approved rides divided by the driver's total number of requests. Drivers are encouraged to maintain a high rate of acceptance of the ride by periodic incentives providing hourly pay that is assured if the driver's acceptance rate is above a certain specified amount. Drivers with low average passenger rating and approval rates are subject to scrutiny. They will also automatically face deactivation on the network. Likewise, riders with a low rating score are at risk of being refused by drivers. Drivers have the option to disregard incoming requests from low-rating passengers below their desired threshold. Drivers that retain strong passenger scores and approval rates are often elevated to advisors or recruiters. In addition to driving for the service, mentors and recruiters recruit new drivers and oversee the application process while receiving additional money for these jobs.

3. Methods

The study is focused on studies conducted between April and August 2019. The study was mostly based on 41 detailed qualitative interviews of those working on the interactive job channels (i.e., Uber and Taxify). We selected interviewees from among those who replied to the broader survey. A non-random survey of 355 Uber and Bolt drivers was performed, and quantitative data from this more general population was also obtained. Following the claims made by [2], we started our research with a web-based survey designed to hit drivers on leading shared riding platforms: Uber, Uber Eats, Bolts, and others. These businesses consider their network staff as independent contractors and not employees.

We recruited drivers to take part in our mobile online survey in two ways. First of all, we aimed ads to Facebook network drivers. The benefits and drawbacks of using Facebook ads for low-wage jobs can not be overemphasized [2]. The prevalent use of Facebook in South Africa and other developing countries makes it a valuable sampling environment. It should be judiciously compared to the telephone-based means of interviewing [3]. We tailored Facebook advertising for our survey to individuals between 18 and 55 who spoke English and reported working on one of the two channels. The commercial was shown to 47,981 users, some of whom were shown advertising several times, and 4869 users clicked on the survey link, just above 10% of those to whom the commercial was shown. Of those who clicked on a survey link shared in the commercial, 476 described themselves as platform staff started the survey, and 355 completed the study.

Respondents to the study were not indicative of the entire employees of the site. Those who marked themselves as platform employees on their Facebook profile, and those who belong to communities connected to the interactive job platform on Facebook – the two networks from which respondents have been drawn – may be more attached to this job than other platform employees and may also vary from the people of other platform workers in different unexpected ways. Given the lack of data on this digital workforce forum's makeup and experience and the difficulties of accessing this sample demographic through other means, the analysis presented here provides a significant, although the non-representative, reflection of this developing market.

One of the questions posed in our online survey was whether the respondent would participate in an hour-long telephone or in-person interview with R120 as an incentive for participation. We called for contact details from the interviewees who showed a desire to schedule the consultation. Of the 355 people who completed the online survey, most (245 or 69%) showed interest in engaging in the follow-up interview. To determine who contacted the 245, we tried to optimize diversity in terms of age, size, ethnicity, geography, political preference, and family household income. We conducted interviews with 41 platform staff or 11 percent of those who showed interest, and interviews lasted at least one hour or less. We use pseudonyms all over to protect the anonymity of the respondents.

Although our survey is not generalizable to the digital network workers' population, interviews have helped us classify processes using reasoning rather than statistical inference and to achieve fullness [4]. Also, we found the same similarities through interviews, which enabled us to trust the findings: that is, identical similarities in the interviewee's account across various platforms. However, we equate different employees' perspectives on other interactive job platforms or the same worker's experience working on multiple platforms. Remarkably, respondents consistently emphasized Uber's higher level of time and task management relative to other networks.

3.1 Passengers interviews

The responses were triangulated by questioning 19 passengers who used or are currently using Uber and Bolt at 6 locations in two cities (Pretoria and Johannesburg). On average, the passengers used the service 4–7 times over three months. Ten of the passengers questioned used both Uber and Bolt, 4 used only Bolt, whereas five used only Uber. The interview aimed to either affirm or disprove the impressions shared by drivers of how passengers use the platform, mainly how drivers are classified and their actions and attitudes towards price increases.

3.2 Analyzing archived data: drivers' online forum

We analyzed the drivers' details on the online forums where all drivers are registered to do so. This is because most drivers said in our interviews that the forums were the primary source of information and places for them to socialize. Two online forums have been observed: one that is not moderated by Uber and Bolt, like different Facebook communities, and the official Facebook pages moderated by Uber and Bolt.

One author who registered as a Bolt driver was granted access to Bolt's (Taxify) and Uber Driver's and Clients 'new driver' Facebook forum, which provides information that is specifically relevant to the business. We also entered other unmoderated private driver groups on Facebook by applying to participate as researchers. This was done to prevent being mistakenly posing as a driver. We have retained a single observation status during this review. To ensure that the method used in [5] was followed, we sampled 142 posts and responses on the online site, noting the algorithmic features chosen from the thousands made over three months.

4. Analysis

We triangulated our observations by conducting and archiving interviews. The interview transcripts [6] were also evaluated qualitatively. The posts excerpts were reviewed, and on the web platform, responses were analyzed using the qualitative data coding program Atlas. The dataset was then classified using three platform algorithmic features, and the data were then opened to the level of the phrase or paragraph around each function. We evaluated the remaining data to identify essential subjects, including social sensibilities and socialization. This lead to the introduction of a total of 289 definitions. The ideas were later grouped into 16 themes, and emerging phenomena were clarified. To understand the connection between categories, we concentrated on eight categories related to our research questions concerning algorithmic management. When checked at 10% of transcripts (Kappa = 0.81), the final coding method demonstrated strong reliability over two coders. We have been able to settle disagreements between coders by debate.

5. Findings

This section discusses how drivers reacted by assigning work algorithms, supporting details, evaluating work success, and how drivers socially make sense of the system's algorithms using online forums.

5.1 Background: driver inspiration

According to the drivers interviewed, one significant advantage of working on digital platforms is the high degree of versatility that the system has when and when it operates on the modern operating network. The low level of commitments and dedication necessary for signing up for the platform is another advantage stated by drivers. Many drivers work on the system full-time, and others do part-time—some work for fun, some of them to feed their curiosity. Many drivers used the forum to coordinate their everyday routine in gaining extra revenue. In addition to the platform's financial versatility, several drivers interviewed said they draw social motivation from the venue. For example., some drivers found meeting enjoyable, connecting with new people, and inspiring the group to participate.

5.2 Algorithmic management: proximity-based passenger-driver pairing

Our findings demonstrate how the openness of algorithmic assignment and the matching of drivers with passengers impact driver communication, workaround, and work strategy. It further describes the potential impacts of computerization decisions used by staff in a particular conventional working environment.

5.3 Collaboration with algorithmic management in terms of passenger allocation

A previous study argues that humans will collaborate less with computer automation activities than human beings. For Uber and Bolt, passengers are allocated to drivers on their applications. There is a guideline on the pace of acceptance and a cut-off period for approval, which motivates drivers to consider as many algorithm assignments as possible. One of the drivers interviewed explained the following: *"passengers may only be refused if the choice can be made within 15 seconds. Both Uber and Bolt find it impossible to deny allocated passengers for different reasons. And if the position on the map where you pick up a passenger is displayed, whether you are unfamiliar with the area, you will not be able to decide within 15 seconds if you want to go to that area."*

Remarkably, one of the reasons that improved drivers' engagement was to figure out whether or not the assignment of passengers made sense to them. While the passengers' assignment was generally based on the similarity between the driver and the passenger location, other considerations led to the assignment. The following variables are the mutual ranking between the driver and the passenger and the driver's login time. This sometimes causes an appeal from remote passengers delegated to drivers who are not closer to passengers. As this situation transpires, several drivers testified that they did not acknowledge the "uncomplimentary" trip task, considering that they must travel a long way (such as more than 10 minutes) to pick up a passenger. Another driver interviewed,

for example, explained: "*I monitor where other drivers are when I'm not with a passenger. So, if I see between two to four drivers in area X and a request is coming from area Y to area X, given that there are at least two to four drivers sitting right there ready to go, then my instinct will tell me that one of two things would have happened. All the drivers transfer the trip request, which is unlikely because they cannot be sitting there and transfer the assigned trip. Or the GPS mapping was not properly coordinated by the system that is assigning the passenger. Then there is a mistake in sending the request to me who is over ten minutes away from the passenger instead of assigning it to a driver who is at least a minute away.*"

The explanation above found it difficult to grasp whether the algorithm assigns the passenger to the driver in error or for legitimate purposes. That's because we could not clarify how the driver app algorithm determined the assignment. Another driver interviewed believed that the task is often made by accident and that he opposes it. However, regardless of how far and inconvenient a passenger assignment might be, drivers can consider the assignment as long as it makes sense to them. For instance, one driver explained as follows: "*When it comes to distance, sometimes I've driven as 15 kilometers, and I know that the fault was not the fault of the passenger. It happened that there weren't many drivers out on the day I happened to be the closest driver at that time.*" This means that the explanation for those passenger assignments might have been significant, but at the moment, this was a feat.

5.4 Workaround strategies for algorithmic assignments

Drivers on the platforms recognize that the classification of passengers is dependent on position proximity. This allowed them to prepare and collaborate to monitor the algorithmic assignment as part of the existing device functionality. They carefully manipulate when and when to work and when to turn to the driver mode on the app so that the kinds of requests and passengers of their choosing can be delegated to them. They limit the location they operate by shutting off the driver feature on the app while returning from a long journey, avoiding low places to prevent dangerous and hazardous conditions, and not going to neighborhoods where the bar is situated to avoid intoxicated riders. They restrict their place to suburban areas to push customers to bars. They get repeat passengers by phone into the travel arrangement, asking passengers to call for a ride while they are in the driver's seat to be allocated.

Some drivers keep away from each other by monitoring other drivers' GPS positions to avoid vying with each other for passenger demands. If drivers take a rest but do not want to turn off their driver app to benefit from the hourly payment promotion, they park their cars in the same place on the GPS, which stops them from receiving any trip order. Both Uber and Bolts express only basic assignment rules, such as "closest drivers are assigned to a ride." This basic understanding allows drivers to maneuver around algorithm assignment strategies. However, the task algorithm's lack of specifics seemed to promote drivers' reluctance to make decisions, often giving them a pessimistic mood to the firms. One of the drivers explained the following.: "*Uber is very tight-lipped on what is going on. What I'm saying is they inform us 'we only assign it to the closest driver,' but we don't understand what is happening behind the scene.*"

5.5 More understanding of the algorithm to the benefit of the driver

Our analysis showed that drivers benefit from a thorough understanding of the algorithmic assignment. Drivers with more experience and knowledge of the method build workarounds to escape unwanted trip assignments. In contrast, those with less knowledge and understanding reject undesirable trip assignments, reducing their approval rates. One of the drivers, for instance, clarified that Bolt's assignment algorithms weigh the number of hours the drivers have been online, and the proximity of the driver's radius to pick up passengers increases the longer they wait for passenger assignments. As a result, he uses his knowledge of the algorithm to turn on and off his driver program periodically when at some stopping point, to disable the device from assigning remote trip requests to him. However, not all drivers have access to this knowledge. Our findings have shown that Bolt drivers who do not understand how the algorithmic assignment functions have attributed the remote assignment to the machine error. Many assumed that higher-ranking drivers would earn priority passenger assignments. These drivers are unable to establish workaround techniques to prevent the assignment of remote trip requests.

5.6 Algorithmic assignment versus preferred pick up

Our results suggest that drivers are usually pleased with the amount of power they have over the assignment algorithms. However, a few drivers have clarified that they are not happy that they have no control over the radius that the assignment algorithm attempts to send passengers to them. One driver interviewed, though, clarified that he wants the assignment algorithm to function to see all the incoming trip requests to pick the one that matches his choice from among them. He also clarified that he had a plan to select the trip requests' position, and he learned the knowledge and comprehension of how best to use them.: *"At some point on some days, there usually are many good trips within some areas. For instance, on Friday, around 6 pm in [area names hidden], there are many trips to and from the airport trips. So, you focus on those.*

Another good thing is that if it's a busy Friday night. And the best way to get that done is to take something that's not in a close area, maybe going far, but then coming back in. It gives us the option for a change of pace." Uber and Bolt's algorithm assignment removed this slight degree of power and predictability. He said that while he tried his hardest to be in those positions in Uber and Bolt's system, there was no assurance that the system would delegate trip requests to him in those places. He's frequently assigned trips outside the area he likes, and he did not want to travel to those places just for a change.

5.7 Algorithmic system does not cater to the driver abilities, feelings, and motivation

Some drivers clarified that hike rates are implemented in some familiar places instead of using the device in any community. They also explained that they frequently travel far to these places as soon as they get alerts about the increased prices. However, half of the drivers interviewed are not inspired by a rise in price notification because the supply-control algorithmic framework caters to their skill, emotions, and inspiration. It was also found that the surge-price shifts too rapidly and unpredictably so that they could not be used in a competitive way to raise their incomes. Furthermore, being in the surge region guarantees that travel requests would be distributed within the area. Any drivers have clarified that the spike price often disappears as they arrive in the spike region.

5.8 Data-driven algorithmic evaluation: performance evaluation through driver rating and acceptance rate

The guidelines used for the approval rate and the driver-passenger rating system benefit the system's overall operation. However, this quantitative method makes drivers responsible for all relationships and is often perceived by drivers as unjust and does not have a meaningful or desirable result, thus negatively affecting drivers.

5.8.1 Treating all assignment rejections equally: an act of unfairness

The acceptance rate regulation requirement is used to encourage drivers to accept most of the requests, which allow more passengers to be allocated to their journeys, arrive at their destination on time, and inspire trust in the use of the platform system. The fact that the degree of approval of the assignment is sustained puts a high burden on drivers to consider most of the trips assigned to them. For example, one of the drivers interviewed explained why he often welcomed a request, even though it was not acceptable for him to consider it.: "*I had no choice but to accept it. I accepted because I want my acceptance rating to go high. We are under a lot of pressure. I cannot give any reason as to why I do not accept it, so I have to accept it.*"

Similarly, Assignment algorithms penalize rejection of passenger requests by drivers, which reduces the driver's approval rate. However, specific drivers often have legitimate reasons, excuses, and circumstances that explain their reluctance to accept travel requests. For example, drivers can prefer not to welcome passengers without photographs at night for safety reasons. Some queries are diverted to the drivers when a few seconds are left for the drivers to approve. This could be attributed to technical issues with the system. Drivers often send e-mails to the company's representative(s) when they believe they have legitimate grounds to refuse the appeal, considering that they will not be penalized for legitimate refusals. Still, most of the time, their e-mail will go without getting a reply from the members.

5.8.2 Using quantitative data only leads to inaccuracy in quality of service

Our study indicates that the Passenger Driver Rating System builds faith and operation behaviors in network work schemes. However, there is not sufficiently metric to be used for driver efficiency assessment. Some drivers rely on passenger ratings as indicators to approve travel requests from passengers or not, believe more in higher-rated passengers, and exercise caution on lower rating passengers. Any of the passengers interviewed clarified that while they ignore driver ratings, the fact that the driver rating system is in place gives them a sense of protection.

The ranking system also allows the drives to provide a feeling of standard service on all their trips. One of the drivers interviewed, for example, explained: "*I want to get five stars. So, I make sure that I am friendly. I relate with the passengers. I ask them their preferences immediately; they step in the vehicle. I ask them if they want me to switch on the Aircon or heater. Or they want me to roll the windows down or up depending on the weather. I also ask them if they will prefer that I play the CD or listen to Radio. And to ones that want the music played, I ask them what type of music they would like to listen to? I sometimes offer them sweets and gum.*"

Drivers take ratings very seriously. High ratings like 4.95 have become the basis upon which some of the drivers pride themselves. Although scores below 4.0 upset some drivers and cause them to fear losing their work on the network, they fear being tracked, measured, and arbitrated by customers. These seem to have adverse psychological effects on those who have not scored close to 4. One of the drivers interviewed explained the following.: "you are forced by *the rating to be careful and cautious. Because it looks like what you are doing is being monitored, rated, and judged. If your rating is low for some time, you could be asked not to drive for some on the platform and reapply for the job later.*"

Some drivers believe that their passenger scores are not a fair representation of their success in terms of operation and driving, as one driver explained: *"It's like in rugby, the line-up does not indicate the specifics of a player. A player could hit three runs, and three tries it doesn't mean that such a player is a productive player."* Some drivers also explained that passengers' psychological and physical conditions, such as being drunk, having a hurry to meet, or catching a late flight, are often reasons they give lower ratings after the trip. They also explained that some passengers interpret the machine anomaly or algorithm functions as bad experiences. This is because they are sometimes unable to manage due to errors (such as GPS errors, traffic jams, rising pricing, etc.) on their side. As a result, they give drivers lower ratings.

5.9 Making sense of the online platform as a forum for discussion

Drivers on the digital work platform function separately in distributed locations and, when they do, make use of internet networks as a significant avenue for their socialization. They use online communities such as Facebook groups, WhatsApp, Telegram, Line, Hangout, etc. as sites for different forums to address their work on the site and the algorithmic site management. One of the constructive aspects of making sense of the online forum is to explore how driver efficiency in terms of ratings and recognition can be strengthened and sustained.

Experienced drivers are often eager to exchange strategies and suggestions with inexperienced drivers who ask questions about boosting their scores. They're willing to do this, relying on the expertise they have gained over time. One of the new drivers in one of the forums, for example, asked how to raise his scores after 74 trips in four weeks. Approximately 120 comments were made within three hours of the publication of the query. Some drivers sympathize with his feelings; some share their everyday encounters as part of the first challenge of getting to the platforms. However, several commenters expressed the specific tricks they use, such as designing a service information manual for their cars' back seat, heading to the Central Business District (CBD) at lunchtime on several short journeys building genuine connections with travelers, etc. Any seasoned drivers often make it clear to beginner drivers that ratings will be steady over time and warn them not to encourage tension to get too much on ratings.

However, in terms of knowledge utility and practices that use assignment algorithms and rising pricing, they tend to have a lesser impact. Most of the posts that raise questions are concerned with how assignment algorithms and surge pricing work, understand the competitive display of surge pricing, and the fields in which higher pricing is often applied. Most of the real-time questions and conversations are about exasperating events — no travel requests in high-priced places or remote travel requests that entail long driving periods.

Much of the online forums' conversation centered on offering emotional and social help instead of informational support. For instance, the driver's post was a matter in which he was irritated at the trip request by the algorithm assigned to him, which forced him to ride from east to west of the area, even though he could see other drivers in the vicinity of the passenger making the request. Much of the comments in response to the post were worried about giving emotional support. However, there were no remarks to justify why the algorithm allocated such a trip to him. Company representatives usually do not appear on the forums to answer the driver's questions.

6. Discussion

This section discusses how our findings will continue to develop and improve the architecture of algorithmic data-driven management.

6.1 Designing and enhancing the algorithmic trip assignment

Myriads of Algorithmic Passenger Trip Requests for Drivers on the Automated Work Network, such as Bolt and Uber, are made automatically within seconds of ride requests. The regular and speedy approval of the drivers' assignments ensures the platforms' operation's reliability and efficacy. Hence, it is the potential to maximize the number of passengers who can make short journeys. Our results indicate that the task is not based purely on the root of the task (i.e., person versus algorithm) in the digital job site's algorithmic management. However, how the assignment is performed and managed, how staff on the site interact and agree with the assignment [7]. According to the study's findings, the details displayed on the computer, the constraint of the time to approve the request for a ride, and the approval rate jointly decides the degree of cooperation between drivers and the assignment algorithm.

Furthermore, our findings indicate that the openness of the algorithmic assignment's method can enable algorithmic management to produce a high degree of coordination and cooperation with drivers. While Uber and Bolt clarify that their algorithmic handling of assignments is based on the proximity of drivers and passengers, our findings suggest that there are other considerations that the algorithm takes into account. This is why passenger demands are often not allocated to drivers who are nearest to travelers. [8] suggests that the art of describing or encouraging staff to ask questions about each trip assignment may help minimize drivers' refusal rate when assigned to a trip that is not close to them, instead of attributing those assignments to a technological glitch to the network work method. This is because clarity and openness will boost drivers' negative emotions or contrasting ideas about businesses that run digital channels. Our results are consistent with previous studies on suggestion networks where exposure has improved users' confidence and adopt suggestions [9]. This study also draws attention to new transparency implications that have gained little attention in previous studies on intelligent systems, i.e., the effect of transparency in algorithmic management and how to open algorithmic assignment leads to improved job strategies address limitations. Drivers who have a thorough understanding of the assignment algorithm can set up workarounds to prevent a less economical journey. In contrast, drivers who only have basic knowledge of proximity-based algorithmic assignment cannot do the same.

It was also discovered that being autonomous contractors on digital work platforms is a crucial factor in the network drivers' preference and stability, which leads to a lack of control over the algorithmic management of trip assignments. Another consideration is the lack of familiarity with such systems. For example, a driver who acts as a driver and a passenger likes a mechanism that allows him to view and select travel requests directly. He assumes that the algorithm is now managing the choices that he will make himself. This is understood as opposition to transition, which often poses versatile, ethical concerns regarding the trend of emerging technologies that are harmful to people's regulation for the sake of overall machine performance and the consequences of learning and growth while at work [10].

6.2 Integrating information support into the algorithm management design

Supply–demand management algorithms were initially developed to solve statistical optimization problems concerning non-human entities. In Bolt and Uber, however, they are used to inspire and regulate human behavior. This poses issues, as the supply–demand management algorithm does not consider the speed at which drivers run. Consistent with previous studies on a smart agent that sought to promote healthy behavior [5], the algorithm struggled to account for people's feelings of inequity towards higher prices and overlooked drivers' social and altruistic motives. This highlights the importance of algorithmic management: (a) the speed and manner in which people work, (b) different forms of inspiration rather than just economic ones, and (c) the feelings people have about the choices that algorithms make. Also, some drivers did not trust the high-priced areas as they had more faith in their expertise. Transparency of how the surge priced region was computed in real-time could increase workers' trust in algorithmic knowledge.

6.3 Integrating data-driven performance evaluation into the algorithm management design

Using driver ratings and approval thresholds, businesses can test drivers on a wide scale. In particular, driver ratings may appear to be a valid assessment tool because customer loyalty is a significant indicator of service performance and human service provider efficiency. Using only the monitored output data in assessing staff, though, has uncovered several problems that may arise if one depends too heavily on quantified metrics without further analysis of their meanings and complexities. Consistent with previous studies on letter-grading schemes or numerical appraisal of teaching skills [11], several random variables beyond drivers' reach affect the way passengers rate drivers. The effectiveness and accuracy of the averaged group assessment, rather than the in-depth holistic review carried out by a human manager or peer, is also at issue. As P18 put it, "you are at the hands of unknown strangers, in [his other work] you are judged by people you meet." Our research also reveals the pitfalls of following a 5-star rating system shared with web goods, content, or business ratings for human employees. Drivers felt that passengers rated conservatively as they did in online reviews; however, interviews with passengers suggest that they are more lenient and positive than drivers think. This misunderstanding indicates that a 5-star ranking metaphor and a heading may have contributed to incorrect comparisons. Finally, the long-term motivational impact of the ranking is still at issue. As the driver ratings were weighted over several journeys, the effect of one positive or negative ride was reduced, and the drivers in our study were less susceptible to changes in their ratings until they were above the minimum threshold.

Effective management offers working procedures and enables improvisation in response to changes and exceptions [12]. On the other hand, task algorithms have penalized all driver rejections of assignments, even though individual drivers had valid motives and situations for doing so. While we have not seen any significant problems with this lack of versatility in our algorithmic analysis assignment, it poses an open challenge in building flexibility in algorithmic management.

An examination is optional in most online rating programs, and many even miss the process. In the ride-sharing service, both riders were urged to score their service experience, and most of them did. Being held responsible for all communications, the drivers were well aware of this external assessment's nature. Trying to offer adequate care with all customer encounters could pose psychological stress to staff. Besides, as comprehensive research on extrinsic incentives' effect on intrinsic motivation indicates, an external device may undermine the innate incentive drivers may have and alter the sense they assign to their behavior. From the passengers' point of view, the uncertainty of the provider's motivation for friendliness and good service risks making the provider's relationship more superficial and perfunctory.

6.4 Designing algorithmic management that supports online forum

Our study found that online forums have been the primary place where drivers socialize, ask questions about each other, and share information and strategies.

In most research on intelligent systems' sensory and mental models, the focus was on individual sensemaking [13–15]. Our research indicates that social sensemaking is another critical task that needs to be correctly understood and endorsed if intelligent systems are effectively implemented. Social sensemaking events at the driver's forums adopted "fragmented social sensemaking" [10]. Many involved contributors have no overarching authority to put together various thoughts and narratives into a cohesive plot. This kind of sensemaking was useful in addressing rating enhancement techniques. There were no accurate or incorrect responses, and employees' knowledge and learned and improvised techniques played a critical role. On the other hand, fractured social sensemaking fell short on topics where only an authority figure had the correct details. This highlights possibilities for creating organized online social sensemaking algorithmic features where individuals can draw on each other's expertise.

7. Scope of the study

As with many research studies, there are certain drawbacks to this study. Our research work's findings are derived from interviews with a sample of drivers from three cities in South Africa, namely Johannesburg, Pretoria, and Durban. We have used the study of archival records. We were unable to perform interviews with the official members or creators of both Uber and Bolt. The organizations that run and maintain the platform's work processes were against their organizations' policies.

Consequently, we conclude that this work's results should be complemented by potential studies that will use various research techniques, such as tests, surveys, and ethnography. This study was performed in the context of the on-demand travel order, which is intended to improve the provision of the future of employment. We also assume that more analysis is needed in various institutional ways, such as with full-time workers.

8. Conclusions

Computer algorithms rapidly assign, refine, and evaluate work. This article discussed the effect of this algorithmic, data-driven management on Uber and Bolt's new ride-sharing services. This study's qualitative research results illustrated possibilities and difficulties in the architecture of human-centered algorithmic job assignment, knowledge and assessment, and the importance of fostering social sensemaking around the algorithmic method. The implications for HCI, CSCW, and Artificial Intelligent Systems research were discussed. We hope that this study will stimulate future work so that we can empower human workers to work with intelligent machines not only in an accurate but also in a rewarding and meaningful way.

Acknowledgements

This study was supported and funded by the research and innovation department of the Tshwane University of Technology. We also thank Mrs. Akinsola Oluwabunmi for reviewing and providing feedback that helped enhance the paper's quality.

Author details

Olalekan Samuel Ogunleye* and Billy Mathias Kalema
Department of Informatics, Faculty of ICT, Tshwane University of Technology, South Africa

*Address all correspondence to: olalekan@olalekansamuel.com

References

[1] J. Berg, M. Furrer, E. Harmon, U. Rani, and M. S. Silberman, "Digital labour platforms and the future of work: Towards decent work in the online world," NY, USA2018.

[2] Bolt. (2019, July 5th). *Improving Your Ratings*. Available: https://support.taxify.eu/hc/en-us/articles/115003396994-Improving-Your-Ratings

[3] Braga M, Paccagnella M, Pellizzari M. Evaluating students' evaluations of professors. Economics of Education Review. 2014;**41**:71-88

[4] J. Burston, D. Nick, and H. Alison, "Digital labour: Workers, authors, citizens," in *Digital Labour: Workers, Authors, Citizens*, ed: Ephemera: Theory & Politics in Organization, 2010.

[5] L. Christian, K. Scott, ., K. Purcell, and A. Smith, "Assessing the Cell Phone Challenge to Survey Research in 2010," ed. Washington, DC: Pew Research Center, 2010.

[6] E. Costanza, J. E. Fischer, J. A. Colley, and N. R. Jennings, "Doing the laundry with agents: a field trial of a future smart energy system in the home," in *CHI*, 2014, pp. 813-822.

[7] S. Maitlis, "The social processes of organizational sensemaking" *Academy of Management Journal,* vol. 48, pp. 21-49, 2005.

[8] H. Mintzberg, *The Nature of Managerial Work*: Harpercollins College Div, 1973.

[9] Mutlu B, Forlizzi J. Robots in organizations: The role of workflow, social, and environmental factors in human-robot interaction. In: *HRI*. 2008. pp. 287-294

[10] N. Srnicek, *Platform Capitalism*: Hoboken, 2016.

[11] Newell S, Marabelli M. Strategic opportunities (and challenges) of algorithmic decision-making: A call for action on the long-term societal effects of 'datification. The Journal of Strategic Information Systems. 2015;**24**:3-14

[12] Vallas SP. Platform capitalism: What's at stake for workers? New Labor Forum. 2019;**28**:48-59

[13] Y. Rayoung and W. N. Mark, "Learning from a Learning Thermostat: Lessons for Intelligent Systems for the Home," in *UbiComp 2013*, Zurich, Switzerland, 2013, pp. 93-102.

[14] Schneider D, Kristen H. Consequences of routine work-schedule instability for worker health and well-being. American Sociological Review. 2019;**84**:82-114

[15] Todd K., Simone S., Margaret B., and I. K., "Tell me more: The Effects of Mental Model Soundness on Personalizing an Intelligent Agent," in *CHI 2012*, Austin, Texas, USA, 2012, pp. 1-10.

Chapter 13

Overview of Some Intelligent Control Structures and Dedicated Algorithms

Kuo-Chi Chang, Kai-Chun Chu, Yuh-Chung Lin and Jeng-Shyang Pan

Abstract

Automatic control refers to the use of a control device to make the controlled object automatically run or keep the state unchanged without the participation of people. The guiding ideology of intelligent control is based on people's way of thinking and ability to solve problems, in order to solve the current methods that require human intelligence. We already know that the complexity of the controlled object includes model uncertainty, high nonlinearity, distributed sensors/actuators, dynamic mutations, multiple time scales, complex information patterns, big data process, and strict characteristic indicators, etc. In addition, the complexity of the environment manifests itself in uncertainty and uncertainty of change. Based on this, various researches continue to suggest that the main methods of intelligent control can include expert control, fuzzy control, neural network control, hierarchical intelligent control, anthropomorphic intelligent control, integrated intelligent control, combined intelligent control, chaos control, wavelet theory, etc. However, it is difficult to want all the intelligent control methods in a chapter, so this chapter focuses on intelligent control based on fuzzy logic, intelligent control based on neural network, expert control and human-like intelligent control, and hierarchical intelligent control and learning control, and provide relevant and useful programming for readers to practice.

Keywords: artificial intelligence algorithm, adaptive fuzzy control, neural network, expert system, learning control

1. Intelligent control based on fuzzy logic

1.1 Basic knowledge of fuzzy logic

A set of things can be distinguished based on binary logic. The essence of the set concept is to classify or divide things according to certain attributes. The whole of all the elements of the research object is called "universe", which is represented by U, also known as "set", "entire domain" or "space". Eigen functions are an important way to represent classical sets. For fuzzy sets and fuzzy concepts, from the perspective of set theory, the connotation of a concept is the definition of a set, and expansion is all the elements that make up a set. In people's mind, there are many concepts that are not explicitly extended, called fuzzy concepts, such as "high" and "short" to describe height [1].

The general approximation analysis of fuzzy systems is a non-linear mapping from input to output. It consists of multiple "If ... Then ... " rules. These rules have simple geometric features in the input and output spaces X × Y, and define many fuzzy blocks on the X × Y space. "If X is a fuzzy set A1, then Y is a fuzzy set B1." This rule corresponds to the A1 × B1 Cartesian product of the input and output spaces. The overlapping function f: X → Y of fuzzy blocks composed of fuzzy rules can be used to approximate the function f: X → Y with the fuzzy system F: X → Y.

Summarizing the control behavior of the above-mentioned people is following the basic idea of feedback control. An experienced operator can summarize the principle of manual operation to control the furnace temperature. The accumulated operating experience accumulated over the years can be summarized into several rules, such as "if the furnace temperature is low, add more fuel", etc., to train young operators that he is qualified for the job and gradually replace. Similarly, several control rules summarized from operating experience can be stored in a computer, allowing the computer to imitate human control decision-making behaviors to automatically control the furnace temperature. This is the basic idea of fuzzy control.

1.2 Components of the fuzzy control system

There is not much difference between a fuzzy control system and a computer digital control system. As shown in **Figure 1**, a fuzzy control system can be divided into four components: fuzzy controller, A/D and D/A interface to U, generalized objects (actuators and controlled objects) and sensors [2].

In the microcomputer fuzzy control system, the sensor replaces the human eye, the fuzzy controller replaces the human brain to control decisions, and the executive mechanism replaces the functions of the human arm and hand.

1.2.1 The basic form of fuzzy control

The decision-making process of fuzzy control is the three basic forms of human fuzzy thinking include fuzzy concept, fuzzy judgment and fuzzy reasoning. In the fuzzy controller, the fuzzy concept is a fuzzy linguistic variable represented by a fuzzy set. For example, the exact amount of error (continuous domain) is converted to the fuzzy quantity on the discrete domain (discourse domain). This process is called fuzzy quantization processing.

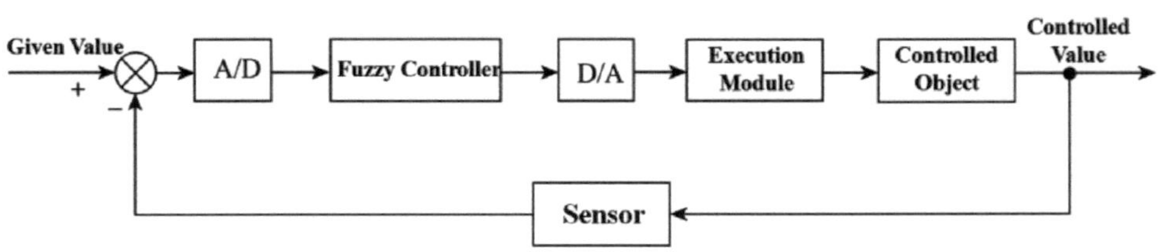

Figure 1.
Block diagram of fuzzy control system.

Human operating experience can be summarized into several fuzzy control rules in language. These rules can be described by a fuzzy relation matrix. It is actually a general principle of the operating process. These fuzzy control rules are also called the language model of the controlled object.

According to the syllogistic fuzzy reasoning synthesis rule, the fuzzy relationship determined by the fuzzy control rule is taken as the major premise of fuzzy reasoning, and the input fuzzy variable is used as the small premise. The known small premise and fuzzy relationship can be concluded by fuzzy inference synthesis.

According to the syllogism fuzzy inference synthesis rule, the fuzzy relationship $\underset{\sim}{R}$ determined by the fuzzy control rule is taken as the major premise of fuzzy reasoning, the input fuzzy variable is $\underset{\sim}{A}$ is taken as the small premise, and the known small premise A and fuzzy relationship R are synthesized by fuzzy relation inference Conclusion $\underset{\sim}{B} = \underset{\sim}{A} \circ \underset{\sim}{R}$.

As shown in **Figure 2**, the schematic diagram of the fuzzy control system is given. For the sake of comparison, the fuzzy logic thinking form of the person is placed above the figure. The three forms of fuzzy logical thinking correspond. Among them, the fuzzy quantization process is to obtain the fuzzy amount of the control variable [3].

For the sake of simplicity, only the error signal is selected as the input variable of the fuzzy controller and abbreviated as e (t) to illustrate the working principle of the fuzzy controller. The microcomputer obtains the precise value of the controlled quantity y by interrupting the sampling, and then compares this quantity with the given quantity to obtain the precise value of the error signal e (e = r-y, here the unit feedback is taken) as the input quantity of the fuzzy controller. The exact amount of the error e becomes the fuzzy amount of the error through the fuzzy quantization process, which can be represented by a subset $\underset{\sim}{e}$ of the corresponding fuzzy language set. Then the fuzzy relationship between the fuzzy amount of the error $\underset{\sim}{e}$ and the fuzzy control rule $\underset{\sim}{R}$ is used to make a fuzzy inference decision. The fuzzy amount of the control amount is shown in Eq. (1).

$$\underset{\sim}{u} = \underset{\sim}{e} \circ \underset{\sim}{R} \tag{1}$$

The fuzzy amount of the control amount cannot be directly sent to the actuator to control the controlled object, the fuzzy amount of $\underset{\sim}{u}$ of the control amount must

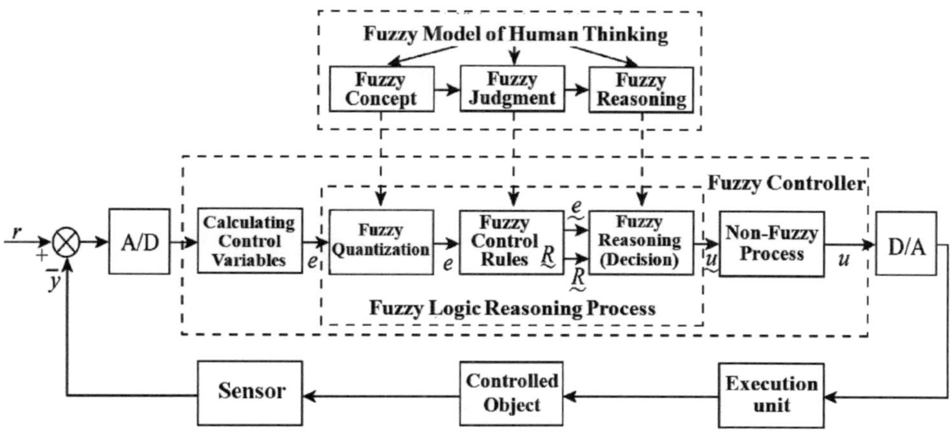

Figure 2.
System principle of fuzzy control.

also be converted into an accurate amount u through non-fuzzy (clarification, deblurring, and defuzzification) processing. After the digital-to-analog conversion into an accurate analog quantity, it is sent to the executive body, which controls the controlled object by one step. Then, it waits for the second sampling and performs the second step control. Continuously controlling in this way will make the actual output of the controlled object approach the expected value with certain accuracy, thereby achieving fuzzy control of the controlled object.

It is not difficult to see that the input quantity e of the fuzzy controller is an accurate quantity, and its output control quantity u is also an accurate quantity. Therefore, the control of the fuzzy controller is not fuzzy, and it can achieve precise control of the controlled object. Only the fuzzy logic reasoning is used in the inference part of the fuzzy controller. The advantages are: first, this reasoning decision does not require an accurate mathematical model of the controlled object; second, this reasoning decision simulates the thinking process of a person, and has intelligent and efficiency.

In the following, a single input single output temperature fuzzy control system is used to specifically explain the working principle of the fuzzy control system. An electric heating furnace is used for the heat treatment of metal parts. According to the requirements of the heat treatment process, the furnace temperature must be kept constant at 600°C. The experience of manual operation to adjust the voltage to control the furnace temperature can be summarized in language as the following control rules: If the furnace temperature is lower than 600°C, the voltage will be increased. When the temperature is lower, the voltage will be higher. If the furnace temperature is equal to 600°C, the voltage will be kept unchanged, as the voltage increases, the voltage will decrease.

According to the above control rules, the application of a microcomputer to achieve fuzzy control of the furnace temperature needs to be designed according to the following steps.

1. Determine the input and output variables of the fuzzy controller.

   Select the difference between the actual value of the furnace temperature and the set value as $e(n) = t0\text{-}t(n)$ as the error input variable, and select the voltage u to adjust the furnace temperature as the output variable of the fuzzy controller.

2. Determine fuzzy language variables of input and output variables.

   First, select a fuzzy subset of the input and output variables as:

   {Negative large, negative small, zero, positive small, positive large} = {*NB, NS, O, PS, PB*}.

Among them, *NB, NS, O, PS, PB* are English abbreviations of negative large, negative small, zero, positive small, and positive large respectively.

Second, the domain X of the selection error e and the domain Y of the control quantity u are both $X = Y = \{-3, -2, -1, 0, 1, 2, 3\}$.

Third, determine the membership functions of the input and output language variables as shown in **Figure 3**. From this, the assignment of fuzzy variables $\underline{e}$ and $\underline{u}$ can be obtained from this, see **Table 1** [4].

Establish fuzzy control rules using the above-mentioned rules for manually adjusting the voltage to control the furnace temperature, using the error as an input variable, and the voltage as an output variable, five rules can be written as follows:

1. If the error is negative, the voltage is positive; If $\underset{\sim}{e} = NB$ then $\underset{\sim}{u} = PB$.

2. If the error is small, the voltage is small; If $\underset{\sim}{e} = NS$ then $\underset{\sim}{u} = PS$.

3. If the error is zero, then the voltage is zero; If $\underset{\sim}{e} = O$ then $\underset{\sim}{u} = O$.

4. If the error is small, then the voltage is small; If $\underset{\sim}{e} = PS$ then $\underset{\sim}{u} = NS$.

5. If the error is positive, the voltage is negative; If $\underset{\sim}{e} = PB$ then $\underset{\sim}{u} = NB$.

In the above rules, the left side is expressed in Chinese, and the right side is written in English if-then fuzzy conditional statements.

1.2.2 Fuzzy matrix representation of fuzzy control rules

A fuzzy control rule is actually a set of multiple fuzzy conditional statements, which can be expressed as a fuzzy relationship from the error domain X to the control quantity domain Y. Because when the universe is limited, fuzzy relations can be represented by fuzzy matrices. In the furnace temperature fuzzy control, the universe of discussion X and Y are limited to 7 levels, so the fuzzy relation matrix can be used to represent the above fuzzy control rules.

The above fuzzy conditional statement can be expressed as a fuzzy relationship as show in Eq. (2).

$$\underset{\sim}{R} = NB_e \times PB_u + NS_e \times PS_u + O_e \times O_u + PS_e \times NS_u + PB_e \times NB_u \quad (2)$$

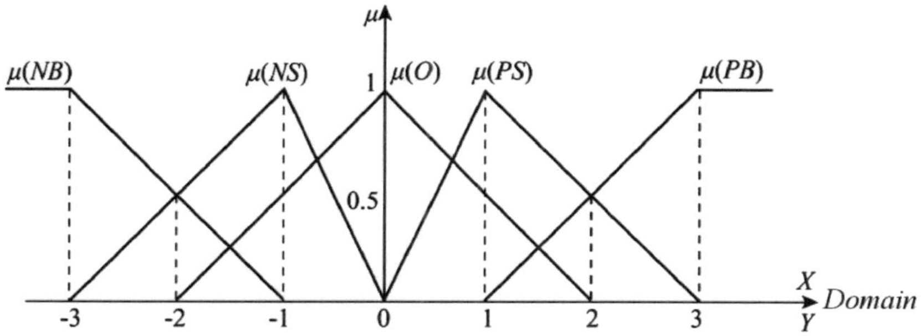

Figure 3.
Membership functions of language variables.

Membership				Language variable			
Quantization level	**−3**	**−2**	**−1**	**0**	**1**	**2**	**3**
PB	0	0	0	0	0	0.5	1
PS	0	0	0	0	1	0.5	0
O	0	0	0.5	1	0.5	0	0
NS	0	0.5	1	0	0	0	0
NB	1	0.5	0	0	0	0	0

Table 1.
Assignment table of fuzzy variables (e, u).

Among them, the subscripts *e* and *u* of language variables *NBe*, *PBu*, etc. indicate that they are language variables of error and control amount, respectively. **Figure 4** shows the fuzzy rule base.

1.3 Adaptive fuzzy control

1.3.1 Components of the fuzzy control system

1.3.1.1 The concept of adaptive control

In the 1950s and 1960s, since classic control was difficult to meet the high control performance requirements of aircraft, rockets, and satellites, a high-performance controller capable of automatically adapting to the changing characteristics of the controlled object, an adaptive controller, was needed.

In order to make the controlled object operate according to the predetermined rules, negative feedback control is used. A natural idea is that when the control performance of the controller does not meet the requirements, the negative feedback control idea is also used to control the controller itself to improve the control performance. This is the basic idea of adaptive control. Therefore, the adaptive controller must have two functions at the same time:

1. According to the operating state of the controlled process, a suitable control amount is given, that is, a control function.

2. According to the control effect of the given control amount, the control decision of the controller is further improved to obtain a better control effect, that is, a learning function.

The adaptive controller performs system identification and control tasks simultaneously. The essence of the adaptive fuzzy controller is to make a control strategy described in language by observing and evaluating the performance of the

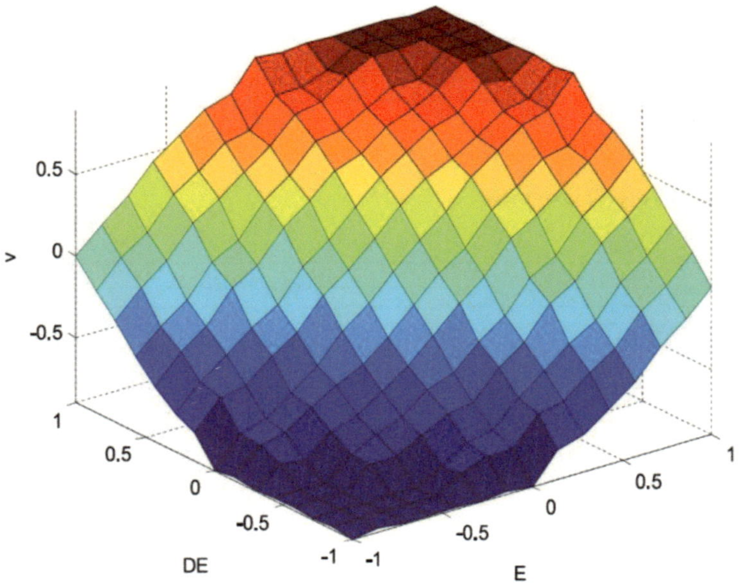

Figure 4.
Control surface of the fuzzy rule base.

controller. There are two types of adaptive control: direct adaptive control and indirect adaptive control. The principle of direct adaptive control is shown in **Figure 5**. It adds an adaptive mechanism to the basic feedback control system. It obtains signals from the original control system. The control performance changes that it can adaptively modify the controller parameters to make the control. Performance remains the same. The principle of indirect adaptive control is shown in **Figure 6**. It uses online identification to identify the parameters of the object, and then uses the identified parameters to adjust the control parameters through the parameter corrector to continuously improve and improve the control performance.

Direct reference adaptive control includes model reference adaptive control (MRAC), while indirect adaptive control is also called self-correcting control (STC).

Introduce fuzzy logic inference system in traditional adaptive control, or act as an adaptive mechanism, or as an object model, or as a controller, or both, forming different forms of adaptive fuzzy control, or fuzzy adaptive control [5].

1.3.1.2. The structure of adaptive fuzzy controller

The adaptive fuzzy controller is based on the basic fuzzy controller, and an adaptive mechanism is added. Its structure is shown in **Figure 7**. The adaptive mechanism in the dashed box in the figure includes three functional blocks, which are:

1. Performance measurement-used to measure the deviation between the actual output characteristics and the expected characteristics in order to provide information for the correction of the control rules, that is, to determine the correction amount P of the output response.

2. Control amount correction—the correction amount of the output response is converted into the correction amount R of the control amount.

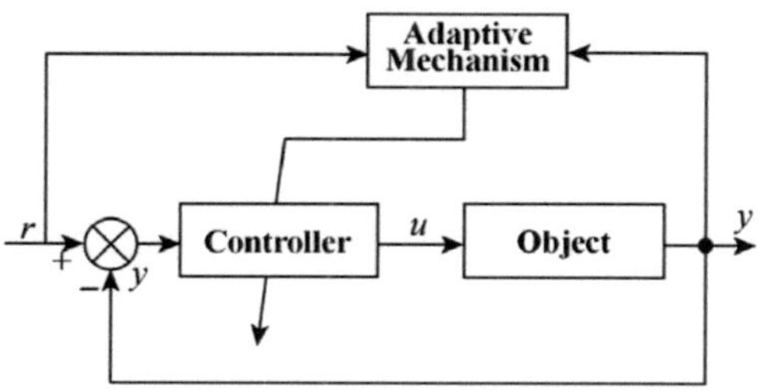

Figure 5.
Block diagram of fuzzy control system structure of direct adaptive control.

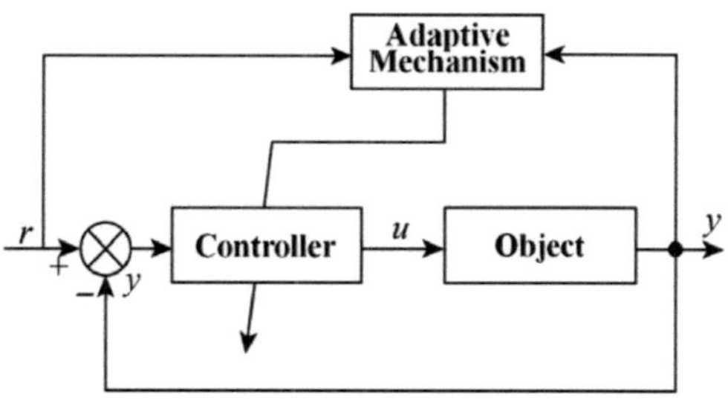

Figure 6.
Structure of indirect adaptive control.

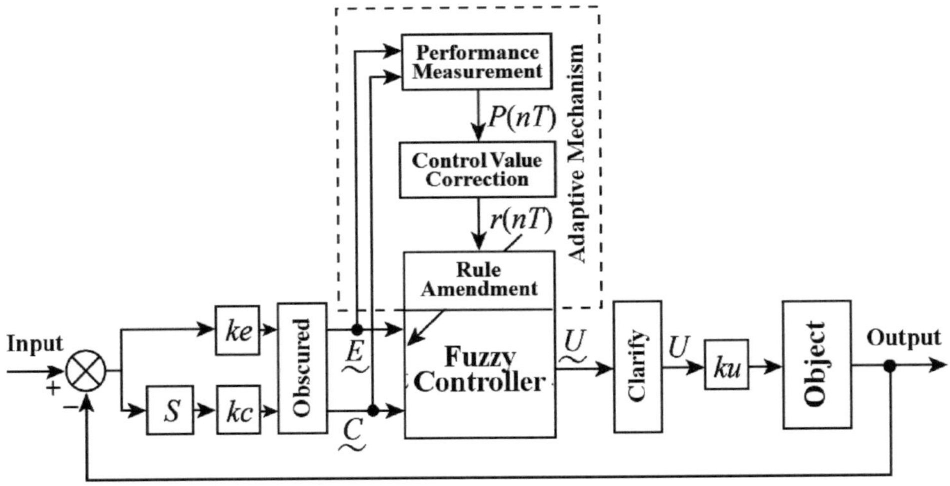

Figure 7.
Structure of the adaptive fuzzy controller.

3. Modification of control rules-correction of control amount is achieved by modifying control rules.

1.3.1.3 Principle of adaptive fuzzy controller

The adaptive fuzzy controller also needs to understand the parameters of the controlled object while controlling the controlled object. Therefore, it is actually a control method that combines fuzzy system identification and fuzzy control. Through identification, we can better "understand" the controlled object so that the controller can "follow" changes in the object and the environment. In this way, the controller itself has a certain ability to adapt to changes, or the adaptive fuzzy controller has higher intelligence.

The three function blocks added by the adaptive fuzzy controller are implemented by software to implement their # functions. The adaptive link can be understood as the introduction of a "soft feedback" inside the fuzzy controller, that is, the feedback of the controller's own performance implemented by software. Through this feedback, the control performance of the controller is continuously adjusted and improved to make the control effect of the control process is sent to the best state.

The above method is still feasible for a system with a single input and single output and which is not critical to the calculation time. The relationship matrix for a multiple input multiple output system is too large for a computer to store and compute.

1.3.2 The principle and method of model reference adaptive fuzzy control

1.3.2.1 Basic principle of model reference adaptive fuzzy control

The model reference adaptive system originates from the concept of self-adaptation of human behavior and causal reasoning (law of cause and effect) being transplanted into the field of control. The causal reasoning model is a general model of the reasoning process that expresses human adaptive characteristics. The causal law model characterizes the qualitative relationship between cause and effect.

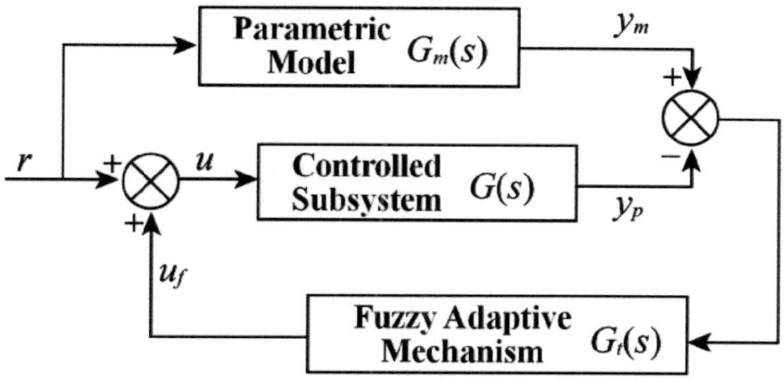

Figure 8.
Structure of model reference adaptive fuzzy controller.

By comparing the model with the real situation, people use adaptive mechanisms instead of people to modify parameters or control strategies to obtain a process that is close to the desired output for control system.

The basic structure of the model reference adaptive fuzzy control system includes three components:

1. Reference model-used to describe the dynamic characteristics of the controlled object or to represent an ideal dynamic model.

2. Controlled subsystem-including the controlled object, feed forward controller and feedback controller, as shown by the dashed box in the Figure.

3. Self-adaptive mechanism—adjust the control parameters of the feed forward controller and feedback controller based on the difference e between the actual output y_P of the controlled object and the reference model output y_m and its changes, so that e = y_m-y_p → 0.

1.3.2.2 Design method of fuzzy adaptive mechanism

The fuzzy adaptive mechanism design method generally has the following two forms:

1. Design fuzzy adaptive mechanism based on fuzzy relation model. The design process based on fuzzy relation model is similar to the design steps of fuzzy control look-up table.

2. Design of fuzzy adaptive mechanism model based on TS fuzzy model, the general structure of the reference fuzzy adaptive system can be expressed in the form of **Figure 8**, where the controlled subsystem is a closed-loop subsystem including the controlled object. The adaptive mechanism generates a fuzzy adaptive signal according to the difference between the output of the reference model and the output of the controlled subsystem, and changes the output of the controlled subsystem to the reference model output [6].

1.3.3 Simulation and programming of adaptive fuzzy control

Adaptive fuzzy control simulation takes the controlled object as Eq. (3).

$$\ddot{x} = -25\dot{x} + 133u \tag{3}$$

Position instruction is sin(πt) take the membership function in Eqs. (4)–(9).

$$\mu_{N3}(x_i) = \frac{1}{(1 + exp\,(5(x + 2)))} \tag{4}$$

$$\mu_{N2}(x_i) = exp\left((x + 1.5)^2\right) \tag{5}$$

$$\mu_{N1}(x_i) = exp\left[((x_i + 0.5))^2\right] \tag{6}$$

$$\mu_{p1}(x_i) = exp\left[-((x_i - 0.5))^2\right] \tag{7}$$

$$\mu_{p2}(x_i) = exp\left[-((x_i - 1.5))^2\right] \tag{8}$$

$$\mu_{p3}(x_i) = \frac{1}{(1 + exp\,(-5(x - 2)))} \tag{9}$$

The initial state matrix of the system is $[\mathbf{1}, \mathbf{0}]$, and the initial values of each element in θ are all taken as 0. Control rule show in Eq. (10) and adaptive rule (11) are adopted. Then take $Q = \begin{bmatrix} 50 & 0 \\ 0 & 50 \end{bmatrix}$, $k_1 = 1, k_2 = 10$, and adaptive parameter γ = 50.

$$u_D = (x|\theta) = \frac{\sum_{l_1=1}^{m_1} \cdots \sum_{l_n=1}^{m_n} \overline{y}_u^{l_1 \ldots l_n} \left(\prod_{i=1}^{n} \mu_{Ai}^{l_i}(x_i)\right)}{\sum_{l_1=1}^{m_1} \cdots \sum_{l_n=1}^{m_n} \left(\prod_{i=1}^{n} \mu_{Ai}^{l_i}(x_i)\right)} \tag{10}$$

$$\dot{V} = -\frac{1}{2} e^T Q e + \frac{b}{\gamma}(\theta^* - \theta)^T \left[\gamma e^T P_n \zeta(x) - \dot{\theta}\right] - e^T p_n b w \tag{11}$$

According to the membership function, write the MATLAB program as follows:

```
% Adaptive fuzzy approximation
clc % Clear screen
clear all; % Remove workplace variables
close all; % Close the display graphics window
L1 = -3;
L2 = 3;
L = L2-L1; % Fuzzy set change range length
T = 0.001;
x = L1: T: L2; % Range of fuzzy set
figure (1);
for i = 1: 1: 6
    if i = 1
        u = 1 ./ (1 + exp (5 * (x + 2)));
    else if i = 6
        u = 1 ./ (1 + exp (-5 * (x-2)));
    else
    u = exp (-(x + 2.5- (i-1)). ^ 2);
end
    hold on;
    plot (x, u, 'r', 'Line Width', 2);
end
```

```
x label ('x'); y label ('Membership function fuzzy set');
grid on
axis tight
```

The running program is shown in **Figure 9** as a membership function graph.

Directly and adaptively control the internal control objects. The MATLAB program is written as follows. Results show in **Figures 10–12**.

```
% S-function for continuous state equation for controlled object
function [sys, x0, str, ts] = s_function (t, x, u, flag)
switch flag,
% Initialization
    case 0,
        [sys, x0, str, ts] = mdlInitializeSizes; % Initialization function
    case 1,
        sys = mdlDerivatives (t, x, u);% Differential function
% Outputs
    case 3,
        sys = mdlOutputs (t, x, u);% Output function
% Unhandled flags
    case {2, 4, 9}
        sys = [];
% Unexpected flags
    otherwise
        error (['Unhandled flag =', num2str (flag)]);
end
function [sys, x0, str, ts] = mdlInitializeSizes
sizes = simsizes;
sizes.NumContStates = 2;
sizes.NumDiscStates = 0;
sizes.NumOutputs = 2; % 2 outputs
sizes.NumInputs = 1; %1 input
sizes.DirFeedthrough = 0;
sizes.NumSampleTimes = 0;
sys=simsizes(sizes);
x0 = [1 0];
```

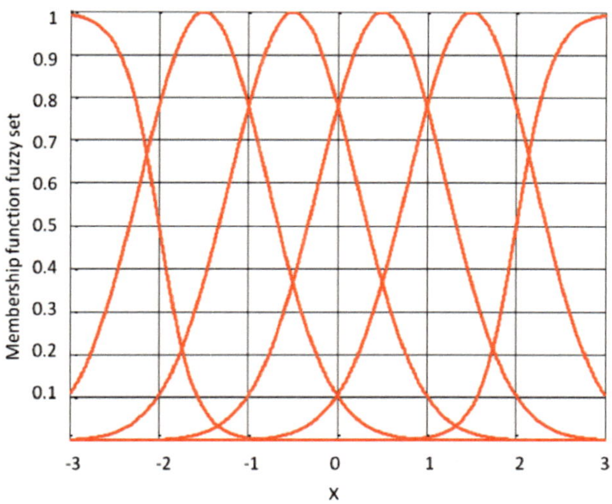

Figure 9.
Membership function graph.

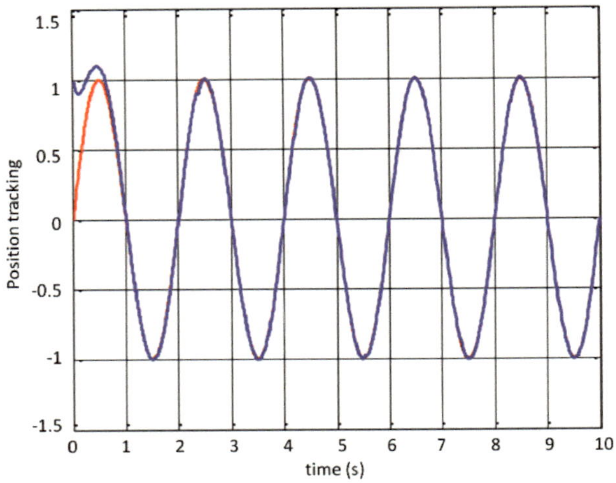

Figure 10.
Simulation result of position tracking.

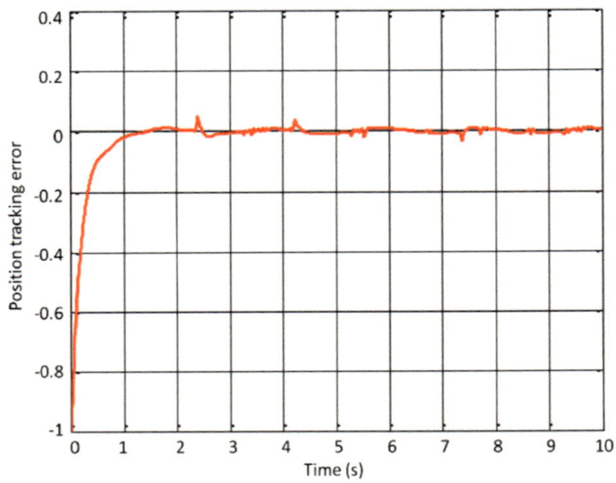

Figure 11.
Simulation result of position tracking error.

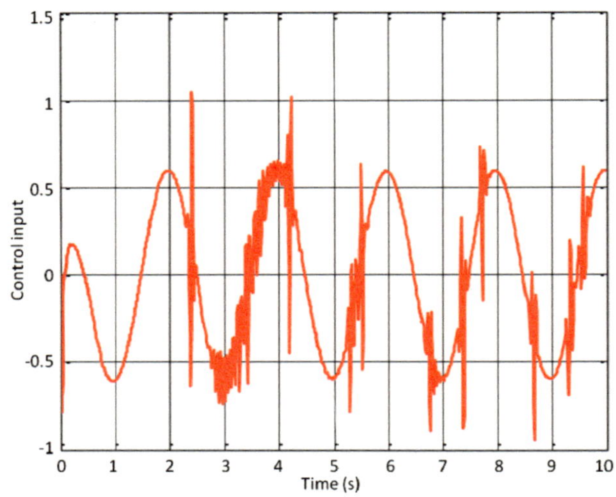

Figure 12.
Simulation result of control input.

DOI: http://dx.doi.org/10.5772/intechopen.91966

```
str = [];
ts=[];
function sys = mdlDerivatives (t, x, u)
% Second-order system
sys (1) = x (2);
sys (2) =-25 * x (2) + 133 * u;
function sys = mdlOutputs (t, x, u)
sys (1) = x (1);
sys (2) = x (2);
% Drawing program writing
close all;
figure (1);
plot (t, y (:, 1), 'r', t, y (:, 2), 'b', 'LineWidth', 2);
xlabel ('time (s)'); ylabel ('Position tracking');
grid on
title ('Location tracking')
figure (2);
plot (t, y (:, 1) -y (:, 2), 'r', 'LineWidth', 2);
xlabel ('time (s)'); ylabel ('Position tracking error');
grid on
title ('Position tracking error')
figure (3);
plot (t, u (:, 1), 'r', 'LineWidth', 2);
xlabel('time(s)');ylabel('Control input');
grid on
title ('Control input signal')
```

2. Neural network

2.1 The connotation of neural network

The neural network model is used to simulate the process of a large number of neurons in the human brain, including information processing, processing, storage, and search. Its main features include (1) the characteristics of distributed storage of information. (2) Information processing and reasoning have the characteristics of parallelism. (3) Information processing has the characteristics of self-organization and self-learning. (4) It has a very strong non-linear mapping capability from input to output [7].

The topology of the neural network connection method is a graph with neurons as nodes and directed connections between nodes as edges. The structure can be divided into two categories: layered and gridded. A neural network with a hierarchical structure consists of several layers. Each layer has a certain number of neurons. Neurons in adjacent layers are unidirectional connected. Normally, neurons in the same layer cannot connect. In a neural network with a network structure, any two neurons may be connected in both directions. The following are several common neural network structures including (1) forward neural networks. (2) Feedback neural network. (3) Integrate neural networks with each other. (4) Hybrid neural network.

Neurons in the human brain pass through the fine structures of many dendrites, collect information from other neurons, and burst electrical activity pulses through the axis. How to adjust the connection weight is reconstructed into different learning algorithms. In order to apply neural networks to solve practical engineering problems, they must be trained. This is neural network teacher learning or supervised learning. And neural network learning usually refers to unsupervised learning of neural networks. In addition, after training the neural network through the sample data set, when new data other than the sample data set appears in the input, the neural network can still obtain new outputs through learning, and can strictly maintain the input–output mapping relationship after the input. The training ability of neural networks is called the generalization ability of neural networks. By changing the structure and parameters of the neural network, you can change the size of the network to make it more suitable for solving specific problems. This process is called the growth and pruning of neural networks [8].

2.2 Types and controls of intelligent control based on neural networks

In the control system, the non-linear mapping capability of neural networks can be used to model complex non-linear objects that are difficult to accurately describe, or to act as controllers, or to optimize calculations, or to perform inference, or fault diagnosis, or both Adaptation of certain functions, etc.

Neural network-based intelligent control this book refers to the collective control of neural network alone control or integration of neural network and other intelligent control methods. The main types of control are the following forms.

1. Neural network direct feedback control. This is a way to directly implement intelligent control using only neural networks. In this control method, the neural network is directly used as a controller, and algorithms such as feedback are used to implement self-learning control.

2. Neural network expert system control. Expert systems are good at expressing knowledge and logical reasoning. Neural networks are better than non-linear mapping and intuitive reasoning. Combining the two to give play to their respective advantages will result in better control results.

3. Neural network fuzzy logic control. Fuzzy systems are good at directly expressing logic and are suitable for directly expressing knowledge. Neural networks are better at learning to express knowledge implicitly through data. The former is suitable for top-down expression, and the latter is suitable for bottom-up learning process. The two are complementary and related. Therefore, their integration can complement each other and better improve the intelligence of the control system. There are three ways to combine neural network and fuzzy logic. First, fuzzy control using fuzzy neural network to drive fuzzy reasoning. This method uses a neural network to directly design multiple membership functions, and combines the neural network as a membership function generator in a fuzzy control system. Second, use neural network to memorize the control of fuzzy rules. An abstract concept value is expressed by a group of neurons with different degrees of excitement, thereby converting abstract empirical rules into input and output samples of a multilayer neural network, and memorizing these samples through a neural network such as a BP network. The use of these experiences for control, in a sense, mimics the way people think about associative memory. Third, the parameters of the fuzzy controller are optimized using a neural network. In addition to the above-mentioned membership functions and fuzzy rules, the factors that affect the control performance in fuzzy control systems also have control parameters such as the quantization factor of error and error change and the output scale factor. These can be optimized using the optimization calculation function of neural network Parameters to improve the performance of the fuzzy control system.

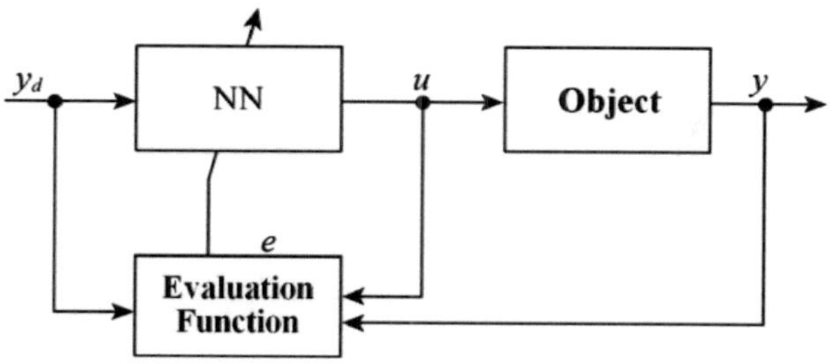

Figure 13.
Structure of model reference adaptive fuzzy controller.

4. Neural network sliding mode control. Variable structure control can be regarded as a special case of fuzzy control, so it belongs to the category of intelligent control. Combining neural network and sliding mode control constitutes neural network sliding mode control. This method classifies the control or state of the system, switches and selects according to changes in the system and the environment, uses the learning ability of the neural network, and improves the sliding mode switching curve through self-learning in an uncertain environment, thereby improving sliding mode the effect of control.

2.3 Neural control based on traditional control theory

The neural network is used as a link or links in a traditional control system to serve as an identifier, controller, estimator, or optimization calculation. There are many ways to do this. Some common ways are summarized as follows.

1. Nerve inverse dynamic control. Let the state observation value of the system be $x(t)$, and its relationship with the control signal $u\ (t)$ is $x(t) = F(u\ (t), x\ (t-1))$. F may be unknown, assuming F is reversible, which can be obtained from $x(t)$, $x(t-1)$, and the dynamic response through training the neural network is $u(t) = H(x(t), x(t-1))$, H is the inverse dynamic of F.

2. Neural PID control. Combining neuron or neural network with conventional PID control, using the learning algorithm of neuron or neural network to optimize and adjust HD control parameters in real-time during the control process according to the dynamic characteristics of the controlled object to achieve online optimization of PID the purpose of controlling performance. Such a composite control form is collectively referred to as neuron PID control or neural HD control.

3. Model reference neural adaptive control. In traditional model reference adaptive control systems, neural networks are used as object models, or as controllers, or as adaptive mechanisms, or to optimize control parameters, or both. Such systems are collectively referred to as model reference neural adaptive control.

4. Nerve self-correcting control. One form of this control structure is the indirect learning control structure of the dual neural network introduced earlier. The control structure of a single neural network is shown in **Figure 13**. The evaluation function is generally taken as $e = yd\text{-}y$, or the following form Eq. (12): [9].

$$e(t) = M_y\left[y_d(t) - y(t)\right] + M_u u(t) \qquad (12)$$

Among them, M_y and M_u are matrices of appropriate dimensions. The effectiveness of this method has been confirmed in the underwater robot attitude control. In addition, the combination of neural network and traditional control, as well as endometrial control, neural predictive control, and neural optimal decision control, will not be described in detail.

2.4 Programming of neural network PID controller

Here is a programming example of neural network PID controller simulation, the simulation results are shown in **Figures 14–18**, write the MATLAB program as follows:

```
% Calculation error
   error = [r1 (k) -y1 (k); r2 (k) -y2 (k); r3 (k) -y3 (k)];
   error1 (k) = error (1); error2 (k) = error (2); error3 (k) = error (3);
   J (k) = 0.5 * (error (1) ^ 2 + error (2) ^ 2 + error (3) ^ 2);% adjust size
   ypc = [y1 (k) -y_1 (1); y2 (k) -y_1 (2); y3 (k) -y_1 (3)];
   uhc = [u_1 (1) -u_2 (1); u_1 (2) -u_2 (2); u_1 (3) -u_2 (3)];
```

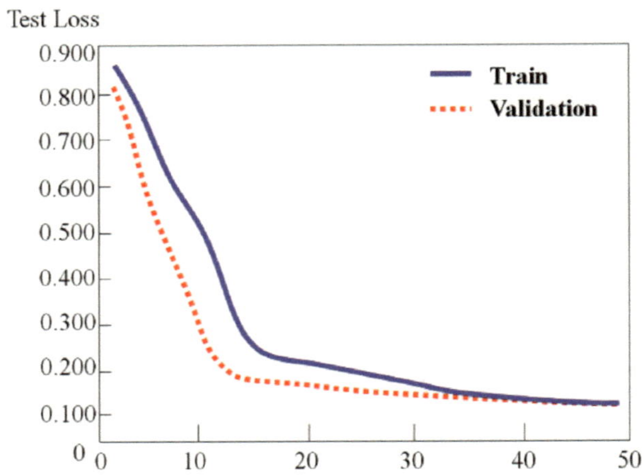

Figure 14.
Simulation result of test loss comparison of train and validation.

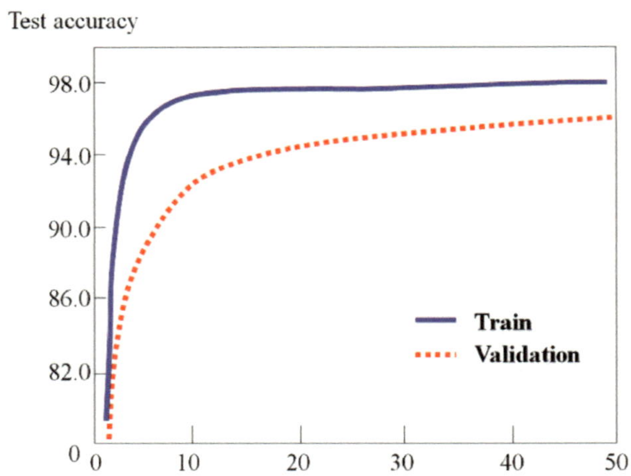

Figure 15.
Simulation result of test accuracy comparison of train and validation.

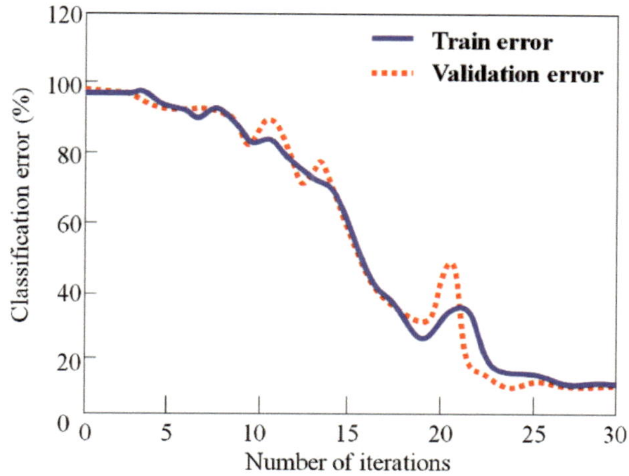

Figure 16.
Simulation result of test error comparison of train and validation.

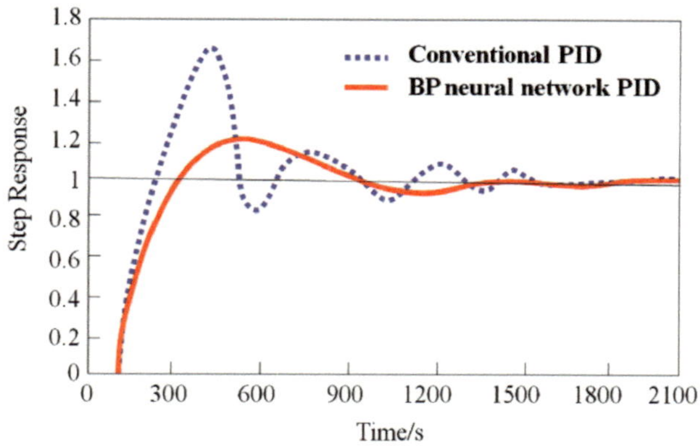

Figure 17.
Step response simulation of conventional PID control and BP neural network PID control.

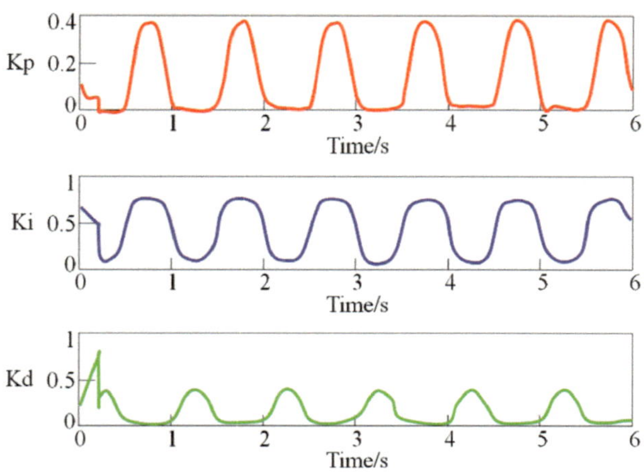

Figure 18.
Change curve of Kp, Ki, Kd parameters.

```
% Hidden layer and output layer weight adjustment
% Adjust w21
Sig1 = sign (ypc ./ (uhc (1) +0.00001));
```

```
dw21 = sum (error. * Sig1) * qo ';
w21 = w21 + rate2 * dw21;
% Adjust w22
Sig2 = sign (ypc ./ (uh (2) +0.00001));
dw22 = sum (error. * Sig2) * qo ';
w22 = w22 + rate2 * dw22;
% Adjust w23
Sig3 = sign (ypc ./ (uh (3) +0.00001));
dw23 = sum (error. * Sig3) * qo ';
w23 = w23 + rate2 * dw23;
% Input layer and hidden layer weight adjustment
delta2 = zeros (3,3);
wshi = [w21; w22; w23];
for t = 1: 1: 3
  delta2 (1: 3, t) = error (1: 3). * sign (ypc (1: 3) ./ (uhc (t) +0.00000001));
end
for j = 1: 1: 3
  sgn (j) = sign ((h1i (j) -h1i_1 (j)) / (x1i (j) -x1i_1 (j) +0.00001));
end
s1 = sgn '* [r1 (k), y1 (k)];
wshi2_1 = wshi (1: 3,1: 3);
alter = zeros (3,1);
dws1 = zeros (3,2);
for j = 1: 1: 3
   for p = 1: 1: 3
      alter (j) = alter (j) + delta2 (p,:) * wshi2_1 (:, j);
   end
end
for p = 1: 1: 3
   dws1 (p,:) = alter (p) * s1 (p, :);
end
w11 = w11 + rate1 * dws1;
% Adjust w12
for j = 1: 1: 3
   sgn (j) = sign ((h2i (j) -h2i_1 (j)) / (x2i (j) -x2i_1 (j) +0.0000001))
end
s2 = sgn '* [r2 (k), y2 (k)];
wshi2_2 = wshi (:, 4: 6);
alter2 = zeros (3,1);
dws2 = zeros (3,2);
for j = 1: 1: 3
for p = 1: 1: 3
alter2 (j) = alter2 (j) + delta2 (p,:) * wshi2_2 (:, j);
end
end
for p = 1: 1: 3
dws2 (p,:) = alter2 (p) * s2 (p, :);
end
w12 = w12 + rate1 * dws2;
% Adjust w13
for j = 1: 1: 3
sgn (j) = sign ((h3i (j) -h3i_1 (j)) / (x3i (j) -x3i_1 (j) +0.0000001));
end
```

```
s3 = sgn '* [r3 (k), y3 (k)];
wshi2_3 = wshi (:, 7: 9);
alter3 = zeros (3,1);
dws3 = zeros (3,2);
for j = 1: 1: 3
for p = 1: 1: 3
alter3 (j) = (alter3 (j) + delta2 (p,:) * wshi2_3 (:, j));
end
end
for p = 1: 1: 3
dws3 (p,:) = alter2 (p) * s3 (p, :);
end
w13 = w13 + rate1 * dws3;
% Parameter update
u_3 = u_2; u_2 = u_1; u_1 = uh;
y_2 = y_1; y_1 = yn;
h1i_1 = h1i; h2i_1 = h2i; h3i_1 = h3i;
x1i_1 = x1i; x2i_1 = x2i; x3i_1 = x3i;
end
time = 0.001 * (1: k);
figure (1)
subplot (3,1,1)
plot (time, r1, 'r-', time, y1, 'b-');
title ('PID neural network control');
ylabel ('Controlled amount 1');
legend ('control the target', 'actual output', 'fontsize', 12);
subplot (3,1,2)
plot (time, r2, 'r-', time, y2, 'b-');
ylabel ('Controlled amount 2');
legend ('control the target', 'actual output', 'fontsize', 12);
axis ([0,0.2,0,1])
subplot (3,1,3)
plot (time, r3, 'r-', time, y3, 'b-');
xlabel ('time / s');
ylabel ('Controlled amount 3');
legend ('control the target', 'actual output', 'fontsize', 12);
print -dtiff -r600
figure (3)
plot (time, u1, 'r-', time, u2, 'g-', time, u3, 'b');
title ('PID control input provided by the neural network to the object');
xlabel ('time'), ylabel ('control law');
legend ('u1', 'u2', 'u3'); grid
figure (4)
plot (time, J, 'r-');
axis ([0,0.1,0,0.5]); grid
title ('network learning objective function J dynamic curve');
xlabel ('time'); ylabel ('control error');
% BPy1 = y1;
% BPy2 = y2;
% BPy3 = y3;
% BPu1 = u1;
% BPu2 = u2;
% BPu3 = u3;
```

```
% BPJ = J
% save BP r1 r2 r3 BPy1 BPy2 BPy3 BPu1 BPu2 BPu3 BPJ
```

3. Expert control and humanoid intelligent control

3.1 Expert control

An expert is someone who has deep theoretical knowledge or rich practical experience in a certain field. Experts' decision-making actions to solve difficult problems can achieve important results because they have accumulated valuable theoretical knowledge and practical experience in their heads. We may use some kind of knowledge acquisition method to store expert knowledge and experience in the professional field into the computer, and rely on its reasoning program to make the computer work close to the level of the expert. It can be based on the special domain knowledge and knowledge provided by one or more human experts. Use experience to reason and judge. An expert system is a computer program system with a large amount of expertise and experience [10].

The basic structure of an expert system usually consists of five parts: a knowledge base, a database, an inference engine, and an interpretation part and knowledge acquisition.

In terms of the main features and structure of the expert system, the industrial production process places several special requirements on the expert control system that are different from general expert systems, including (1) high reliability and long-term continuous operation. (2) Real-time nature of online control. (3) Excellent control performance and anti-interference. (4) Flexible and easy to maintain [11].

Because industrial process control has the aforementioned special requirements for expert control systems, expert control systems control process objects, and domain expert knowledge is usually represented by production rules. Generally speaking, the expert control system consists of the following parts: (1) Database. (2) Rule base. (3) Inference engine. (4) Human-machine interface. (5) Planning.

Constructing an expert control system requires not only complex design and long commissioning cycles, but also a large amount of human, material and financial resources. Therefore, for some controlled objects, considering the control performance indicators, reliability, real-time performance, and performance/price ratio requirements, the expert control system can be simplified.

Expert controller is usually composed of four parts: knowledge base, control rule set, reasoning mechanism and information acquisition and processing. The scale of the knowledge base and control rule base of the expert controller is small, and the reasoning mechanism is simple. Therefore, it can be controlled by microcontroller, programmable controller (PLC), etc. to realize.

According to the characteristics of industrial process control, production rules are used to describe the causality of the controlled process, and control rule sets can be established through fuzzy control rules with adjustment factor analysis and description [12].

Set the input set E and output set U of the expert controller to be Eq. (13) and (14)

$$\begin{aligned} E &= \{-e_n, -e_{n-1}, \cdots, -e_1, 0, e_1, e_2, \cdots, e_n\} \\ U &= \{-u_n, -u_{n-1}, \cdots, -u_1, 0, u_1, u_2, \cdots, u_n\} \end{aligned} \tag{13}$$

$$f(E) = U \tag{14}$$

The control rule set is summarized and summarized based on the knowledge set. It reflects the expertise and experience of experts, and reflects the intelligent control decision-making behavior of people in the operation process. The design control rule set includes the following 6 rules:

1. IF $E > E_{PB}$ THEN U=U_{NB}.

2. IF $E < E_{NB}$ THEN U=U_{PB}.

3. IF $C > C_{PB}$ THEN U=U_{NB}.

4. IF $C < C_{NB}$ THEN U=U_{PB}.

5. IF $E \bullet C < 0$ OR $E = 0$ THEN U=INT[αE+(1-α)C].

6. IF E·C > 0 OR $C = 0$ AND $E \neq 0$ THEN U = INT[βE+(1-β)C+$\gamma \sum_{i=1}^{k} E_i$].

Among them, E, C, and U are fuzzy variables of error, error change, and control amount, respectively, and the quantization level of C is selected exactly the same as E, U; and the maximum positive values of E, C, and U, respectively, and E_{NB}, C_{NB} and U_{NB} are negative maximums of E, C, and U, respectively; α, β, and γ factors to be adjusted are determined by empirical rules of knowledge concentration; $\sum_{i=1}^{k} E_i$ intelligent integration terms for errors are used to improve the stability of the control system State performance; the symbol INT〔a〕means to take an integer closest to a.

Considering that the control decision of the expert controller completely depends on the characteristics of the input data, the controller adopts a data-driven forward reasoning method to sequentially determine the conditions of each rule. If the conditions are met, the rule is executed, otherwise the search is continued. Since there are corresponding controls rules for each of the control input variables £: and C, the target can be searched.

Simulation and practical application show that the above-mentioned expert controller not only has the characteristics of fast dynamic response, small overshoot, and high steady-state accuracy, but also has simple control algorithm programming, flexible control rule modification, good real-time performance, and changes in the parameters of the controlled object. Has strong robustness.

3.2 Human-like intelligent control

Conventional PID control controls the controlled object based on the linear combination of the proportional, integral, and derivative of the controlled system error. According to the mathematical models of different control objects, the three control parameters Kp, Ki, and Kd of the PID are appropriately set to obtain a satisfactory control effect. Linear PID control cannot solve the problem that increasing the control amount can reduce the steady-state error and improve the accuracy, but it will reduce the stability. Physically speaking, the control process is the process of information processing and energy transfer. Therefore, the information processing ability is improved, a more reasonable control law is designed, and the energy transmission of the controlled system is achieved too quickly, stably, and accurately in the shortest time and/or the lowest cost. This is the key problem to be solved in the control system design [13].

It is not enough to use only linear control methods in PID control. It is necessary to introduce some non-linear control methods as needed. The system's dynamic process and transient process, according to the needs of the system's dynamic characteristics, behavior and control performance, use variable gain (gain adaptive), intelligent integration (non-linear integration) and intelligent sampling. This requires expert control experience, heuristics Intuitive judgment and intuitive reasoning rules. Such control decisions are conducive to solving the contradiction between fastness, stability and accuracy in the control system, and can enhance the adaptability and robustness of the system to uncertain factors.

Intelligent control basically imitates human intelligent behavior for control and decision-making. Some scholars have found through experiments that after obtaining the necessary operational training, the artificially implemented control method is close to optimal. This method does not require knowledge of the structural parameters of the object, nor does it require the guidance of an optimal control expert. In the following analysis of the step response characteristics of the second-order system, we can see the basic idea of implementing human-like intelligent control (see **Figure 19**) [14].

It can be found in **Figure 14** (1) that variable gain control should be used in the OA segment. Use a larger gain in the initial section and increase it to a certain stage to reduce the gain so that the system continues to run through the inertia rise. (2) In section AB, the control function shall try its best to reduce the overshoot. In addition to proportional control, the integral control function should be added to enhance the control function through the integral error and make the system output return to the steady state value as soon as possible. (3) In the BC segment, the error starts to decrease, and the system shows a steady state change trend under control. At this time, no integral control operation should be added. (4) The system output decreases in the CD segment, the error changes in the opposite direction, and reaches the maximum value (positive) at point D. At this time, proportional plus integral control should be used. (5) In the DE segment, the system error gradually decreases, and the control effect should not be too strong, otherwise overshoot will occur again.

The basic idea of human-like intelligent control is to use computer to simulate the artificial control behavior in the control process, to maximize the identification and use of the characteristic information provided by the dynamic process of the control system, to make heuristic judgment and intuitive reasoning. This can effectively control objects that lack accurate models.

3.2.1 Characteristic variables of system dynamic behavior

In order to use a computer to automatically realize human-like intelligent control, the system must be able to automatically recognize the dynamic behavior of the control system through some characteristic variables in order to mimic human intelligent control decision-making behavior. In fuzzy control, the error e and the

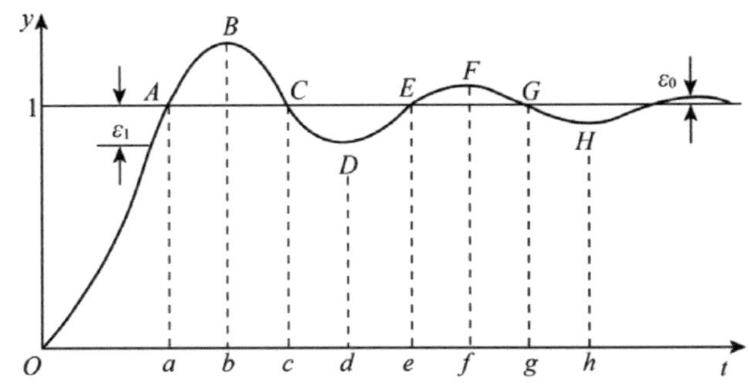

Figure 19.
Unit step response curve for second-order system.

error change Δe are usually selected as the input variables of the fuzzy controller. Generally, the output u of the fuzzy controller can be expressed as Eq. (15).

$$u = f(e, \Delta e) \tag{15}$$

If the control is based on the magnitude of the error e, it is difficult to obtain satisfactory control results for some complex systems. For example, when the controlled system has a large error and it is changing rapidly in the direction of reducing the error, if only based on the large error and not taking into account the rapid change of the error, it is necessary to increase the control amount so that the system eliminates large errors as soon as possible. Error, such control will inevitably lead to the negative consequences of over-regulation and reverse error. When two input variables e and Δe are used for control, the above-mentioned blindness can be avoided. Therefore, for a complex system under manual control, the more people learn about the state, dynamic characteristics, and behavior of the controlled system during the control process, the better the control effect will be.

How to identify the state, dynamic characteristics and behavior of the controlled system according to the input and output information is the first problem to be solved by human-like intelligent control. To this end, starting from the two basic variables of error e and error change Δe, a characteristic variable is designed to identify the characteristic mode of the dynamic process.

3.2.1.1 Characteristic variable $e_n \bullet \Delta e_n$

The product of the error e and the error change Δe constitutes a characteristic variable describing the dynamic process of the system, and whether the value of the characteristic variable is greater than zero can describe the trend of the system dynamic process error change. Let e_n and e_{n-1} denote the error values of the current and previous sampling moments respectively, then $\Delta e_n = e_n - e_n - 1$. For different stages of the dynamic system response curve shown in **Figure 20**, the values of the characteristic variables $e_n \bullet \Delta e_n$ are shown in **Table 2**.

When $e_n \bullet \Delta e_n > 0$, as shown in the *AB* and *CD* sections in **Figure 20**, it shows that the dynamic process of the system changes in the direction of increasing error, that is, the absolute value of the error gradually increases. When $e_n \bullet \Delta e_n < 0$, as shown in **Figure 20**, *BC* and *DE*, it shows that the dynamic process of the system changes in the direction of decreasing error, that is, the absolute value of the error gradually decreases. In the control process, the computer can easily recognize the symbol of $e_n \bullet \Delta e_n$, so as to grasp the behavior characteristics of the dynamic process of the system, so as to better formulate the next control strategy.

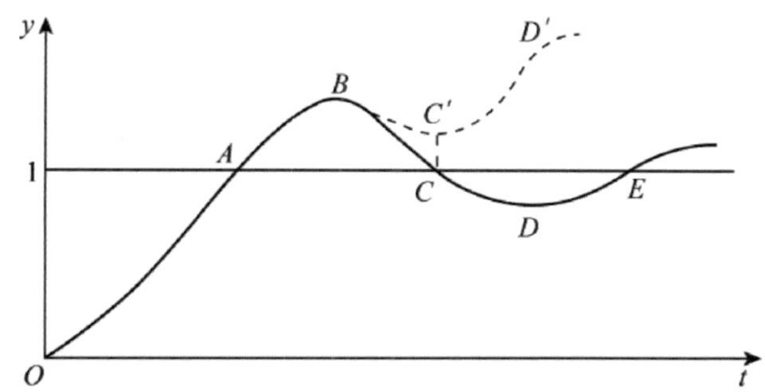

Figure 20.
Curve of dynamic process.

	OA part	AB part	BC part	CD part	DE part
e_n	>0	<0	<0	>0	>0
Δe_n	<0	<0	>0	>0	<0
$e_n \bullet \Delta e_n$	<0	>0	<0	>0	<0

Table 2.
Sign change of characteristic variables.

3.2.1.2 Characteristic variable e•Δe

The absolute value of the ratio of the error change Δe to the error e is defined as the characteristic variable describing the error change trend in the dynamic process of the system. The combined use of $|\Delta e/e|$ and $e_n \bullet \Delta e_n$ can further divide the characteristics of the dynamic process and facilitate the capture of different modes of the dynamic process. For example, in **Figure 19**, the *AB* segment of the curve can be subdivided into the following three cases:

1. The segment is close to the point A: $e_n \bullet \Delta e_n > 0$ and $|\Delta e/e| > \alpha$ indicate a mode in which the dynamic process presents a small error and a large error change.

2. The AB segment is near the middle part: $e_n \bullet \Delta e_n > 0$ and $\beta < |\Delta e/e| < \alpha$ indicates that the magnitude of the error and the variation of the dynamic process are in a medium state.

3. Segment AB is close to point B: $e \bullet \Delta e > 0$ and $|\Delta e/e| < \beta$ indicates that the dynamic process presents a mode with large errors and small error changes.

The above-mentioned α and β are constants set according to the needs of control, and there are α > β. Similarly, readers of the BC, CD, and DE segments of the curve in **Figure 15** can perform similar analysis.

3.2.1.3 Characteristic variable $\Delta e_n \bullet \Delta e_n - 1$

The product of two adjacent error changes is defined as the characteristic variable that characterizes the extreme state of the error. If $\Delta e_n \bullet \Delta e_n - 1 < 0$, it means that the extreme value appears; if $\Delta e_n \bullet \Delta e_n - 1 > 0$, it means that there is no extreme value. The combination of the characteristic variables $\Delta e_n \bullet \Delta e_n - 1$ and $e_n \bullet \Delta e_n - 1$ can determine the change trend of the dynamic process when the error has an extreme value. As shown in **Figure 19**, extreme values appear at points B and C ‘, But their $e_n \bullet \Delta e_n$ values have opposite signs like below:

Point B: $\Delta e_n \bullet \Delta e_n - 1 < 0$, the error of $e_n \bullet \Delta e_n$ tends to decrease after point B.
Point C′: $\Delta e_n \bullet \Delta e_n - 1 < 0$, $e_n \bullet \Delta e_n > 0$, the error gradually increases after point C′.

3.2.1.4 Characteristic variables $|\Delta e_n \bullet \Delta e_n - 1|$

The magnitude of the absolute value of the ratio of the error change at the current moment to the error change at the previous moment is defined as the characteristic variable describing the local change trend of the system error. It also indirectly indicates the effect of early control. If the ratio is large, it indicates that the effect of early control is not significant; if the ratio is small, it indicates that the effect of early control is significant.

3.2.1.5 Characteristic variable Δ (Δe)

The sign of the change rate (secondary difference) of the error change is defined as a characteristic quantity that describes the dynamic process as overshoot or callback. For example, for the curve shown in **Figure 19**, there are two cases:

1. ABC segment: $\Delta(\Delta e) > 0$, it is in the overshoot segment.

2. CDE segment: $\Delta(\Delta e) < 0$, which is in the callback segment.

The essential characteristics of the above-designed characteristic variables are that they are not an absolute quantity, but a symbol variable, or a relative quantity. Symbol variables are used to characterize the direction of the dynamic process change trend, and relative quantities are used to characterize the speed of the dynamic process change. The above-mentioned symbol variables and characteristic variables (relative quantities) that characterize the degree of change in a dynamic process are collectively referred to as qualitative variables.

In order to use computers to realize human-like intelligent control, it is necessary to try to teach human operation experience, qualitative knowledge and intuitive reasoning to the computer, and let it apply this knowledge through flexible and flexible judgment, reasoning and control algorithms to perform human-like intelligent control. The main source of online information obtained by a computer is the input R and output Y of the system, from which the error e and the error change Δe can be calculated. Through e and Δe, the characteristic quantities that characterize the dynamic characteristics of the system can be further obtained. The computer can capture the characteristic information of the dynamic process with the aid of the above-mentioned characteristic quantities, and recognize the dynamic behavior of the system as a basis for control decisions. According to the dynamic characteristics and dynamic behavior of the system, the most effective control form is selected from a variety of control modes to precisely control the controlled object. Computers can use qualitative knowledge and intuitive reasoning in the control process. This is fundamentally different from traditional control theories, and it is precisely this point that embodies human intelligence. This method solves the contradiction of speed, stability and accuracy in the control process very well.

3.2.2 Humanoid intelligent control principle

The composition of human-like intelligent controller is similar to the basic structure of an expert controller, which consists of the following four parts.

1. Acquisition and processing of characteristic information. According to the input and output sampling data, the current time error and error change are calculated, and then the characteristic variables necessary to identify the controlled dynamic process mode are obtained.

2. Feature pattern set. The feature pattern set stores certain feature pattern classes, which also include necessary parameters, thresholds, empirical data, and control parameters. It is similar to the knowledge base in an expert controller.

3. Pattern recognition. The pattern recognition plays the role of an inference mechanism. According to the obtained feature variables at the current time, it searches for feature pattern classes that match the constraints provided, and provides prerequisites for control decisions.

4. Control rule set. The control rule set is actually a rule-based controller. The process of control decision is to implement a mapping from the feature pattern set to the control rule set. In general, the number of feature patterns is greater than or equal to the number of control rules.

The working process of the human-like intelligent controller can be summarized into three steps: First, the system judges the characteristic mode of the dynamic process according to the calculated characteristic variables; second, the inference mechanism searches for a matching control rule according to the characteristic mode class; third, the controller executes the above control rules to control the controlled object [15].

This completes a step-by-step intelligent control algorithm, and then cyclically controls step by step until the error of the controlled system reaches the desired index.

3.3 Multiple modes of human-like intelligent control

A variety of human-like intelligent control modes have been formed to imitate human control and decision-making processes: humanoid intelligent switch control, humanoid proportional control, humanoid intelligent integral control, humanoid intelligent sampling control, and humanoid extreme value sampling and control. In addition, in the human-like intelligent control, a combination of variable gain proportional control, proportional differential control, and open-loop and closed-loop control is also used.

3.3.1 Human intelligence integration principle

3.3.1.1 Human-like intelligent integration principle

The introduction of integral control in the control system is an important way to reduce the steady-state error of the system. **Figure 21(c)** shows the integral process of the integral control action on the error in conventional PID control. This integral effect simulates human memory characteristics to a certain extent. It "remembers" all the information about the existence and changes of errors. The disadvantages of the integral control function based on this integral form are: first, the integral control function is not targeted, and sometimes does not meet the objective needs of the control system; second, because this integral effect is always integrated as long as the error exists, it is easy to cause integral saturation in the application of actual practice, which will reduce the rapidity of the system. Third, the integral parameters of this integral control are not easy to select, and improper selection will cause the system to oscillate.

The reason why the integral control function is not good is that the integral control function does not well reflect the intelligent control decision-making thoughts of experienced operators. In the integral curve interval (a, 6) in **Figure 21(c)**, the integral effect is opposite to the control effect of an experienced operator. At this time, the system has overshoot. The correct control strategy should be to add a negative control value to the constant value to reduce the overshoot and reduce the error as soon as possible. However, the integral control effect in this interval increases a positive control amount. This is because the integral result in the $(0, \alpha)$ interval is difficult to be offset and the sign is changed, so the integral control amount remains positive. As a result, the system overshoot cannot be reduced quickly, which prolongs the transition process time of the system.

In the $(6, c)$ section of the above integral curve, the system error changes from a maximum value to a decreasing direction, and there is a trend of steady state change. At this time, a certain proportional control effect should be added, but the integral control effect should not be added. Otherwise, it will cause system callback.

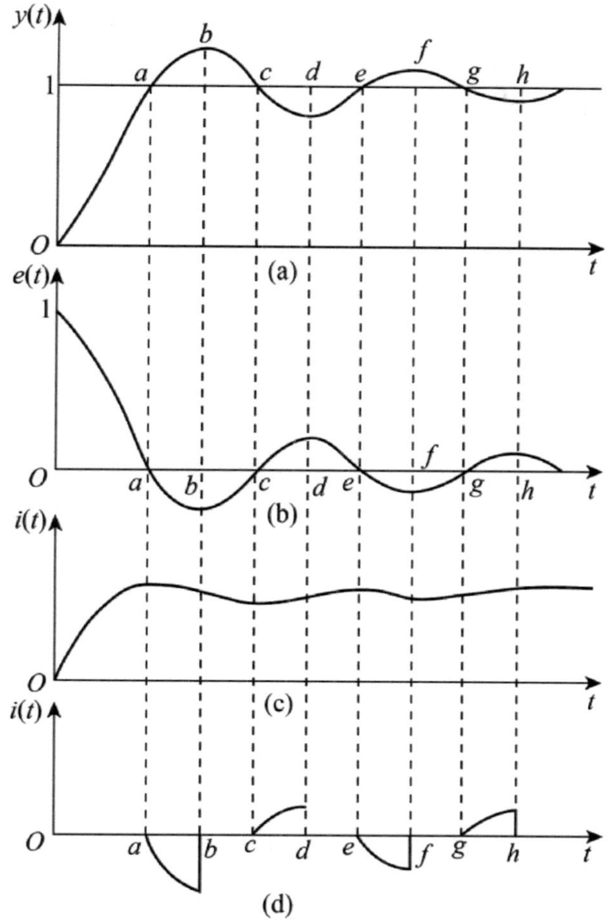

Figure 21.
Error and error integration curve. (a) (y (t)) Unit step response curve for second-order system. (b) (e (t)) Unit step response curve for second-order system. (c) In conventional PID control Integral control is the process of integrating errors. (D) Integrate on equal intervals.

In order to overcome the shortcomings of the integral control function described above, the integral curve shown in **Figure 21(d)** is used, that is, the integration is performed in the intervals (a, b), (c, d), and (e, f). The integral can provide the correct additional control amount for the integral control function in a timely manner, and can effectively suppress the increase of system error; while in the interval (0, a), (b, c), and (d, e), stop integral role to facilitate the system to transition to a steady state by virtue of inertia. At this time, the system is not in a state of out of control, it is also restricted by control functions such as proportion.

This integral function better simulates human memory characteristics and human-like intelligent control strategies. It selectively "remembers" useful information and "forgets" useless information, so it can overcome the shortcomings of general integral control. It has the characteristics of non-linear integration of human-like intelligence, which is called such integration of human-like intelligence.

3.3.1.2 Human-like intelligent integration control algorithm

In order to introduce the function of intelligent integration into the control algorithm, we must first solve the problem of logical judgment of introducing intelligent integration.

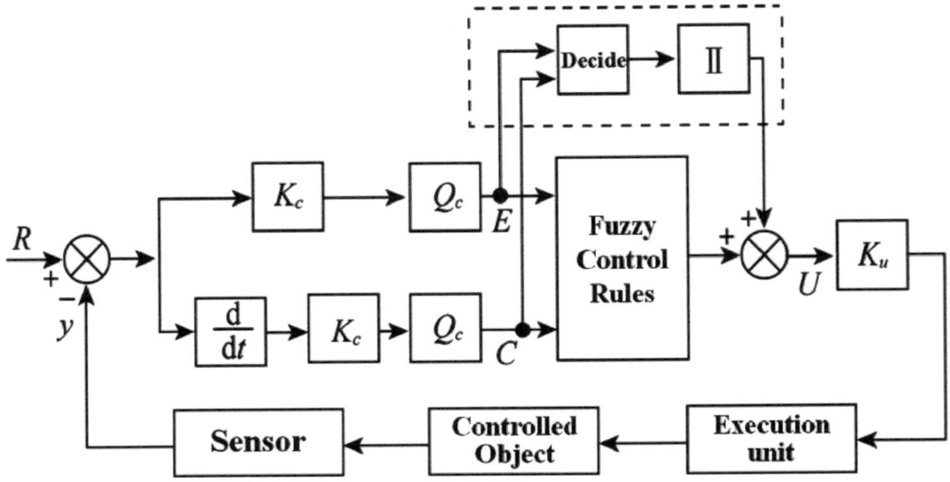

Figure 22.
Structure of a humanoid intelligent controller.

This condition can be determined by comparing the intelligent integration curve in **Figure 21** with **Figure 22** and **Table 2** [16].

When the error e_n and the error change Δe_n at the current sampling time have the same sign, that is, $e_n \bullet \Delta e_n > 0$, the error is integrated; on the contrary, the error e_n and the error change Δe_n have different signs, that is, when $e_n \bullet \Delta e_n < 0$, errors are not integrated. This is the basic condition for introducing intelligent integration. Considering the extreme points of errors and error changes, that is, the boundary conditions, the conditions for introducing intelligent integration and not introducing intelligent integration can be synthesized as follows:

When $e \bullet \Delta e > 0$ or $\Delta e = 0$ *and* $e \neq 0$, the error is integrated, that is, intelligent integration; when $e \bullet \Delta e < 0$ *or* $e = 0$, the error is not integrated, that is, no integration effect is introduced.

The digital simulation results show that the human-integrated intelligent integral control algorithm significantly improves the steady-state accuracy of the fuzzy control system due to the introduction of intelligent integral control. Compared with ordinary fuzzy controllers, the human-integrated intelligent integral control algorithm has the advantage of high steady-state accuracy. Compared with conventional PID control, this control algorithm has the advantages of fast response speed, small overshoot, or no overshoot. Therefore, this is a control algorithm with simple structure and good control performance for intelligent control.

3.3.2 Multiple modes of human-like intelligent control

The computer control systems of most production processes are continuous discrete hybrid systems. In such a system, the detection of time-continuous signals and the output of computer-controlled quantities, considering the problem of signal reproduction, require the correct selection of discrete-time sampling periods. In the control process, the main consideration is to help improve the control quality as much as possible.

3.3.2.1 Effect of sampling period on digital control

The upper limit of the sampling period is selected, but the choice of the lower limit of the sampling period is restricted by many factors. The smaller the sampling period is better to reproduce the signal. However, if the sampling period is too small, the signal-to-noise ratio is low, the quantization noise is large, and it is easy to be interfered, which affects the control performance. In process control, the lower limit of the control cycle selection is limited by the control algorithm operation time. Therefore, the sampling period and the control period are not both as small as possible.

As for the analog design method of the digital controller, the smaller the sampling period, the closer the characteristics of the digital controller are to the characteristics of the analog controller. As for the discrete design methods of digital controllers, most are based on the discretized object model. The discretization model of the object depends on the selection of the sampling period, so the sampling period not only affects the distribution of the zero and pole positions of the model, but also affects the accuracy of the model. Too long a sampling period may even lead to the loss of useful high-frequency information, thereby reducing the model order.

3.3.2.2 Human-like intelligent sampling control for lag process

A large number of controlled processes have varying degrees of time lag, which brings difficulties to the process system. The ratio of the lag time to the capacity lag time constant T reflects the difficulty of control. As the τ/T value increases, the difficulty of control increases accordingly. When approaching or exceeding T, the effect of using ordinary PID control is very poor, and Smith predictive control must be used. However, Smith control requires an accurate controlled process model, and complex controlled processes are often difficult to establish accurate mathematical models. Therefore, many improvements have been made to Smith control, and some control algorithms have emerged to overcome lag. Nevertheless, it should be said that the problem of large lag process control is still a topic of great concern in the control field.

As we all know, for an object with a pure lag time τ, its control effect must be reflected in time τ. Therefore, control within time is of no value, so the sampling control shown in **Figure 23** is generated. The sampling period T_s is slightly larger than τ, and the control time (on time) Δt is about $1/10\ T_s$. This choice will bring two disadvantages: first, the interference and sampling will be seriously out of sync. Because the pure lag time of the controlled process is generally large, T_s is chosen to be large, and the control time Δt is very small. In this way, the system is in an open-loop state during the $T_s - \Delta t$ time of each sampling cycle, and some urgent needs cannot be obtained. Useful information such as changes in output $y(t)$ caused by fixed-value disturbances or a given input that requires $y(t)$ to track as quickly as possible. Second, the feedback information obtained is too small and untargeted, which makes the control in a blind state, resulting in a long transition process.

The disadvantages of the above sampling control are passive waiting and blind control. In short, such control lacks the intelligent sampling characteristics of the lag

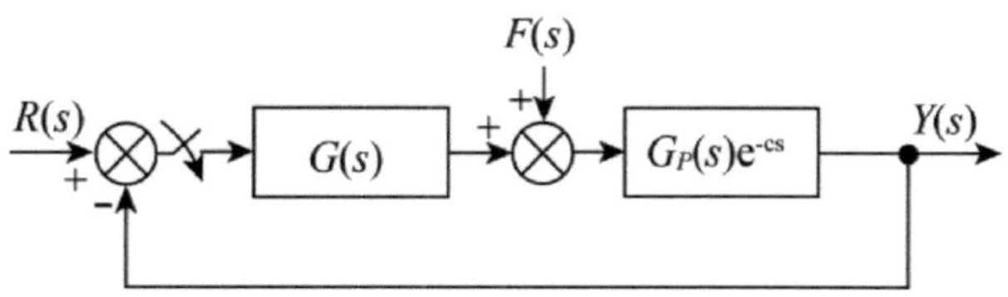

Figure 23.
Sampling control principle.

process of manual control. The basic strategy of manual sampling control to overcome large lags can be described as follows:

$$\text{Wait} \rightarrow \text{look} \rightarrow \text{tune} \rightarrow \text{wait again} \rightarrow \text{look again} \rightarrow \text{adjust again} \ldots\ldots$$

According to the above control strategy, the principle of a human-like intelligent sampling control system is shown in **Figure 24**. Among them, INT indicates intelligent device; $F(s)$ indicates fixed value interference; $G_{cj}(s)$ indicates controller; $Y(s)$ indicates controlled variable; $G_p(s)e^{-\tau s}$ indicates controlled object; B indicates intermediate feedback coefficient; $R(s)$ indicates interference; P takes the constant "1" or "0".

The function of the intelligent device is to control an intelligent sampling switch K, which "opens" or "closes" when certain conditions are met, that is shown in Eq. (11),

$$K = \begin{cases} 0 \ (\text{Disconnect}) & e \cdot \dot{e} < 0 \text{ or } |e| < \delta \\ 1 \ (\text{Closed}) & e \cdot \dot{e} > 0 \text{ or } \dot{e} = 0, |e| \geq \delta \end{cases} \tag{16}$$

In the formula, e, $\dot{e}$ are the error and the first derivative of the error; δ is the insensitive region.

When the controlled variable deviates from the expected value, the smart device sends a signal that the K switch is closed for sampling, and the controller controls in time until the controlled variable has a tendency to return to a balanced position; when the controlled system error value is within the allowable range, switch K disconnect, the system is in an open-loop working state. At this time, the energy required to be maintained by the object is supplied by the controller or the stored energy of the object.

In **Figure 24**, $G_{cj}(s)$ represents the jth controller $j \in (1,2)$, which is attractive considering that the given interference and fixed value interference often have different control laws and effects. Its selection is made automatically by the logical relationship of the design like Eq. (9):

$$G_{cj}(s) = \begin{cases} G_{c1}(s), & dc(t)/dt \neq 0 \\ G_{c2}(s), & dc(t)/dt = 0 \end{cases} \tag{17}$$

Among them, $c(t) = L^{-1}\left[\frac{1}{T_s+1} R(s)\right]$, T is determined as needed. In this way, the adaptability and effectiveness of the controller are enhanced, and it is ensured that

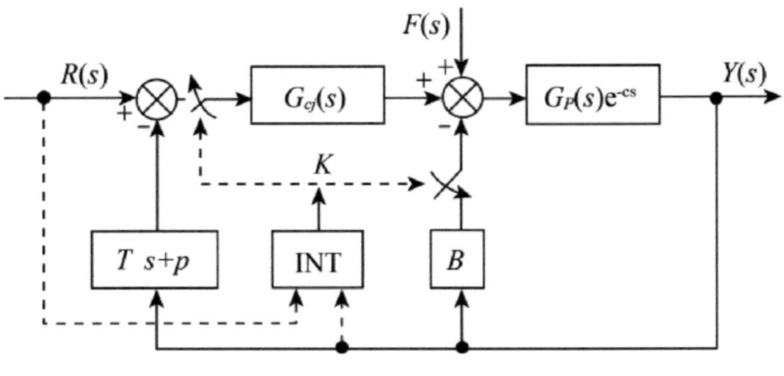

Figure 24.
Intelligent sampling control schematic diagram.

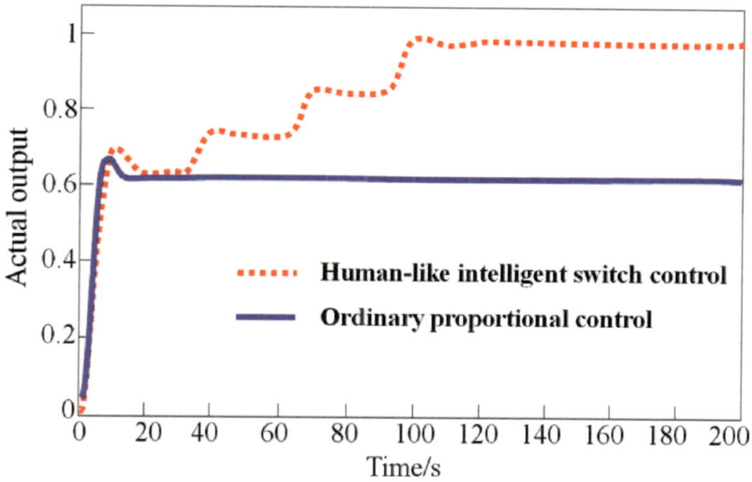

Figure 25.
Human-like intelligent switch control simulation result.

the corresponding controller is selected with different interference. The introduction of intermediate proportional feedback *B* is mainly for design convenience.

Through further analysis of the above-mentioned intelligent sampling control mechanism, it can be seen that when the switch *K* is closed and the system is in closed-loop control, whether the system is a follow-up system or a fixed value system, $e^{-\tau s}$ will appear in the characteristic equation, but by setting appropriately due to the logic judgment function of the smart device, the control parameters of the smart device can be established only in a short period of time, so that the unstable factors are eliminated. When the switch is open and the system is in open loop, $e^{-\tau s}$ has no effect on stability. In the above-mentioned intelligent mining control scheme, a smart device is used to determine whether the system is open-loop or closed-loop. In the open-loop process, the controller is in the active waiting phase with an observation function. The purpose of this waiting is to prepare for better control. In the closed-loop process, the controller is in the control phase, which is a manifestation of waiting action for strongly targeted. This control method cleverly avoids the adverse effects brought by $e^{-\tau s}$, and successfully solves the stability problem of the system.

Intelligent sampling control is a novel control method with open loop in the closed loop and closed loop in the open loop. Its entire working process is similar to an experienced operator. It can continuously observe and perform real-time correction as required. Therefore, this control has strong robustness and fastness, and can be easily realized by a microcomputer program.

3.3.3 Programming of human-like intelligent control

Here is a programming example of human-like intelligent switch control simulation, the simulation results are shown in **Figure 25**, write the MATLAB program as follows:

```
k = 1;% scaling factor K
g1 = tf (1, [8 6 1]);% continuous system model G (s)
g2 = feedback (k * g1,1);
tt = 1;
y = [0 0.0329];% output matrix initialization
u = ones (1,100);% input matrix initialization
```

```
u (1) = 0;
e = [0 0.9671];% deviation matrix initialization
for n = 3: 1: 200
   e (n) = k * (u (n-1) -y (n-1));
   if (e (n) -e (n-1) <= 0.0001) & (y (n-1) <= 1)% Judge whether the steady-
state and boundary conditions are satisfied
   tt = tt-1;
      if tt == 0% judging whether it is within a switching action period
         u = (u (n) + 0.31 * e (n)) * ones (1,200);% Switch action content
         tt = 40;% set smart switch action period
      end
   end
   e (n) = k * (u (n-1) -y (n-1));
   y (n) = 1.489 * y (n-1) -0.549 * y (n-2) + 0.0329 * e (n-1) + 0.0269 * e (n-2);%
difference equation
end
t = 0: 1: n-1;
g3 = step (g2, t);
graphical output of plot (t, y, t, g3)% response curve
```

Here is a programming example of human-like intelligent integral control simulation, the simulation results are shown in **Figure 26**, write the MATLAB program as follows:

```
y = [0 0.132];% output matrix initialization
u = ones (1,1000);% input matrix initialization
u (1) = 0;
e = [0 0.868];% deviation matrix initialization
c = [0 0];% process matrix initialization
kp = 0.803;% PID scale factor
ki = 0.282;% PID integration coefficient
kd = 0.02;% PID differential coefficient
esum = e (2);
for n = 3: 1: 100% conventional PID control system for reference
   y (n) = 1.559 * y (n-1) -0.559 * y (n-2) + 0.3 * c (n-1) + 0.1 * c (n-2);
   e (n) = u (n) -y (n);
```

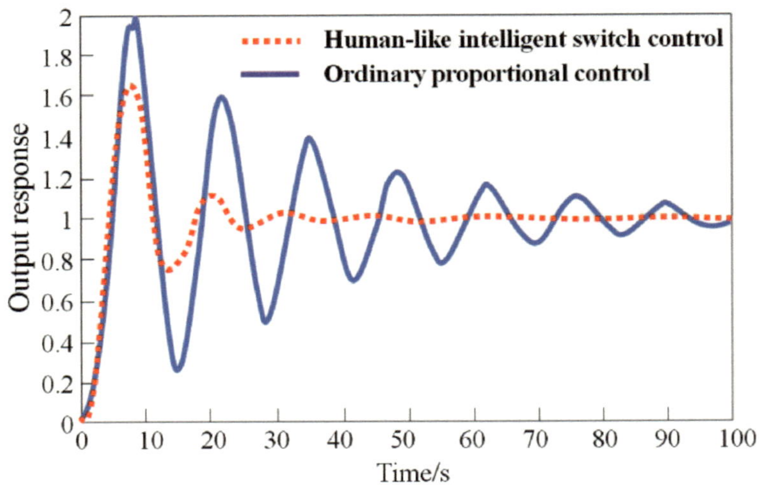

Figure 26.
Intelligent sampling control schematic diagram.

```
        esum = esum + e (n);% deviation summation
        c (n) = kp * (e (n)) + kd * (e (n) -e (n-1)) + ki * esum;% PID link output
        y (n) = 1.559 * y (n-1) -0.559 * y (n-2) + 0.3 * c (n-1) + 0.1 * c (n-2);% c (n)
signal drives the differential equation
    end
    t = 0: 1: n-1;
    plot (t, y)% response curve output
    for n = 3: 1: 100% humanoid intelligent integral control
        y (n) = 1.559 * y (n-1) -0.559 * y (n-2) + 0.3 * c (n-1) + 0.1 * c (n-2);
        e (n) = u (n) -y (n);
        esum = esum + e (n);
        edrta = e (n) * (e (n) -e (n-1));
        if (edrta> 0) & (e (n) ~ = 0)
            c (n) = kp * (e (n)) + kd * (e (n) -e (n-1)) + ki * esum;% for integration
        else
            c (n) = kp * e (n) + kd * (e (n) -e (n-1));
        end
        y (n) = 1.559 * y (n-1) -0.559 * y (n-2) + 0.3 * c (n-1) + 0.1 * c (n-2);
    end
    t = 0: 1: n-1;
    plot (t, y, 'r')% response curve output
```

4. Hierarchical intelligent control and learning control

Large systems usually have the following characteristics: the high-level order of the system, a large number of subsystems and interrelationships, a large number of system evaluation goals, and conflicts between different goals. People studying complex problems usually deal with them at different levels. Similarly, more complex large-scale system control problems are usually broken down into several interrelated subsystem control problems to deal with. Large-scale complex control systems use multi-level and multi-objective control to form a pyramid-like hierarchical control structure. Aiming at the large system control form, according to the information exchange method and related processing methods, it is generally divided into three basic forms: decentralized control, distributed control and hierarchical control. The main structure of the large-scale system control hierarchy includes multiple descriptions, multi-level descriptions, and multi-level descriptions. According to the number of decision-making objectives, the system can be divided into single-stage single-objective systems, single-stage multi-objective systems, and multi-stage multi-objective systems [17].

Regarding the hierarchical control principle of the aforementioned pyramid control structure, the configured controller receives information from an upper-level controller (or a decision unit) and is used to control the controller (or a subsystem) at a lower level. The possible conflicts between controllers depend on the coordination of the superior controller (or coordinator). There are many methods for coordination, but most of them are based on the two basic principles of association prediction coordination and association balance coordination.

4.1 Structure and basic principle of hierarchical intelligent control

The human central nervous system is organized according to a multilayer structure. Therefore, the multi-level hierarchical control structure has become a typical structure of intelligent control. Multilevel hierarchical intelligent control system is a branch of intelligent control. It was first applied in industrial practice and it played an important role in the formation of intelligent control systems. Hierarchical intelligent control structure, according to the intelligence level, it is divided into three levels: organization level, coordination level and control level. Hierarchical intelligent control principle uses the principles and methods of human intelligence, such as human organizers and coordinators, to have the ability to use and process knowledge, and have different degrees of self-learning ability to achieve control purposes. The principle of multi-level and multi-level hierarchical intelligent control of large-scale systems has three characteristics: (1) The more high-level units, the larger the scope of influence on system behavior, and therefore requires higher decision-making intelligence. (2) The decision cycle of the high-level unit is longer than the decision cycle of the lower unit, which mainly deals with factors that involve system behavior and change slowly. (3) The higher the level, the more uncertain the description of the problem, and the more difficult it is to formulate quantitatively. Therefore, the hierarchical intelligent control system is based on the aforementioned principle: accuracy increases as intelligence decreases. Under the unified organization of high-level organizers, multi-layer intelligent control systems can achieve optimal control of complex systems [18].

4.2 Hierarchical learning control system

4.2.1 The concept of learning control

Learning is one of the basic intelligences of people. Learning is to gain knowledge. Therefore, learning control that simulates human learning intelligent behavior in control belongs to the category of intelligent control.

Learning control means that if a system can learn the information inherent in the unknown characteristics of a process or its environment, and use the obtained experience for further estimation, classification, decision-making and control, so that the quality of the system can be obtained. Improve, and then call this system a learning system. The learning information obtained by the learning system is used to control the process with unknown characteristics. Such a system is called a learning control system.

The unknown environment in the learning control system includes the controlled dynamic process and its interference. The learning control law can be different learning control algorithms. The memory is used to store the control information and related data in the control process. The performance index evaluation is to control the learning control. The experience gained in the process is used to continuously estimate the characteristics of unknown processes for better decision-making control.

Because there are many ways to implement the learning control algorithm, the composition of the learning control system will also have different structural forms due to the different learning algorithms.

4.2.2 The main form of learning control

First, adaptive control with learning function. The adaptive control system should have two functions: one is a conventional control function, which is implemented by a closed-loop feedback control loop; the other is a learning function, which is implemented by another feedback control loop composed of an adaptive mechanism, and its control object is the controller itself.

The adaptive control system has a learning function, but the structure of this learning is different from the structure of the learning control system. The learning function of adaptive control is the feedback and evaluation of control performance of conventional controllers, and then the parameters or structure of controllers are adjusted or corrected online through adaptive mechanisms, so that the next step of control performance is better than the previous step. Yes, this is learning. It can be considered that the adaptive learning system is a two-level hierarchical control structure composed of a double closed-loop control system. The conventional control loop is a low-level form of a hierarchical structure, which completes the direct control of the controlled object. The second loop including the adaptive mechanism and the conventional controller is a high-level form of the hierarchical structure. The feedback control form completes the learning function of the controller control behavior.

The iterative learning control system and the repeated learning control system do not have a two-level hierarchical structure like the adaptive control system, but only increase the memory to remember the past control experience. The learning in iterative control is realized through the empirical memory of "weighted sum of control action and error" in the past. The assumption that the system is not deformed and the intermittent repetitive training of the memory unit are the essential characteristics of iterative learning control. The memory function of the repeated learning control is completed by the repeated controller, and its correction of the control effect is not realized intermittently but continuously.

Second, learning control based on neural reasoning. In a neural control system where the neural network directly acts as a controller, the neural network actually changes the connection weight between neurons in the network through a learning algorithm, thereby changing the non-linear mapping relationship between the input and output of the neural network, and gradually approaching the controlled dynamics. The inverse model of the process is used to achieve the task of control. This learning of neural networks is different from the learning forms in iterative learning control and repetitive learning control. The former study is based on the idea of approximation, while the latter uses the previous control experience of the control system to find an ideal input characteristic curve based on the actual output signal and the expected signal of the measurement system, so that the controlled object can produce the desired motion. The process of "finding" is the process of learning control.

Third, learning control based on pattern recognition. The first problem encountered when applying the principles and methods of pattern recognition to control systems is how to describe the dynamic characteristics of the controlled object. For controlled objects with a reference model, pattern recognition is mainly used as a signal processing method, but it has no practical value due to the large amount of calculation. For some complex production processes that cannot be modeled or parameter estimated, pattern recognition has become an important means to obtain working condition information and knowledge.

5. Conclusion and future prospects

In **Table 3**, we introduce the advantages and disadvantages of Artificial Neural Networks, Machine learning control, Bayesian probability control, Fuzzy control, Expert system and Genetic algorithm Control.

Advanced intelligent control is mainly used in comprehensive application scenarios such as computer technology, GPS positioning technology, or precision

Category	Description	Advantage	Disadvantage
Artificial Neural Networks [19, 20]	A mathematical or computational model that mimics the structure and function of a biological neural network and is used to estimate or approximate functions. In most cases, artificial neural networks can change the internal structure based on external information. It is an adaptive system, that is, it has learning functions.	(1) Non-linear, neural network can fully approximate any non-linear function in theory. (2) Parallel distributed processing, neural network has a high degree of parallel structure and parallel implementation capabilities, making it have a greater degree of fault tolerance and strong data processing capabilities. (3) Learning and self-adaptability can learn and remember the information provided by the knowledge environment. (4) Multi-variable processing. The neural network can naturally process multiple input signals and has multiple outputs. It is very suitable for multi-variable systems.	(1) The solution provided by the neural network is still a "black box". It can neither read out the appropriate cause for such a specific behavior, nor can it manually modify the neural network to change a specific expected behavior. (2) For most general mass-market products, the computing power of pure neural networks is limited. (3) Choosing the appropriate network model and set the parameters of the learning algorithm is still a "black art" and requires more experience.
Machine learning control [21, 22]	Machine learning control is programming computers to optimize a performance criterion using example data or past experience. The main applications are complex non-linear systems that are not suitable for control system methods.	(1) Very strong learning ability. (2) The neural network has many layers and a wide width, and can theoretically map to any function, so it can solve very complicated problems. (3) Highly dependent on data, the larger the amount of data, the better his performance. (4) There are many frameworks that can be used and portability is good.	(1) Large amount of calculation and high cost. Many applications are not suitable for use on mobile devices. (2) High computing power requirements. The mainstream computing power uses GPU and TPU, so the hardware requirements are high and the cost is high. (3) Model design is very complicated and requires a lot of human and material resources and time to develop new algorithms and models. (4) Because learning is dependent on data and is not highly interpretable. In the case of imbalanced training materials, problems such as gender discrimination and racial discrimination will occur, which is prone to bias.
Bayesian probability control [23, 24]	Bayesian probabilistic control should measure the confidence of an individual for an uncertain proposition and use this property to control it, so it is subjective in this sense. Using the probability theory proposed by Bayes, we can examine the sensitivity of decision-making. Bayes proposed the concepts of prior and posterior probability: the prior probability can be modified according to new information to obtain the posterior probability. Therefore, Bayesian theory is used to incorporate new information into analysis.	(1) The Bayesian model has stable classification efficiency. (2) It performs well on small-scale data, can handle multi-class tasks, and is suitable for incremental training, especially when the amount of data exceeds the memory, you can go for incremental training in batches. (3) Not very sensitive to missing data, and the algorithm is relatively simple, often used for text classification.	(1) Bayesian model has the smallest error rate compared with other classification methods. However, this is not always the case. This is because given the output category of the Naive Bayes model, the attributes are assumed to be independent of each other. This assumption is often not true in practical applications. When the correlation is large, the classification effect is not good. (2) You need to know the prior probability, and the prior probability often depends on the hypothesis. There can be many hypothetical models, so the prediction effect is not good due to the hypothetical prior model. (3) Since we determine the classification by using the prior and the data to determine the posterior, the classification decision has a certain error rate. (4) Very sensitive to the form of input data.

Category	Description	Advantage	Disadvantage
Fuzzy control [25, 26]	Computer intelligent control based on fuzzy set theory, fuzzy linguistic variables, and fuzzy logic reasoning. The input of this mechanism is to change the original 0 and 1 data into a value between 0 and 1 through fuzzification, which is closer to human thinking than the original nonzero or one dichotomy. In the process of inference, the data is fuzzy, but through the steps of defuzzification, the output can be made to an accurate value. Fuzzy control is often used in intelligent computing, expert system construction, and neural network-like applications.	(1) Fuzzy control is a rule-based control. It directly uses language-based control rules. The starting point is the control experience of field operators or the knowledge of relevant experts. It is not necessary to establish an accurate mathematical model of the controlled object in the design. Therefore, the control mechanism and strategy are easy to accept and understand, and the design is simple and easy to apply. (2) Starting from the qualitative understanding of industrial processes, it is relatively easy to establish language control rules, so fuzzy control is very suitable for those objects whose mathematical models are difficult to obtain, whose dynamic characteristics are not easy to grasp or whose changes are very significant. (3) Model-based control algorithms and system design methods can easily lead to large differences due to different starting points and performance indicators; but a system's language control rules are relatively independent. It is easy to use fuzzy connections between these control rules. Finding a compromise option makes the control effect better than conventional controllers. (4) Fuzzy control is designed based on heuristic knowledge and language decision rules, which is helpful for simulating the processes and methods of artificial control, enhancing the adaptive ability of the control system, and making it have a certain level of intelligence. (5) The robustness of the fuzzy control system is strong, and the influence of interference and parameter changes on the control effect is greatly reduced, which is especially suitable for the control of nonlinear, time-varying and purely lagging systems.	(1) The design of fuzzy control still lacks systematicness, which is difficult to control complex systems. So how to establish a system of fuzzy control theory to solve a series of problems such as the mechanism of fuzzy control, stability analysis, and systematic design methods. (2) How to obtain fuzzy rules and membership functions, that is, the design method of the system is currently based on experience. (3) Fuzzy processing of information will cause the control accuracy of the system to decrease and the dynamic quality to deteriorate. If you want to improve the accuracy, you will inevitably increase the number of quantization levels, which will lead to the expansion of the rule search range, reduce the speed of decision-making, and even fail to control in real time. (4) There is still room for discussion on how to ensure the stability of fuzzy control systems, that is, how to solve the problems of stability and robustness in fuzzy control.
Expert system [27]	This is a programming system with expert-level problem-solving capabilities in a specific field. It can effectively use the effective experience and expertise accumulated by experts for many years to solve problems that require experts to solve by simulating the expert's thinking process.	(1) The expert system can work efficiently, accurately, thoughtfully, quickly and tirelessly. (2) The expert system is not affected by the surrounding environment when solving practical problems, and it is impossible to forget and forget. (3) The expertise of experts can be freed from the constraints of time and space in order to promote valuable and scarce	(1) Its design cost and construction cost are extremely high, so it is rare in the industry today. (2) The ability to deal with common sense is low and the requirements for experts are very high. (3) There is still a need to develop deep inference systems. (4) The ability to interpret at different levels is poor. (5) It is difficult to make the

Category	Description	Advantage	Disadvantage
		expert knowledge and experience. (4) Expert systems can promote development in various fields. (5) Expert systems can bring together the knowledge and experience of experts in multiple fields and their ability to collaborate to solve major problems. (6) The level of military expert system is one of the important signs of national defense modernization. (7) The development and application of expert systems have huge economic and social benefits. (8) The research expert system can promote the development of the whole science and technology.	expert system have the ability to learn. (6) It is costly and difficult to achieve a decentralized expert system. (7) Poor ability to easily acquire and update knowledge.
Genetic algorithm control [28]	The genetic algorithm is a computational model that simulates Darwin's genetic selection and natural evolutionary biological evolution process. Its idea originates from the natural laws of biological genetics and survival of the fittest, and it is a search algorithm with an iterative process of "survival + detection". The genetic algorithm targets all individuals in a population, and uses randomization techniques to guide an efficient search of an encoded parameter space. Among them, selection, crossover and mutation constitute the genetic operation of the genetic algorithm; the five elements of parameter coding, initial population setting, fitness function design, genetic operation design, and control parameter setting constitute the core content of the genetic algorithm.	(1) produce interpretable results. (2) Results are easy to apply. (3) The range of data types that can be processed is extremely large. (4) Can be used for optimization. (5) Easy integration with neural network.	(1) Many problems have encoding difficulties. (2) No guarantee of optimization. (3) High computing cost. (4) Not many commercial software packages available.

Table 3.
Development trend of intelligent control technology.

sensing technology. With the increasingly fierce competition in the product market, intelligent products have obtained good application advantages in practical operations and applications. The main results include greatly improving the operator's operating efficiency, improving the quality of work in some dangerous places, and reducing work intensity, solving application of some dangerous and key constructions jobs, enhanced machine automation and intelligence, improved equipment reliability and reduced maintenance costs, and intelligent fault diagnosis, environmental protection and energy saving, etc.

According to the, the integration is shown in **Table 4** [29, 30].

Category	Item	Description
Performance development	High speed, high precision and high efficiency.	Speed, accuracy and efficiency are the key performance indicators of machinery manufacturing technology. Due to the use of high-speed CPU wafers, RISC wafers, multi-CPU control systems, and AC digital servo systems with high-resolution absolute detection elements, effective measures to improve the dynamic and static characteristics of the machine tools are also taken. Has been greatly improved.
	Flexible	It includes two aspects: the flexibility of the CNC system itself, the CNC system adopts a modular design, and the function coverage is large. Strong cutting ability, easy to meet the needs of different users; the flexibility of group pull system, the same group control system can automatically adjust the material flow and information flow according to the requirements of different production processes, so as to maximize the use of group control system efficacy.
	Process composite and multi-axis	Composite processing with the main purpose of reducing process and auxiliary time. It is developing in the direction of multi-axis and multi-series control functions. The process compounding of NC machine tools means that after the work piece is clamped on a machine tool at one time, the multi-process and multi-surface composite processing is completed through various measures such as automatic tool change, rotating spindle head or turntable.
	Real-time intelligence	Early real-time systems were usually aimed at relatively simple and ideal environments, and their role was to schedule tasks to ensure that tasks were completed within prescribed deadlines. Artificial intelligence attempts to implement various intelligent behaviors of humans with computational models. To date, science and technology have developed. Real-time systems and artificial intelligence are combined. Artificial intelligence is developing in a more realistic field with real-time response, and real-time systems are also developing in more complex applications with intelligent behavior. This has created a new field of real-time intelligent control.
Functional development	Graphical user interface	The user interface is the interface between the CNC and the user. Because different users have different requirements for the interface, the workload of developing the user interface is huge, and the user interface has become one of the most difficult parts in computer software development. Current Internet, virtual reality, scientific computing visualization, and multimedia technologies also place higher demands on user interfaces. The graphical user interface greatly facilitates the use of non-professional users. People can operate through windows and menus to facilitate the realization of blueprint programming and fast programming, 3D color stereo dynamic graphic display, graphic simulation, dynamic tracking and simulation of graphics, different directions of view and partial display scaling.
	Visualization of scientific computing	Visualization in scientific computing can be used to efficiently process and interpret data, so that information exchange is no longer limited to words and

Category	Item	Description
		language education, but visual information such as graphics, images, and animation can be used directly. The combination of visualization technology and virtual environment technology has further broadened the application fields, such as design without drawings, virtual prototype technology, etc., which is of great significance for shortening product design cycles, improving product quality, and reducing product costs. In the field of numerical control technology, visualization technology can be used for CAD / CAM, such as automatic programming and design, automatic parameter setting, dynamic processing and display of tool compensation and tool management data, and visual simulation of machining processes.
	Diversification of interpolation and compensation methods	Multiple interpolation methods such as linear interpolation, circular interpolation, cylindrical interpolation, spatial elliptical surface interpolation, thread interpolation, polar coordinate interpolation, 2D + 2 spiral interpolation, NANO interpolation, NURBS interpolation (non-uniform Rational B-spline interpolation), polynomial interpolation, etc. Various compensation functions such as clearance compensation, verticality compensation, quadrant error compensation, pitch and measurement system error compensation, speed-related feed forward compensation, temperature compensation, tool radius compensation with smooth approach and exit, and opposite point calculation.
	Built-in high-performance PLC	The high-performance PLC control module is built in the CNC system, which can be directly programmed with trapezoidal circles or high-level languages. It has intuitive online debugging and online help functions. The programming tool contains the real side of the standard PLC user program for lathe and milling machine. Edit and modify based on PLC user program, so as to create your own application program conveniently.
	Application of multimedia technology	Multimedia technology integrates computer, audio-visual, and communication technologies, so that the computer has the ability to comprehensively process sound, text, image and video information. In the field of CNC technology. The application of multimedia technology can achieve comprehensive and intelligent information processing, which has great application value in real-time monitoring systems and fault diagnosis of production field equipment and monitoring of production process parameters.
Architecture development	Integrated	Adopting highly integrated CPU, RISC wafers and large-scale programmable integrated circuits FPGA, EPLD, CPLD, and ASIC wafers for special integrated circuits, which can improve the integration of the CNC system and the speed of hardware and software. Applying LED flat panel display technology Improve display performance. Flat panel displays have the advantages of high technology content, light weight, small size, low power consumption, and easy portability. Achieve oversized display. Apply advanced packaging and interconnect technologies to integrate semiconductor and surface-mount technologies. By

Category	Item	Description
		increasing the density of integrated circuits, reducing the length and number of interconnects to reduce product prices, improve performance, reduce component size, and increase system reliability.
	Modular	The hardware modularity is easy to realize the integration and standardization of the numerical control system. According to different functional requirements, the basic modules, such as CPU, memory, position servo, PLC, input and output interface, and communication modules, are made into standard serial products. The building block method is used to cut functions and increase or decrease the number of modules to form different grades of CNC systems.
	Networking	Machine tool networking can perform remote control system and unmanned operation. Networking can program, set, operate and run other machine tools on any one machine tool. The pictures of different machine tools can be displayed on the screen of each machine at the same time.

Table 4.
Development trend of intelligent control technology.

At present, the neural fuzzy network has its input and output space divided into a checkerboard pattern. Although it is very easy to implement in hardware, as the input and output variables increase, the number of this checkerboard pattern increases. This leads to an unrealistic increase in the number of required memory or hardware, because more training data is needed for better spatial segmentation than the method in the learning process, otherwise insufficient learning will occur. In complex systems, in order to avoid an increase in the number of partitions, it is a research direction to find a more flexible and irregular partitioning method. The implementation of fuzzy controllers with neural networks such as learning capabilities has become a very interesting research area in recent years. However, many neural fuzzy networks with learning capabilities often require expert knowledge before application. In order to improve its performance, it is also the research direction to automatically generate fuzzy rules and adjust the attribution function only from the training data. In recent years, because genetic algorithms have global optimization capabilities, genetic algorithms have become another useful tool. Currently, genetic algorithms are used to adjust the fuzzy controller's attribution function and neural network-like weighting values. The combination of genetic algorithms and neural-like fuzzy networks to accelerate their learning speed is also worth exploring.

Other intelligent control methods such as machine learning control, Bayesian probability control, expert system and genetic algorithm control, etc., or the aforementioned artificial neural networks and fuzzy control, will be from communication technology, manufacturing technology, construction technology, transportation technology and the integration of energy technology and other different levels is also an important task for us to continue to make good use of the advantages of intelligent control methods and eliminate the disadvantages in the future.

Author details

Kuo-Chi Chang[1]*, Kai-Chun Chu[1], Yuh-Chung Lin[1] and Jeng-Shyang Pan[2]

1 Fujian University of Technology, China

2 Shandong University of Science and Technology, China

*Address all correspondence to: albertchangxuite@gmail.com

References

[1] Takagi T, Sugeno M. Derivation of fuzzy control rules from human operator's control actions. IFAC Proceedings Volumes. 1983;**16**(13): 55-60. DOI: 10.1016/S1474-6670(17) 62005-6. ISSN: 1474-6670

[2] Ying H, Siler W, Buckley JJ. Fuzzy control theory: A nonlinear case. Automatica. 1990;**26**(3):513-520. DOI: 10.1016/0005-1098(90)90022-A. ISSN: 0005-1098

[3] Sastry S, de Silva C. Intelligent Control. Boca Raton: CRC Press; 1995. DOI: 10.1201/9780203750513

[4] Xiang X, Yu C, Lapierre L, et al. Survey on fuzzy-logic-based guidance and control of marine surface vehicles and underwater vehicles. International Journal of Fuzzy Systems. 2018;**20**: 572-586. DOI: 10.1007/s40815-017-0401-3

[5] Lin HR, Cao BY, Liao Y. Fuzzy control. In: Fuzzy Sets Theory Preliminary. Cham: Springer; 2018. DOI: 10.1007/978-3-319-70749-5_3

[6] El-samahy AA, Shamseldin MA. Brushless DC motor tracking control using self-tuning fuzzy PID control and model reference adaptive control. Ain Shams Engineering Journal. 2018;**9**(3): 341-352. DOI: 10.1016/j.asej.2016.02.004 ISSN: 2090-4479

[7] Zhang X, Zhou X, Lin M, Sun J. The IEEE Conference on Computer Vision and Pattern Recognition (CVPR). 2018. pp. 6848–6856

[8] Rajendra Acharya U, Oh SL, Hagiwara Y, Tan JH, Adeli H. Deep convolutional neural network for the automated detection and diagnosis of seizure using EEG signals. Computers in Biology and Medicine. 2018;**100**: 270-278. DOI: 10.1016/j.compbiomed.2017.09.017. ISSN: 0010-4825

[9] Nagabandi A, Kahn G, Fearing RS, Levine S. Neural network dynamics for model-based deep reinforcement learning with model-free fine-tuning. In: 2018 IEEE International Conference on Robotics and Automation (ICRA), Brisbane, QLD. 2018. pp. 7559-7566. DOI: 10.1109/ICRA.2018.8463189

[10] Wu M, She J-H, Nakano M, Gui W. Expert control and fault diagnosis of the leaching process in a zinc hydrometallurgy plant. Control Engineering Practice. 2002;**10**(4): 433-442. DOI: 10.1016/S0967-0661(01) 00122-8 ISSN: 0967-0661

[11] Liu J. Expert PID control. In: Intelligent Control Design and MATLAB Simulation. Singapore: Springer; 2018. DOI: 10.1007/978-981-10-5263-7_2

[12] Wang M, Yang Z, Xiangchao S, Zhang B, Wu X, Shi Y. Simulation and experimental research of engine rotary speed for gas engine heat pump based on expert control. Energy and Buildings. 2013;**64**:95-102. ISSN: 0378-7788. DOI: 10.1016/j.enbuild.2013.04.003

[13] Yu H, Tseng HE, Langari R. A human-like game theory-based controller for automatic lane changing. Transportation Research Part C: Emerging Technologies. 2018;**88**: 140-158. ISSN: 0968-090X. DOI: 10.1016/j.trc.2018.01.016

[14] Jin J, Ma X, Kosonen I. An intelligent control system for traffic lights with simulation-based evaluation. Control Engineering Practice. 2017;**58**: 24-33. DOI: 10.1016/j.conengprac.2016.09.009 ISSN: 0967-0661

[15] Rath AK, Parhi DR, Das HC, Muni MK, Kumar PB. Analysis and use of fuzzy intelligent technique for navigation of humanoid robot in obstacle prone zone. Defence Technology. 2018;**14**(6):677-682. ISSN:

2214-9147. DOI: 10.1016/j.dt.2018.03.008

[16] Chen J, Wang G, Hu X, et al. Lower-body control of humanoid robot NAO via kinect. Multimedia Tools and Applications. 2018;**77**:10883-10898. DOI: 10.1007/s11042-017-5332-3

[17] Xiao L, Li M, Wang F, Li Y. A hierarchical framework for intelligent launch site system. In: 2018 IEEE International Conference on Information and Automation (ICIA), Wuyishan, China. 2018. pp. 814-818. DOI: 10.1109/ICInfA.2018.8812548

[18] Pham DB, Lee SG. International Journal of Precision Engineering and Manufacturing. 2018;**19**:1291. DOI: 10.1007/s12541-018-0153-5

[19] Tan C, Sun F, Kong T, Zhang W, Yang C, Liu C. A survey on deep transfer learning. In: Kůrková V, Manolopoulos Y, Hammer B, Iliadis L, Maglogiannis I, editors. Artificial Neural Networks and Machine Learning– ICANN 2018. ICANN 2018. Lecture Notes in Computer Science. Vol. 11141. Cham: Springer; 2018

[20] Karatzoglou A, Schnell N, Beigl M. A convolutional neural network approach for modeling semantic trajectories and predicting future locations. In: Kůrková V, Manolopoulos Y, Hammer B, Iliadis L, Maglogiannis I, editors. Artificial Neural Networks and Machine Learning– ICANN 2018. ICANN 2018. Lecture Notes in Computer Science. Vol. 11139. Cham: Springer; 2018

[21] Holmes L, LaHurd A, Wasson E, McClarin L, Dabney K. Racial and ethnic heterogeneity in the association between total cholesterol and pediatric obesity. International Journal of Environmental Research and Public Health. 2016;**13**:19. Available from: https://www.mdpi.com/1660-4601/13/1/19

[22] Granda JM, Donina L, Dragone V, et al. Controlling an organic synthesis robot with machine learning to search for new reactivity. Nature. 2018;**559**: 377-381. DOI: 10.1038/s41586-018-0307-8

[23] Han W, Wang W, Li X, Xi J. Statistical-based approach for driving style recognition using Bayesian probability with kernel density estimation. IET Intelligent Transport Systems. 2019;**13**(1):22-30. DOI: 10.1049/iet-its.2017.0379

[24] Wang J, Yan W, Xu H, et al. Investigation of the probability of a safe evacuation to succeed in subway fire emergencies based on Bayesian theory. KSCE Journal of Civil Engineering. 2018;**22**:877-886. DOI: 10.1007/s12205-018-0620-7.

[25] Sun K, Mou S, Qiu J, Wang T, Gao H. Adaptive fuzzy control for nontriangular structural stochastic switched nonlinear systems with full state constraints. IEEE Transactions on Fuzzy Systems. 2019;**27**(8): 1587-1601. DOI: 10.1109/TFUZZ.2018.2883374

[26] Xia J, Zhang J, Sun W, Zhang B, Wang Z. "finite-time adaptive fuzzy control for nonlinear systems with full state constraints," in IEEE transactions on systems. Man, and Cybernetics: Systems. July 2019;**49**(7):1541-1548. DOI: 10.1109/TSMC.2018.2854770

[27] Son LH, Tuan TM, Fujita H, Dey N, Ashour AS, Ngoc VTN, et al. Dental diagnosis from X-ray images: An expert system based on fuzzy computing. Biomedical Signal Processing and Control. 2018;**39**:64-73. DOI: 10.1016/j.bspc.2017.07.005

[28] Reynolds J, Rezgui Y, Kwan A, Piriou S. A zone-level, building energy optimisation combining an artificial neural network, a genetic algorithm, and model predictive control. Energy.

2018;**151**:729-739. DOI: 10.1016/j.energy.2018.03.113

[29] Wang T, Tao Y, Liu H. Current researches and future development trend of intelligent robot: A review. International Journal of Automation and Computing. 2018;**15**:525-546. DOI: 10.1007/s11633-018-1115-1

[30] Wang X, Zhang Y, Xu B, Wang Y, Wang Y. Development status and trend of control system and intelligent equipment for greenhouse. In: 2018 20th International Symposium on Electrical Apparatus and Technologies (SIELA), Bourgas. 2018. pp. 1-4. DOI: 10.1109/SIELA.2018.8447105

Section 7

Electric Motor Drives

Chapter 14

DC Motor Synchronization Speed Controller Based on Microcontroller

Aamir Shahzad, Muhammad Kashif, Tariq Munir and Maogang He

Abstract

In this chapter, we report the design and fabrication of an improved speed synchronizer device in which two dc motors has been controlled on different sequences programmed by microcontroller. Depending on the programmed software, the device is used to command a rolling of machines, synchronizes the dc motors speed, and displays the result on liquid crystal display (LCD). Flash memory of the microcontroller is used to program for controlling this device where permanent memory is needed to store different parameters (codes for motor speed, LCD display, ratio control, and rotary encoder's feedback). The present simulation gives new reliable results with better performance for the speed and direction than the earlier available synchronizers. It has been shown that the speed and direction are dependent on both the ratio setting and frequency of encoder in two dc motors speed synchronizer. It is shown that this device is applicable for controlling, monitoring, and synchronizing identical processes and can be implemented in multiple domains, from textile industry and home control applications to industrial instruments.

Keywords: speed synchronizer, liquid crystal display, microcontroller, dc motors speed

1. Introduction

Nowadays, the advancements in industrial automation research technology and the desires on performance and feature in manufacture processes have become in complex and integrated innovation systems. Speed synchronizers have various applications in many control industrial processes (textile, chemical, marine, etc.) and have shown a vital role in control system design and optimization. The speed synchronizer is cost-effective and highly efficient device and it works as a bridge between two (or multiple) independent drives for many control applications with a suitable value of ratio control. The synchronization units are appropriate for different type of drives (DC, AC, Servo, etc.) that are adjustable in speed and direction under influence of speed reference signal of the 0–5 volts [1]. The significant results of using synchronizer is to reduce the working and fabricating cost in designing new control devices in terms of reducing the time and minimizing the ambiguity in going process modernization at the small level to manufacturing scale. Recent advances in microelectronics and algorithm design including information technology have brought significant changes in speed and motion control technology. As a result, the predictive advancement, broadly useful, practical approach for the speed and motion control is an enduring target of greatly of the industrial automation technologist both in science and engineering developments [2].

The earlier variable speed drive was mechanical type and was used a variable pitch diameter pulley. This drive is still in operation but for clear reasons are not in common uses in practical and industrial applications nowadays. Recently, the three significant kinds of electronically variable speed drives are used in present industrial machines. Firstly, the brush style dc motor is almost the totally used adjustable speed drive on blow molding machineries. Alternative current (AC) adjustable frequency and AC motor drives for blow molding technologies are used and replaced which is second kind of variable speed drives. Thirdly, the brush dc drive has been replaced with Brushless dc drive [3]. The direction, speed and power of an AC or dc motor can be controlled with the help of electronic device called as variable speed drive. Variable frequency AC drives provide processing requirements and several familiar industrial appliances for example pumps, fans and conveyors in a variety of working situations. The dc drives have various development applications and they can used to command material handling and processing machine such as metals, mining, printing, and other industries. In industrial automation increases gradually and many applications need directional and positional control simultaneously [3, 4].

Over the past two decades, numerous measurements of speed and direction control synchronization have been studied [5–11]. The revolving mechanical arrangements form a significant and singular class of systems that, with or without the interaction through some coupling, demonstrate synchronized motion (or speed motors), for instance the case of rotating machinery in manufacture plants, power generators, wrapping of cloth in many textile industrial applications and unbalanced rotors in milling machines [12]. The synchronization of revolving systems and electromechanical arrangements was reported by Blekhman et al. [13]. Subsequently, the teleoperated master-slave systems were developed by Dubey et al. [14], and by Lee and Chung [15]. Shortly after, the Brunt [16] and Liu et al. [17] were designed multifinger robot-hands, multirobot systems and multi-actuated platforms. Only a few years later, Huijberts et al. calculated synchronization of rotating bodies and electromechanical systems [18]. For the medical applications, the master-slave teleoperated systems were used in surgical procedure giving rise to more accurate and less enveloping surgical treatment measures [19, 20]. In case of aerospace applications, the synchronization approaches were employed to reduce the error of the relevant manner in developments of satellites [21, 22]. The case of group formation of multiple robotic vehicles is addressed in [23]. Mirza and Hussain published their work of speed asynchronization mode of dc motors in the regime of nonlinear process through selective position of feedback and integral control for textile industry [1]. Since then, speed synchronizers have provided a key automation tool in process industries, medicine, material science, plasma process engineering, and aerospace science and multiphase and thermal transport processes [9–11].

In this study, a speed synchronizer based on microcontroller for controlling and monitoring the speed of two dc motors, which is our particular motivation. Embedded system is a system that has a microcomputer or microcontroller inside which can reads the input, process them and gives the feedback according to the preprogram condition. Embedded systems are designed to do some specific tasks and have minimal requirements for memory and program length [24]. The application of microcontrollers with designed algorithm, dc drives, and complex programmable logic devices (CPLD) in factory automation motivates the development of computer integrated manufacturing systems. In textile industry many processes require speed synchronization of more than one motors involved in the process. Rolling of cloth should be synchronized with the speed of weaving spindle to avoid damage and similarly motors-speed synchronization is vital in a conveyor-belt driven by multiple motors. The digitally controlled dc machines (or motors) can have much aggravated phenomena owing to poor sampling period selection [1, 7]. The role of synchronization control not only involves in practical and industrial applications, however, there are many physical phenomena that may be explored through synchronization control theory [12].

The objective of presented article is to design and implement of an advanced digital speed control synchronizer using ATMEL-based microcontroller and dc drives. A digital speed (or motion) control algorithm has been developed for the synchronization control of two motors, which can easily be extended to control of multiple motors. The designed speed and motion control system is based on a digital torque drive for dc motors with encoder interface. This article presents the algorithm for design implementation of a speed and motion control and to examine the understanding of synchronization efficiency for industrial application. Additionally, this design and numerical approach is different from those used in the earlier studies. The calculations are performed for a wide domain of synchronization control parameters for the drives. In order to characterize the performance of the project, the different parameters of synchronization control have been studied and analyzed, so that the systems working may be enhanced.

2. Methodology and design

2.1 The PROTOS framework

The circuitry was designed for the synchronization control of two dc motors using commercially available devices and components. There are two main parts in the system, which are hardware and software development. The schematic design for the system and components testing in standalone state are include in hardware development. While the software skill developed including drive circuit connection establishment and improvements to algorithm. Some assumptions are made in the prototype system and recommendations or future improvements are suggested [25]. In order to utilize facilities and advantages of both microcontrollers along with their standard features and programming resources and CPLD along with their flexibility to implement additional hardware arrangements for interfacing or progress of computational performance, we need to set a number of this design system purposes the PROTOS system should fulfill.

An experimental dc motor synchronization control system based on microcontroller has been developed and its design implemented for control automation of two dc motors as shown in **Figure 1**. This controller can command two drives one byte external port; these outputs take values depending of the inputs states. One port states are displaying on a 16 characters, 4/2 lines material liquid crystal display (LCD) display. The ATmega8535 has a 13-bit program counter capable of addressing an 8 k × 14 program memory space [26]. In this experiment an attempt has been made to fabricate and study a circuit that is able to control and monitor the synchronization control of two dc motors. The basic proposal was to design a synchronizer to ensure its performance equivalent to dedicated industrial synchronizers designed by marketable big companies. The synchronizer based on ATmega8535 microcontroller hardware was divided into six major modules; power supply module, dc motor speed and direction control module, encoder interface module for two dc

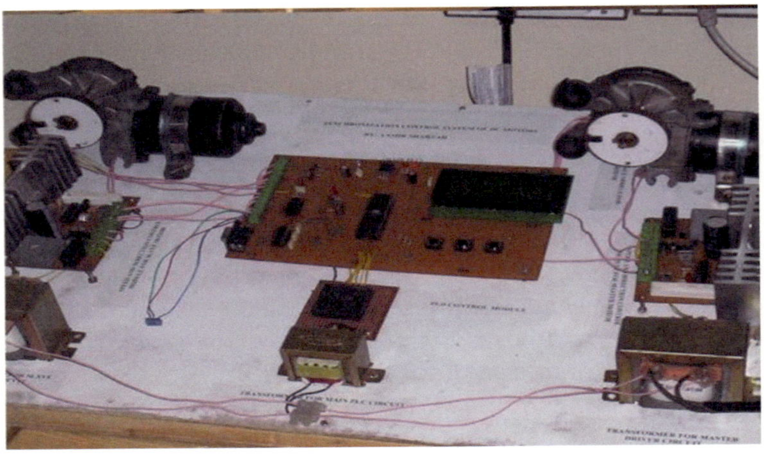

Figure 1.
Prototype snap for synchronization control system for two dc motors.

motors, central processing unit (CPU) and control module, speed synchronization graphic interface module and LCD display module. In order to develop the project the following methodology has followed: the theory of microcontroller, logic control devices, encoder, dc motor, circuit diagrams, operation and its function has been studied in detail [7–15, 26–29]. The first step was to draw a schematic diagram of the hardware in a software package called Protel, according to circuit diagrams [5, 6]. The encoder circuit design has been made on computer and its film developed in printing press and finally its hard fitting with sensor has been made accordingly. A fitting of dc motor shaft coupled with encoder has been made for sensing motor position, direction and whole system fitted on separate wooden board. After completing the circuit of encoder interface unit, the program has been written in VHDL programming language of CPLDs family. The written program has been debugger using active VHDL simulator and downloaded according to the application with the help of programmer such as control sensing and monitor motion or position of motors [29, 30].

After completing the circuit of CPU and control unit, the program has been written in C-51 programming language of MCS-51 microcontroller family. The written program has been debugger using simulator AVS-51 and downloaded according to the application with the help of downloader ATmega8535 such as synchronization of two dc motors. Check the performance and error of microcontroller, and CPLD ICs has debugger the written program and circuits through respective simulators and check timing signals on different pins and finally monitor programs of both ICs.

2.2 Measurement scheme using microcontroller and CPLD

The schematic circuit diagram (CPU and control module) of the microcontroller scheme is given in **Figure 2**. The hardware and software for CPU and control unit are to be contained within the microcontroller. Design of the CPU and control hardware was tested using the ATmega8535 development board and breadboard. The power supply was then tested and found to be functional. After the first programming the microprocessor was configured to use the external oscillator [26–30]. Testing of the communication between the Atmel chip and other chips in the circuit design was successful using the communication protocol designed. Switching between two voltage levels required the use of the microcontroller on board comparator. It was

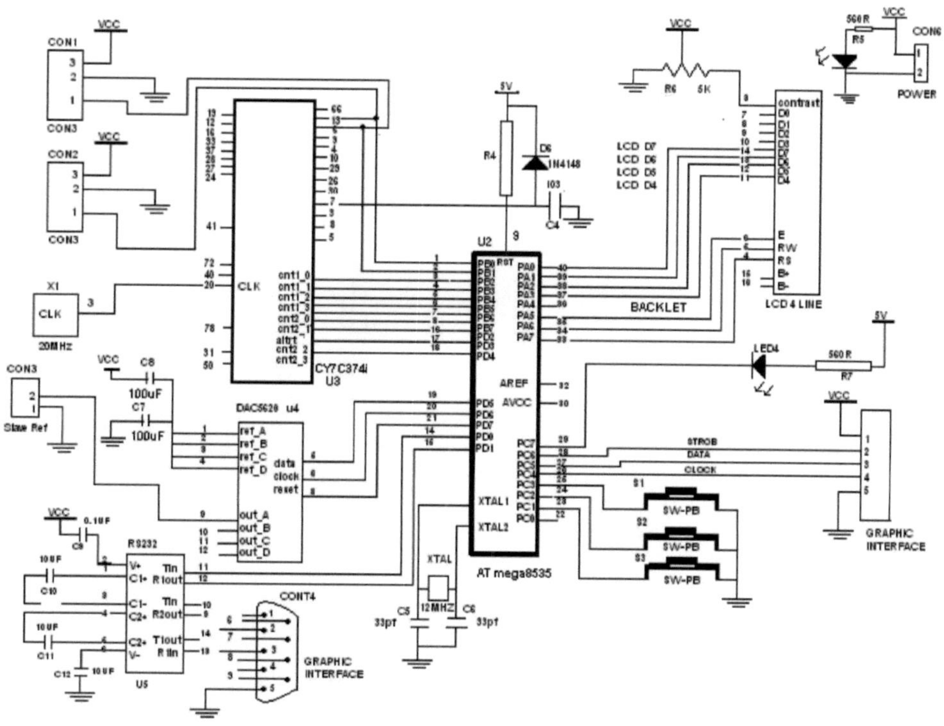

Figure 2.
Schematic circuit diagram of CPU and control unit.

important that the circuit was reverted to switching between 5 and 0 V; therefore, a standard I/O pin on the Atmel could be confined to detect the communications signal. This test avoided the complication of noise interfering with the synchronization signal [26]. The Vero board circuitry was tested successfully. The CY7C374 is a flash erasable complex programmable logic device and is part of the FLASH370 family of high-density, high-speed CPLDs [29, 30].

2.3 Measurement scheme for speed and direction control module

The schematic diagram for speed and direction control module for two dc Motors with built-in power supply is given in **Figure 3**. In dc motor speed and direction control module, the device L6203 combines a full H-bridge in a single package and it was noted that this device is perfectly matched for controlling dc motors. The power stage was carried out this device which is essential for both direction and speed control for dc motor control. A current regulator (L6506) along with this device (L6203) provides a constant current drive for dc motors that control the speed of the system [7]. The value of peak current may be obtained as:

$$I_{peak} = \frac{V_{ref}}{R_{sense}} \quad (1)$$

The voltage across the sense resister (R_S) was detected by the current regulator L6506 in order to control the motor current, and it evaluates detected voltage to monitor the speed of the motor during the brake of the motor. A suitable resistor was used between the R_S and each sense input of the L6506. It was noted that the relations between the inputs of the device of L6203 and the outputs of the L6506 require a lengthy path. It was noted that a resistor should be connected between each input

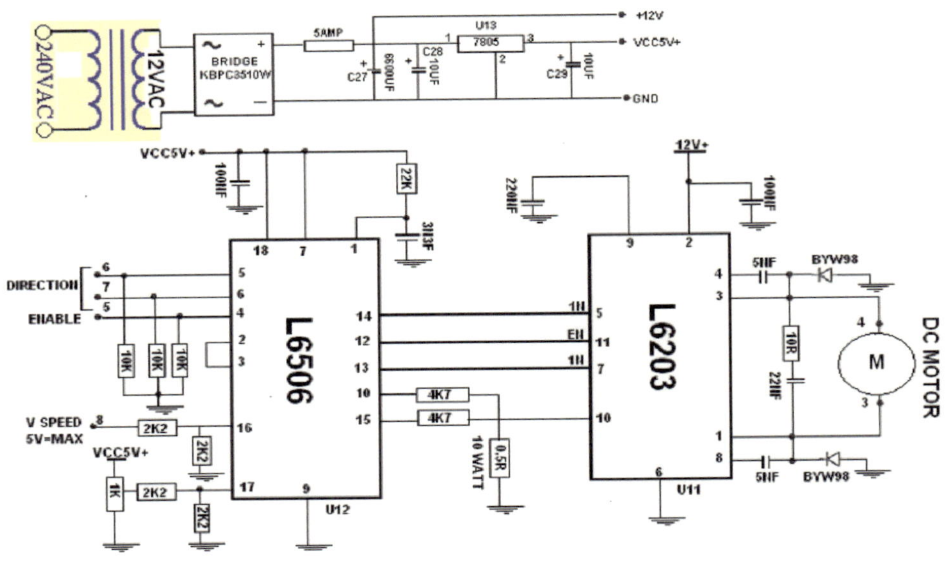

Figure 3.
Schematic diagram for speed and direction control module for two dc motor.

of the L6203 and ground its terminal. We have used the sunbber circuit very close to the output connections of the device L6203 which was fabricated by the series of resister (R) and capacitor (C). Moreover, each power output connection and ground terminal was connected with the help of one diode (BYW98). An appropriate big capacitor was used and linked close to the supply pin of the device L6203, when the power supply cannot sink current properly. It is important that a capacitor at pin 17 of the L6506 allows the system application work smoothly. The L6506 was proposed for work with double bridge drivers, such as the device L6203 to drive dc motors. The major purpose of the device (L6203) was to detect and monitor the current in each of the load windings (dc motors). The dual chopper was run using a general on chip oscillator. An ordinary on chip oscillator was used to drive the dual chopper and it wad regulated the working frequency for the pulse width modulated drive. The sunbber circuit on pin 1 sets the working frequency which was calculated as:

$$f = \frac{1}{0.69RC}, \text{ where R > 10 K} \tag{2}$$

The oscillator gives signals to set the two flip flops which in turn cause the outputs to trigger the drive. The current in the load winding (dc motor) attained the computerized programmed peak point, then the voltage across the sense resister becomes equivalent to the reference voltage (V_{ref}) signal which leads to the comparator readjusts its flip flop interrupting the drive current until the next oscillator signal appears. A suitable choice of the value of reference voltage (V_{ref}) and sense resister were used to program the peak current for each dc motor. It was examined that the ground noise issues in multiple arrangements has been avoided by synchronizing the oscillators. This was done by the oscillator output of the master device was connected to the sync pins of each of the devices and the R/C pin of the unused oscillators was grounded [7–11].

3. Results and discussion

A control system is designed and fabricated using microcontroller and CPLD from the point of simplicity and cost effectiveness to act as speed synchronization and direction control for two (or multiple) dc motors, already shown in **Figure 1**. The presented system was tested running the two motors for the long period of time and in different environments. It is observed that the synchronization system was worked to expectation. Some calibration of the analogy to digital conversion (ADC) in the control circuit module was necessary and this was done using software. Calibration had to account for the voltage drop across the resister in series with the control trigger and motor response. The calibration improved the speed range available to the driver circuit of dc motors, confirming the earlier results [1, 7–11]. Once dc motors were running on synchronization mode it was expected that contact noise between the motor brushes and due to the power signals would cause incorrect speeds. Therefore, the error checking software has been included in order to diagnose the erroneous signals, and it is important that this was found not to be a problem. It was noted that there were periodic speed control but unusual cases of erroneous speed behavior. The graphic interface unit has two zones of synchronization, one for when graphic interface unit has its middle value of indications (green-indications), the two motors has same speed, this mode was called synchronization of motors. On the other hand, the second zone of graphic interface has its end value of indications (red-indications), shows that the speeds of two motors were not in same phase. This feature adds realism to the synchronization control system of dc motors by using software.

3.1 Principle of operation: Positional synchronization and ratio control

All operation was based on setting an analogue synchronization between the speed and direction control modules first. This has been achieved by feeding a common speed reference voltage to the drives and tuning the drive speeds in order to get them into an approximate synchronism. A ratio agreement was significant for the slave drive. This analogue pre-synchronization was matched the two speeds within an error range of approximately 0.8%, confirming the earlier results of Ref. [31]. Presently, the digital synchronization has to balance for the analogue speed errors in order to get an absolute, angular and positional synchronization with no drift and no cumulative displacement of the motor shafts. This required a digital feedback of the angular shaft position of the drives. In our case, incremental shaft encoders (e.g., encoder simulation from a resolver system) were used for this feedback purpose.

The synchronizer (CPU and encoder modules) was continuously checked the two shaft positions and immediately responded by an analogue correction signal when an angular error started to appear. This analogue correction signal, added to the slave's reference with the correct polarity, was maintained the shaft positions of master and slave inline. The synchronized speed of developed system was monitored by using LCD. It was noted that the synchronizer responds within only microseconds to each individual encoder pulse and the slave was practically have no chance to drift away. We have been adopted the way to generate feed forward signal (FFS) given in Ref. [31]. Currently, this FFS was generated internally from the frequency of the master encoder and no external voltage was applied to the analogue input. It was observed that the encoder frequency always represents the real actual speed of the master. Moreover, a FFS was needed to run the drives, and no external correction voltage was added to receive the total slave speed reference. It was significant that the digital FFS used here was higher than at maximum speed, for encoder frequencies [7, 8, 31].

3.2 Simulation results

In this section, the simulation measurements of encoder interface with motors by using the Cypress and Wrap software were presented. The calculations of simulated waveforms with each module for implementation of CPLD (CY7C374i) were shown. Also, this subsection describes the implementation of C-code of AVR ATmege8535 microcontroller. The compiler software simulator has been used for both functional and timing simulation [29, 30]. For functional simulation, the written code was simulated before synthesis. After fitting (CPLDs) or place, simulate the design using the same original test bench as a test fixture, but with logic and routing delays added. In case of any major problems which have made an assumption on the device specification that was incorrect, or has not measured some aspect of the signal required to/ from the programmable logic device. Consequently, we have collected data on the problem and go back to the drawing (or behavioral) board. The output master up, reset, slave up and mixed simulation waveforms for encoder interface with motor for synchronization control of dc motors are shown in **Figures 4–7**. These presented timing waveforms show a successful prototype of a synchronization control system for dc motors. The designed system was

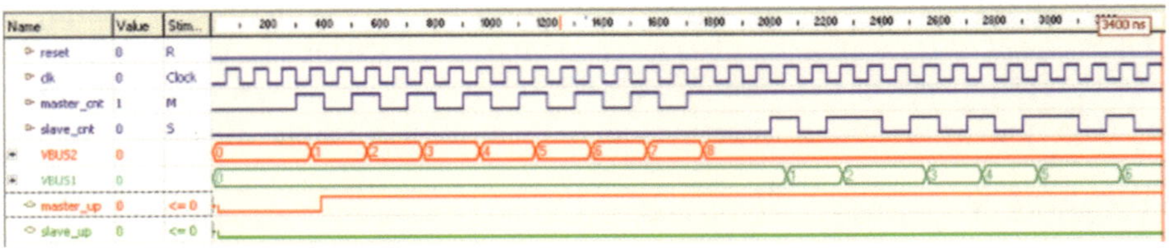

Figure 4.
Master up timing simulated wave form for encoder interface with motors.

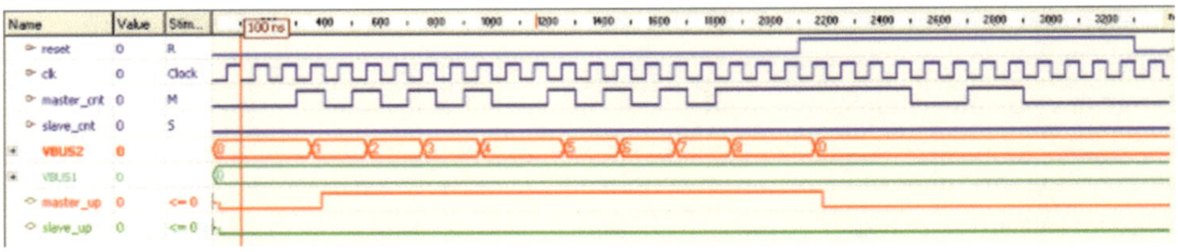

Figure 5.
Reset timing simulated wave form for encoder interface with motors.

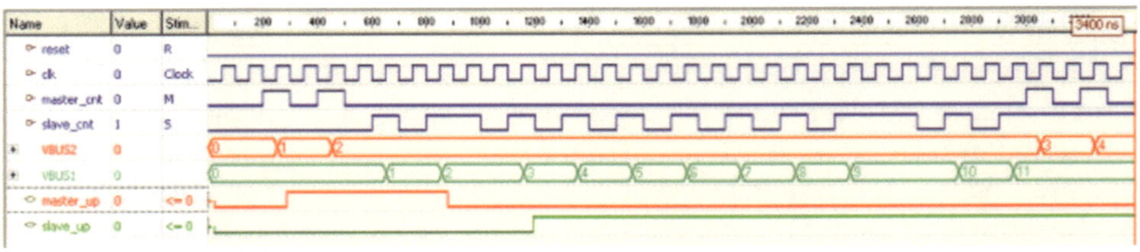

Figure 6.
Slave up timing simulated wave form for encoder interface with motors.

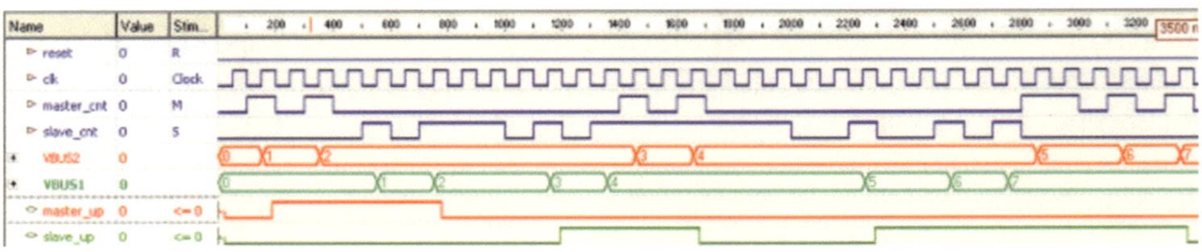

Figure 7.
Mixed timing simulated wave form for encoder interface with motors.

demonstrated the advantages of a synchronization control system over an analog system and improved the possibility for security when data mentioned on graphic interface module, challenge and competition when synchronization of dc motors speed required in industry, especially in textile.

4. Summary

An experimental dc motor synchronization control based on microcontroller has been developed and its design implemented for the purpose of simplicity and cost effectiveness to act as power converter, master and slave controller. The main functions that each drive performs were readily programmable. The drives were forced into coordination and synchronization by some sort of software mechanism which was also implemented by programming. The speed synchronization was provided with the help of the FFS technique on the LCD. The designing of a synchronizer system to control the direction and speed of a geared dc motor has been successfully employed in this chapter. The dc motors have speed control facilities which show that speed and direction of rotation can be changed at any time to meet new situation. The hardware of the proposed system was interfaced with computer using RS232 serial communication port, for future communications. It is concluded that the dc motors were ideally suited to a multitude of industrial including textile and marine applications in which variable speed high torque are required. In addition, systems present in other applications, which include industrial equipment such as fan drives, ship propulsion and steel rolling mills, can be synchronized by a proposed design with proper modifications. This paper gives a platform for further development in the field of industrial use of dc motors.

Acknowledgements

The authors thank Ramiza FARRUKH (MPhil Supervisor, UAF Pakistan) for providing his valuable suggestions for MPhil thesis completion and useful discussions.

Abbreviations

LCD	liquid crystal display
AC	alternative current
CPLD	complex programmable logic devices
CPU	central processing unit
ADC	analog to digital conversion
FFS	feed forward signal

Author details

Aamir Shahzad[1,2]*, Muhammad Kashif[1], Tariq Munir[1] and Maogang He[2]

1 Molecular Modeling and Simulation Laboratory, Department of Physics, Government College University Faisalabad (GCUF), Faisalabad, Pakistan

2 Key Laboratory of Thermo-Fluid Science and Engineering, Ministry of Education (MOE), Xi'an Jiaotong University, Xi'an, P.R. China

*Address all correspondence to: aamirshahzad_8@hotmail.com; aamir.awan@gcuf.edu.pk

References

[1] Mirza M, Sarfraz H. DC-motors speed asynchronization in nonlinear process by selective state feedback and integral control. Advances in Modelling and Analayis. 2002;**57**:45-61

[2] Bates MP. PIC Microcontrollers: An Introduction to Microelectronics. 2nd ed. Amsterdam, Neatherland: Boston-Newnes; 2004

[3] Allen B. Guide to Drives: Introduction to DC Drives. London: Elsevier; 2005. pp. 1-20

[4] Mohan N. Power Electronics. New York, NY, USA: John Wiley & Sons; 1995

[5] Shahzad A, Farooq H, Akbar S, Yousaf M, Hafeez K, Hanif M. Hybrid progranmmable logic controller for load automation. Journal of Applied Sciences. 2007;7:20904-20907

[6] Shahzad A, Islam-ul Q, Akhtar S, He MG. Cost-effective micro programmable logic controller for system automation. Journal of Basic and Appilied Sciences. 2007;**9**:368-372

[7] Singh R, Kumar S, Kumar R, Kumar V. Synchronisation speed and direction control of DC motor. International Journal of Electronic and Electrical Engineering. 2014;**7**:943-952

[8] Chauhan SJ, Semwal S. Microcontroller based speed control of DC geared motor through RS-232 Interface with PC. International Journal of Engineering Research and Applications. 2013;**3**:778-783

[9] Bashi SM, Aris I, Hamad SH. Development of single phase induction motor adjustable speed control using M68HC11E-9 microcontroller. Journal of Applied Sciences. 2004;**5**:249-252

[10] Tipsuwan Y, Chow M-Y. Fuzzy logic microcontroller implementation for DC motor speed control. IEEE Transactions on Power Electronics. 1999;**11**:1271-1276

[11] Ganiyu RA, Shoewu O, Olatinwo SO, Omitola OO. Development of a microcontoller-based motor speed control system using Intel 8051. Journal of Advanced Engineering Techologies. 2014;**12**:1-6

[12] Blekhman II. Synchronization in Science and Technology. New York, NY, USA: ASME Press Translations; 1988

[13] Blekhman II, Landa PS, Rosenblum MG. Synchronization and chaotization in interacting dynamical systems. ASME Applied Mechanics Reviews. 1995;**48**:733-752

[14] Dubey RV, Chan TF, Everett SE. Variable damping impedance control of a bilateral telerobotics system. IEEE Control Systems. 1997;**17**:37-45

[15] Lee HK, Chung MJ. Adaptive controller of a master-slave system for transparent teleoperation. Journal of Robotic Systems. 1998;**15**:465-475

[16] Brunt M. Coordination of redundant systems [PhD Thesis]. The Netherlands: Technical University Delft; 1998

[17] Liu YH, Xu Y, Bergerman M. Cooperation control of multiple manipulators with passive joints. IEEE Transactions on Robotics and Automation. 1999;**15**:258-267

[18] Huijberts HJC, Nijmeijer H, Willems RMA. Regulation and controlled synchronization for complex dynamical systems. International Journal of Robust and Nonlinear Control. 2000;**10**:336-377

[19] Hills JW, Jensen JF. Telepresence technology in medicine: Principles and

applications. Proceedings of the IEEE. 1998;**86**:569-580

[20] Guthart GS, Salisbury JK Jr. The intuitive telesurgery system: Overview and application. Proceedings of the IEEE International Conference on Robotics and Automation. 2000:618-621

[21] Wang PKC, Sparks A, Banda S. Co-ordination and control of multiple microspacecraft moving in formation. The Journal of the Astronautical Sciences. 1996;**44**:315-355

[22] Kang W, Yeh H. Coordinated attitude control of multi-satellite systems. International Journal of Robust and Nonlinear Control. 2002;**12**:185-205

[23] Yamaguchi H, Arai T, Beni G. A distributed control scheme for multiple robotic vehicles to make group formations. Robotics and Autonomous Systems. 2001;**36**:125-147

[24] Iovine J. PIC Microcontroller Project Book. New York, NY, USA: McGraw-Hill; 2000

[25] Shen SK. Remote control using infrared with message recording [PhD Thesis]. Universiti Teknologi Malaysia; 2008

[26] Gadre DV. Programming and Customizing the AVR Microcontroller. New York, NY, USA: McGraw-Hill; 2001

[27] Geller DA. Programmable Controllers Using the Allen-Bradley SLC-500 Family. Upper Saddle River, NJ: Prentice Hall; 2000

[28] Wilson RJ, Adler R, Brad F. Determining DC motor rotation direction. Automotive Engineering International. 1998;**1064**:184-187

[29] Skahill K. VHDL for Programmable Logic. 1st Ed. Sunnyvale, California, USA: Cypress Semiconductor, Addison-Wesley Longman Publishing; 1996

[30] Brown S, Vranesic Z. Fundamentals of Digital Logic with VHDL Design. New York, NY, USA: McGraw-Hill; 2000

[31] Corporation MKS. Low Cost Synchrocontroller. BY125 Product Operating Instruction. Germany: MKS Instruments, Inc; 2005

Chapter 15

Wavelet Neural Networks for Speed Control of BLDC Motor

Ameer L. Saleh, Adel A. Obed, Hamza H. Qasim, Waleed I.H. Breesam, Yasir I.A. Al-Yasir, Naser Ojaroudi Parchin and Raed A. Abd-Alhameed

Abstract

In the recent years, researchers have sophisticated the synthesis of neural networks depending on the wavelet functions to build the wavelet neural networks (WNNs), where the wavelet function is utilized in the hidden layer as a sigmoid function instead of conventional sigmoid function that is utilized in artificial neural network. The WNN inherits the features of the wavelet function and the neural network (NN), such as self-learning, self-adapting, time-frequency location, robustness, and nonlinearity. Besides, the wavelet function theory guarantees that the WNN can simulate the nonlinear system precisely and rapidly. In this chapter, the WNN is used with PID controller to make a developed controller named WNN-PID controller. This controller will be utilized to control the speed of Brushless DC (BLDC) motor to get preferable performance than the traditional controller techniques. Besides, the particle swarm optimization (PSO) algorithm is utilized to optimize the parameters of the WNN-PID controller. The modification for this method of the WNN such as the recurrent wavelet neural network (RWNN) was included in this chapter. Simulation results for all the above methods are given and compared.

Keywords: BLDC motor, particle swarm optimization (PSO), wavelet neural network (WNN), speed control

1. Introduction

Brushless DC (BLDC) motors have a wide application in our life due to their high-power density and high dynamic response. In addition, the BLDC motor is utilized with constant loads, varying loads, and position applications with high accuracy. This motor is generally controlled utilizing electronically commutation by three-phase power semiconductor bridge inverter with rotor position sensors that are required for starting and providing proper firing sequence to turn on the power devices in the inverter bridge. Based on the rotor position, the power devices are commutated sequentially every 60° [1, 2]. The mathematical model and the Simulink model of BLDC motor to control the speed of a BLDC by using conventional methods are introduced in Refs. [2–6]. The DC-DC converter technique is utilized to control the speed of the motor [5, 6].

In the past decade, artificial intelligence techniques such as neural networks, fuzzy-neural networks, and wavelet neural networks control have been utilized to control the speed of the BLDC motor [7–10]. Since BLDC motor is a multivariable and nonlinear system, it is complex to obtain high performance by applying classical PID control. The main objective of this chapter is to develop wavelet neural networks (WNNs) to control the speed of the BLDC motor, and the recurrent wavelet neural network (RWNN). These methods lead to an enhanced dynamic performance of the system of motor drive and are resistant to load perturbations. The learning strategy for the wavelet neural network and PID controller is developed based on PSO algorithm.

2. Wavelet networks

The combination of wavelet principle and neural networks has led to producing new representing network of wavelet neural network (WNN). Wavelet networks are feedforward networks utilizing wavelets as activation functions. Wavelet networks substitute the sigmoid activation components of the classical feedforward artificial neural networks (ANNs) with wavelets transform function. In wavelet neural networks, both the translation (position) and the dilation are tuning besides weights. The utilization of wavelet node outcomes in efficient networks are optimally approximated and estimated for nonlinear and nonstationary functions [11, 12]. There are two main types to construct the wavelet neural network:

- **Wavenet (fixed grid WNs):** in this type, the neural network and the wavelet processing are accomplished separately. The input signal is first decomposed utilizing some wavelet bases by neurons in the hidden layer with fixed wavelet bases; positions and dilations of the wavelets are preset and only the weights have to be adjusted by learning the network. The main problem is the choice of wavelet frames/bases [12, 13].

- **Wavelet network (adaptive WNs):** this type merges the two theories, which are the dilation and the translation of wavelets along with the summer weights that are adjusted in conformity with some learning procedures. Generally, the modeling of the wavelet network involves two steps: determining the network construction (the number of neurons in each layer, the number of layers, and the type of activations function (wavelet transform)) and modifying the wavelet network parameters by some optimizing algorithm method [12, 13].

3. Structure of wavelet neural network (WNN)

The arrangement of WNN is similar to that of the neural network. The hidden layer contains neurons, whose activation functions are driven from a wavelet basis. These wavelet neurons are generally referred to as wavelons, whose parameters of the inputs contain the wavelet dilation and translation elements [12, 14]. Wavelet networks can be categorized into recurrent and nonrecurrent (feedforward) types.

3.1 Feedforward wavelet neural network (FFWNN)

The FFWNN has no feedback connection. That is, the output is calculated straightly from the input with feedforward connections [12]. There are two arrangements of feedforward wavelet networks:

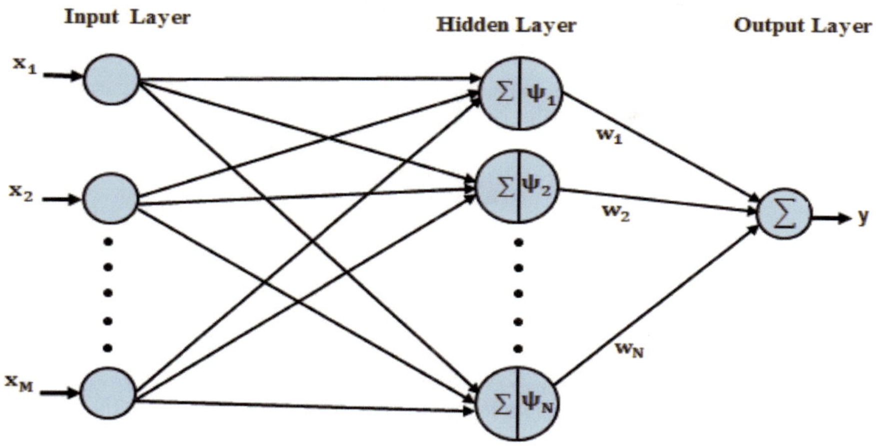

Figure 1.
The building of radial basis wavelet neural network (RBWNN).

3.1.1 Radial basis wavelet neural network (RBWNN)

Radial basis wavelet neural network (RBWNN) is the simplest form of the wavelet network [15, 16]. The arrangement of radial basis wavelet neural network (RBWNN) is shown in **Figure 1**. This network approaches any required signal f(t) by simplifying a linear combination of a group of daughter wavelets $\psi_{a,b}$, where $\psi_{a,b}$ are created by dilation a and translation b from mother wavelet ψ [11, 17].

$$\psi_{a,b} = \psi\left(\frac{x-b}{a}\right) \tag{1}$$

The network output is specified as follows [10, 18]:

$$y = \sum_{n=1}^{N} w_N \psi_{a_N,b_N} \tag{2}$$

where x is the input signal, N is the number of neuron in the hidden layer, and w_N is the weights of the output. The network parameters w_N, a_N, and b_N can be training and optimizing by any optimization technique. In this chapter, the PSO algorithm is used to minimize the error according to the fitness function as will be demonstrated later.

3.1.2 Conventional wavelet neural network

The conventional WNN is a general form of radial basis wavelet neural network [19]. **Figure 2** depicts the building of the conventional wavelet network, the number of hidden layers and neurons that are selected to create an appropriate WNN, and the parameters that are optimized by the PSO algorithm. The input layer can be represented by a vector $x = [x_1, x_2, \ldots, x_M]$, the output layer represented by a vector $y = [y_1, y_2, \ldots, y_K]$, and the activation function of hidden layer is the wavelet basis function. The output Y_j can be given as follows [11, 19]:

$$Y_j = \sigma(u_j) = \sigma\left[\sum_{n=1}^{N} w_{j,n}\, \psi_{a_n b_n}\left(\sum_{m=1}^{M} v_{n,m}\, x_m\right) + g\right] \tag{3}$$

where, $j = 1, 2, 3, \ldots, K$; M is the number of inputs; K is the number of output layers; N is the number of hidden layers; g is the bias; and $\sigma(u)$ is the activation function of the output layer, the most common formula of activation function being sigmoid function which can be illustrated as follows [12]:

$$\sigma(u) = \frac{1}{1 + e^{-u}} \tag{4}$$

3.2 Recurrent wavelet neural networks (RWNNs)

In recurrent WNNs, the output depends not only on the present inputs of the network but also on the prior outputs or conditions of the network [12, 15]. Recurrent networks have feedback and are also known as feedback networks. There are several types of recurrent networks that depend on the feedback connection [12–22].

In the recurrent wavelet network structures, the wavelet network input involves delayed samples of the system output y(k). The number of inputs increases with the order of the system actuality demonstrated. **Figure 3** depicts the structure of recurrent wavelet network. Hence, the output for each layer can be calculated as [20, 23, 24]:

$$\psi_N = \psi\left(\frac{u_N - b_N}{a_N}\right) \tag{5}$$

where a_N and b_N are translation and dilation parameters of wavelets. The inputs of this layer for time n can be denoted as:

$$u_N(n) = x_N(n) + \psi_N(n-1) \times \varnothing_N \tag{6}$$

where $\varnothing_N$ represents the weight of the self-feedback loop. The output of the network is given as follow:

$$y = \sum_{N=1}^{N} w_N \psi\left(\frac{u_N - b_N}{a_N}\right) \tag{7}$$

$$u(n) = x(n - D_i) + y(n - D_0) \times r_N \tag{8}$$

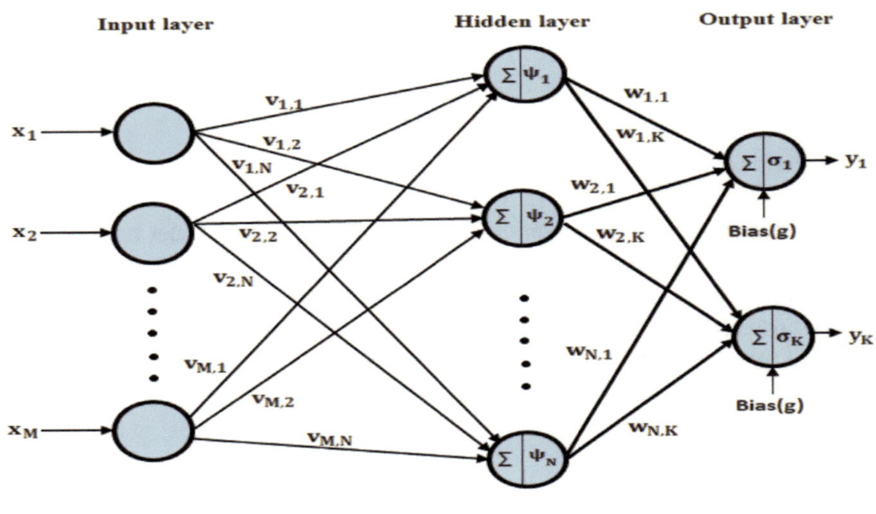

Figure 2.
The building of conventional wavelet neural network.

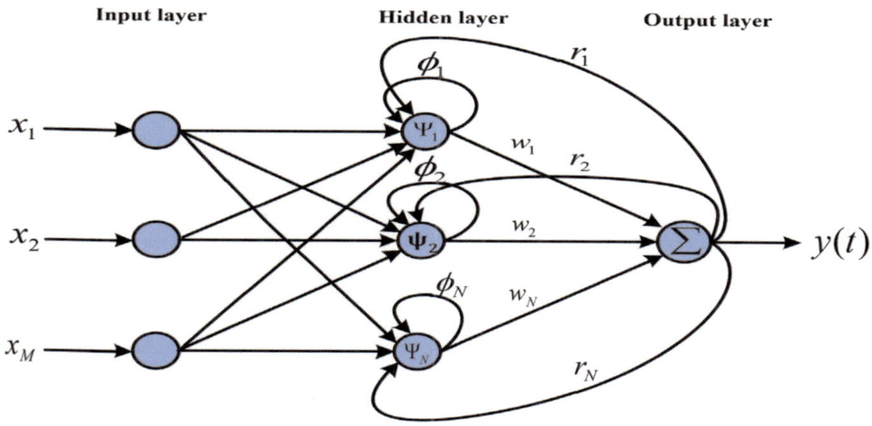

Figure 3.
The recurrent wavelet neural network.

where x is the input signal, N is the number of neuron in the hidden layer, w_N is the output weight, D_i, D_0 is the number of delay for the input and output network, and r_n is the weight of the output feedback loop.

4. Particle swarm optimization

Particle swarm optimization is an inhabitance-based computational procedure motivated from the simulation of gregarious behaviors (social-psychological): fish schooling, bird flocking, and swarm theory. PSO was firstly invented and established by Eberhart and Kennedy [25, 26]. In the PSO algorithm, in place of utilizing evolutionary operators such as mutation and crossover to operate algorithms, the population dynamics emulates a "bird flock" behavior, where social sharing of information takes place and individuals can yield from the finds and prior experience of all the other companions through the search for food. Therefore, each companion, called particle, in the population, which is called swarm intelligence as shown in **Figure 4**, is assumed to fly in several directions over the search space to meet the request fitness function [27, 28].

4.1 Particle swarm optimization algorithm

The PSO algorithm is one of the evolutionary computation techniques to solve optimization troubles. In this algorithm, a swarm of individuals or entities called particles flies over the exploration space [29, 30]. Each particle acts as a probable

Figure 4.
Swarm intelligence.

solution to the optimization troubles. The position of a particle is influenced by the best position visited by itself, i.e., its own knowledge or experience and the position of the best particle in its knowledge of neighboring particles. When the neighborhood is the entire swarm, the best position in the neighborhood of the particle is denoted as the global best position and the resulting algorithm is referred to as the global best position PSO, where the finest prior position that gives the minimum fitness value of any particle is called local best position (lbest). The index of the best particle of all particles in the population is called global best position (gbest). The algorithm is generally referred to as the lbest PSO when smaller neighborhoods are used. For each particle, the performance is measured utilizing an objective function that differs depending on the optimization challenge. The basic PSO algorithm is given below according to the flow chart which is shown in **Figure 5** [31–35].

Step 1. Generation of population particles

Create particles regularly distributed over x, then choose the number of particles, number of iterations, modification accelerating coefficients c_1 and c_2, the inertia weight (w) and random numbers R_1, and R_2 to start the optimum searching.

Step 2. The initialization for each particle

Initialize the present position $x_i(t)$ and the velocitiy $v_i(t)$ for each particle. The particles are randomly produced among the minimum and maximum limits of parameter values. Each particle is treated as a point in a D-dimensional space. The ith particle is denoted as $x_i = (x_{i1}, x_{i2}, \dots, x_{iD})$. The velocity for the particle i is

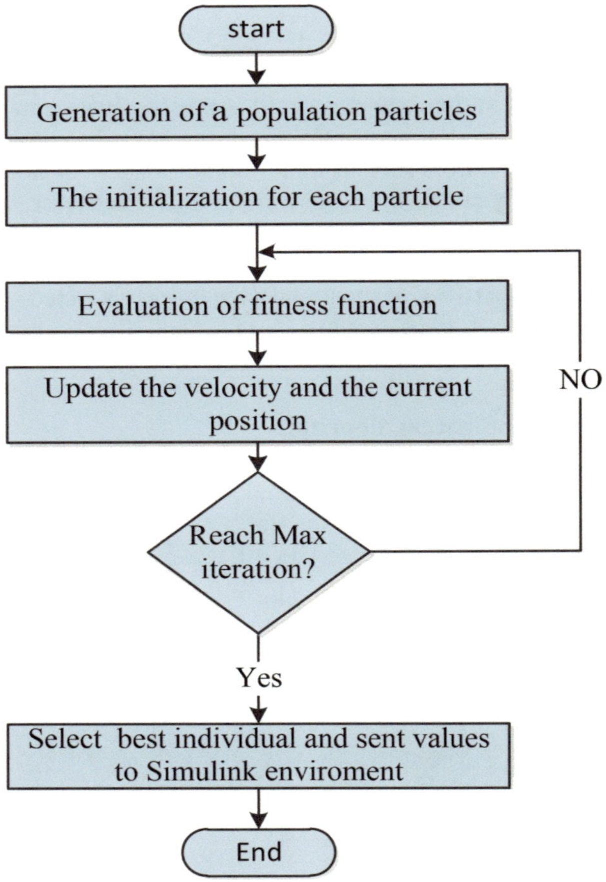

Figure 5.
General flow chart of PSO.

represented as $v_i = (v_{i1}, v_{i2}, \dots, v_{iD})$, then the local best position (lbest) and the global best position (gbest) are initialization.

Step 3. Evaluation of fitness function

The overall performance (speed of convergence, efficiency, and optimization accuracy) of the PSO algorithm counts on the objective function that observes the optimization search. The objective function is chosen to minimize the reference constraints. The popular performance standards based on the error condition are integrated absolute error (IAE), integrated of time weight square error (ITSE), and integrated of square error (ISE) that can be estimated theoretically in the frequency domain [31, 32, 36]. In this chapter, multiobjective functions are utilized based on the integral of the squared error (ISE) criterion and overshoot (M_p) criterion as follow [37, 38]:

$$\text{fitness function} = \min(\text{ISE}) + \min(M_p) \tag{9}$$

where

$$\text{ISE} = \int e^2(t)dt \tag{10}$$

$$M_p = \max(n) - (n_{ref}) \tag{11}$$

$$e(i) = D(i) - y(i) \tag{12}$$

where y(i) is the system output and D(i) is the desired output, while n is the actual speed and n_{ref} is the desired speed.

Step 4. Update the swarm

The updating of the velocity $v_i(t)$ and the present position $x_i(t)$ for each particle in the swarm is done according to Eqs. (13) and (14). Then the main loop and the objective function are calculated for updating positions of particles. If the new value is improved than the previous lbest, the new value is fixed to lbest. Similarly, gbest value is also updated as the best lbest. Velocity of each agent can be modification by the following:

$$v_i^{k+1} = w * v_i^k + c_1 * R_1 * (\text{lbest}_i - x_i^k) + c_2 * R_2 * (\text{gbest}_i - x_i^k) \tag{13}$$

And, the present position can be modification by the following:

$$x_i^{k+1} = x_i^k + v_i^{k+1} \tag{14}$$

where x_i^k is the present position of particle i at iteration k, v_i^k is the velocity of particle i at iteration k, w is the inertia weight which can be represented in Eq. (15), c_1, c_2 represent positive acceleration constants and R_1, R_2 are random variables uniformly distributed in the range [0; 1].

$$w = w_{max} - \frac{(w_{max} - w_{min})}{\text{iter}_{max}} \tag{15}$$

where, w_{min} is the inital weight, w_{max} is the final weight, iter_{max} is the maximum iteration number.

Step 5. Stopping criteria

If the current iteration number reaches the predetermined maximum iteration number, then exit. Otherwise, execute another initialization for each particle and reiterate the process.

5. Speed control of BLDC motor based on wavelet neural network

The WNN-PID controller based on PSO is proposed in this section, which combines the ability of the artificial neural networks for learning with the ability of wavelet for identification, control of dynamic system, and also having the capability of self-learning and adapting [10, 11, 19, 37]. Two types of wavelet network are modified in this section, feedforward WNN and proposed recurrent WNN with online tuning optimization using PSO algorithm [22–24].

5.1 WNN-PID controller based on PSO

In this type of controller, the WNN is utilized with PID controller based on PSO algorithm. WNN-PID controller utilizes online learning by PSO algorithm, where the PSO learning algorithm is used to train the translation parameters a_k and b_k, weights connection in the WNN, and the parameter (k_p, k_i, k_d) of PID controller on-line with the model of BLDC motor to control the speed at the desired value. There are two major issues to implement any wavelet neural networks. First, the network architecture is used and second, the algorithm is used to learn the network by the PSO algorithm. **Figure 6** depicts the block diagram of the BLDC motor with WNN-PID based on PSO algorithm. The structure and the design of the WNN-PID controller will be given in the next subsection.

5.2 Design of the structure of WNN-PID controller based on PSO training algorithm

1. **Design of PSO algorithm:** the PSO algorithm is discussed in Section 5, where each particle parameters are initiated to make a population and then the algorithm is accomplished according to the flow chart given in **Figure 6**, which includes training the parameters of this controller to guarantee the minimization of an objective function. The objective fitness is evaluated as follows:

$$\text{fitness function} = \min(\text{ISE}) + \min(M_p) \tag{16}$$

where ISE is the integrated of square error and M_p is the maximum peak overshoot.

2. **Design of WNN-PID controller**: to design the WNN-PID controller, the type of WNN must be selected as shown in Section 3 and also the number of layers and neurons and the wavelet function type must be selected [16]. In this chapter, the input layer has two inputs: the speed error and the change of this error. One hidden layer with four neurons is used. Three types of mother wavelet functions are used and they are: the Mexican hat function is [10, 11, 22, 37, 39]

$$\psi(x) = (1 - x^2)e^{\frac{-x^2}{2}} \tag{17}$$

The first partial derivative Mexican hat is

$$\psi(x) = (-x)e^{\frac{-x^2}{2}} \tag{18}$$

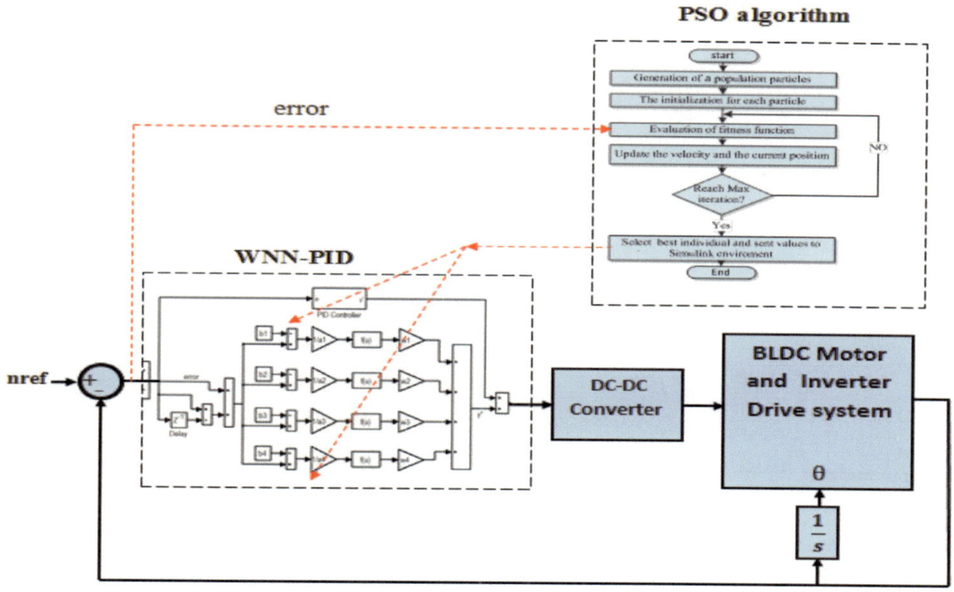

Figure 6.
Block diagram of the BLDC motor with WNN-PID controller based on PSO algorithm.

The Morlet's basic wavelet function is

$$\psi(x) = \mathrm{Cos}(w\,x)e^{\frac{-x^2}{2}} \tag{19}$$

where x is the desired signal and w is a variable value, which was adopted to satisfy the admissibility condition. w = 5 is chosen. The output layer contains one output which is the sum of PID controller and WNN outputs. The parameters values of the WNN-PID controller, such as the dilation factors $a_{k's}$ and the translation factors $b_{k's}$ of the mother wavelet function, the weights connection $w_{k's}$ of the WNN, and PID parameters (k_p, k_i, k_d), are optimized online in PSO algorithm. The results given in this chapter are for Mexican hat function only. The results for the rest functions are similar to that in Mexican hat and are not given.

6. Simulink implementation and results for a BLDC motor drive based on WNN-PID controller

6.1 Speed control based on feedforward WNN-PID controller

The feedforward WNN with PID controller (FWNN-PID) is utilized to control the speed of the BLDC motor as shown in the Simulink model in **Figure** 7. The inputs of the WNN are the speed error and the change of this error, while the hidden layer has four neurons and one output in the output layer. The translation and dilation factors, weights connection for WNN, and PID parameters are learning on-line in PSO algorithm. The output of WNN is given by Eq. (2).

The PSO parameters are given in **Table 1**. These parameters are chosen to get optimal parameters for the PID controller and the wavelet neural network; when it is tuned on-line in PSO algorithm and BLDC motor drive, the optimal values for the PID controller parameters and the WNN parameters (a's, b's, w's) are given in **Tables 2** and **3**, respectively.

The BLDC motor drive is implemented in Simulink/Matlab program as shown in **Figure 6** with the optimal values of PID controller parameters and the optimal

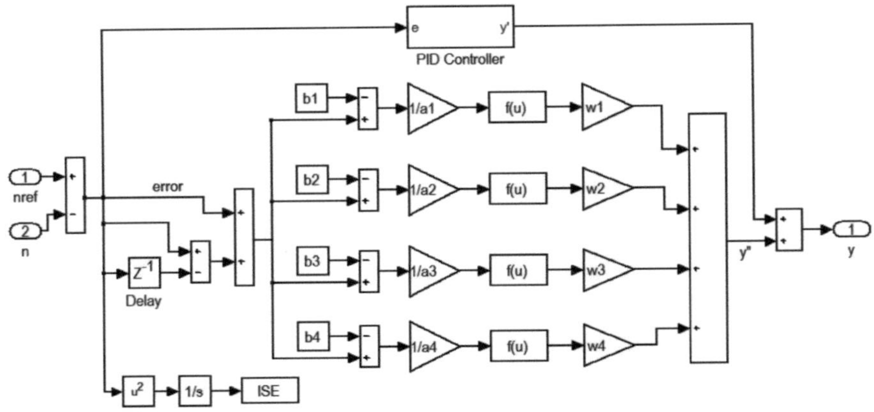

Figure 7.
Simulink model of WNN-PID controller.

PSO_Parameters	Value
Size of the swarm "no of birds"	50
Maximum iteration number	50
Dimension	15
PSO parameter c_1	1.2
PSO parameter c_2	1.2
W_{max}	0.9
W_{min}	0.3

Table 1.
PSO parameter values.

Parameters	K_p	K_i	K_d
Values	6.958	3.241	0.006274

Table 2.
PID parameters tuned using PSO for WNN-PID controller.

WNN dilation parameters				WNN translation parameters				WNN weights parameters			
a_1	a_2	a_3	a_4	b_1	b_2	b_3	b_4	w_1	w_2	w_3	w_4
4.535	1.227	3.013	3.141	1.248	0.314	2.576	1.352	3.632	3.321	2.2843	2.4722

Table 3.
WNN parameters tuned using PSO.

values of WNN parameters. **Figure 8** shows the speed response of the BLDC motor due to change in reference speed. The motor is started at a speed of 500 rpm and then is changed in step to a speed of 500 rpm for every 0.2 s. The actual speed of the motor is tracking the desired speed with a good response. The system starts at no load and suddenly a torque 2 N m (full load) is added at t = 0.4 s. **Figure 9** shows the speed response of the BLDC motor at 2000 rpm during no load and load condition. The developed torque during no load and load condition is shown in **Figure 10**. The position signal, the torque-speed characteristics, the phase current i_a, Phase Back-emf e_a voltage, and line voltage v_{ab} are given in **Figures 11–15**, respectively.

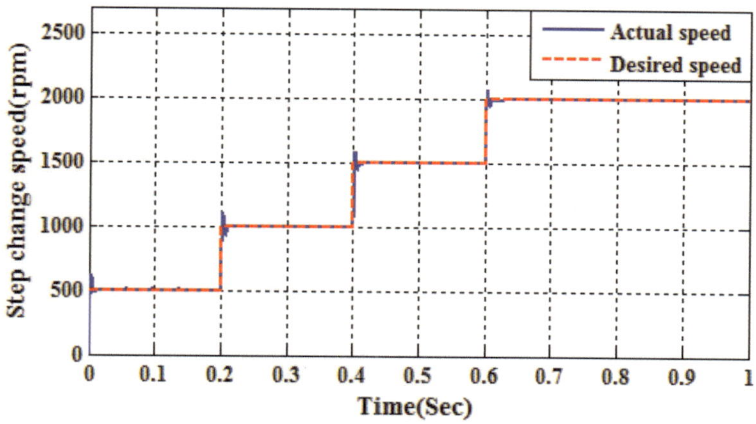

Figure 8.
Step change in speed response with feedforward WNN-PID controller.

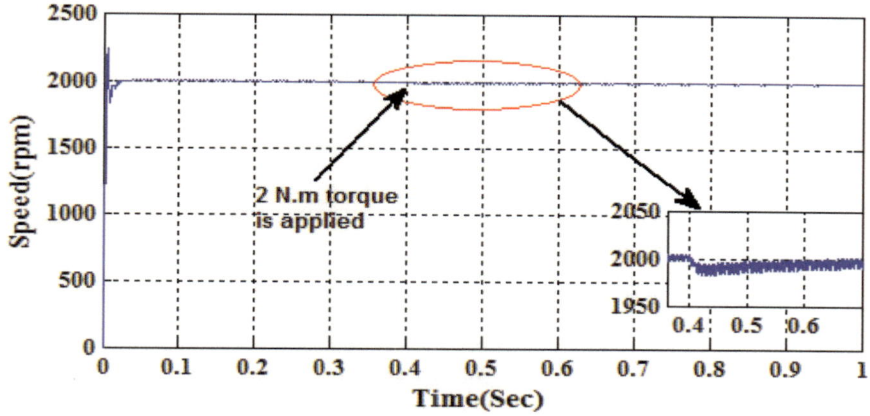

Figure 9.
Speed response of the BLDC motor with WNN-PID controller.

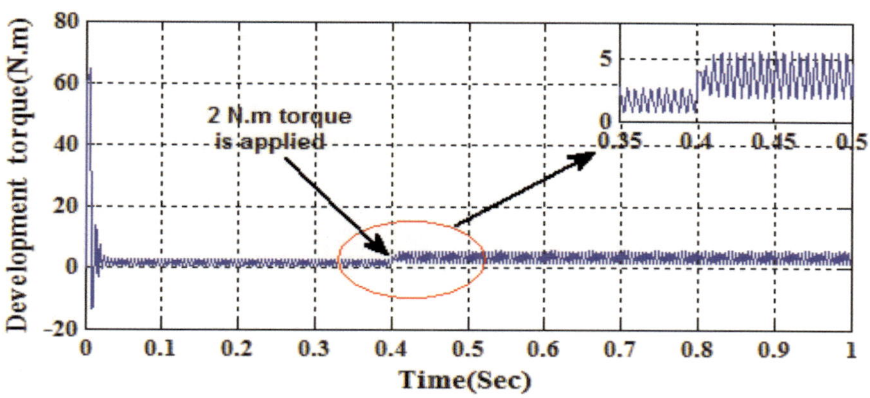

Figure 10.
Developed torque of BLDC motor with feedforward WNN-PID controller.

6.2 Speed control based on proposed recurrent WNN-PID controller

The RWNN that is proposed here is similar to that of feedforward WNN with feedback connections. The RWNN consists of three layers, with two inputs in the

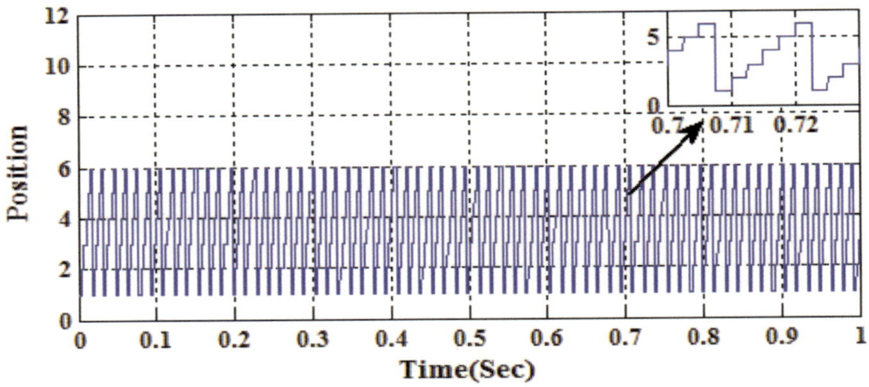

Figure 11.
Position stair signal of BLDC motor with WNN-PID controller.

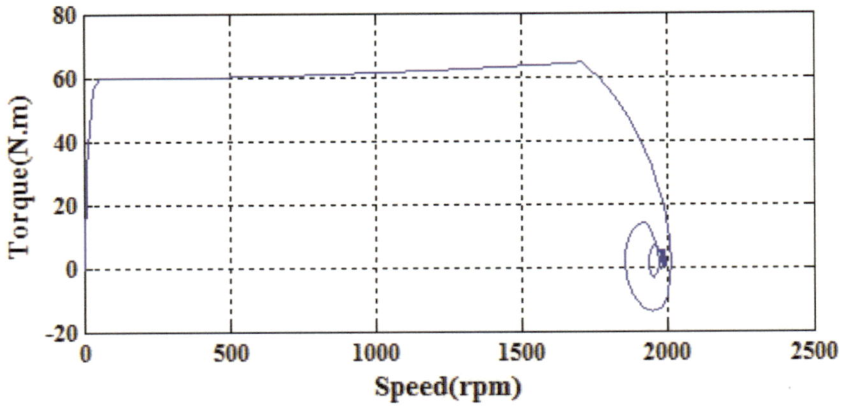

Figure 12.
Torque-speed characteristics of BLDC motor with feedforward WNN-PID controller.

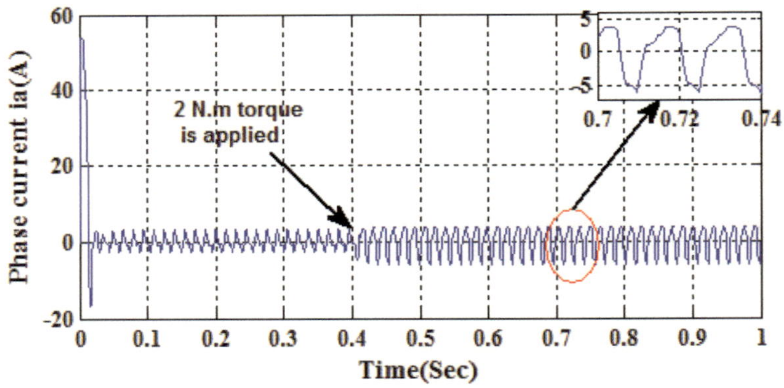

Figure 13.
Phase current i_a of BLDC motor with feedforward WNN-PID controller.

input layer, the hidden layer has four neurons, with one output in the output layer and feedback connection for each layer. In this section, the feedback connection is called "Fully feedback." Besides, the RWNN contains a number of delay samples in the input and output layers as shown in **Figure 16**. The translation and dilation factors, weights and PID parameters are learning on-line to utilize PSO method in

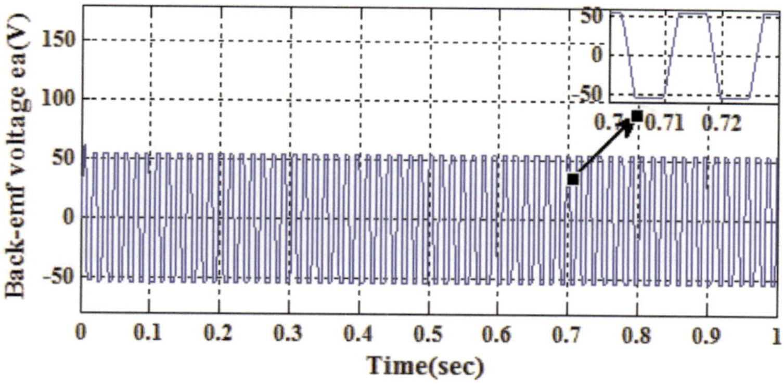

Figure 14.
Phase back-emf e_a voltage for BLDC motor with WNN-PID controller.

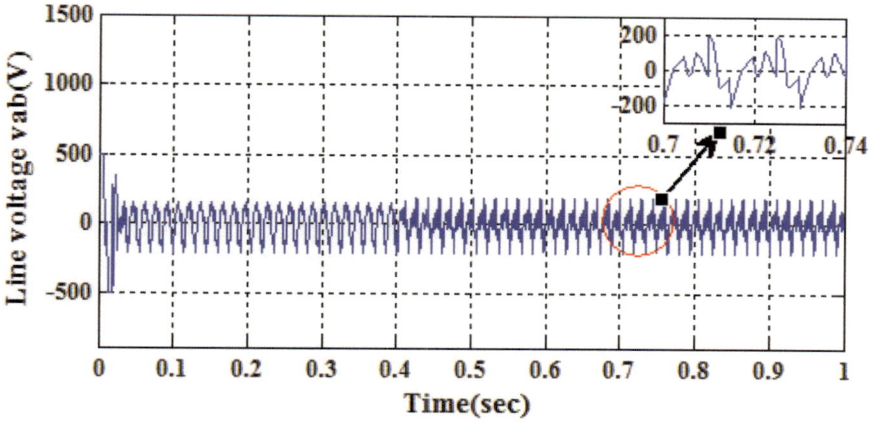

Figure 15.
Line voltage v_{ab} for BLDC motor with feedforward WNN-PID controller.

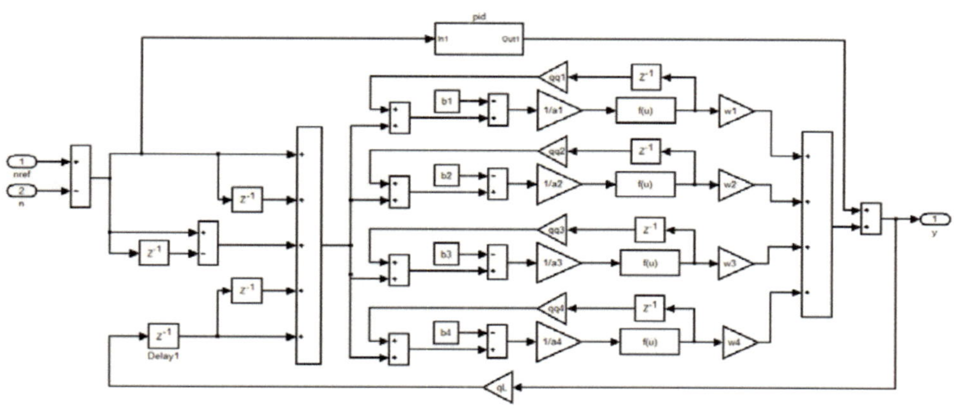

Figure 16.
Simulink model for a proposed recurrent WNN-PID controller.

the same manner used in the previous subsection and the results are given in **Tables 4–6**. The output of WNN is described by Eqs. (7) and (8).

The BLDC motor drive system with RWNN-PID controller is simulated in Matlab/Simulink program as shown in **Figure 6**. The time period that is assumed in

Parameters	K_p	K_i	K_d
Values	4.8256	2.6003	0.0105

Table 4.
PID parameters tuned using PSO for RWNN-PID controller.

WNN dilation parameters				WNN translation parameters				WNN weights parameters			
a_1	a_2	a_3	a_4	b_1	b_2	b_3	b_4	w_1	w_2	w_3	w_4
2.1781	3.0484	3.0181	1.9509	0.1266	5.3694	1.8668	3.2328	0.8769	3.2828	2.7151	0.04124

Table 5.
RWNN parameters tuned using PSO.

Feedback parameters of RWNN				
qq1	qq2	qq3	qq4	qL
0.2957	0.4738	1.0581	4.2460	3.6623

Table 6.
Feedback parameters of RWNN tuned using PSO.

this model is 1 s. The WNN-PID controller can be utilized for speed control in a wide range between 0 and the rated value, with better performance and more flexibility in the controller. **Figure 17** depicts the step change in speed of the BLDC. The motor is started at a speed of 500 rpm and then is changed in step to 500 rpm each 0.2 s. The actual speed of the motor is tracking the desired speed with a good response. The system starts at no load and suddenly a torque 2 N m (full load) is added at t = 0.4 s. **Figure 18** shows the speed response of the BLDC motor at 2000 rpm during no load and load conditions. The developed torque during no load and load conditions is shown in **Figure 19**. The position signal, the torque-speed characteristics, the phase current i_a, Phase Back-emf e_a voltage and line voltage v_{ab} are given in **Figures 20–24**, respectively.

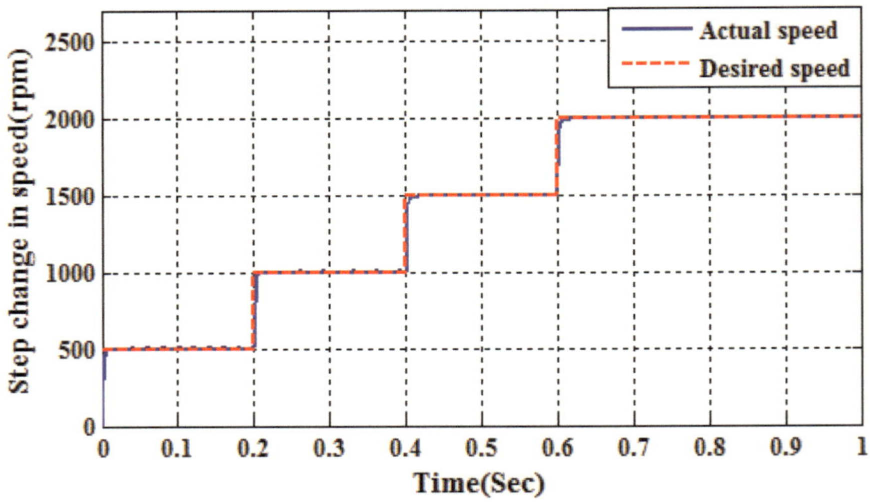

Figure 17.
Step change in speed of BLDC motor with RWNN-PID controller.

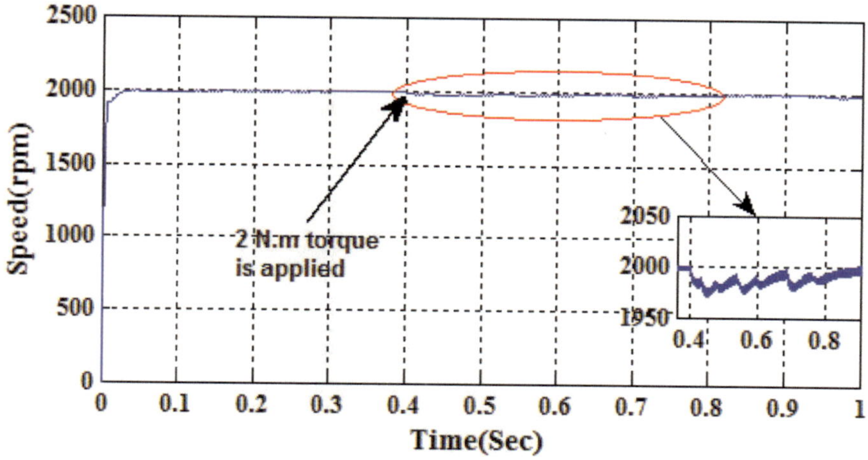

Figure 18.
Speed response of the BLDC motor with RWNN-PID controller.

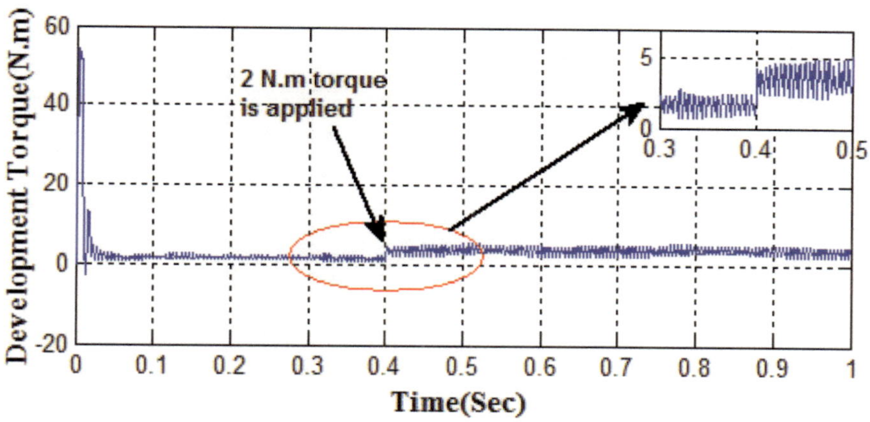

Figure 19.
Development torque of BLDC motor with RWNN-PID controller.

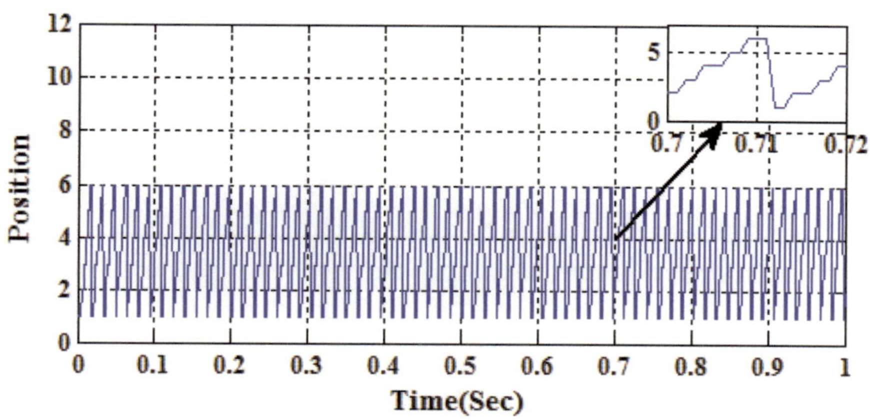

Figure 20.
Position stair signal of BLDC motor with RWNN-PID controller.

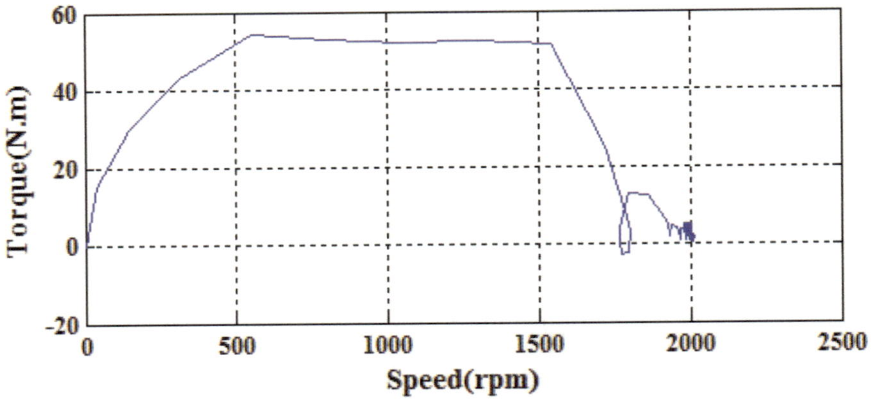

Figure 21.
Torque-speed characteristics of BLDC motor with RWNN-PID controller.

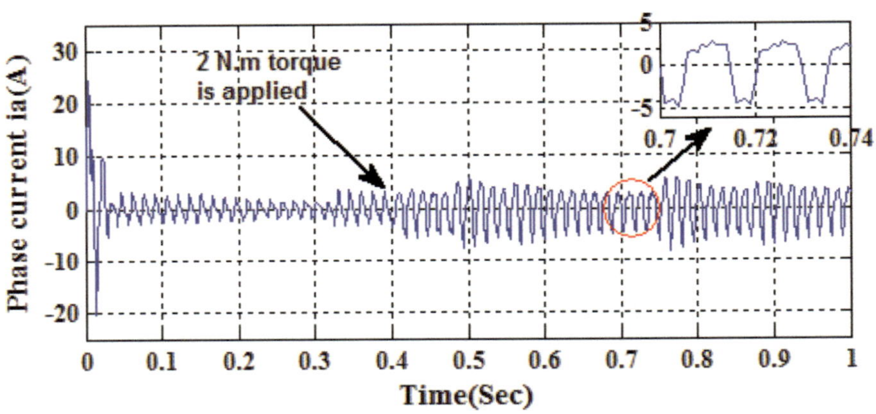

Figure 22.
Phase current i_a of BLDC motor with RWNN-PID controller.

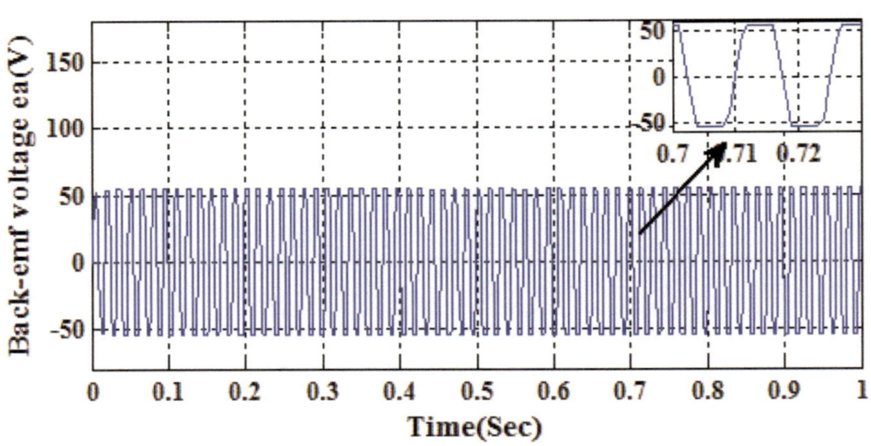

Figure 23.
Phase back-emf e_a voltage of BLDC motor with RWNN-PID controller.

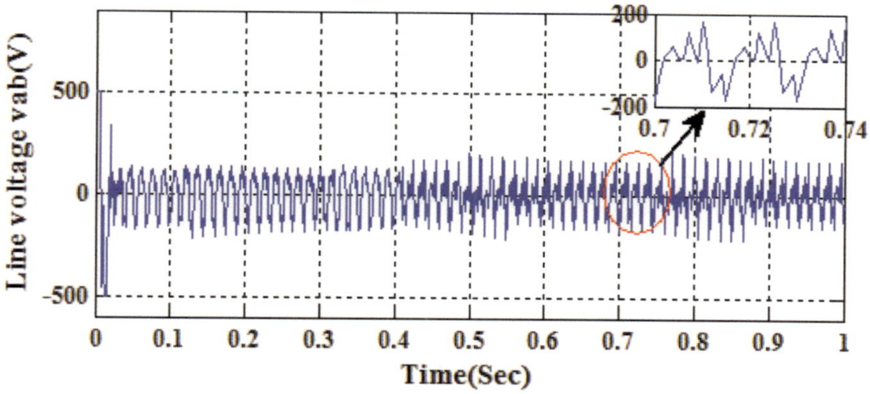

Figure 24.
Line voltage v_{ab} for BLDC motor with RWNN-PID controller.

7. Comparison of two methods for speed control of BLDC motor drive

Comparing among various wavelet neural network schemes shows that the WNN-PID is a preferable method to overcome the nonlinearity in this model with high reliability, more robustness and being good with better performance than the RWNN method as shown in **Table 7**. In addition, the performance of the WNN controller in the real application tends to make the system more robust and less sensitive as well as high precision and excellent flexibility.

Performance	WNN-PID	RWNN-PID
Rise time (s)	0.0035	0.0038
Settling time (s)	0.03	0.04
Steady state error	3×10^{-3}%	2×10^{-3}%
Overshoot	0.12%	Approximately 0%

Table 7.
Performance of speed control of BLDC motor in each methods.

8. Conclusion

In this chapter, the WNN is used with the PID controller to make an adapted controller named as the WNN-PID controller. This controller is utilized to control the speed of BLDC motor in an extensive range and can stock preferable performance than a traditional controller. Two schemes of wavelet neural network are modified for speed control of BLDC motor such as WNN and RWNN. PSO algorithm is utilized for tuning and learning the parameters of the two controllers. The two methods are implemented and tested for different conditions and the performance is compared as shown in **Table 7**. From the simulation results, one can conclude that the proposed WNN controller with PID controller is the best scheme in performance and stability. In addition, using the proposed WNN controller to control the speed of BLDC motor gives better results compared to traditional methods.

Author details

Ameer L. Saleh[1], Adel A. Obed[2], Hamza H. Qasim[3], Waleed I.H. Breesam[4], Yasir I.A. Al-Yasir[5]*, Naser Ojaroudi Parchin[5] and Raed A. Abd-Alhameed[5]

1 Department of Electrical Engineering, College of Engineering, University of Misan, Misan, Iraq

2 Department of Electrical Engineering Technical College, Middle Technical University, Baghdad, Iraq

3 Department of Communication Engineering, Universiti Tun Hussein Onn Malaysia, Johor, Malaysia

4 Electrical Department, Basra Oil Training Institute, Basra, Iraq

5 Faculty of Engineering and Informatics, University of Bradford, Bradford, UK

*Address all correspondence to: y.i.a.al-yasir@bradford.ac.uk

References

[1] Padmaraja Y. Brushless DC (BLDC) Motor Fundamentals. Microchip Technology Inc.; 2003

[2] Stefan B. BLDC motor modeling and control—A MATLAB/SIMULINK implementation [master thesis]. Gothenburg, Sweden: Electrical Power Engineering, Chalmers University of Technology; 2005

[3] Rambabu S. Modeling and control of a brushless DC motor [master thesis]. In: Power Control and Drives Technology. Rourkela: National Institute of Technology; 2007

[4] Ulasyar A, Zad HS, Zohaib A. Intelligent speed controller design for brushless DC motor. In: 2018 International Conference on Frontiers of Information Technology (FIT); January; Islamabad, Pakistan. 2019

[5] Saleh AL, Obed AA. Speed control of brushless DC motor based on fractional order PID controller. International Journal of Computer Applications. 2014; **95**:1-6

[6] Obed AA, Kadhim AK. Speed and current limiting control strategies for BLDC motor drive system: A comparative study. International Journal of Advanced Engineering Research and Science. 2018;**5**(2):119-130

[7] Niasar AH, Vahedi A, Moghbelli H. Speed control of a brushless DC motor drive via adaptive neuro-fuzzy controller based on emotional learning algorithm. In: IEEE Proceedings of the Eighth International Conference on Elec trical Machines and Systems; vol. 1. 2005. pp. 230-234

[8] Awadallah MA, Bayoumi EHE, Soliman HM. Adaptive deadbeat controllers for brushless DC drives using PSO and ANFIS techniques. Journal of Electrical Engineering. 2009;**60**(1):3-11

[9] Kumar NS, Kumar CS. Design and implementation of adaptive fuzzy controller for speed control of brushless DC motors. International Journal of Computer Applications. 2010;**1**(27): 36-41

[10] Maohua Z, Changliang X, Yang T, Dan L, Zhiqiang L. Speed control of brushless DC motor based on single neuron PID and wavelet neural network. In: IEEE International Conference on Control and Automation Guangzhou, China; June. 2007. pp. 617-620

[11] Hameed WI, Sawadi BA, Al-Kamil SJ, Al-Radhi MS, Al-Yasir YIA, Saleh AL, et al. Prediction of solar irradiance based on artificial neural networks. Inventions. 2019;**4**:45

[12] David V. Wavelet neural networks [dissertation submitted for the Master Science]. In: Data Analysis, Networks and Nonlinear Dynamics. Department of Mathematics, University of York; 2005

[13] Billings SA, Hua-Liang W. A new class of wavelet networks for nonlinear system identification. IEEE Transactions on Neural Networks. 2005;**16**:862-874

[14] Dhiraj A, Sunil K. Adaptive fuzzy wavelet network control design for nonlinear systems. International Journal of Advanced Technology & Engineering Research (IJATER). Jan. 2013;**3**:148-157

[15] Cheng-Jian L, Hung-Ming T. FPGA implementation of a wavelet neural network with particle swarm optimization learning. Mathematical and Computer Modelling. 2008;**47**:982-996

[16] Gaviphat L. Adaptive self-tuning neuro wavelet network controllers [doctor of philosophy thesis in Electrical Engineering]. Virginia Polytechnic Institute and State University; March 1997

[17] Bhowmik PS, Pradhan S, Prakash M, Roy S. Investigation of wavelets and radial basis function neural network for incipient fault diagnosis in induction motors. In: IEEE International Conference on Circuits, Controls and Communications (CCUBE); December. 2013. pp. 1-5

[18] Shouxin R, Ling G. Application of a wavelet packet transform based radial basis function neural network to analyze overlapping spectra. In: IEEE Congress on Image and Signal Processing; China; May. 2008. pp. 228-232

[19] Hamza MI. A wavelet network control scheme for path tracking of mobile robot [master thesis in Electrical Engineering]. University of Basrah, Aug. 2013

[20] Sung Jin Y, Park JB, Choi YH. Direct adaptive control using self recurrent wavelet neural network via adaptive learning rates for stable path tracking of mobile robots. In: IEEE American Control Conference, Vol. 1. 2005. pp. 288-293

[21] Hameed WI, Saleh AL, Sawadi BA, Al-Yasir YIA, Abd-Alhameed RA. Maximum power point tracking for photovoltaic system by using fuzzy neural network. Inventions. 2019;**4**:33

[22] Obed AA, Saleh AL. Speed control of BLDC motor based on recurrent wavelet neural network. Iraqi Journal of Electrical and Electronic Engineering. 2014;**10**(2):118-129

[23] Saleh AL, Obaid BA, Obed AA. Motion control of linear induction motor based on optimal recurrent wavelet neural network-PID controller. International Journal of Engineering & Technology. 2018;**7**(4):2028-2034

[24] Obed AA, Saleh AL, Kadhim AK. Speed performance evaluation of BLDC motor based on dynamic wavelet neural network and PSO algorithm. International Journal of Power Electronics and Drive System (IJPEDS). 2019;**10**(4):1742-1750

[25] Mehdi N, Hossein N, Malihe M. A PSO-based optimum design of PID controller for a linear brushless DC motor. Proceedings of World Academy of Science, Engineering and Technology. 2007;**20**:211-215

[26] Reddy MB, Obulesh YP, Raju SS. Particle swarm optimization based optimal power flow for volt-var control. ARPN Journal of Engineering and Applied Sciences. 2012;**7**:20-25

[27] Effatnejad R, Bagheri S, Farsijani M, Talebi R. Economic dispatch with particle swarm optimization and optimal power flow. International Journal on "Technical and Physical Problems of Engineering" (IJTPE). 2013;**5**:9-16

[28] Mohammed HJ, Abdullah AS, Ali RS, Abd-Alhameed RA, Abdulraheem YI, Noras JM. Design of a uniplanar printed triple band-rejected ultra-wideband antenna using particle swarm optimisation and the firefly algorithm. IET Microwaves, Antennas and Propagation. 2016;**10**(1):31-37

[29] Portillo AA, Michael F, Chunjiang Q. Particle swarm optimization for PID tuning of a BLDC motor. IEEE International Conference on Systems, Man, and Cybernetics; October. 2009. pp. 3917-3922

[30] Sivanandam SN, Visalakshi P, Bhuvaneswari A. Multiprocessor scheduling using hybrid particle swarm optimization with dynamically varying inertia. International Journal of Computer Science & Applications. 2007;**4**:95-106

[31] Soni YK, Rajesh B. BF-PSO optimized PID controller design using ISE, IAE, IATE and MSE error criteria. International Journal of Advanced

Research in Computer Engineering & Technology (IJARCET). 2013;**2**: 2333-2336

[32] Mohammed HJ, Abdullah AS, Ali RS, Abdulraheem YI, Abd-Alhameed RA. Performance comparison of particle swarm optimization, and genetic algorithm in the design of UWB antenna. Journal of Telecommunications. 2014;**27**(2):22-26

[33] Taeib A, Ltaeif A, Chaari A. A PSO approach for optimum design of multivariable PID controller for nonlinear systems. In: International Conference on Control, Engineering & Information Technology (CEIT'13) Proceedings Engineering & Technology. Vol. 2. 2013. pp. 206-210

[34] Mohammed HJ et al. Evaluation of genetic algorithms, particle swarm optimization, and firefly algorithms in antenna design. In: 2016 13th International Conference on Synthesis, Modeling, Analysis and Simulation Methods and Applications to Circuit Design (SMACD), Lisbon. 2016. pp. 1-4

[35] Sharaf AM, El-Gammal AAA. A novel particle swarm optimization PSO tuning scheme for PMDC motor drives controllers. In: IEEE International Conference on Power Engineering, Energy and Electrical Drives; March. 2009. pp. 134-139

[36] Xie W, Wang J-S, Wang H-B. PI controller of speed regulation of brushless DC motor based on particle swarm optimization algorithm with improved inertia weights. In: Mathematical Problems in Engineering. 2019. pp. 1-12

[37] Saleh AL, Mohammed MJ, Kadhim AS, Raadthy HM, Mohammed HJ. Design fuzzy neural petri net controller for trajectory tracking control of mobile robot. International Journal of Engineering & Technology. 2018;7(4):2256-2262

[38] Saleh AL, Obed AA, Al-Yasir YIA, Elfergani ITE, Rodriguez J, Clarke RW, et al. Anti-windup scheme based on 2DOF-PIλDμ controller for velocity tracking of linear induction motor. International Transactions on Electrical Energy Systems. 2019;**40**:1-17

[39] Obed AA, Kadhim AK. Multi-resolution wavelet PID speed and current controllers of BLDC motor based on invasive weed optimization technique. International Journal of Applied Engineering Research. 2018; **13**(8):6234-6243